A Dynamic Theory of Space-Time

A Matter of Waves

First Edition

Dr. Robert Nieves

Library of Congress Control Number: 2020914889

ISBN 9798667276289

A Dynamic Theory of Space-Time: A Matter of Waves

Covering novel theoretical ideas in contemporary research style, this book is ideal for students, researchers, and enthusiastic readers in all areas of cosmology and theoretical physics. There are four main parts of the book focusing on waves of space, kinematics, space-time and gravitation, and waves of the fields of force, to discuss the greatest questions and challenges to modern physics. A dynamic theory of space-time is proposed based upon well accepted concepts of physics and as a foundation to The General Theory of Relativity.

The structure of the book follows the gradual path of research and investigation into our physical reality that culminates in the development of avant-garde notions, ideas, and realizations that are described and presented mathematically as expansions to current concepts of modern physics. The last four chapters of the book describe the multidimensionality of space-time, the natures of gravitation and light, black holes, and an analysis of the present gravitational theory, among other advanced topics, some of which are covered in-depth and in a straightforward style, for the general advancement of science and technology.

Robert Nieves has a diversified professional experience in engineering, teaching, international business administration, and physics and cosmology research. Dr. Nieves holds a Bachelor of Science in Electrical Engineering from the Illinois Institute of Technology and an MBA and a DIBA from Nova Southeastern University in Florida.

Dedicated to my family and friends

CONTENTS

PART I

WAVES OF SPACE-TIME

PART II

KINEMATICS

PART IV

WAVES OF THE FIELDS OF FORCE

PART I

WAVES OF SPACE-TIME

Chapter 1

The Dynamic Theory of Space-Time

§ *1. On the dynamic theory of space-time*

As a photon surfs the wave front of its time field, it also propagates at a velocity of *c*, the photon also expands radially if unobstructed, in isotropic and homogeneous space-time, in an expanding scale at a velocity of *c*, manifesting a probabilistic wave in its future time, for the massless photon with its associated energy and momentum. The emergent motion of the photon surfing the wave front of time is relative to the three spatial dimensions, for the photon is stationary with respect to its position in its temporal location. Thus, such photon exists in that instant of time, not moving through its temporal dimension, even though the photon's imaginary clock would not tick-tock, the photon has an apparent measurable velocity with respect to space. (Born, 1999)

An object that is moving through time would have to travel at a velocity less or greater than *c* through space-time. If the object were to travel with a velocity faster than *c* through subluminal or luminal space-time, it would reverse direction in time and travel backwards in time through space-time, because there would not be any space to travel into except the space that existed in its past time. Thus, as the object travels backward through time offsetting spatial divergence, the object travels through space. If particles were to travel faster than *c* as tachyonic particles, or anti-matter, the momentum of these particles would be transferred to motion in the backward temporal direction or anti-time. If the object travels at a velocity less than *c* through space-time, it would travel relativistically through space-time or with slower velocity. If the object travels at a velocity equal to *c* through space-time, it is not moving in its time field, only relative to space; such object may be referred to as timeless or existing at tick-tock-zero. It perpetually exists in a single cycle of time.

Time expands in all directions and increases entropy since every particle in space-time has a greater probability distribution to be located elsewhere from its present location in space-time. The expansion of time is the framework of the second law of thermodynamics. Time may be viewed as an intrinsic relative movement, and an emergent property, of the underlying dimensions of space. Time may be viewed as having both magnitude and direction. Hence, time may be a vector in any of its directions, or a vector field, or just a magnitude, a zeroth order tensor (scalar) of a measured interval.

The expansion of time is fundamental in nature; it endows a particle or photon at a local point in the present time-scape with an advanced nonlocal probability distribution function that may travel at a velocity of *c* through its temporal dimension in the opposite direction to the arrow of time. If obstructed by a physical object or photon during observation, the probability nonlocal function would collapse localizing the particle or photon at its present local point in space-time. Hence, the expansion of time is the framework underlying the nonlocality of physical particles and the wave-like quality of light (or photon). All matter exists in space-time. (Taylor, 1966)

The equation $\delta x_4/\delta t = ic$ is the spatiotemporal velocity equation for spatiotemporal distance X_4 relative to coordinate time along the temporal *i*-axis where *t* is a resultant coordinate time, X_4 is a spatiotemporal distance equal to *ict*, and *r* is a distance of a resultant coordinate space. The spatiotemporal velocity with respect to coordinate space may be less, equal, or greater than *c*. The spatiotemporal distance X_4 with respect to coordinate space may expand faster than the velocity of light. Space is expanding faster at the outer boundary of our universe. Therefore, the time field at the outer boundary of our universe is creating new instants of space at a spatiotemporal velocity faster than *c*. The spatiotemporal velocity with respect to coordinate time is equal to *ic*. However, the spatiotemporal velocity with respect to coordinate space *r* is equal to $\delta x_4/\delta r = icv_t$. If $\delta x_4/\delta r$ is greater than *c*, then v_t is greater than unity. Hence, coordinate time is expanding faster than coordinate space in $\delta t/\delta r$. Thus, time is accelerating with respect to space. Accelerated inflation happens!

The time-scape is conjugate to space and it expands at a velocity icv_t

relative to the three spatial dimensions. The time-scape is three-dimensional with three temporal axes. Since time may expand differently relative to space, and space may expand or contract differently relative to time, the speed of light depends on the relative reciprocal relationship between space and time at the locality of the observation. This assures all observers at all inertial frames at the locality of observation that the laws of physics behave the same.

The equation $E = m_0 v^2$ exemplifies and implies the effect of the velocity of time on rest mass, since v is equal to the speed of light c in the well-known equation, giving a particle at rest in space, but still moving in time, a temporal momentum. (Einstein, 1952)

§ 2. On the Eddington's postulate

If we consider the following postulate (Eddington's Postulate): 'Every point in space-time expands freely in all directions unless obstructed.' We may hypothesize that this postulate is supported by Huygens Principle.

Every point on a local spatiotemporal wave-front in isotropic homogeneous space-time may be considered a source of secondary spherical spatiotemporal wavelets which spread in the outward direction at the speed of time (light). The new spatiotemporal wave-front is the tangential surface to all, or of all, of these secondary spatiotemporal wavelets.

The principle that any point on a spatiotemporal wave-front may be regarded as the source of secondary spatiotemporal wavelets and that the surface that is tangent to the secondary spatiotemporal wavelets, the envelope, can be used to determine the future position of the spatiotemporal wave-front supports Eddington's Postulate.

If we consider an extended line of space-time points, the resultant spatiotemporal wave will consist of an infinite number of space-time points and may be thought of as generating a plane spatiotemporal wave front.

If a space-time locality is isotropic and homogeneous, allowing time to expand with the same speed regardless of its direction of propagation, the three-dimensional spatiotemporal envelope of a space-time point will be spherical.

When a spatiotemporal wave expands in a single space-time locality at a constant speed, the Huygens' wave construction preserves the general form of the wave-front. That is, spheres propagate and become larger spheres as shown in figure 1. (Baker, 1987)

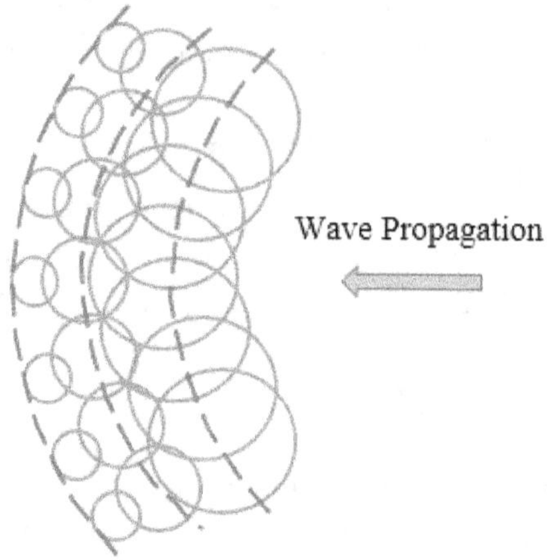

Figure 1.

The distance that the temporal wave travels is *ict*. We call this distance X_4.

A temporal wavelet in isotropic and homogeneous space-time centered at (x, y, z, *ict*) with constant velocity has a future wave-front at $ic(t - \Delta t)$, opposite to the direction of the arrow of time. Its past wave-front is at $ic(t + \Delta t)$ as shown below. The future wave-front is contained within the present wave-front, and the present wave-front is contained within the past wave-front.

This temporal wave behavior exemplifies why time travels through a physical object in its world tube in an inertial frame, the present encompasses the future probability distribution of its existence, and its past existence will encompass both its present and future probability distributions.

Time emerges counterintuitively to our current perception of the past, present, and future of linear time. The arrow of time for expanding space-time points from the future to the past. Space-time expands, increasing entropy, from the future, through the present, to the past.

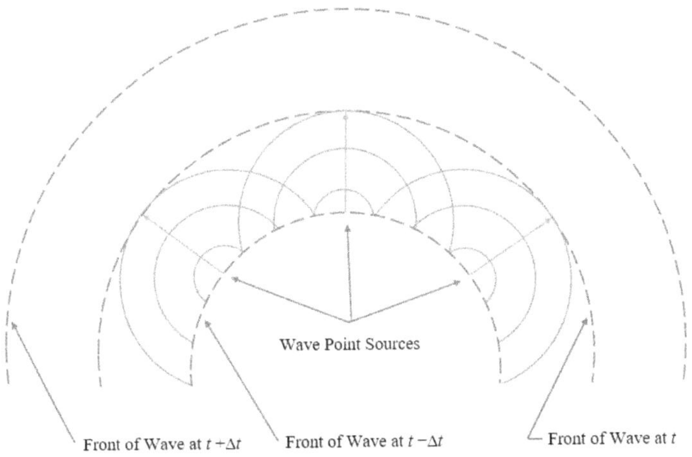

Figure 2.

Every point in space-time expands in all directions unless obstructed. Therefore, there is spatiotemporal wavelet interference when two spatiotemporal wavelets meet while traveling along the same isotropic and homogeneous space-time. The spatiotemporal wavelet interference causes space-time to take on a form that results from the net effect of the two individual spatiotemporal wavelets between the center points of the two wavelets.

§ 3. On the principle of superposition of spatiotemporal waves

The principle of superposition of spatiotemporal waves may be stated as follows:

'When two spatiotemporal wavelets interfere, the resulting displacement of space-time at any locality is the algebraic sum of the displacement of the individual spatiotemporal wavelets at the same locality in space-time.'

For instance, if two spatiotemporal wavelets have a displacement in the same direction at any locality along space-time, constructive interference will occur between the spatiotemporal wavelets. If two spatiotemporal wavelets have a displacement in the opposite direction at any locality along space-time, destructive interference will occur between the spatiotemporal wavelets. The principle of superposition prevents objects of mass embedded in isotropic and homogeneous space-time from moving away from each other due to the expansion of every point in isotropic and homogeneous space-time if there is destructive interference between adjacent spatiotemporal wavelets.

Consequently, objects retain their proportional scale and distances with respect to one another in isotropic and homogeneous space-time, while time passes as every point in space-time expands in all directions unless obstructed.

A photon surfing on the wave-front of a spatiotemporal wavelet can still transfer its momentum and energy (momenergy) to the wave-front of the adjacent wavelet, surf the adjacent wavelet in the direction of the arrow of time that points to its direction of motion. Hence, light has the speed of the time field at the locality of measurement, and a photon is timeless. (Born, 1999)

3.1. The Spatiotemporal Wave Equivalence Principle

As the key idea of the General Theory of Relativity, the spatiotemporal equivalence principle reinforces that if a gravitational field of a body of mass is pulling in one direction, then that effect is completely equivalent to an acceleration in the opposite direction due to the spatiotemporal wave expansion. During launch from earth, a rocket ship usually accelerates vertically pushing any object of mass into the floor of the control deck. A rocket ship accelerating forward in space-time feels like a gravitational field pushing the astronaut back against his or her seat.

Hence, a gravitational field is equivalent to the spatiotemporal wave acceleration, they both affect spatial and temporal extents in the direction of motion during the interference of spatiotemporal waves as they expand or contract. According to the General Theory of Relativity, covariant equations always satisfy the principle of spatiotemporal wave equivalence.

The metric of nearly flat space-time may be regarded as still having some cosmological curvature. Consequently, the nearly flat metric may be regarded as a special case of the spatiotemporal field, not the absence of the field. Our universe is inherently curved by the nature of its spatiotemporal divergence. The traditional idea of flat space-time serves as a useful contrast to the natural and universal concept of curvature. The gravitational field emerges within space-time as spatiotemporal waves interact, not as an extrinsic attribute of space-time. Gravity is a measurable quality of the geometry of space-time. The geometry of space-time emerges from the spatiotemporal wave medium. Multidimensional Riemannian geometry has provided a natural spatiotemporal framework where the elementary laws of physics may be formulated.

In a sense, the spatiotemporal geometry tells charges where to go; the spatiotemporal field tells the field potential of charges how to change. The gradient(s) of a spatiotemporal field potential is (are) the paths of greatest change.

3.2. The polarizable vacuum analogy

The exponent of the spatiotemporal growth factor may be expressed as the square of the ratio of a velocity, v_r, to the velocity of light in free space-time, c_0. The spatiotemporal expansion or contraction is equivalent to the concept of polarizable vacuum since spatiotemporal properties are functions of position and the spatiotemporal dimensions. As spatiotemporal wavelets expand or contract in a region of space-time, the spatiotemporal properties may increase, decrease, or remain the same. (Wilson, 1921), (Dicke, 1957)

$$\ln e^{\frac{gr}{c^2}} = \left(\frac{v_r}{c_0}\right)^2 \qquad (3.1)$$

$$e^{\left(\frac{v_r}{c_0}\right)^2} \equiv \frac{\mu'}{\mu_0} \equiv \frac{\varepsilon'}{\varepsilon_0} \equiv \frac{m'}{m_0} \equiv \frac{E_0}{E'} \equiv \frac{\partial t'}{\partial t_0} \equiv \frac{\partial x_0}{\partial x'} \qquad (3.2)$$

The constancy of the speed of light may be expounded through the application of spatiotemporal properties of permeability, permittivity, and the relativistic effects on mass and energy as a function of velocity according to the General Theory of Relativity. (Puthoff, 2002) The relativistic ratios of the spatiotemporal properties change together with the factor of spatiotemporal growth, to preserve the ratio of the impedance of free space-time, $\sqrt{\mu'/\varepsilon'} = \sqrt{\mu_0/\varepsilon_0}$, to maintain the magnetic field-to-electric field energy ratio stable during the adiabatic translation of an atomic structure and its particles from one spatiotemporal region to another of differing spatiotemporal growth.

The energetic state(s) of a physical system, E_0/E', may vary as the system translates to a spatiotemporal region of differing spatiotemporal growth. Moreover, the law of conservation of charge, $\partial q \equiv \partial x' \cdot \partial t' \equiv \partial x_0 \cdot \partial t_0$, is an outcome of the relativistic changes of the spatiotemporal properties.

The Lorentz factor (gamma) is the factor in General Relativity by which length, time, mass, and energy change for an object while that object is moving relativistically in free space-time. Let us express the Lorentz factor in terms of the exponential growth factor and the electromagnetic properties of permeability or permittivity.

$$\gamma = \frac{1}{\sqrt[2]{1-\left(\frac{v_r}{c_0}\right)^2}} = \frac{1}{\left(\frac{e^{\frac{0}{c^2}} - e^{\frac{gr}{c^2}}}{e^{\frac{0}{c^2}}}\right)^{\frac{1}{2}}} = \frac{1}{\sqrt[2]{\ln e^{\frac{c^2}{c^2}} - \ln e^{\frac{gr}{c^2}}}} \tag{3.3}$$

$$\gamma = \frac{1}{\sqrt[2]{\ln \frac{e^{\frac{c^2}{c^2}}}{e^{\frac{gr}{c^2}}}}} = \frac{1}{\sqrt[2]{\ln e^{\frac{c^2-gr}{c^2}}}} \equiv \frac{1}{\sqrt[2]{\ln \frac{e\mu_0}{\mu'}}} \equiv \frac{1}{\sqrt[2]{\ln \frac{e\varepsilon_0}{\varepsilon'}}} \tag{3.4}$$

Hence, the above results for the Lorentz factor allow us to contemplate relativistic effects from an electromagnetic point of view.

3.3. The spatial divergence about a charged body of mass and its implications.

The spatial divergence about a charged body of mass is given by the spatiotemporal growth factor of its spatiotemporal surroundings which is raised to twice the exponential power of the ratio of two spatiotemporal accelerations. The exponential growth factor can be written as a Taylor series.

$$e^{2\frac{gr^2}{rc^2}} \approx 1 + 2\frac{gr^2}{rc^2} + \ldots \tag{3.5}$$

The exponential growth factor of the space-time about a charged body of mass is the ratio of the spatiotemporal acceleration $\ddot{a}_m$ about the mass to the spatiotemporal acceleration $\ddot{a}_{st}$ of free space-time.

$$\frac{gr^2}{rc^2} \equiv \frac{\ddot{a}_m}{\ddot{a}_{st}} \tag{3.6}$$

The exponential growth factor represents the growth factors of two spatiotemporal wavelets between two points in the space-time near the charged body of mass. The square root of the growth factor represents the spatial divergence of a single wavelet. The exponential growth factor is a solution to the above equation that satisfies the Newtonian limit.

$$\sqrt[2]{e^{2\frac{gr^2}{rc^2}}} = e^{\frac{gr^2}{rc^2}} = e^{\frac{gr}{c^2}} = \sigma \tag{3.7}$$

The ratio of electric field E_Q of the electrostatic charge to the alternating electric field E of a body of mass has the effect of modulating both curvature and spatial divergence about the body. If $E_Q > E$, the amplitude of spatial divergence increases, if $E_Q < E$, the amplitude of spatial divergence decreases.

Neither the spatiotemporal expansion or contraction in a gravitational field, nor an alternating or dynamic electric field, is static, but it is possible to consider a snapshot of a temporal instant to be static for analytical purposes. (Puthoff, 2002)

$$\frac{d^2 e^{\frac{gr}{c^2}}}{dr^2} + \frac{2}{r}\frac{de^{\frac{gr}{c^2}}}{dr} = \frac{1}{e^{\frac{gr}{c^2}}}\left[\left(\frac{de^{\frac{gr}{c^2}}}{dr}\right)^2 - \frac{E_Q}{E}\frac{g^2}{c^4}\right] \tag{3.8}$$

The exponential growth factor is a solution to the above equation that satisfies the Newtonian limit. If curvature $\left(g^2/c^4\right)$ increases, on the right-hand side of the above equation, the amplitude of spatial divergence would decrease. Conversely, if curvature decreases, the amplitude of spatial divergence would increase.

If curvature increases toward infinity, or $E >> E_Q$, we have

$$\frac{d^2 e^{\frac{gr}{c^2}}}{dr^2} + \frac{2}{r}\frac{de^{\frac{gr}{c^2}}}{dr} \approx \frac{1}{e^{\frac{gr}{c^2}}}\left(\frac{de^{\frac{gr}{c^2}}}{dr}\right)^2 \tag{3.9}$$

Let us consider an electrostatically charged body of mass that also has an alternating electric field *E*. We may express the electrostatic charge as *Q*.

$$Q = 4\pi\varepsilon_0 r^2 E\sigma^2 \tag{3.10}$$

Where the exponent of the spatiotemporal factor of growth is the ratio of an electrostatic field to an alternating electric field.

$$\frac{\frac{Q}{4\pi\varepsilon_0 r^2}}{E} = \frac{E_Q}{E} \tag{3.11}$$

$$\frac{E_Q}{E} = \frac{1}{2}\ln\sigma^2 = \frac{1}{2}\ln e^{2\frac{E_Q}{E}} \tag{3.12}$$

$$e^{\frac{gr}{c^2}} = \mathrm{Cosh}\sqrt[2]{\frac{g^2r^4}{c^4r^2} - \frac{E_Q}{E}} + \sqrt[2]{\frac{\frac{g^2r^4}{c^4}}{\frac{g^2r^4}{c^4} - \frac{E_Q r^2}{E}}}\,\mathrm{Sinh}\sqrt[2]{\frac{g^2r^4}{c^4r^2} - \frac{E_Q}{E}} \tag{3.13}$$

Simplifying the above equation, we obtain

$$e^{\frac{gr}{c^2}} = \mathrm{Cosh}\sqrt[2]{\frac{g^2r^2}{c^4} - \frac{E_Q}{E}} + \sqrt[2]{\frac{\frac{g^2r^4}{c^4}}{\frac{g^2r^4}{c^4} - \frac{E_Q r^2}{E}}}\,\mathrm{Sinh}\sqrt[2]{\frac{g^2r^2}{c^4} - \frac{E_Q}{E}} \qquad (3.14)$$

for $g^2r^4 / c^4 - E_Q r^2 / E \neq 0$.

As the gravitational field of the charged body of mass approaches zero at a distance r from the body, then the factor of spatiotemporal growth approaches one, i.e. space-time is nearly flat, in the absence of any other gravitational fields. As the factor of spatiotemporal growth differs from one, the curvature of space would change, and the spatiotemporal metric, $ds^2 e^{\pm\sigma}$, would also be affected. The exponential factor of spatiotemporal expansion modulates the measurable spatiotemporal metric, $g_{\mu\nu}$. However, the speed of light if measured would be the same. The straightedges and clocks would measure the spatiotemporal metric.

Therefore, it is possible to suggest that electromagnetism may have an inherent gravitational nature, and vice versa. Such a gravitational nature provides an alternative approach for the possible theoretical manipulation of electromagnetism to affect the metric of the spatiotemporal medium about a charged body of mass.

The motion of the body of mass in four spatiotemporal dimensions proceeds against the gradient of the scalar field ϕ given by

$$\Box\phi = \sum_{0}^{3} \frac{\partial \phi}{\partial x_n} = -\frac{i}{c}\frac{\partial \phi}{\partial x_0} + \frac{\partial \phi}{\partial x_1} + \frac{\partial \phi}{\partial x_2} + \frac{\partial \phi}{\partial x_3} \qquad (3.15)$$

and the potential field is described by the d'Alembertian as follows:

$$\Box^2\phi = -\frac{\partial^2 \phi}{\partial x_0{}^2} + \frac{\partial^2 \phi}{\partial x_1{}^2} + \frac{\partial^2 \phi}{\partial x_2{}^2} + \frac{\partial^2 \phi}{\partial x_3{}^2} = \nabla^2\phi - \frac{1}{c^2}\frac{\partial^2 \phi}{\partial t^2} = \nabla^2\phi - \frac{\partial^2 \phi}{\partial r^2} \qquad (3.16)$$

Let us imagine a thought experiment of a moving charged body of mass with a strong electromagnetic field that modulates the spatiotemporal curvature of the medium about the body of mass to provide a contracting spatiotemporal bridge, which may be directed and traversed in the direction of the advanced wave between two spatiotemporal points. The spatiotemporal region around the moving body of mass would contract as outer unaffected space-time expands if unobstructed. The object would move through previous spatiotemporal structures toward its destination. Certainly, an equation that represents such a fantastic scenario would include terms of curvature, kinetic energy, and electromagnetic energy.

Let us examine an equation for the electromagnetic energy, kinetic energy, and the spatiotemporal curvature of the medium, for the above thought experiment. (Puthoff, 2002)

$$\nabla^2 e^{\frac{gr}{c^2}} - e^{4\frac{gr}{c^2}} \frac{d^2 e^{\frac{gr}{c^2}}}{dr^2} = -\frac{e^{\frac{gr}{c^2}}}{\frac{c^4}{8\pi G}} \left\{ \left[\frac{m_0 c^2}{2e^{\frac{gr}{c^2}}} \left(\frac{1+\frac{v^2}{c^2} e^{2\frac{gr}{c^2}}}{\sqrt[2]{1-\frac{v^2}{c^2} e^{2\frac{gr}{c^2}}}} \right) \delta^3(r) \right] + \right. \tag{3.17}$$

$$\left. \frac{1}{2}\left(\frac{\frac{B^2}{\mu_0}}{e^{2\frac{gr}{c^2}}} + \varepsilon_0 E^2 e^{2\frac{gr}{c^2}} \right) - \frac{\frac{c^4}{32\pi G}}{e^{4\frac{gr}{c^2}}} \left[\left(\nabla e^{2\frac{gr}{c^2}} \right)^2 + e^{4\frac{gr}{c^2}} \left(\frac{de^{2\frac{gr}{c^2}}}{dr} \right)^2 \right] \right\} \tag{3.18}$$

$$-\frac{\nabla^2 e^{\frac{gr}{c^2}}}{e^{\frac{gr}{c^2}}} + e^{3\frac{gr}{c^2}} \frac{d^2 e^{\frac{gr}{c^2}}}{dr^2} = \frac{8\pi G}{c^4} \left\{ \left[\frac{1}{e^{\frac{gr}{c^2}}} \frac{1}{2} m_0 c^2 \left(\frac{1+\frac{v^2}{c^2} e^{2\frac{gr}{c^2}}}{\sqrt[2]{1-\frac{v^2}{c^2} e^{2\frac{gr}{c^2}}}} \right) \delta^3(r) \right] + \right. \tag{3.19}$$

$$\left. \left(\frac{1}{2}\frac{B^2}{\mu_0} e^{-2\frac{gr}{c^2}} + \frac{1}{2}\varepsilon_0 E^2 e^{2\frac{gr}{c^2}} \right) - \frac{1}{4}\left(\frac{c^4}{8\pi G} \right) \left[\left(\frac{\nabla e^{2\frac{gr}{c^2}}}{e^{2\frac{gr}{c^2}}} \right)^2 + \left(\frac{de^{2\frac{gr}{c^2}}}{dr} \right)^2 \right] \right\} \tag{3.20}$$

Simplifying all the curvature terms to the left side of the equation we get

$$\left(\frac{c^4}{8\pi G}\right)\left\{-\frac{\nabla^2 e^{\frac{gr}{c^2}}}{e^{\frac{gr}{c^2}}}+e^{3\frac{gr}{c^2}}\frac{d^2 e^{\frac{gr}{c^2}}}{dr^2}+\frac{1}{4}\left(\frac{\nabla e^{2\frac{gr}{c^2}}}{e^{2\frac{gr}{c^2}}}\right)^2+\frac{1}{4}\left(\frac{de^{2\frac{gr}{c^2}}}{dr}\right)^2\right\}= \tag{3.21}$$

$$=\frac{1}{2}m_0 c^2 e^{-\frac{gr}{c^2}}\left(\frac{1+\frac{v^2}{c^2}e^{2\frac{gr}{c^2}}}{\sqrt[2]{1-\frac{v^2}{c^2}e^{2\frac{gr}{c^2}}}}\right)\delta^3(r)+\frac{1}{2}\frac{B^2}{\mu_0}e^{-2\frac{gr}{c^2}}+\frac{1}{2}\varepsilon_0 E^2 e^{2\frac{gr}{c^2}} \tag{3.22}$$

After differentiating terms of curvature, where $r = f(x, y, z)$, we obtain

$$\left(\frac{c^4}{8\pi G}\right)\left\{-\frac{g^2}{c^4}\frac{e^{\frac{gr}{c^2}}}{e^{\frac{gr}{c^2}}}+\frac{g^2}{c^4}e^{3\frac{gr}{c^2}}e^{\frac{gr}{c^2}}+\frac{1}{4}\left(\frac{4g^2}{c^4}\frac{e^{4\frac{gr}{c^2}}}{e^{4\frac{gr}{c^2}}}\right)+\frac{1}{4}\left(\frac{4g^2}{c^4}e^{4\frac{gr}{c^2}}\right)\right\}=\left(\frac{c^4}{8\pi G}\right)\frac{\nabla^2 e^{2\frac{gr}{c^2}}}{2e^{-2\frac{gr}{c^2}}}= \tag{3.23}$$

$$=\frac{1}{2}m_0 c^2\left(\frac{e^{-\frac{gr}{c^2}}+\frac{v^2}{c^2}e^{\frac{gr}{c^2}}}{\sqrt[2]{1-\frac{v^2}{c^2}e^{2\frac{gr}{c^2}}}}\right)\delta^3(r)+\frac{1}{2}\frac{B^2}{\mu_0}e^{-2\frac{gr}{c^2}}+\frac{1}{2}\varepsilon_0 E^2 e^{2\frac{gr}{c^2}} \tag{3.24}$$

$$\left(\frac{c^4}{8\pi G}\right)2\frac{g^2}{c^4}e^{4\frac{gr}{c^2}}=\left(\frac{g^2}{4\pi G}\right)e^{4\frac{gr}{c^2}}=\frac{1}{2}m_0 c^2\left(\frac{e^{-\frac{gr}{c^2}}+\frac{v^2}{c^2}e^{\frac{gr}{c^2}}}{\sqrt[2]{1-\frac{v^2}{c^2}e^{2\frac{gr}{c^2}}}}\right)\delta^3(r)+\frac{1}{2}\frac{B^2}{\mu_0}e^{-2\frac{gr}{c^2}}+\frac{1}{2}\varepsilon_0 E^2 e^{2\frac{gr}{c^2}} \tag{3.25}$$

Let us express the factor of spatiotemporal expansion as $\sigma = e^{gr/c^2}$ to simplify the above equation. The spatial Delta function $\delta^3(r)$ has units of $1/m^3$. It is interesting to note how the electromagnetic field in this equation becomes less magnetic and more electrical with spatiotemporal divergence. During universal spatiotemporal expansion the opposite effect is predicted. Thus, the equation for the resultant four-dimensional scalar field ϕ is given by

$$\Box\phi = \frac{g^2}{4\pi G}\sigma^4 = \frac{1}{2}m_0c^2\left(\frac{\frac{1}{\sigma}+\frac{v^2}{c^2}\sigma}{\sqrt[2]{1-\frac{v^2}{c^2}\sigma^2}}\right)\delta^3(r)+\frac{1}{2}\frac{B^2}{\mu_0}\frac{1}{\sigma^2}+\frac{1}{2}\varepsilon_0E^2\sigma^2 \qquad (3.26)$$

$$\Box\phi = 4\lambda G_{\mu\nu} = \left(\frac{c^4}{8\pi G}\right)\frac{\nabla^2 e^{2\frac{gr}{c^2}}}{2e^{-2\frac{gr}{c^2}}} \qquad (3.27)$$

Where the constant 4λ is equal to the reciprocal of Einstein's constant for energy density, $c^4/8\pi G$.

Multiplying top and bottom of the fraction by the factor of spatiotemporal expansion, we obtain

$$\frac{\left(\sqrt[2]{1+\frac{v^2}{c^2}\sigma^2}\right)^2}{\left(\sqrt[2]{(1)\sigma^2-\frac{v^2}{c^2}\sigma^4}\right)} = \frac{\left(\sqrt[2]{\left(\frac{e^{\frac{gr}{c^2}}+e^{-\frac{gr}{c^2}}}{e^{\frac{gr}{c^2}}+e^{-\frac{gr}{c^2}}}\right)^2+\left(\frac{e^{\frac{gr}{c^2}}-e^{-\frac{gr}{c^2}}}{e^{\frac{gr}{c^2}}+e^{-\frac{gr}{c^2}}}\right)^2 e^{2\frac{gr}{c^2}}}\right)^2}{\sqrt[2]{\left(\frac{e^{\frac{gr}{c^2}}+e^{-\frac{gr}{c^2}}}{e^{\frac{gr}{c^2}}+e^{-\frac{gr}{c^2}}}\right)^2 e^{2\frac{gr}{c^2}}-\left(\frac{e^{\frac{gr}{c^2}}-e^{-\frac{gr}{c^2}}}{e^{\frac{gr}{c^2}}+e^{-\frac{gr}{c^2}}}\right)^2 e^{4\frac{gr}{c^2}}}} \qquad (3.28)$$

If we examine the content of the fraction, the numerator illustrates the additive effect of the retarded and advanced wavelets that is augmented at a greater rate than the denominator. The denominator illustrates the subtractive effect of the retarded and advanced wavelets. Thus, the resultant effect is an increase in kinetic energy at a velocity that is faster than the speed of light.

The above simplified equation relates to the General Theory of Relativity for celestial bodies of mass, but it represents the warping of space-time in the direction of propagation of a charged body of dilated mass moving faster than the speed of light. The electromagnetic energy warps the spatiotemporal curvature to tell the body of mass how to move faster than the speed of light relative to the direction of the retarded spatiotemporal wavelet. However, the body of mass is moving relativistically at less than the speed of light in the direction of the advanced spatiotemporal wavelet.

Simplifying the fraction further to re-introduce it into the equation,

$$\frac{g^2}{4\pi G}\sigma^4=\frac{1}{2}m_0c^2\left(\frac{2}{\sqrt[2]{1-\frac{v^2}{c^2}\sigma^2}}-\sqrt[2]{1-\frac{v^2}{c^2}\sigma^2}\right)\frac{\delta^3(r)}{\sigma}+\frac{1}{2}\frac{B^2}{\mu_0}\frac{1}{\sigma^2}+\frac{1}{2}\varepsilon_0E^2\sigma^2 \quad (3.29)$$

$$\frac{g^2}{4\pi G}\sigma^4=\left(\frac{m_0c^2}{\sqrt[2]{1-\frac{v^2}{c^2}\sigma^2}}-\frac{1}{2}\left(m_0\sqrt[2]{1-\frac{v^2}{c^2}\sigma^2}\right)c^2\right)\frac{\delta^3(r)}{\sigma}+\frac{1}{2}\frac{B^2}{\mu_0}\frac{1}{\sigma^2}+\frac{1}{2}\varepsilon_0E^2\sigma^2 \quad (3.30)$$

$$\frac{g^2}{4\pi G}\sigma^4=\left(\frac{1}{2}\frac{m_0c^2}{\sqrt[2]{1-\frac{v^2}{c^2}\sigma^2}}+\frac{1}{2}\frac{m_0v^2\sigma^2}{\sqrt[2]{1-\frac{v^2}{c^2}\sigma^2}}\right)\frac{\delta^3(r)}{\sigma}+\frac{1}{2}\frac{B^2}{\mu_0}\frac{1}{\sigma^2}+\frac{1}{2}\varepsilon_0E^2\sigma^2 \quad (3.31)$$

$$\frac{g^2}{4\pi G}\sigma^4=\left(\frac{1}{2}\frac{m_0\left(c^2+v^2\sigma^2\right)}{\sqrt[2]{1-\frac{v^2}{c^2}\sigma^2}}\right)\frac{\delta^3(r)}{\sigma}+\frac{1}{2}\frac{B^2}{\mu_0}\frac{1}{\sigma^2}+\frac{1}{2}\varepsilon_0E^2\sigma^2 \quad (3.32)$$

It is interesting to note that the rest mass dilates, but the mass of the kinetic energy contracts while it subtracts from the rest mass, as the charged body of mass traverses the spatiotemporal vortex, a non-trivial topology. Hence, it is possible to suggest that the rest mass dilates as the body of mass moves at less than c as it would be expected, while the kinetic mass contracts as the charged body of mass moves at greater than c in the direction of the advanced spatiotemporal wavelet. Therefore, for relativistic motion in homogeneous and isotropic space-time in either the retarded or the advanced direction, mass and time dilate and space contracts to obey the laws of conservation of spatial momenta.

Moreover, the rest mass dilates as a result of the velocity of time decreasing as the object speeds up through space, and this effect increases the momentum of the object and its kinetic energy, as the object travels against the increasing gravitational field about the mass opposite to the direction of motion, in the direction of the advanced spatiotemporal wavelet.

The above hypothetical scenario resembles a charged body of mass traversing a spatiotemporal rift, or an Ellis-Bronnikov bridge.

$$\Box\phi = 4\lambda G_{\mu\nu} = \frac{1}{\sigma}\nabla_\mu\phi_a\nabla_\nu\phi_a g_{\mu\nu} - \frac{1}{2\sigma}\nabla_\mu\phi_b\nabla_\nu\phi_b g_{\mu\nu} + \frac{1}{2\sigma^2}\nabla_\mu\phi_\beta\nabla_\nu\phi_\beta g_{\mu\nu} + \frac{\sigma^2}{2}\nabla_\mu\phi_\varepsilon\nabla_\nu\phi_\varepsilon g_{\mu\nu} \qquad (3.33)$$

The above tensor field equation corresponds to two asymptotically flat spatiotemporal regions joined at a singularity-free two-sphere, geodesically complete, without one-way event horizons. The bridge connects the spatiotemporal regions that are asymptotically flat in each direction of recession from the tunnel in the middle. The bridge is gravitationally attractive on one side and strongly repulsive on the other side. The spatiotemporal vector field is the velocity field of the spatiotemporal flow from the attractive side to the repulsive side of the bridge, with an accelerating gravitational field all the way through. The bridge is traversable in either direction by photons, particles, or signals.

How would the resultant curvature of the scalar field of a body of mass be affected by the mass of the body, the cosmological matter, and the spatiotemporal-pressure-modifying electromagnetic field?

From previous research we have

$$G_{\mu\nu}+\bar{G}_{\varepsilon\beta}=\frac{8\pi G}{c^4}\left(T_{\mu\nu}-\Lambda_{\mu\nu}+\Phi_{\varepsilon\beta}\right) \tag{3.34}$$

$$G_{\mu\nu}+\bar{G}_{\varepsilon\beta}=\kappa\left(T_{\mu\nu}-\Lambda_{\mu\nu}+\Phi_{\varepsilon\beta}\right) \tag{3.35}$$

$$\hat{R}_{\mu\nu}-\frac{1}{(n-1)}g_{\mu\nu}\hat{R}=\kappa\left(T_{\mu\nu}-\Lambda_{\mu\nu}+\hat{\Phi}_{\mu\nu}\right) \tag{3.36}$$

Denoting the stress-energy-momentum tensors we have

$$T_{\mu\nu}=-\frac{2\rho\phi}{m}\left(\nabla_{\mu}\phi\nabla_{\nu}\phi-\frac{1}{2}g_{\mu\nu}\nabla^{\alpha}\phi\nabla_{\alpha}\phi\right)=-\frac{2\rho\phi}{m}\left(\nabla_{\mu}\phi\nabla^{\mu}\phi-2\rho^2\phi^2\right)g_{\mu\nu} \tag{3.37}$$

$$\Lambda_{\mu\nu}=2\lambda\phi g_{\mu\nu} \tag{3.38}$$

$$\hat{\Phi}_{\mu\nu}=-\frac{\rho}{\kappa}\left(\nabla_{\mu}\phi\nabla_{\nu}\phi-\frac{1}{2}g_{\mu\nu}\nabla_{\mu}\phi\nabla^{\mu}\phi\right)-\frac{g_{\mu\nu}}{\kappa}\Box\phi=\frac{1}{\kappa}\left(\rho\phi\nabla_{\mu}\nabla^{\mu}\phi-n\delta_{\mu\nu}\Box\phi\right) \tag{3.39}$$

The trace of the resultant curvature from the local, cosmological, and electromagnetic field energy densities is $\hat{R}$, and κ is Einstein's constant for energy density, λ is a cosmological term for pressure, and n is the number of spatiotemporal dimensions.

$$\hat{R}\phi=-\frac{2\kappa\rho\phi}{m}\left[\nabla_{\mu}\phi\nabla^{\mu}\phi-2\rho^2\phi^2\right]+\rho\phi\nabla_{\mu}\nabla^{\mu}\phi-2\kappa\lambda\phi-n\delta_{\mu\nu}\Box\phi \tag{3.40}$$

Multiplying the entire equation by the field ϕ to simplify terms,

$$\hat{R}\phi^2=-\frac{2\kappa\rho\phi^2}{m}\left[\nabla_{\mu}\phi\nabla^{\mu}\phi-2\rho^2\phi^2\right]+\rho\phi^2\nabla_{\mu}\nabla^{\mu}\phi-2\kappa\lambda\phi^2-n\delta_{\mu\nu}\Box\phi^2 \tag{3.41}$$

$$\hat{R}\phi^2=-\frac{2\kappa\rho\phi^2}{m}\left[\nabla_{\mu}\phi\nabla^{\mu}\phi-2\nabla^{\alpha}\phi\nabla_{\alpha}\phi+\frac{m}{2\kappa}\nabla_{\mu}\nabla^{\mu}\phi\right]-2\kappa\lambda\phi^2-n\delta_{\mu\nu}\Box\phi^2 \tag{3.42}$$

$$\hat{R}\phi^2 = -\frac{2\kappa\rho\phi^2}{m}\left[\nabla_\mu\phi\nabla^\mu\phi + \frac{m}{2\kappa\phi}\nabla_\mu\phi\nabla^\mu\phi - 2\nabla^\alpha\phi\nabla_\alpha\phi\right] - 2\kappa\lambda\phi^2 - n\delta_{\mu\nu}\Box\phi^2 \quad (3.43)$$

Substituting the scalar field $\phi = m/\rho$ for pressure to subtract terms,

$$\hat{R}\phi^2 = -\frac{2\kappa\rho\phi^2}{m}\left[\nabla_\mu\phi\nabla^\mu\phi + \frac{\rho}{2\kappa}\nabla_\mu\phi\nabla^\mu\phi - 2\nabla^\alpha\phi\nabla_\alpha\phi\right] - 2\kappa\lambda\phi^2 - n\delta_{\mu\nu}\Box\phi^2 \quad (3.44)$$

Substituting terms and simplifying,

$$\hat{R}\phi^2 = -\frac{2\kappa\rho\phi^2}{m}\left[\left(1 + \frac{\rho}{2\kappa}\right)\nabla^\alpha\phi\nabla_\alpha\phi - 2\nabla^\alpha\phi\nabla_\alpha\phi\right] - 2\kappa\lambda\phi^2 - n\delta_{\mu\nu}\Box\phi^2 \quad (3.45)$$

Since $|\rho| >> |\kappa|$, where $|\kappa| = 8\pi G/c^4 \approx 10^{-43}$, we obtain

$$\left|1 + \frac{\rho}{2\kappa}\right| = \left|\frac{2\kappa + \rho}{2\kappa}\right| \approx 1 \quad (3.46)$$

$$\hat{R}\phi^2 = -\frac{2\kappa\rho\phi^2}{m}\left[\nabla^\alpha\phi\nabla_\alpha\phi - 2\nabla^\alpha\phi\nabla_\alpha\phi\right] - 2\kappa\lambda\phi^2 - n\delta_{\mu\nu}\Box\phi^2 \quad (3.47)$$

The exponential factor of spatiotemporal expansion modulates the measurable spatiotemporal metric, $g_{\mu\nu}$, so, the speed of light if measured would be the same, and *n* is the number of spatiotemporal dimensions. In that sense, an electromagnetic field in free space-time is a gauge field. Light may be considered a universal gauge.

$$\hat{R}\phi^2 = -\frac{2\kappa\rho\phi^2}{m}\left[-\nabla^\alpha\phi\nabla_\alpha\phi\right] - 2\kappa\lambda\phi^2 - g_{\mu\nu}\Box\phi^2 \quad (3.48)$$

$$2\kappa\lambda\phi^2 = 2\kappa\lambda\left(\frac{m}{\rho}\right)\phi = 2\kappa\left[\phi^3\left(\sqrt[2]{\rho}\right)\right]\left(\frac{\lambda}{\sqrt[2]{\rho}}\right)\left(\frac{1}{\phi}\right) = 2\kappa\phi\left(\frac{\phi m}{\sqrt[2]{\rho}}\right)\left(\frac{\lambda}{\sqrt[2]{\rho}}\right)\left(\frac{1}{\phi}\right) \quad (3.49)$$

Substituting with the quantum mechanical wave equation describing the physical changes for the field ϕ over space and time, we have

$$2\kappa\lambda\phi^2 = 2\kappa\phi\left(\frac{\phi m}{\sqrt[2]{\rho}}\right)\left(\frac{\lambda}{\sqrt[2]{\rho}}\right)\left(\frac{1}{\phi}\right) = 2\left(\frac{\phi\kappa m}{\sqrt[2]{\rho}}\right)\left\{-\frac{1}{2}\frac{\hbar^2}{m}\left[\Box\left(\frac{\lambda}{\sqrt[2]{\rho}}\right)+\left(\frac{\lambda\Box\sqrt[2]{\rho}}{\rho}\right)\right]\right\} \quad (3.50)$$

Since $\Box \equiv \kappa/\phi = \kappa\rho/m$ we obtain

$$\frac{2\kappa\lambda\phi^2}{\phi^2} = 2\left(-\frac{1}{2}\right)\left(\frac{\nabla_\mu\phi\nabla_\nu\phi}{\phi^2}\right) = -\left(\frac{\nabla_\mu\phi\nabla_\nu\phi}{\phi^2}\right) \quad (3.51)$$

Denoting the resultant curvature using the above terms, we get

$$\hat{R}\phi^2 = \frac{2\kappa\rho\phi^2}{m}\left[\nabla^a\phi\nabla_a\phi\right] - \frac{2\phi\kappa m}{\sqrt[2]{\rho}}\left\{-\frac{1}{2}\frac{\hbar^2}{m}\left[\Box\left(\frac{\lambda}{\sqrt[2]{\rho}}\right)+\left(\frac{\lambda\Box\sqrt[2]{\rho}}{\rho}\right)\right]\right\} \quad (3.52)$$
$$-\frac{\kappa\rho}{m}g_{\mu\nu}\phi^2$$

$$\hat{R} = \frac{2\kappa\rho\phi^2}{m\phi^2}\left(\nabla^\mu\phi\nabla_\nu\phi\right) - \left(-\frac{\nabla_\mu\phi\nabla_\nu\phi}{\phi^2}\right) - \frac{\kappa\rho}{m}g_{\mu\nu} \quad (3.53)$$

Therefore, the resultant curvature is given by

$$\hat{R} = \frac{2\kappa\rho}{m}\left(\nabla^\mu\phi\nabla_\nu\phi\right) + \left(\frac{\nabla_\mu\phi\nabla_\nu\phi}{\phi^2}\right) - \frac{\kappa\rho}{m}g_{\mu\nu} \quad (3.54)$$

Similarly, in terms of the Lagrangian action *S*, and the Hamiltonian $\hat{H}$, for the relativistic motion of the mass traveling through an Ellis-Bronnikov bridge, we find

$$S = \int_{t_1}^{t_2} L\,dt = \int_{t_1}^{t_2} -mc^2\left(\frac{d\tau}{dt}\right)dt = \int_{t_1}^{t_2} -mc^2\left(\sqrt[2]{1-\frac{v_r^{\ 2}}{c^2}}\right)dt \quad (3.55)$$

$$p = \frac{\partial L}{\partial v_r} = -\frac{1}{2}\left[\frac{mc^2(-2v_r)}{c^2\sqrt[2]{1-\frac{v_r^{\,2}}{c^2}}}\right] = \frac{mv_r}{\sqrt[2]{1-\frac{v_r^{\,2}}{c^2}}} = \frac{mv_r}{v_\tau} \tag{3.56}$$

Through pair production we obtain

$$\hat{H} = v_r p_r - L = \frac{m_0 v_r^{\,2}}{\sqrt[2]{1-\frac{v_r^{\,2}}{c^2}}} - \left[-m_0 c^2 \sqrt[2]{1-\frac{v_r^{\,2}}{c^2}}\right] = \frac{m_0 c^2}{\sqrt[2]{1-\frac{v_r^{\,2}}{c^2}}} = \frac{m_0 c^2}{v_\tau} \tag{3.57}$$

It is possible to consider the case of the Hamiltonian of the system as a two-body problem, or pair production for the body of mass, where the Hamiltonian may be expressed as the sum of the dilating Hamiltonian and the contracting Hamiltonian, in the directions of the retarded and the advanced spatiotemporal waves. It is possible to hypothesize that the body of mass is projected on the retarded wave onto the spatiotemporal dimensions of the future as it dilates, whereas the body of mass is projected on the advanced wave onto the spatiotemporal dimensions of the past as it contracts. Thus, it is possible to consider the projections only as projections onto spatial dimensions, or spatial layers, by factoring out the temporal derivatives involved. It has been observed that the Hamiltonian energy of a system allows for pair production for a particle of mass. Would pair production or pair annihilation be observable for a system of particles?

The two-body problem approach predicts the individual motions of each object of mass interacting with each other gravitationally and regulating the spatiotemporal pressure.

$$\hat{H}_d = \hat{T}_r + \hat{U} \tag{3.58}$$

$$\hat{H}_c = -\hat{L} - \hat{U} \tag{3.59}$$

$$\hat{H} = \hat{H}_d + \hat{H}_c \tag{3.60}$$

Where $T_r = v_r p_r$ is kinetic energy, $-L$ is potential energy, and $\pm U$ is self-potential energy or one-body potential energy.

Let us denote the Hamiltonian as a spatial three-dimensional integral of the dilating Hamiltonian and the contracting Hamiltonian as a formulation of the General Theory of Relativity. The spectrum of the formulation is the set of possible outcomes when one measures the total energy of a system.

$$\hat{H} = \int d^4x \left(\frac{mc^2}{v_\tau} \right) = \hat{H}_D + \hat{H}_C = \int d^3r \left(W^\perp \hat{H}^\perp + Wi\hat{H} \right) \tag{3.61}$$

$$d^4x = d^3r \cdot dt \tag{3.62}$$

$$W = \left(-g^{00}\right)^{-\frac{1}{2}} = -\frac{i}{\sqrt[2]{g^{00}}} \tag{3.63}$$

$$Wi = \left(g^{00}\right)^{-\frac{1}{2}} = \frac{1}{\sqrt[2]{g^{00}}} \tag{3.64}$$

The $W-$denominator removes all temporal derivatives to yield the spatial derivatives only. The "i" symbol represents the three imaginary directions onto a surface. The projection operator on the Hamiltonian operator, $\hat{H}^\perp$, or on $W^\perp$, denotes the four-dimensional, or higher dimensional, spatiotemporal mapping to its three-dimensional form. It may be helpful to think of the way a lighted three-dimensional object would project onto the ground plane as a two-dimensional shadow.

The above formalism proposes that space-time is foliated into a family of space-like hypersurfaces, labeled by their constant coordinate time t^i, and where each spatial slice has coordinates given by x^i. The metric tensor of the three-dimensional spatial slices and their conjugate momenta are the dynamic variables. A Hamiltonian may be defined from these variables, and thereby the equations of motion may be written for the General Theory of

Relativity in the form of the equations of Hamilton. (Arnowitt, 1959) As the body of mass is projected on the advanced wave, onto the spatiotemporal dimensions of the past as it contracts into the Ellis-Bronnikov bridge, the electromagnetic field force of the body of mass pulsates as a wave with a magnetic field component and an electric field component given by

$$\left(\frac{c^4}{8\pi G}\right)2\frac{g^2}{c^4}\sigma^4 = \frac{1}{2}\frac{B^2}{\mu_0}\frac{1}{\sigma^2}+\frac{1}{2}\varepsilon_0 E^2\sigma^2 \tag{3.65}$$

$$\frac{g^2}{8\pi G}\sigma^4 = \frac{1}{4}\left(\frac{B^2}{\mu_0}\frac{1}{\sigma^2}+\varepsilon_0 E^2\sigma^2\right) \tag{3.66}$$

The torsion and modulation of the electromagnetic field has the effect of shaping, widening, and stiffening, the spatiotemporal walls (surfaces) of the vortex in the tunnel of the bridge. As the body of mass moves through the bridge, the spatiotemporal pressure is lower behind the body of mass, and higher in front of the body of mass in the direction of propagation.

What is the force of the electromagnetic field onto the initial opening surface of the hypothetical Ellis-Bronnikov bridge?

$$\frac{g^2}{8\pi G}\sigma^4 = \frac{1}{4}F_\beta R_\beta\frac{1}{\sigma^2}+\frac{1}{4}F_\varepsilon R_\varepsilon\sigma^2 \tag{3.67}$$

Where R is the trace of the curvature tensor and σ is the factor of spatiotemporal growth.

$$\frac{g^2}{8\pi G}\sigma^4 = \frac{1}{4}\frac{\vec{B}^2}{\mu_0}\frac{1}{\sigma^2}+\frac{1}{4}\varepsilon_0\vec{E}^2\sigma^2 = \frac{1}{4}\frac{c^2\vec{E}_\beta{}^2}{\mu_0\sigma^2}+\frac{1}{4}\varepsilon_0\vec{E}_\varepsilon{}^2\sigma^2 \tag{3.68}$$

Let us consider an instant $(\partial t'/\partial t_0)$ in the temporal domain,

$$\frac{g^2}{8\pi G}\sigma^4 \rightarrow \frac{1}{4}\frac{c^2\vec{E}_\beta{}^2}{\mu_0}\frac{1}{\sigma^2}+\frac{1}{4}\varepsilon_0\vec{E}_\varepsilon{}^2\sigma^2 \tag{3.69}$$

$$\frac{g^2}{8\pi G}\left(\frac{\partial t'}{\partial t_0}\right)^4 = \frac{1}{4}\frac{c^2\vec{E}_\beta^{\,2}}{\mu_0}\frac{1}{\left(\frac{\partial t'}{\partial t_0}\right)^2}+\frac{1}{4}\varepsilon_0\vec{E}_\varepsilon^{\,2}\left(\frac{\partial t'}{\partial t_0}\right)^2 \quad (3.70)$$

Substituting from electromagnetic terms to equivalent electrogravitic terms in the frequency domain on the right side of the equation.

$$\frac{g^2}{4\pi G}e^{4\frac{\mu'}{\mu_0}} = \frac{1}{2}\left[q_1V_1\left\langle\delta^3(r)\right\rangle e^{-2\frac{\mu'}{\mu_0}}+q_2V_2\left\langle\delta^3(r)\right\rangle e^{2\frac{\mu'}{\mu_0}}\right] \quad (3.71)$$

Where the electromagnetic term $q_1V_1\delta^3(r)$ denotes magnetic field variables, and the term $q_2V_2\delta^3(r)$ denotes electric field variables.

The voltage is given by Faraday's voltage law,

$$V = N\frac{\partial\phi}{\partial t} \quad (3.72)$$

$$\phi = \int B\, dA \quad (3.73)$$

$$\frac{\partial\phi}{\partial t} = A\cdot\frac{\partial B}{\partial t} \quad (3.74)$$

Where N is the number of turns (loops) in a coil, ϕ is the magnetic flux per turn, or magnetic lines through a turn per square meter, dA is the area of one turn of the coil, and $\partial\phi/\partial t$ is the velocity of the magnetic flux through one turn, or how fast the magnetic flux is changing through one turn with respect to time.

Let us introduce two rotating perpendicular field forces $(F_1 \perp F_2)$, the variable F_1 is for a magnetic field force, F_2 is for an electric field force, and let us add a negative sign on the electric field acceleration.

$$\frac{\theta}{2} = \frac{90^0}{(nN)} \tag{3.75}$$

$$\theta = \frac{180^0}{(nN)} \tag{3.76}$$

Where n is the number of Newtons of the force that rotates about the other force's axis and determines the degrees per second of rotation, and r is equal to zero for the factor of spatiotemporal expansion σ.

$$\frac{g^2}{4\pi G} = \left(\frac{F_1}{x_1}\frac{1}{v^2\lambda^2}\right)\left(-\frac{1}{2}\frac{F_2}{x_2}v^2\lambda^2\right)(\theta) = \left(\frac{F_1}{x_1}\frac{1}{\lambda}\right)\left(-\frac{1}{2}\frac{F_2}{x_2}\frac{1}{\lambda}\lambda^2\right)(\theta) \tag{3.77}$$

$$\frac{g^2}{4\pi G} = \left(\frac{F_1}{x_1}\frac{1}{\lambda}\right)\left(-\frac{1}{2}\frac{F_2}{x_2}\frac{1}{\lambda}\lambda^2\right)(\theta) = -(\mu_0)(\mu_0\cdot\lambda^2)\left(\frac{\theta}{2}\right) = -(4\pi \text{ x } 10^{-7})^2\cdot\lambda^2\cdot\left(\frac{\theta}{2}\right) \tag{3.78}$$

$$\frac{g^2}{4\pi G} = -1.6\pi^2 \text{ x } 10^{-13}\ \frac{N^2}{m^4}\cdot\lambda^2\cdot\left(\frac{\theta}{2}\right) = -1.6\pi^2 \text{ x } 10^{-13}\ \frac{N^2}{m^2}\cdot\left(\frac{\theta}{2}\right) \tag{3.79}$$

$$\frac{g^2}{4\pi G} = \left(\frac{F_1}{x_1}\frac{1}{v^2}\right)\left(-\frac{1}{2}\frac{c^2}{x_2}\frac{m_2\cdot v^2}{x_2}\right)(\theta) = (m_1)\left(\frac{c^2}{x_2}\right)\left(-\frac{1}{2}\frac{m_2\cdot v^2}{x_2}\right)(\theta) \tag{3.80}$$

$$\frac{g^2}{4\pi G} = (m_1)\left(\frac{c^2}{x_2}\right)\left(\frac{1}{\lambda^2}\right)\left(-\frac{1}{2}\frac{m_2\cdot v^2}{x_2}\right)(\lambda^2)(\theta) = \left(\frac{F_1}{\lambda}\right)\left(\frac{F_2}{\lambda}\right)\left(\frac{1}{2}\frac{180^0}{(nN)}\right) \text{ x } 10^{-13}\frac{N}{m^2} \tag{3.81}$$

$$\frac{g^2}{4\pi G} = \left(\frac{F_1}{\lambda}\right)\left(\frac{F_2}{\lambda}\right)\left(\frac{\theta}{2}\right) \text{ x } 10^{-13}\frac{N}{m^2} \tag{3.82}$$

Therefore, denoting the field equation in terms of energy density, or pressure, we find the force of the electromagnetic field onto the initial opening surface of the hypothetical Ellis-Bronnikov bridge.

$$\frac{g^2}{4\pi G} = \frac{1}{2}\cdot q_1 V_1 R_1 \frac{1}{v^2}\cdot q_2 V_2 R_2 v^2 \cdot \theta \cdot 10^{-13}\frac{N}{m^2} \tag{3.83}$$

Where q is charge, V is voltage, v is frequency, and R is the trace of curvature.

What is the gravitational constant in terms of the permeability of free space-time?

Simplifying terms, the gravitational constant is given by

$$\frac{1}{G} = \frac{-\mu_0^{\,2}\cdot\lambda^2}{g^2}\cdot\frac{\theta}{2} = \frac{-6.4\pi^3 \text{ x } 10^{-13}\, N^2/m^2}{g^2}\cdot\left(\frac{\theta}{2}\right) \tag{3.84}$$

$$G = \frac{g^2}{-6.4\pi^3 \text{ x } 10^{-13}\,\frac{N^2}{m^2}\cdot\left(\frac{\theta}{2}\right)}\ \frac{m^3}{Kg\cdot s^2} \tag{3.85}$$

The gravitational constant illustrates a direct connection between electromagnetism and space-time, as space-time expands or contracts. In other words, the spatiotemporal medium has an electrogravitic nature, i.e. an electromagnetic aspect and a spatiotemporal aspect.

An analysis of Alcubierre's warp drive

Let us consider the theoretical proposal for a spatiotemporal warp drive by the eminent physicist Miguel Alcubierre. (Alcubierre, 1994)

$$ds^2 = -d\tau^2 = g_{\alpha\beta}dx^\alpha dx^\beta \tag{3.86}$$

$$ds^2 = -\left(\alpha^2 - \beta_i\beta^i\right)dt^2 + 2\beta_i dx^i dt + \gamma_{ij}dx^i dx^j \tag{3.87}$$

Where β^i is a shift vector for spatial coordinate hypersurfaces which vanishes at $r_s < R$, γ_{ij} is the three-metric of the hypersurfaces, and α is the lapse function that gives the interval of proper time between nearby hypersurfaces as measured by the Eulerian observers, i.e. those observers whose four-velocity is normal to the hypersurfaces. R is a radial distance from the warp drive. The following metric will propel the spaceship along a trajectory described by an arbitrary function of time, $x_s(t)$.

$$ds^2 = -dt^2 + \left[dx - v_s f(r_s)dt\right]^2 + dy^2 + dz^2 \qquad (3.88)$$

The metric is expressed in the language of a (3 + 1) formalism. The equation of the metric describes the spatiotemporal region about the spaceship that includes the region of influence of the warp drive, as the warp drive distorts the spatiotemporal medium.

Since the motion is linear in the x-direction, we have

$$dy^2 = 0 \qquad (3.89)$$

$$dz^2 = 0 \qquad (3.90)$$

The escape velocity of Alcubierre's spaceship is given by

$$v_s = \frac{dx_s(t)}{dt} = \sqrt[2]{\frac{2GM}{r_s}} = c \qquad (3.91)$$

What would be the centripetal force (toward center) as the spaceship rotates?

$$F_c = \frac{GME}{r_s^2 c^2} = \frac{E}{r_s} \qquad (3.92)$$

The warp drive would create a linear distortion in the direction of propagation given by

$$ds^2 - dt^2 + \left[dx - f(r_s)cdt\right]^2 \qquad (3.93)$$

The initial conditions for the warp drive equation are

$$x_s(t)=\begin{cases} D & @\ t>T \\ 0 & @\ t<0 \quad (\textit{Spaceship starts moving at } t_0) \end{cases} \tag{3.94}$$

Where distance $x_s(t)$ is an arbitrary function of time that describes a trajectory. D is a proper spatial distance between departure and destination points. The warp drive is centered at spatial coordinates $(x_s(t),0,0)$ in the cockpit of the spaceship. Space-time is nearly flat within a radial distance of R on the x-y plane. The distance r_s, or the distance R, is positive toward the bow (front) and negative toward the stern (rear).

The range of the distance parameter r_s is given by

$$f(r_s)=\begin{cases} 1 & r_s \in (-R+d,\ R-d) \\ 0 & r_s \in (-R,\ R) \end{cases} \tag{3.95}$$

$$\lim_{\frac{gr}{c^2}\to\infty} f(r_s)=\begin{cases} 1 & r_s \in (-R+d,\ R-d) \\ 0 & \text{otherwise} \end{cases} \tag{3.96}$$

The function $f(r_s)$ is one when the distortion, $\pm d$, is subtracted from the distance R, or $-R$, e.g. $-R-(-d)$, during spatiotemporal compression (bow) or expansion (stern), and zero when the distortion is not present. Hence, $f(r_s)$ is defined as a function describing the spatiotemporal distortion.

The distance parameter r_s ranges from $R-d$ in the bow distortion, or the direction of propagation, to $-R+d$ in the stern distortion, when $f(r_s)=1$.

Since the distance r_s is in the x-direction and $x=0$ at $t<0$,

$$r_s(t)=\sqrt[2]{(x-x_s(t))^2+y^2+z^2}=\sqrt[2]{(x-x_s(t))^2} \tag{3.97}$$

$$r_s(t)=\sqrt[2]{\left(0-x_s(t)\right)^2}=x_s(t) \tag{3.98}$$

The function $f(r_s)$ is the spatiotemporal function of the distortion about the spaceship with $R>0$, and $d>0$.

$$f(r_s)=\frac{\tanh\left(\frac{g}{c^2}(-R+d)\right)-\tanh\left(\frac{g}{c^2}(R-d)\right)}{2\tanh\left(\frac{g}{c^2}(-R+d)\right)} \tag{3.99}$$

	$d=0$	$d\neq 0$	$R>d$	$R<d$
$f(r_s)$	N/A	1	1	1

Figure 3.

Where $R>0$ and $\frac{gr}{c^2}>0$, as the exponent of the factor of spatiotemporal expansion σ.

$$f(r_s)=\frac{1}{2}\frac{\frac{e^{\frac{g}{c^2}(-R+d)}-e^{-\frac{g}{c^2}(-R+d)}}{e^{\frac{g}{c^2}(-R+d)}+e^{-\frac{g}{c^2}(-R+d)}}-\frac{e^{\frac{g}{c^2}(R-d)}-e^{-\frac{g}{c^2}(R-d)}}{e^{\frac{g}{c^2}(R-d)}+e^{-\frac{g}{c^2}(R-d)}}}{\frac{e^{\frac{g}{c^2}(-R+d)}-e^{-\frac{g}{c^2}(-R+d)}}{e^{\frac{g}{c^2}(-R+d)}+e^{-\frac{g}{c^2}(-R+d)}}} \tag{3.100}$$

By using an identity of the hyperbolic function,

$$f(r_s)=\frac{1}{2}\frac{\frac{e^{\frac{2g}{c^2}(-R+d)}-1}{e^{\frac{2g}{c^2}(-R+d)}+1}-\frac{e^{\frac{2g}{c^2}(R-d)}-1}{e^{\frac{2g}{c^2}(R-d)}+1}}{\frac{e^{\frac{2g}{c^2}(-R+d)}-1}{e^{\frac{2g}{c^2}(-R+d)}+1}} \tag{3.101}$$

$$f(r_s)=\frac{1}{2}\left(\frac{e^{\frac{2g}{c^2}(-R+d)}-1}{e^{\frac{2g}{c^2}(-R+d)}+1}-\frac{e^{\frac{2g}{c^2}(R-d)}-1}{e^{\frac{2g}{c^2}(R-d)}+1}\right)\left(\frac{e^{\frac{2g}{c^2}(-R+d)}+1}{e^{\frac{2g}{c^2}(-R+d)}-1}\right) \quad (3.102)$$

As the distortion grows larger, $d >> R$, the exponents may be substituted with a dummy square velocity ratio variable ℓ^2, to lower the factor of contraction in the bow (front) to the denominator of the fraction, to clarify the spatiotemporal pressurization.

$$f(r_s)=\frac{1}{2}\left[1-\left(\frac{e^{\frac{2g}{c^2}(R-d)}-1}{e^{\frac{2g}{c^2}(R-d)}+1}\right)\left(\frac{e^{\frac{2g}{c^2}(-R+d)}+1}{e^{\frac{2g}{c^2}(-R+d)}-1}\right)\right] \quad (3.103)$$

$$f(r_s)=\frac{1}{2}\left[1-\left(\frac{\frac{1}{e^{2l_b^2}}-1}{\frac{1}{e^{2l_b^2}}+1}\right)\left(\frac{e^{2l_s^2}+1}{e^{2l_s^2}-1}\right)\right] \quad (3.104)$$

It is interesting to note that the volumetric elements are contracting in the bow of the spaceship (front) and expanding in the stern (rear). This specific theoretical warp drive has negligible linear distortion effects on the starboard (right), or port (left), but it distorts the spatial and temporal directions of aloft and underside, as it propagates. This spatiotemporal condition creates the effect of a gravitational slope where the spaceship is falling downhill as it propagates.

The frontal contraction creates a region of higher spatiotemporal pressure whereas the rear expansion creates a region of lower spatiotemporal pressure.

This condition resembles the Law of Inertia for moving bodies of mass when acted upon by an unbalanced force. The spatiotemporal wavelets are closer together in the front and farther apart in the rear of the spaceship. To reverse the direction of propagation, it is only necessary to reverse the polarity of the distortion of the warp drive, so the spaceship, depending on design, may not have to turn around.

It is possible for the spatiotemporal warp drive to provide linear motion, curvilinear motion, or volumetric motion through a spatiotemporal region or a spatiotemporal bridge.

Let us consider the Alcubierre's warp drive for curvilinear motion on three spatial planes.

For curvilinear motion on the x-y plane with $dz^2 = 0$,

$$ds^2 = -dt^2 + \left[dx - v_x f\left(r_x\right)dt\right]^2 + \left[dy - v_y f\left(r_y\right)dt\right]^2 + dz^2 \qquad (3.105)$$

$$ds^2 = -dt^2 + \left[dx - f\left(r_x\right)cdt\right]^2 + \left[dy - f\left(r_y\right)cdt\right]^2 + dz^2 \qquad (3.106)$$

For curvilinear motion on the x-z plane with $dy^2 = 0$,

$$ds^2 = -dt^2 + \left[dx - v_x f\left(r_x\right)dt\right]^2 + dy^2 + \left[dz - v_z f\left(r_z\right)dt\right]^2 \qquad (3.107)$$

$$ds^2 = -dt^2 + \left[dx - f\left(r_x\right)cdt\right]^2 + dy^2 + \left[dz - f\left(r_z\right)cdt\right]^2 \qquad (3.108)$$

For curvilinear motion on the y-z plane with $dx^2 = 0$,

$$ds^2 = -dt^2 + dx^2 + \left[dy - v_y f\left(r_y\right)dt\right]^2 + \left[dz - v_z f\left(r_z\right)dt\right]^2 \qquad (3.109)$$

$$ds^2 = -dt^2 + dx^2 + \left[dy - f\left(r_y\right)cdt\right]^2 + \left[dz - f\left(r_z\right)cdt\right]^2 \qquad (3.110)$$

and for volumetric motion,

The function $f\left(ct_s\right)$ is the spatiotemporal function of the distortion about the spaceship with $\left(R/c\right) > 0$ and $\left(d/c\right) > 0$, expressed with respect to time.

$$f\left(ct_s\right)=\frac{\tanh\left(\frac{g}{c^2}\left(-ct_R+ct_d\right)\right)-\tanh\left(\frac{g}{c^2}\left(ct_R-ct_d\right)\right)}{2\tanh\left(\frac{g}{c^2}\left(-ct_R+ct_d\right)\right)} \tag{3.111}$$

The six-dimensional metric is given by

$$ds^2=-c^2dt^2+dr^2 \tag{3.112}$$

$$ds^2=-\left\{\left[cdt_x-f\left(ct_x\right)cdt\right]^2+\left[cdt_y-f\left(ct_y\right)cdt\right]^2+\left[cdt_z-f\left(ct_z\right)cdt\right]^2\right\}+ \tag{3.113}$$

$$\left\{\left[dx-\left(v_x\right)f\left(r_x\right)dt\right]^2+\left[dy-\left(v_y\right)f\left(r_y\right)dt\right]^2+\left[dz-\left(v_z\right)f\left(r_z\right)dt\right]^2\right\} \tag{3.114}$$

$$ds^2=-\left\{\left[cdt_x-f\left(ct_x\right)cdt\right]^2+\left[cdt_y-f\left(ct_y\right)cdt\right]^2+\left[cdt_z-f\left(ct_z\right)cdt\right]^2\right\}+ \tag{3.115}$$

$$\left\{\left[dx-f\left(r_x\right)cdt\right]^2+\left[dy-f\left(r_y\right)cdt\right]^2+\left[dz-f\left(r_z\right)cdt\right]^2\right\} \tag{3.116}$$

$$ds^2=-\left\{\left[cdt_x-f\left(ct_x\right)cdt\right]^2+\left[cdt_y-f\left(ct_y\right)cdt\right]^2+\left[cdt_z-f\left(ct_z\right)cdt\right]^2\right\}+ \tag{3.117}$$

$$\left\{\left[dx-f\left(r_x\right)dr\right]^2+\left[dy-f\left(r_y\right)dr\right]^2+\left[dz-f\left(r_z\right)dr\right]^2\right\} \tag{3.118}$$

The six-dimensional metric for a symmetrical volumetric drive is given by

$$ds^2=-\left[cdt-f\left(ct_s\right)cdt\right]^2+\left[dr-f\left(r_s\right)dr\right]^2 \tag{3.119}$$

The six-dimensional metric for an asymmetrical volumetric drive in the syntax of a (3 + 3) formalism is given by

$$ds^2 = -\left\{\left(cdt_x \cdot cdt_y \cdot cdt_z\right)^{\frac{1}{3}} - \left[f\left(ct_x\right)\cdot f\left(ct_y\right)\cdot f\left(ct_z\right)\right]^{\frac{1}{3}} cdt\right\}^2 + \quad (3.120)$$

$$\left\{\left(dx \cdot dy \cdot dz\right)^{\frac{1}{3}} - \left[f\left(r_x\right)\cdot f\left(r_y\right)\cdot f\left(r_z\right)\right]^{\frac{1}{3}} dr\right\}^2 \quad (3.121)$$

A propulsion mechanism based on such a modifiable six-dimensional spatiotemporal distortion just begs to be given the well-known name of the "hyperdrive" of science fiction.

As the spatiotemporal medium about the hyperdrive expands or contracts, how much positive or negative work is being done by the expanding or contracting spatiotemporal medium per unit of volume?

$$\pm\frac{dF \cdot d^2}{da} = \pm\frac{8\pi mc}{\Lambda^2 eL_p} = \pm\frac{8\pi mca}{\Lambda eL_p{}^2} \quad (3.122)$$

To enlarge and maintain a spatiotemporal bridge would be positive work, to reduce it would be negative work since the spatiotemporal medium itself would be exerting the force.

As the spaceship drive starts, if it were possible for the hyperdrive to produce a rotating electron-positron pair to initiate a distortion, what is the tangential momentum of the distortion?

$$p_T = \frac{\left(\phi_B \cdot r_P{}^2\right)}{A \cdot f_P} = \sqrt[2]{\frac{k_e{}^4 q^4 r_0}{c^4 \varepsilon_0{}^2 m_P r_P{}^2 f_P}} \approx \frac{61}{40} \quad \text{atto Newtons/second} \quad (3.123)$$

Where r_P, m_P, f_P are Planck units of distance, mass, and frequency, q is the charge of an electron or positron, ε_0 is the permittivity of free space-time at a radial distance of r_0, A is the surface area of a turn, and ϕ_B is the rotating magnetic flux through A.

As the body of mass, or spaceship, rotates, charges flow as a current $\vec{J}$ through the bottom of the body of the spaceship. As the bottom of the body of the spaceship acts a conductor, the electric field would establish a corona discharge around it if the spaceship is within the atmosphere of a planet, and a rotating electromagnetic field in the direction of the opening distortion. The radial component of the tangential momentum of the electromagnetic field would reinforce the opening and enlargement of the spatiotemporal bridge.

Let us consider Ampere's circuital law in differential form,

$$\nabla \times \vec{B} = \mu_0 \left(\vec{J} + \varepsilon_0 \frac{\partial \vec{E}}{\partial t} \right) = \mu_0 \vec{J} + \varepsilon_0 \mu_0 \frac{\partial \vec{E}}{\partial t} = \mu_0 \vec{J} + \frac{1}{c^2} \frac{\partial \vec{E}}{\partial t} \quad (3.124)$$

$$\mu_0 \vec{J} = \left(\nabla \times \vec{B} \right) - \frac{1}{c^2} \frac{\partial \vec{E}}{\partial t} = \left(\nabla \times \vec{B} \right) - \mu_0 \varepsilon_0 \frac{\partial \vec{E}}{\partial t} \quad (3.125)$$

$$\vec{J} = \frac{1}{\mu_0} \left[\left(\nabla \times \vec{B} \right) - \frac{1}{c} \frac{\partial \vec{B}}{\partial t} \right] \quad (3.126)$$

Hence, the current $\vec{J}$ may be expressed entirely as a function of the magnetic field $\vec{B}$. The rotating charges springs the magnetic field that exerts pressure on the spatiotemporal surfaces, or walls, of the opening bridge.

What is the spatiotemporal area that the gravitational force is acting on during the opening of the spatiotemporal bridge?

Starting with the six-dimensional EFE,

$$R_{\mu\nu} - \frac{1}{(n-1)} g_{\mu\nu} R = -\frac{8\pi G}{c^4} T_{\mu\nu} \quad (3.127)$$

$$\left(\frac{c^4}{8\pi G} \right) \left(R_{\mu\nu} - \frac{1}{(n-1)} g_{\mu\nu} R \right) = -T_{\mu\nu} \quad (3.128)$$

Reducing tensorial terms of the six-dimensional EFE and substituting area as the reciprocal of curvature,

$$\left(\frac{c^4}{8\pi G}\right)\left(-\frac{R}{5}\right)=\left(\frac{c^4}{40\pi G}\right)\left(-\frac{1}{A}\right)=-T \tag{3.129}$$

$$-\left(\frac{c^4}{40\pi G}\right)\left(\frac{c^4}{4^2\pi G^2 M^2}\right)=-T \tag{3.130}$$

Squaring terms of energy density, area, and curvature,

$$\left(\frac{c^8}{5^2\cdot 4^3\pi^2 G^2}\right)\left(\frac{c^8}{4^4\pi^2 G^4 M^4}\right)=T^2=\left(3\rho+3p\right)^2=9\left(\rho+p\right)^2 \tag{3.131}$$

$$\left(\frac{c^8}{5^2\cdot 4^3\pi^2 G^2}\right)\left(\frac{c^8}{4^4\pi^2 G^4 M^4}\right)=9\left(\rho+p\right)^2 \tag{3.132}$$

The initial area A of the Schwarzschild spatiotemporal bridge is given by

$$A=4\pi\left(\frac{2GM}{c^2}\right)^2=\frac{4^2\pi G^2 M^2}{c^4} \tag{3.133}$$

Squaring the area of a curved funnel leading to an infinitesimal point since curved funnels have double curvature,

$$A^2=\frac{4^4\pi^2 G^4 M^4}{c^8} \tag{3.134}$$

The cosmological constant has the same effect as an intrinsic energy density ρ_{vac} of space-time. The cosmological constant is given by

$$\Lambda=\frac{8\pi G\rho}{c^2} \tag{3.135}$$

$$\Lambda^4 = \frac{4^6 \pi^4 G^4 \rho^4}{c^8} \qquad (3.136)$$

Substituting for the area, α , of the opening spatiotemporal bridge,

$$\left(\frac{c^8}{5^2 \cdot 4^3 \cdot \pi^2 \cdot G^2}\right)\left(\frac{1}{\alpha^2}\right) = 9(\rho + p)^2 \qquad (3.137)$$

As the initial area α expands, it quadruples to the square area. Balancing terms on both sides of the equation for a quarter term of square area expansion,

$$\left(\frac{c^8}{5^2 \cdot 4^3 \cdot \pi^2 \cdot G^2}\right)\left(\frac{4^2}{5^2 \alpha^2}\right) = 9(\rho + p)^2 \qquad (3.138)$$

$$\left(\frac{4^2 \cdot c^8}{5^4 \cdot 4^3 \cdot \pi^2 \cdot G^2}\right) = 9\alpha^2 (\rho + p)^2 \qquad (3.139)$$

$$\left(\frac{c^8}{5^4 \cdot 4 \cdot \pi^2 \cdot G^2}\right) = 9\alpha^2 (\rho + p)^2 \qquad (3.140)$$

Taking the square root of both sides,

$$\left(\frac{c^8}{2500 \cdot \pi^2 \cdot G^2}\right) = 9A^2 (\rho + p)^2 \qquad (3.141)$$

$$\sqrt{\left(\frac{c^8}{2500 \cdot \pi^2 \cdot G^2}\right)} = \sqrt{9\left(\frac{\Lambda^4}{16\pi^2}\right)(\rho + p)^2} \qquad (3.142)$$

Let us express the cosmological curvature as the area grows from the infinitesimal size of a naked singularity, a gravitational singularity without an event horizon, that may be generated by a local energy density which would extend the Schwarzschild radius out beyond its initial size, to a size A^2 in one fourth of the distance,

$$\frac{c^4}{50\cdot\pi\cdot G}=\frac{3\cdot\Lambda^2}{4\cdot\pi}(\rho+p) \tag{3.143}$$

$$\frac{c^4}{50\cdot G}=\frac{3\cdot\Lambda^2}{4}(\rho+p) \tag{3.144}$$

A standard form for the six-dimensional equation of the spatiotemporal bridge may be expressed as

$$\frac{1}{\Lambda^2}=\frac{75\cdot G}{2\cdot c^4}(\rho+p) \tag{3.145}$$

Let us consider three theoretical configurations of spatiotemporal bridges that may provide a path through a naked singularity. First, let us imagine a spatiotemporal bridge that would be mostly spatial, with temporal coordinate changes that are almost negligible. Both gates (portals) on the bridge would be spatially separated, but nearly simultaneous on the temporal medium, e.g. a spatiotemporal bridge from Earth's orbit to the orbit of Mars. If it were possible for a signal or a spaceship to enter that bridge with an escape velocity of c, the time elapsed on the clock of the spaceship to that of an observer on Earth would be negligible. Category-one is a spatial bridge through six-dimensional space-time.

Second, let us envision a spatiotemporal bridge that would be both spatial and temporal, but the change in temporal coordinates are toward the future, e.g. a spatiotemporal bridge from Earth's orbit to the orbit about an exoplanet. If it were possible for a signal or a spaceship to enter that bridge with an escape velocity of c, the spatial distance and time elapsed on the clock of the spaceship to that of an observer on Earth may be considerable compared to the previous trip to the orbit of Mars, depending on the curvature of the bridge or the capacity of the hyperdrive. Category-two is a spatial and temporal bridge through six-dimensional space-time.

Third, let us visualize a spatiotemporal bridge that would be both spatial and temporal, but the change in temporal coordinates are toward the far past, e.g. after a long stay, a spatiotemporal bridge is re-opened from the orbit about the exoplanet to the orbit of Earth to arrive one hour after the spaceship, or the signal, left, to protect the timeline. Interdimensional signals may be carefully encrypted and spatiotemporally stamped for access according to chronological law.

The latter theoretical scenario is partial to the view of the existence of parallel universes and represents a form of interdimensional temporal travel. The opposite direction of travel of the same spatiotemporal bridge would be to the far future. Category-three is a spatial, temporal, and interdimensional bridge through six-dimensional space-time.

It is interesting to note that if a combination of at least two of these forms of travel were possible, a traveler would be able to arrive at any destination instantly, or simultaneously, according to the clock at the inertial frame of reference of departure. In such scenario, as soon as the spaceship launched, it would be landing at the destination. Moreover, if the gates (portals) of the bridges are farther apart in the same spatiotemporal plane, the spatial distortion effect is greater than the temporal toward positive infinity-eternity, but if the gates are concentric, the temporal distortion effect is greater than the spatial effect on the same spatiotemporal plane, toward negative infinity-eternity. Hence, a spatiotemporal bridge may be directed, and later re-directed, aligned, or later re-aligned, in the direction of the resultant spatiotemporal metric toward the future, present, or past.

A micro spatiotemporal bridge may be useful to an advanced civilization to cloak communication signals between sender and receiver, in a private channel, to evade less-technologically-developed unintended receivers or eavesdroppers.

A category-three micro spatiotemporal bridge would provide a useful channel for a chronoviewer which allows for monitoring spatiotemporal coordinate locations and events, or intercepting broadcasted signals from either the past, present, or future under strict chronological law. A cloaked signal may not be broadcasted in a conventional way, e.g. coded signals may be point-to-point low-energy pulse-modulated discrete signals, or very compacted digital images, that are directional in nature and do not reflect, scatter, or transmit radially out as a wave.

3.4. The Bianchi Identities: Curvature and torsion in the spatiotemporal wave medium.

The Bianchi identities, developed by the eminent mathematicians Gregorio Ricci-Curbastro and his brilliant student Tullio Levi-Civita, and later simplified by the prominent mathematician Elie Cartan, may be interpreted as a conservation theorem for the curvature and torsion of a Riemannian or pseudo-Riemannian manifold. The second Bianchi identity is typically used for the study of manifolds that have more than three spatial dimensions.

The Bianchi identities demonstrate how curvature and torsion behave on a six-dimensional tesseract. The orientation of each pair of opposite sides on the inner cube of a tesseract represents a spatial direction of a dimension. The first Bianchi identity is given by

$$R_{\tau xyz} + R_{\tau yzx} + R_{\tau zxy} = 0 \qquad (3.146)$$

Let us describe each of the four covariant indices as a spatial or temporal direction of a spatial or temporal dimension. Each spatiotemporal direction has two senses. Three-dimensional time is represented by the covariant index, τ , which is orthogonal to each spatial dimension, spatial width is represented by x, spatial depth is represented by y, and spatial height is represented by z.

In the first Bianchi identity, the temporal index, τ , stays in the same orientation, but every spatial index permutes to the next spatial direction. Each permuted spatial index is rotated in the direction of torsion to an adjacent spatial direction; first by 90 degrees for $R_{\tau yzx}$, and then by another 90 degrees for $R_{\tau zxy}$, with respect to the starting Riemannian tensor, $R_{\tau xyz}$. As each of the three spatial direction indices are permuted on a Riemannian or pseudo-Riemannian tensor, the inner cube of the tesseract is rotated. So, after the permutation of the last Riemannian tensor, $R_{\tau zxy}$, each permuted spatial index of the starting Riemannian tensor has shifted through each direction to a final direction that is orthogonal to the starting orientation. Thus, the sum of the permutations of the first Bianchi identity may not be without curvature and torsion.

Let us visualize three separate parallel transports on three separate sides of the tesseract in a Cartesian coordinate system, first, a vertical z-direction vector is parallel transported along the two possible paths of directions x and y on a horizontal plane, secondly, a vertical x-direction vector is parallel transported along the two possible paths of directions y and z on a horizontal plane, and thirdly, a vertical y-direction vector is parallel transported along the two possible paths of directions x and z on a horizontal plane. If there is no curvature or torsion in each case, the parallel transported vector for each case would end up in the same orientation. However, as it has been noted above, after parallel transport there might be curvature and torsion. Moreover, the orientation of curvature and torsion on each vertex of the inner cube of the tesseract changes as the tesseract is rotated.

When there is curvature and torsion as the above parallel transport starts at a vertex of the tesseract, each of the three pairs of paths would end up at a distinct point at a diagonally opposite vertex, to form three separate points. The three points may be visualized as the three corners of a triangle. The curvature and torsion of vertices of the tesseract involved in parallel transport may represent positive, zero, or negative curvature and torsion. If the torsion and curvature on a vertex is zero, all three points will coincide, the vertex of the tesseract would close, and all sides of the triangle would add up to zero.

In the case of a six-dimensional sphere, the curvature is distributed evenly over its entire surface. All three directions have equal positive intrinsic curvature and torsion. The sum of curvature and torsion of each pair of sides adds up to zero. So, if directions are rotated to an orthogonal orientation, all curvature and torsion in each direction stays the same. The same is true for the tesseract only if each vertex has the same intrinsic curvature and torsion. The surface of any side of the inner cube of a tesseract has no intrinsic curvature and torsion. There is intrinsic curvature and torsion at the eight vertices, where the curvature is singular. The point of singular curvature is equivalent to the apex of a cone in space.

The curvature of the inner cube of a tesseract is concentrated in equal parts at each of the eight evenly spaced vertices rather than being distributed over the entire surface as with a sphere. This implies that straight lines can be drawn unambiguously on the surface of one side of the tesseract to the surface of another side, if the line doesn't pass precisely through a vertex. Therefore, the first Bianchi identity represents the parallel transport case in which the sum of the intrinsic curvature and torsion of each vertex of the tesseract involved in parallel transport adds up to zero along the same starting direction for each of the three permutations. Then, in that case, the first Bianchi identity is validated.

This conclusion would imply that the intrinsic curvature and torsion of each vertex of the tesseract on each of the Riemannian tensors is covariantly conserved. If one vertex, or two vertices, has positive intrinsic curvature and torsion, the other two vertices, or one vertex, would offset that curvature and torsion. This inherent property of space-time exemplifies the constructive or destructive interference of six-dimensional spatiotemporal waves.

The second Bianchi identity is given by

$$R_{\tau xyz;r} + R_{\tau xzr;y} + R_{\tau xry;z} = 0 \qquad (3.147)$$

Again, the temporal index, τ , stays in the same orientation, but the indices for the directions of spatial dimensions are rotated in the direction of torsion by 90 degrees. The covariant derivative is taken on each term of the second Bianchi identity with respect to spatial directions *r*, *y*, and *z*. The covariant derivative expresses the application of parallel-transport in each of the above spatial directions. The curvature and torsion measured by parallel transport at a vertex of a tesseract cancel out with the curvature and torsion measured by parallel transport at a diagonally opposite vertex. The covariant derivative of the starting Riemannian tensor, $R_{\tau xyz;r}$, taken in the spatial direction of *r*, represents intrinsic curvature and torsion on a vertex of the tesseract with τxyz indices. The spatial direction, *r,* may be interpreted as orthogonal to *x, y, and z,* toward the center of the inner cube. The inner cube of the tesseract, corresponding to the starting Riemannian tensor, $R_{\tau xyz}$, may be interpreted to be oriented in the spatial directions of *x*, *y* and *z*.

After the application of parallel transport on each vertex of the tesseract, the intrinsic curvature and torsion is summed over each spatial direction index, where each index represents *n* dimensions, and the sum of the spatiotemporal perturbations in the second Bianchi identity is null. Therefore, the first Bianchi identity expresses the causal and geometrical relationship of the constructive or destructive interference of six-dimensional spatiotemporal waves, while the second Bianchi identity describes how to measure the effect of the constructive or destructive interference. Thus, the second Bianchi identity motivates local energy conservation in the EFEs, allowing the covariant derivative of the Einstein tensor, $G_{\mu\nu}$, or the covariant derivative of the stress-energy-momentum tensor, $T_{\mu\nu}$, of the General Theory of Relativity, to be set to zero to be covariantly conserved.

§ *4. On the Thomson scattering cross section*

The effect of the expansion and acceleration of space-time on a Thomson scattering cross section:

Thomson scattering is the low-energy limit of the scattering of electromagnetic radiation by electrons when an incident wave of photons accelerates the electrons to cause radiation at the same frequency as the incident photonic wave, scattering the electromagnetic wave, whenever the photonic energy is much less than the mass energy of the electrons, and the speed of the electrons is non-relativistic. (Weinberg, 2003)

Let us imagine that the components of the electric field of the incident photonic wave and the radiated wave are radial and perpendicular (tangential) to the plane of observation in space-time.

The Thomson scattering cross section is the integrated area of scattering of the wave in the directions of the radial and perpendicular planes of the electric field components.
The Thomson scattering cross section measures the ability of a particle, such as an electron, to remove photons from a directed beam and send them into new directions.

Let us consider the classical Thomson scattering effect of electrons on low energy incident photons E_i which may be calculated by dividing the emitted scattering power of electrons, $-\partial E_S/\partial t$, the average power radiated to all angles, by the average incoming power per unit area, the Poynting flux *S*, of the photonic electromagnetic field.

$$\sigma_T = \frac{-\dfrac{\partial E_S}{\partial t}}{|S|} = \frac{\dfrac{4\pi}{3}\dfrac{e^4|E_i|^2}{4^2\pi^2\varepsilon_0 m_e^2 c^3}}{\dfrac{1}{2}c\varepsilon_0|E_i|^2} \tag{4.1}$$

The Thomson scattering cross section for low energy incident photons is given by

$$\sigma_T = \frac{8\pi}{3}\left(\frac{e^2}{4\pi\varepsilon_0 m_e c^2}\right)^2 = \frac{8\pi}{3}\left(\frac{e}{Q_p}\right)^4\left(\frac{\ddot{a}}{c^2}\right)^2 \quad \text{meters}^2 \tag{4.2}$$

$$m_e = \frac{\hbar c}{\ddot{a}} \tag{4.3}$$

$$\left(\frac{\ddot{a}}{c^2}\right)^2 = r_a^2 \tag{4.4}$$

Where the classical electron radius is shown above as equivalent to the product of the electron-Planck charge ratio raised to the fourth power and the Thomson space-time spatiotemporal distance r_a squared. The scattering of the incident photonic wave is the result of the expansion and acceleration of space-time through the volume of space encompassing the integrated Thomson cross section area that may be represented by a square area σ_T that is proportional to an area r_a^2. The expansion of space-time through the scattering volume results in the uniform distribution of the charges of electrons, about the surface area of the scattered waves, and is proportional to the Planck charge.

§ 5. The temporal wavelength and period of time fields

Intensity of a temporal cubic wavelength on adjacent spatiotemporal cube, where E is energy from Work and F is force:

$$I_T = \text{Pressure x Velocity} = \frac{F}{\lambda^2} \cdot \frac{\lambda}{T} = \frac{\langle E \rangle}{\lambda^2 T} = \frac{F}{\lambda T} = \frac{F}{q} = \frac{\text{Power}}{\lambda^2} \qquad (4.5)$$

From previous research, q represents a unit of spatiotemporal charge.

Velocity and acceleration of temporal wavelet:

$$\partial v = \frac{\partial \lambda}{\partial T} \qquad (4.6)$$

$$\partial a = \frac{\partial \lambda}{\partial T^2} \qquad (4.7)$$

For three adjacent temporal cubes:

Contracting time field:

$$\frac{\partial \lambda_{n+1}}{\partial T_{n+1}} > \frac{\partial \lambda_n}{\partial T_n} > \frac{\partial \lambda_{n-1}}{\partial T_{n-1}} \qquad (\lambda_{n+1})^3 > (\lambda_n)^3 > (\lambda_{n-1})^3 \qquad (4.8)$$

Flat time field:

$$\frac{\partial \lambda_{n+1}}{\partial T_{n+1}} = \frac{\partial \lambda_n}{\partial T_n} = \frac{\partial \lambda_{n-1}}{\partial T_{n-1}} \qquad (\lambda_{n+1})^3 = (\lambda_n)^3 = (\lambda_{n-1})^3 \qquad (4.9)$$

Expanding time field:

$$\frac{\partial \lambda_{n-1}}{\partial T_{n-1}} < \frac{\partial \lambda_n}{\partial T_n} < \frac{\partial \lambda_{n+1}}{\partial T_{n+1}} \qquad (\lambda_{n-1})^3 < (\lambda_n)^3 < (\lambda_{n+1})^3 \qquad (4.10)$$

Where Lambda (λ) is the wavelength of the temporal wavelet during period (T). The ratio $\delta\lambda/\delta T$ is the temporal velocity of the temporal wavelet. The temporal intensity is the power developed per area of the temporal wave-front. As the temporal velocity decreases, the amplitude of the temporal wavelet increases, increasing the power per unit of area of the wave-front (intensity). Greater intensity increases potential energy across the temporal wavelength distance within the locality of the time field. This greater potential energy across the wavelength of the temporal wavelet (*icT*), and its conjugate spatial wavelet, in the presence of mass invokes spatiotemporal (gravitational) potential energy and gravitational acceleration.

§ 6. The spatiotemporal deceleration or gravity near celestial bodies

The following equations define the spatiotemporal (gravitational) deceleration near celestial bodies, *g*, the wavelength, frequency, and period of the time field at the planetary radius, and the speed of time as the speed of light *c*.

Kepler's 3rd Law resembles the temporal deceleration equation in the format of the proportional equivalence between time-like (temporal) and space-like (spatial) distances.

$$g = -\left(\frac{1}{8\pi r^2}\right)\frac{d^2V}{dt^2} \tag{4.11}$$

$$-\frac{d^2V}{dt^2} = 8\pi g r^2 = 8\pi G m_0 \tag{4.12}$$

$$-\frac{d^2V}{dt^2} = \frac{\lambda^3}{T^2} = 8\pi g r^2 \tag{4.13}$$

$$c = \lambda f \tag{4.14}$$

$$\lambda = \frac{c}{f} = cT \tag{4.15}$$

$$\lambda^3 f^2 = c^2 \lambda \tag{4.16}$$

$$-\frac{d^2V}{dt^2} = \frac{\lambda^3}{T^2} = c^2 \lambda \tag{4.17}$$

$$\lambda = \frac{8\pi g r^2}{c^2} \tag{4.18}$$

$$c = \sqrt[2]{\frac{8\pi g r^2}{\lambda}} \tag{4.19}$$

At the Earth's radius,

$$-\frac{\lambda^3}{T^2} = 8\pi g r^2 = 8\pi\left(9.8\frac{m}{s^2}\right)(6{,}731{,}000m)^2 \approx 10^{16}\frac{m^3}{s^2} \tag{4.20}$$

At the Moon's radius,

$$-\frac{\lambda^3}{T^2} = 8\pi g r^2 = 8\pi\left(1.622\frac{m}{s^2}\right)(1{,}737{,}400m)^2 \approx 10^{14}\frac{m^3}{s^2} \tag{4.21}$$

Similarity to Kepler's 3rd Law,

$$P^2 \propto a^3 \tag{4.22}$$

where P is the period T of the orbit, and distance a is the perihelion or shortest radius of orbit.

$$T^2 \approx \left(10^{-16}\right)\lambda^3 = \kappa\lambda^3 \tag{4.23}$$

where κ is Kepler's constant for Earth.

§ 7. The pressure of space-time on an object under water

We may express the velocity of light, c, as the product of the acceleration of space-time, δ, near the surface of an object of mass like the earth, multiplied by the wavelength of space-time, λ. The square of the speed of light is equal to the product of the acceleration of space-time, δ, times its wavelength, λ. If the spatiotemporal wave dilates, its period and wavelength would increase, and its acceleration would stay the same, maintaining the constant acceleration and speed of light.

$$c^2 = \delta\lambda = \frac{\lambda^2}{T^2} \qquad (4.24)$$

$$c = \sqrt[2]{\delta\lambda} \qquad (4.25)$$

where $\delta = \frac{\lambda}{T^2}$ is the acceleration of space-time and $\frac{\lambda}{T}$ is its velocity c.

Pressure on an object, such as a submarine under the waters of an ocean, has been calculated as the product of the density of the liquid (water), the gravitational acceleration, and the depth of the water.

Pressure may be defined as the force per unit area on the surface of an object under water.

$$P = \rho g h \qquad (4.26)$$

Hence, the following equations will show how these variables are related to the expansion of space-time under water to produce a force per unit area on an object that is under water.

Where h is the depth of the water to a point under water where the pressure is being applied.

Then, the product of the gravitational acceleration, g, and the depth of the water, h, is equal to the Pressure, P, divided by the density of the liquid, ρ, at the locality where the pressure is being applied on the object under water.

$$P = \rho g h \rightarrow \frac{\hbar \nu_{st}}{a} \qquad (4.27)$$

Where $\hbar$ is Planck's constant, ν_{st} is the spatiotemporal frequency, and a is the spatiotemporal volume of a quantum of energy.

$$\frac{P}{\rho} = gh \tag{4.28}$$

In terms of units,

$$\frac{P}{\rho} = \frac{\frac{N}{m^2}}{\frac{Kg}{m^3}} = \frac{N}{m^2} \cdot \frac{m^3}{Kg} = \frac{N \cdot m}{Kg} = \frac{Kg \cdot m \cdot m}{s^2 \cdot Kg} = \frac{m^2}{s^2} = \left(\frac{m}{s}\right)^2 \leftrightarrow c^2 \tag{4.29}$$

Hence,

$$\frac{P}{\rho} = gh \quad \left(\frac{m}{s}\right)^2 \tag{4.30}$$

$$c^2 = \delta\lambda = \left(\frac{\lambda}{T}\right)^2 \tag{4.31}$$

Therefore, we can see that the pressure felt by an object under a liquid, like ocean water, follows the format of the mathematical expression of the product of the acceleration of the spatiotemporal field times the wavelength of the spatiotemporal field at the locality where pressure is being applied on the object under water.

The intensity of the gravitational field is directly proportional to the wavelength and acceleration of the spatiotemporal field. Water waves are influenced by spatiotemporal waves about a massive celestial body. Pressure is exerted by spatiotemporal waves and spatiotemporal acceleration.

The passage and contraction of spatiotemporal waves near a massive celestial body like the Earth at a significant depth above the ocean floor, or ground surface, underlie the circular orbital motion of deep-water particles as shown below.

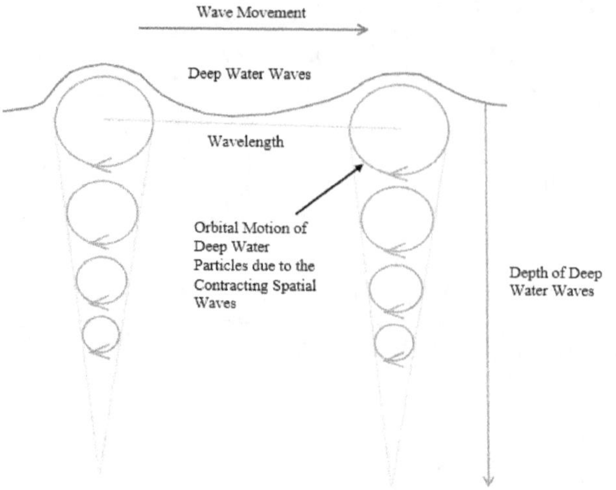

Figure 4.

Similarly, the passage and contraction of flatter and smaller spatiotemporal waves near the ocean floor, or ground surface, underlie the elliptical orbital motion of shallow water particles as shown below.

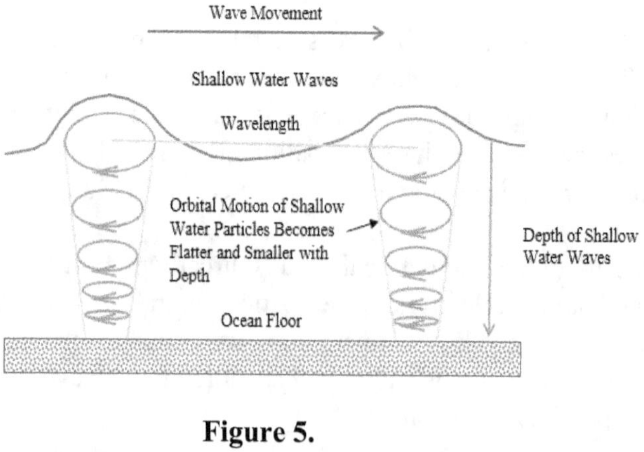

Figure 5.

§ 8. *On epistemological assumptions regarding the nature of time and space*

- There exists a cause and effect relationship between measurable time and observable change.

- Time endows the cause, and change is the effect, for any observable event.

- Any and all observable change is ultimately the endowed effect of time.

- The Principle and Scope of Temporal Effect:

  The quantity of time at any observable event that is available to effect any and all changes in any and all attributes, characters, properties, and qualities of all virtual or physical realities of the observable event is finite.

- Every event has a boundary of time.

- Motion is a perceivable change and change is the endowed effect of space-time. Therefore, space-time ultimately endows the cause of all changes that result in Motion, and Changes that result in Motion are ultimately the endowed effects of Space-time.

- Expanding space-time is that which makes motion possible.

- Any observable change is associated with a temporal change and any temporal change may or may not be associated with an observable change.

- Time endows the cause of change, but the observable change on the hands of a self-moving clock is the measurable effect.

§ 9. *Time Quotes*

- Time creates the space to allow things to happen elsewhere.

- Time prevents the omnipresence of things.

- What then is time? Time is what it is even when there is no one asking about it. If I were to explain time to someone like Saint Augustine, I would say emerging time is diverging space-time. Space endows time and time endows space.

- The difference between proper and coordinate time is all curved.
- To an artist everything you can imagine may be real, but to a physics researcher, everything that may be real you can imagine. Space-time is both real and imaginary, it is complex.
- Time brings a fresh beginning to everything.
- Time is neither friend nor foe, it is just change.
- Time is the shortest distance between two points at the speed of light.
- Why is time emergent? Because time comes to us even when we are not going anywhere.
- Time is the stream of consciousness.
- The present used to be the future, but now I realize that it has become the past.
- Time is amazingly fair and generous, we all get the same amount every day.
- If the sands of time get in your shoe, time may allow you to stay a little longer.
- Time is a great teacher, but you better be punctual, time waits for no one.
- Space can't wait for time to make more space, while time is always waiting for space to get out of its way.
- Time is a kind of place like space, but a place perceived as frequencies and amplitudes.
- In some sense, space-time is the illusion that realizes gravity which does not exist. In some way, gravity comes from the curvature of space-time which is part of the illusion.

- A Quantum Limerick:

  'Once upon a Planck time, there was a tachyon named Sprite,
  whose speed was a lot faster than light;
  A particle sent Sprite out one day,
  to collapse its quantum function wave,
  whose qubits were handed over with delight!'

- Praise Stanza to Einstein:

  'One stone' was thrown upon the water,
  making relative waves near and far,
  across the seven seas of energy and matter,
  riding endlessly on a light ray from a star.

§ *10. Conclusion*

As a photon surfs the wave of its time field, the spatiotemporal waves are preemptive waves which create space for the motion of objects of mass in space-time. Space-time is an emergent and an intrinsic relative wave field that is fundamental to space-time-mass. Spatial or temporal waves expand or contract reciprocally as a result of the constructive or destructive interference of spatiotemporal waves at a spatiotemporal locality. Every point in space-time expands according to Huygens' Principle. Electrons scatter incident low-energy photons through the uniform distribution of electronic charges during the expansion and acceleration of space-time. The spatiotemporal expansion plays a key fundamental role in physical processes. The increase in the potential of spatiotemporal waves results in a gravitational field near an object of mass or an object in relativistic motion. The pressure exerted on an object of mass under water is due to the contraction of spatial waves and the dilation of temporal waves at any point near the surface of the object, or near the ground surface under water. Time endows the cause and change is the effect of any observable event. Motion, gravitational fields, inertia, physical fields are all endowed effects of the spatiotemporal field.

Chapter 2

The Anatomy of Chronos

§ 1. On the anatomy of chronos

The enigma of time through the ages of history has been a topic often considered too scientifically vague for philosophers, and too philosophically ambiguous for scientists. However, time waits for no one and exists for its own purpose. Time has existed without observation or chronology before biological life itself.

Time flows equally at points in the universe sharing equal characteristics of space-time-energy. In the vastness of space-time, this attribute of time may be useful as a cycle-counter or clock of the universe, even though space-time may not follow a universal time flow. In relativity, an event and its observer may or may not have different clocks that run at different paces. Moreover, the difference in the flows of time, or in the tick-tock of different clocks, one stationary, the other moving at relativistic speed, becomes more noticeable and measurable.

By nature of the speed of time and the expansion of space-time, there is always a temporal distance between an event and its observer. The speed of time would be the same in inertial frames of reference for one or more observers at observer locations in the universe sharing equal characteristics of space-time-energy. The passage of time is part of an event itself. Other events involving matter or energy, whether observed or not, even though measured or measurable, occur in a time-like distance or time duration at the space-time locations of the events. The evolution of biological life is a good example of this principle. The time-like distance may be measured by the cycles of the wavelengths of time using a clock if the event and the observer share equal characteristics of the surrounding space-time in the same frame of reference, or by the time-like distance difference measured between the observer's clock in an inertial frame of reference and the event-referenced clock.

Time-like distance, spatial distance, and their units are universal. In homogenous and isotropic space-time at the current cosmic microwave background baseline temperature of 2.735 Kelvin above zero, the time-like distance that time flows through during a spatial distance of *ct* (a metron) is a second, and the spatial distance that time flows in 1/*ct* of a second (a centon) is a meter. These relativistic types of measurement units are helpful in the universal comparison of gravitational fields, and space-time metrics. Time is an intrinsic dimension of an event or between two or more events, and stands as a dynamic property of the spatiotemporal dimensional structure of space-time-energy. Time is Energy. Time is a form of compressed energy that expands into space-time. Time transfers Energy to both Matter and Space. Time is itself an unfinished event, changing the universe in its process, according to its speed at any universal point of reference.

1.1. Temporal Wave Theory

Temporal waves are dispersive in nature at their source and experience damping as they pass through a medium other than vacuum. Primary temporal waves passing through a medium have two associated velocities: the group velocity, and the individual phase velocity of each temporal wave crest. Any individual temporal wave in the temporal wave group travels fast, but if we take a wider view of the wave group as it travels outward from the temporal point of source we see that the wave group travels slower than a particular individual temporal wave. The new temporal ripples rise up, rush forward, and fade away at the front of the temporal wave group. The increase in the speed of temporal waves is possible through the reversal of the behavior of the waves in the temporal wave group. Light rides on the back of temporal waves and spatial waves through the spatiotemporal medium, so the velocity of light increases as the spatiotemporal group velocity is increased over a particular phase velocity.

The amplitude of a temporal wave decreases as the atomic structure in the medium creates outgoing temporal ripples when the primary temporal wave sweeps by the atoms in the structure. These rings of temporal ripples can be refocused backwards to a single point by exciting the atoms in the medium. When these temporal ripples overlap and combine with the primary temporal wave, they cancel the temporal wave racing on ahead, suppressing the fast-moving temporal wave front and slowing down the temporal wave group.

The equations governing the behavior of temporal waves say that it is possible for reverse temporal waves to converge exactly on their starting point, creating a real time-reversal towards the sink (reverse source) that absorbs all the temporal waves. Temporal waves exhibit universal time-reversal symmetry. The spatiotemporal wave may be expressed as

$$\frac{1}{c}\frac{\partial x'}{\partial t'} = \frac{\left(e^{ix_0} + e^{-ix_0}\right)}{2} + \frac{\left(e^{ix_0} - e^{-ix_0}\right)}{2i} \tag{1.1}$$

Where the second term is imaginary or temporal in nature.

It is interesting to note that the first spatial term of the above spatiotemporal wave equation also generates a temporal component, and similarly, the second temporal term generates a spatial component. Therefore, time is emergent and creates more space, which in turn allows the emergence of more time.

The velocity of a temporal wave depends on the time differential across the temporal wave. The time differential is the same principle as the principle of inertia of moving bodies in space-time. If the temporal region (timescape) in front of the wave is less compressed than the temporal region behind the temporal wave, the temporal wave will travel forward. If conditions are reversed, then the temporal wave would travel backward. When both the front and rear temporal regions are equally compressed or dilated, the temporal wave would freeze or not travel. As the temporal wave freezes at a location in expanding space-time, the spatiotemporal points at that location will continue expanding, leading to a split of the temporal wave into a retarded temporal wave (forward) and an advanced temporal wave (backward). As the split occurs, there is a dilated temporal region that develops in the middle of the waves that is more compressed than the front of the retarded temporal wave or the front of the advanced temporal wave, which creates a time differential across the temporal waves. Then, the temporal waves move in opposite direction.

As the amplitude of a temporal wave increases (amplification), a temporal wave will increase its velocity. This reaction of a temporal wave may be employed to speed up the velocity of light over the speed of light *c* (celeritas) in vacuum through certain atomic mediums. By exciting the atoms of the medium, the rings of temporal ripples from the atomic structure can be refocused backwards to freeze and augment the fast-moving temporal wave fronts, and to speed up the resulting temporal wave groups of the retarded and advanced temporal waves. As an electromagnetic wave, light rides on spatiotemporal waves that propagate in vacuum at the speed of light. The light barrier has been a tenet of physics for over a century, but the speed of spatiotemporal waves needs to be considered a new tenet. Nothing travels faster than the waves of space-time.

If the temporal wave group distortion of the light source gets very extensive, the temporal group velocity becomes negative, and there is an image pair creation event. When a light source breaks the light barrier through an excited atomic medium, the light source would travel superluminally to a point of reference in the medium where there would be an image pair creation event of the light source splitting into two images traveling in opposite directions, as the light source continues to speed up faster than the speed of light *c* in the direction of propagation.

An observer in a subluminal inertial frame of reference, or a subluminal free-falling frame of reference, would observe a retarded image of the light source traveling forward in time in the direction of propagation and an advanced image of the light source traveling backwards in time that precedes the temporal wave group of the light source before it enters the medium.

As the light source slows down to luminal and subluminal speed through the medium, an observer in a subluminal inertial frame of reference, or a subluminal free-falling frame of reference, would observe two light sources coming together from opposite directions, merging into a single light source in an image pair annihilation event, and a single light source speeding away.

Gravitational acceleration is the product of a spatiotemporal differential. As an object of mass enters the gravitational field of a celestial body, the side of the object closer to the center of gravity of the celestial body has more dilated temporal waves, and compressed spatial waves, than the opposite side of the object facing away from the center of gravity. This results in a gravitational acceleration towards the center of gravity of the celestial body. As the temporal wave dilates, clocks tick tock slower. Hence, the velocity of a temporal wave, the guiding spatial curvature, the principle of inertia of a moving body, and gravitational acceleration, are all the result of the spatiotemporal differential principle. Thus, a temporal wave does not accelerate, or decelerate, due to gravitational acceleration; a temporal wave changes its velocity due to a spatiotemporal differential across the spatiotemporal wave. The spatiotemporal differential changes radially from the center of gravity of the source of gravitation across the opposite ends of the body of mass being gravitationally accelerated.

1.2. Faster-Than-Space-time: The Super Spatiotemporal Speed Conjecture

A light source traveling at luminal speed rides on the spatiotemporal wave in the direction of propagation between two distant points in space-time and returns back to its origin. An observer at an inertial frame of reference, or at a free-falling frame of reference, near the starting point, would see the image of the light source through a powerful telescope only half-way through the round trip before the light source re-appears at the starting point. The light source has traveled back at luminal speed on the temporal wave in the opposite direction, and the observer can see the light source back and the images of its return trip simultaneously. The light source has traveled instantaneously through space, but not through time, between the two distant points, from the perspective of its own inertial frame of reference. The light source has traveled from the past to the future of the observer with respect to the observer's inertial frame of reference, or free-falling frame of reference, at no time at all from the perspective of the light source inertial frame of reference. For the light source, the journey is luminal travel through retarded temporal waves in opposite directions, but it is not time travel.

Contrary to expectation, a superluminal source of light can be devised. Consequently, if a light source travels at luminal speed, and changes its speed to a superluminal speed to cross the light barrier, it may be speculated that such light source may travel into the propagation region of an advanced temporal wave, or in the reverse direction into the propagation region of its retarded temporal wave. The light source may travel into existing space or past space. As the light source travels at luminal speed, and changes its speed to a superluminal speed at the starting point, an observer in a subluminal inertial frame of reference, or in a subluminal free-falling frame of reference, sometime later, would see the real light source back at the starting point and an image pair creation event. The observer would see an image pair creation event of the light source through a powerful telescope pointed in the direction of propagation. One image of the light source is from the outbound trip and the other image is from the return trip as a simultaneous movie of the round trip. Eventually, the observer would see the retarded and advanced images meet and disappear in an image pair annihilation event.

From a different spatial perspective, as the superluminally traveling light source approaches an observer in a subluminal inertial frame of reference, or in a subluminal free-falling frame of reference, the observer would see an image pair annihilation or creation event depending on the direction of crossing. The observer would see two light source images coming together into a single light source as an image pair annihilation event, and the real single light source speeding away. The two light source images exemplify a light source image riding on each temporal wave, a retarded and an advanced temporal wave, at the locality of observation. At the above superluminal speed, an image pair creation or annihilation event does not represent time travel from the perspective of the observer at a subluminal inertial frame of reference, or in a subluminal free-falling frame of reference, either at the starting or arriving point.

If the light source travels a lot faster than light, it is possible to speculate that the light source may travel in reverse through its retarded temporal wave into the propagation region of a prior retarded temporal wave. As the light source travels from the existing past to the farther future of time, the light source may arrive at the future inertial frame of reference, or free-falling frame of reference, of an observer. The flow of time is from the future, through the present, to the past. Even though, the observer may regard that future inertial frame of reference as a past event. As the light source travels to a farther future of time, the space-time within or around the light source would contract, making the light source seem to disappear in a flash into a point of light, at the starting point in the subluminal inertial frame of reference, or in the subluminal free-falling frame of reference, of a local observer.

Since closed time-like curves are not excluded by the General Theory of Relativity, what are probable hypothetical outcomes in time travel?

First, let us consider the probability that there are no parallel universes only a single reality. As the light source travels a lot faster than light to its destination, two real light sources appear in the inertial frame of reference, or in a free-falling frame of reference, of an observer prior to start. One of the two light sources with exotic negative mass takes off in reverse and eventually meets up with the outbound light source from the starting point with positive mass to annihilate each other in a real pair annihilation event. Sometime after, the outbound light source leaves the starting point traveling a lot faster than light. Consequently, there is only one light source left that arrived prior to the outbound light source leaving the starting point. Hence, the possibility of paradoxes exists in time travel within a single reality in accordance with certain solutions permitted theoretically by the General Theory of Relativity that contain closed time-like curves. The Novikov self-consistency conjecture asserts that if an event exists that would cause a paradox or any change to the past whatsoever, then the probability of that event is zero. In that situation, it would be impossible to create time paradoxes. Second, let us consider the probability that there are parallel universes. As the light source travels a lot faster than light to its destination, a single real light source appears in the inertial frame of reference, or in a free-falling frame of reference, of an observer prior to start, because the light source has traveled to a parallel universe. Moreover, there are two identical light sources existing in the parallel universe for a real alternate pair creation event. Events between the two universes may happen differently from this point forward. The possibility of paradoxes does not exist in time travel between alternate realities of parallel universes.

Third, let us consider the probability that the physical laws of the universe allow time travel within a single reality and between alternate realities of a multiverse. According to the chronology protection conjecture of the eminent physicist Stephen Hawking, the laws of physics are such as to prevent time travel on all but submicroscopic scales. In this situation, time travel is permissible through the existence of closed time-like curves in some exact solutions of the General Theory of Relativity. In the far future of time, space-time and light exist in submicroscopic scales as space and time are infinitesimally contracted. Will the infinitesimally contracted spatiotemporal structure and energy of a light source, that has traveled to the far future of time, be allow to time travel by the physical laws of the universe?

Information is knowledge in the form of a message, data, or illustration, which may be transmitted, to rearrange or re-sequence a previous construct in the physical world. If a source of light travels superluminally from the far future of time to the past, information may also travel superluminally. This last assertion is commonly rescinded by the axiom that information cannot travel faster than light per the Special Theory of Relativity. That axiom exemplifies an electromagnetic waveform, such as a light signal, traveling between a transmitter and a receiver, that may be moving away from each other subluminally, at a total speed greater that the speed of light. In such case, no existing mode can make up for the growing distance between sender and receiver. However, the above axiom does not preclude that information may not travel superluminally, but that the transmission and receipt of information through space-time via an electromagnetic waveform is relative.

Would information be allowed to time travel to the past from the future of time under the constraint of the chronology conjecture?

Information may be coded and compressed into submicroscopic scales in a way that can be later unfolded upon arrival in the past. Even though the General Theory of Relativity predicts paradoxes for time travel to the past, these paradoxes are known to go away when they are considered in quantum mechanical terms in mathematical simulations. An illustration may be coded or encrypted to send messages to the past from the future of time. These messages may be sequenced through time to speed up the technological progress of a society in areas of slow progress, to guide areas that are misdirected, or to forewarn of imminent peril. Messages may also be helpful to reassure existing knowledge, encourage novel scientific theories in the past to proceed at a steady step, or to increase overall acceptance and implementation of new knowledge and theories. The effectiveness of the time travel messages may depend on accessibility, temporal global positioning, and method of delivery to the greatest number of observers at the destination's frame of reference. Responsible information time travel to the past may be advantageous to society as our knowledge may be considerably advanced. The achievements of a scientist are built upon the prediscoveries of other scientists. Moreover, the observers, or watchers, of temporal exploration may advance future education, historical accuracy, and social development through the eyes of the future. Nonetheless, it would be wise if projects for the research and advancement of humanity are closely supervised and managed under the protection and guidance of widely-accepted chronological laws.

§ 2. On the gravitational spatiotemporal field

As space-time flows creating on its course the dimensions of space and time, the flow of time, from one present location of space-time to another in its future, may change unequably, curving space, which in turn would manifest a gravitational field commensurate with the acceleration or deceleration of time and the curvature of space. The gravity of the Earth is a phenomenon that results from changes in the speed and structure of space-time due to the deceleration of time and the contraction of space. These changes may result from the presence of mass or energy in space-time and the inability of time, or space, to flow at free speed and free acceleration at the boundary between mass, or energy, and the surrounding space-time. (Taylor, 1966)

The temporal present is the boundary between past space-time and future space-time. The present boundary expands into the past as time flows outwardly forward from every point in space-time, which is opposite to the arrow of time (arrow of causality), and as it reaches a time-like distance in its past, then all that created space-time becomes part of the past, and a new time-like distance future lies ahead. (Feynman, 1964)

§ 3. The postulate of metric point expansion (Eddington's Postulate)

"Every point in space-time expands freely in all directions unless obstructed." Thus, time is a fundamental part of the structure of space-time from its beginning and it is space-time's driving force during its continuous expansion. Time is able to expand at the speed of light, or faster than light at superluminal speed. At the outer boundary of our universe, space-time is expanding faster than light. Space-time is of the universal essence, without space-time there would not be space, time, gravity, the formation of matter, or the evolution of life.

Time, or space, expands as a matter of scale as it exists. The physical relationship between the macro-scale and the micro-scale of reality that determine quantum gravitational fields (time fields) are dependent on the scale of time, or space. The laws of physics depend on the scale of time, space, or mass, of the universe, for space-time expands freely in all directions if not obstructed. Time is relative depending on the scale of the observer, the motions of the observer's frame of reference and that of the observed, and the observer's perspective.

A gravitational field is a spatiotemporal field; it has been proven by the General Theory of Relativity that in the gravitational field of a large symmetrical mass, such as the earth, time slows down or speeds up in a normal direction to the surface of the mass, and space contracts. Thus, the gradient of the spatiotemporal field normal to the mass invokes Gravity. The gradient of the spatiotemporal acceleration near a mass is the gradient of the gravitational acceleration that may be felt by physical objects near the mass. The closer to the surface of the mass the stronger the gravitational field and the slower a clock would tick-tock. The farther, in a normal direction from the surface of the mass, the weaker the gravitational field becomes, and the faster a clock would tick-tock. The speed of time would have its greatest rate of change along the gradient of the spatiotemporal field.

Potential Energy is the difference in the acceleration of space-time between two points or locations in a gravitational field. A mass moving between these two locations would experience a change in its potential energy. The mass would gain potential energy if it moves from the point of lower acceleration of space-time to the point of higher acceleration of space-time, and it would lose potential energy, ceteris paribus, if it moved in the opposite direction between the same two points. Thus, the difference between accelerations of space-time between spatiotemporal localities constitutes potential energy.

Space, Time, and Energy are the pillars of the structure of our universe where the bases are symbolized by the yardstick and the clock, except the measuring yardstick may be curved, contracting, or expanding, and the clock may slow down, speed up, accelerate, or decelerate during its tick-tock, in a relativistic way, not in an absolute way, depending on the local attributes of space-time-energy. (Einstein, 1952)

§ 3.1. The matter-antimatter asymmetry of the universe

It is possible to argue that the whole universe, as presently understood, is out of balance. Accordingly, whenever or wherever matter was manifested, there should have been an equal quantity of antimatter created. Antiparticles have qualities such as electric charge that are opposite to those of particles. Antimatter can also be made of neutral antiparticles with zero overall electric charge. Antineutrons are made of antiquarks which have opposite electric charge to quarks which make up neutrons. Antimatter was discovered in 1932 by the renowned physicist Carl David Anderson. Afterwards, it has been discovered that large quantities of antimatter are produced in lightning strikes, and that interactions of cosmic ray flux in the upper atmosphere produce antimatter that is trapped in the magnetosphere of the earth. However, antimatter is almost entirely absent from the present universe in comparison to the predominance of matter according to observers. So, where is the antimatter created at the Big Bang? During the formation and expansion of the universe did something favor matter over antimatter?

The equations of particles of the Special Theory of Relativity have dual solutions, this condition is a feature of the expansion of every point in space-time. It is possible to hypothesize the dual solution as the solution for matter where particles move on an advanced wave and the solution for antimatter where particles move on a retarded wave. Hence, when matter and antimatter are manifested in space-time, after any possible oscillation of state of matter or antimatter, some of each manifestation of energy, even if it is a tiny quantity, may move on either the advanced or the retarded spatiotemporal wave through the spatiotemporal continuum. Did the movement of particles of matter through the spatiotemporal continuum toward the past interfere with the probability of pairs to cause the oscillating particles to decay slightly more as matter toward the past or very near past?

For example, if matter moves on the advanced spatiotemporal wave, antimatter would move on the retarded spatiotemporal wave. As matter moves on the advanced spatiotemporal wave, it would travel towards the past as antimatter travels toward the future on the retarded spatiotemporal wave. A present observer only able to observe the matter that remained from the past or very near past which is observable. This hypothesis would explain how matter eventually became more common in our universe than antimatter. Then, what was the fate of the antimatter moving on the retarded wave?

As antimatter arrives in the future, there would be a greater quantity of matter than antimatter from the earlier universe to annihilate with. It is possible to hypothesize from the above matter-antimatter model that the stable structures of matter of the early universe may have been more common than the stable antimatter structures. As the universe expanded over time, the physical structures became more based on matter and less on antimatter. This hypothesis is supported by the observation of some matter structures from the early universe toward the farthest observable regions of the universe. The opposite hypothesis is not supported which would have resulted in antimatter structures from an early universe which have not been observed so far. It has also been observed that a pair of particle and antiparticle may annihilate to release energy very efficiently.

What causes matter to be attracted toward the past or very recent past, and antimatter toward the future? Current of charges may be of either polarity. The dissimilarity of the polarity of current for opposite particles of matter, or opposite antiparticles of antimatter, may go as deep as the dissimilarity of the past and future of space-time-mass. Every particle or antiparticle produced may annihilate with a hole or antihole, as allowed by the equation of particles of the Special Theory of Relativity. It is possible to hypothesize that as a pair of particle and antiparticle is produced, the particle is attracted to the total charge of the holes in matter of its past or very near past, while its antiparticles would be attracted to the total charge of the antiholes in antimatter of the future, through the spatiotemporal continuum. The polarity of the total charge of a hole or antihole, for either a particle or an antiparticle, may differ from the very near past to the past, or from the future to the past, depending on the polarity of the individual particle or antiparticle. We may ask the rhetorical questions, is it possible for other universes of the multiverse to have an opposite development of particles and antiparticles? Would an opposite development in another universe be as stable as this universe has been?

As our universe expands infinitely and eternally, there will always be a future space-time, somewhere or somewhen, for the antiparticles to travel to before every pair of particle and antiparticle ever manifested may annihilate. On the other hand, if our universe were finite and self-contained, there would be a future space-time, where and when, all matter and antimatter may ultimately annihilate.

§ 4. On Einstein's own Mach's Principle

The physical law of the expansion of space-time is the underlying framework for the concept of centrifugal force that a body in angular motion experiences as the body interacts with the extending wavelets of isotropic and homogeneous space-time. Inertia is the interaction of mass and the spatiotemporal wave field at a local scale, in isotropic and homogeneous space-time, in the structure of the universe. The conservation of gravitational angular momentum results from the action and reaction of spatiotemporal waves about a body in motion along its path through space-time.

If you stand in a field upon the earth looking at the seemingly stationary distant stars, with your arms resting freely at your side, and then you start spinning faster and faster, the physical law of the expansion of space-time, would make you feel what appears to be a centrifugal force, and your arms are pulled away from your body. You would realize, before eventually feeling very dizzy, that the whirling stars around you are too far away from you, even at the speed of light, to pull your arms away.

The law of inertia, or Newton's first law, is a consequence of spatiotemporal wave theory in the weak-field regime, the constructive or destructive interference of spatiotemporal waves, and of Newton's third law, for every spatiotemporal wave action there is an equal and opposite spatiotemporal wave reaction, about the mass of a body that is at rest, which will stay at rest, or in motion, which will stay in motion, without being obstructed or acted upon by an unbalanced external force, or by the actions of an external mass, or energy field.

The law of inertia is a direct result of the law of gravity for bodies at rest or in motion. The balance or equilibrium of gravitational acceleration forces, or the balance of the acceleration forces of space-time about a body of mass, at rest or in motion, is altered when an external force acts upon the body of mass and impacts an acceleration that breaks the balance of the gravitational acceleration forces, or the acceleration forces of space-time. The body of mass at rest is put into motion in the direction of the external force, or the body of mass in motion may be redirected in a new direction of motion. At relativistic speeds, bodies of mass moving through isotropic and homogeneous space-time experience dilation, or contraction, as a direct result of the law of inertia, when the acceleration forces of space-time about the body of mass, or the gravitational acceleration forces, are unbalanced or no longer in inertial equilibrium. The gravitational forces or acceleration forces of space-time are the acceleration forces of the spatiotemporal waves as they emerge and expand about nodes of mass or energy.

The relativity of inertia is exemplified on objects of mass moving at relativistic speeds through space-time by the relativistic effects on mass, time, and length, where the unbalance of inertial frame forces about the object are directly responsible for dilation, extension, and contraction. Gravitational fields are spatiotemporal fields of space-time-mass.

The inertia of an isolated body or mass is determined by the interaction of the acceleration and pressure of space-time with the physical state of the body or mass at a locality, regardless of the inertia of all other external bodies or masses. Mach's principle manifests itself universally through the equivalence principle, and upholds the constancy of the speed of light.

The principle of equivalence for gravitational effects or inertial effects, as a result of the acceleration and pressure of space-time about bodies of mass, are locally indistinguishable.

The Lense-Thirring effect, a gravitomagnetic frame dragging effect, is a predictive effect for the precession of a particle orbiting near a body of mass, as well as a physical expression of Mach's principle on the dragging of inertial frames near rotating bodies of mass. A clock revolving closely around a body of mass in the direction of rotation will tick slower than tock, while a clock orbiting opposite to the direction of rotation will experience time dilation, as viewed by a distant observer in an inertial frame of reference. Moreover, the clock may experience either a repulsive or an attractive force, among other effects, as it revolves around the body of mass depending on its angular speed, radial distance, and sense of rotation. (Lense, 1918) Mach's Principle may be interpreted as a particle which is completely on its own in the universe having a state of motion that is as much a meaningful consequence of the equilibrium of space-time wave forces as it is about the conservation of energy and momentum throughout space-time.

Consequently, the acceleration and pressure of space-time on the distribution of matter and field-energy momentum, at a particularly chosen Cauchy surface, or moment, in our universe, determines the inertial frame at each point in the same universe.

§ 5. Comments on the dimensions of time

- Richard Feynman allegedly quipped, "Time is what happens when nothing else does." Hence, it is worth adding that 'Time is the first thing to happen and the last.' (Feynman, 1998)

- Julian Barbour allegedly wrote "Physics must be recast on a new foundation in which change is the measure of time, not time the measure of change." Then, it is reasonable to also ask: Can there be motion without time? Would a mechanical clock tick-tock if its hands are not moving at all? Motion is change and time invokes change. (Barbour, 1999)

- Contrary to Ernst Mach, every day we can measure the changes of things because of time. Therefore, the perception of time is a construct created in our minds through our senses based on the changes of things through the passage of time. Time is the framework of perception. Thus, the existence and passage of time promotes change. Motion is change.

- Motion exists in space-time as comparative change as long as there are two or more things to compare.

- Every point in space-time, including the center point of the light-cone in the world history of a particle, expands in all directions of space-time with the passage of time. This is the divergence of space, and the foundation of the postulate of metric point expansion in cosmology.

- Reality is situated in Time. Each moment of reality exists as a moment in a continuum preceding or succeeding another moment of time. Time exists timelessly. Time is perpetual. Time creates instants of space; these instants of space are what we sense as moments in time.

- The existence of time implies eternity, [illegible] , and the existence of space implies infinity, ∞ , as time and its conjugate space may have no beginning or end.

- In order to envision the unusual reality of space-time, one must embrace the unrecognized non-linearity of time. To measure time passing non-linearly is half the achievement.

- Meaning is ascribed to random events through time, but nature is an objective reality. Existence has meaning, and 'meaning and free will' exist in the self-awareness of a mind. Thus, meaning may be ascribed by the outcomes of chance of random events, but nature, space, time, and physical properties are objective realities of the wave medium. The rational and emotional mind benefits from the equilibrium of evidence and compassion.

- There is conformity and consistency in reality, regardless of human acknowledgement or ignorance. Things are the way they are because of cause and effect, not because of premises for the benefit of human reason.

- Each instant of space invoked by time is a Now. Each now is a distance in space that can be measured with a yardstick, but a future now, or then, is a distance in time that is measurable by a clock. A Now is experienced in the very near past, not the present or future. Our senses are not able to perceive the present due to a biological time delay in perception; what we sense with any of our human senses is always the very near past. Our Nows are an illusion of the present, only how things were. Reality may metaphorically be imitating surreal art, future Daliesque clocks melting into yardsticks in the present with our senses only able to observe these bases and events in the very near past.

- Each moment in reality, each now, is a configuration of instants of space involved in the process of change of space-time, and energy, or space-time-energy, which precedes future configurations of moments in time.

- Truth of existence is a prominent part of the process of change during cause and effect between the past, present, and future. In our current temporal paradigm, things that exist have a causal relation to other things in the past, and if those things persist in the present, they invoke causal relation with things in the future. *Things that will be, are related to the way things are, because of the way things were*. However, if we consider the counterintuitive paradigm shift that time flows from the future, through the present, to the past, then our perspective would be that things that will be have a causal relationship to the way things are, because of the way things were. In a sense, every outcome of chance of every random event will, can, and has happened in objective realities of the multiverse of creation.

- In our current paradigm, cause and effect (arrow of causality) follow the arrow of time. Cause precedes effect in time, but the moment of cause is space-like formed a priori while the moment of effect is time-like. If time is perceived to flow from the future, through the present, to the past, the arrow of time is opposite to the arrow of causality of our current perception. In the case of the latter, or if time were to run backward in case of the former, the arrow of time, not of causality, would point from effect to cause, and time would precede space. If Time is Cause, Motion is Effect.

- The length of time, or of a time interval, is the temporal distance difference as measured by the clock of a local observer between the instant of Cause and the instant of Effect of a process or action. Causality enables an observer in space-time to record a temporal interval.

- If time, space, and energy are removed from our experience of recurring observations and conceptualizations of reality, the only framework remaining would be the mathematics of physics.

- Space, time, and energy in all its forms are fundamental properties of objective reality. Motion is emergent. Can there be motion without space or time? Would subatomic particles move in space if time is not passing? Motion of particles is emergent upon the existence of space and time. Space-time endows motion. Under such scenario, the electron surfs on the wave front of space-time at a past time-like distance, slightly into the past from the future scale of time of the nucleus of its atom. The potential difference created by quantum gravitational spatiotemporal fields may be the source of propulsion of the electron and other subatomic particles in the atom. The orbital movement of the particles may depend upon position and momentum within the gravitational spatiotemporal field. The perpetual flow of space-time may propel the perpetually emergent motion of the fundamental particles of matter.

- Time stands still at the event horizon of a black hole; space-time stays equally expanded in all directions if unobstructed. Time expands beyond existing space-time outside of the event horizon, curves, and contracts inward inside the event horizon of the black hole. Outside the event horizon, time expands outwardly toward the past, away from the event horizon, curves, and contracts inward inside the event horizon toward the future, and toward the singularity of the black hole. Hence, the arrows of time inside and outside the event horizon are in opposite direction. The spherical sector inside the black hole from the imaginary surface of the event horizon to the center point of the singularity resembles a space-time reflection of the past space-time-energy spherical sector outside of the black hole. However, at the event horizon, the arrow of time does not point in any direction if unobstructed, and time is not expanding or contracting. Thus, if an imaginary particle were to travel "only" onto the surface, or "only" within the layer, or boundary of that surface if that may be, of the event horizon, with an imaginary internal clock, that would not tick tock, time would not pass, expand or contract, if unobstructed, at the event horizon of a black hole. Such imaginary particle on "only" the event horizon experiences an equivalent temporal condition to an identical imaginary particle if it were possible for the imaginary particle to travel at the speed of light in isotropic and homogenous space-time outside the event horizon. In both scenarios, time is not passing, and the effect of not moving through time for both imaginary particles, is indistinguishable. This effect may be respectfully referred to as the Hawking equivalence principle of luminal motion.

- Gravity tells Space and Time how to go, Space and Time tell Gravity how to pull.

- Gravitational dynamics may typically refer to the motion of planets about a star. Nevertheless, when one considers the spatiotemporal metric describing such a weak-field system $\left(\ddot{a}/8\pi << rc^2\right)$, the spatial derivatives are smaller than the temporal derivatives. From the perspective of free and unobstructed space-time, spatial and temporal derivatives are proportional, this condition is nearly quiescent. If one considers strong-field gravitational systems $\left(\ddot{a}/8\pi \leq rc^2\right)$ such as black holes or neutron starts, then one finds that the spatial derivatives are far smaller to the temporal derivatives. Hence, as temporal derivatives become proportional to spatial derivatives, the spatiotemporal medium commences to exhibit faster wave motion. This strong-field state of the spatiotemporal wave medium propagates forward through the dimensions of free and unobstructed space-time. Gravitational dynamics is analogous to the propagation, production, and back-reaction of gravitational waves.

- The oscillating spatiotemporal wave influences the motion of particles and realizes the quantum mechanical realm. The expansion or contraction of space-time at all scales sustains the fundamental particles and their interactions through the quantum wave medium of space-time and the passage of time. The expansion or contraction of space-time permeates everything at all levels and causes time to proceed as the eternal energy source of space-time-mass. The modulation of the spatiotemporal dimensions produces a quantum mechanical wave function similar to a de Broglie matter wave. As space-time curves, it tells the mass of a particle how to move on the path of a geodesic, through the curvature realized by the expansion, or contraction, and modulation, of quantum space-time.

§ 6. Conclusion.

A temporal distance is the spatiotemporal distance between the location of two events at the speed of light. Time is an intrinsic energy of space-time that is transferable to matter and space. Time invokes space and space invokes time. The deceleration of spatiotemporal fields invokes gravitational fields and the curvature of space-time. The gradient of a spatiotemporal field invokes a gravitational field. In that sense, time is cause, and space is effect. Every point in space-time expands freely in all directions unless obstructed. The effects of the laws of physics depend on the scale of space-time, the properties of space-time depend on the scale of objective reality, and the laws of physics depend on the scale of the universe.

Potential energy is the difference in the acceleration of a spatiotemporal field between two spatiotemporal locations in the field potential. Inertia is the interaction of mass and the spatiotemporal wave field at a local scale about an object while centrifugal force is the effect of the spatiotemporal wave field on a body in angular motion. Thus, if the spatiotemporal wave field is the cause, inertia, or centrifugal force, may be the effect.

The principle of equivalence for gravitational effects or inertial effects is locally indistinguishable. Mach's principle may be interpreted by the meaningful consequences of the equilibrium of space-time as it is about the conservation of energy and momentum through space-time for a particle completely on its own in its universe. Space-time is the framework of perception. Space-time is cause, and change is effect. Space and time endow motion.

Chapter 3

The Ontological Reality of Spatiotemporal Expansion

§ 1. The Spatiotemporal Expansion from an Eternal Singularity.

In the field of cosmology, it is probable that researchers have pondered the following questions:

What is the physical reality of the expanding vacuum? What exactly in space-time is expanding? Why does the expansion phenomenon occur? What physical processes are causing the expansion and acceleration of space-time?

One possible way we may think about the spatiotemporal waves is as expanding densely-packed spherical waves that start as a compacted multi-dimensional manifold that emerges from the tightly compacted dimensions of inner infinity toward the unfolded dimensions of outer infinity as shown in figure 1.

These ontological singularity wavelets lie between the ends of infinity at a scale that allows time to pass, to be sensed, to be measured, through the realms and scales of particles, molecules, objects of mass, sentient beings, celestial bodies and the expanding universe.

A typical question to ask of any theory of the universe that involves a singularity is: what existed or happened before the singularity? Hence, is it possible for singularities to be eternal? Since at a singularity it is reasonable to assume that time is not passing as it does in our universe and physical dimensions are unified. At tic tac zero, there is no universal expansion.

Certainly, the quintessence of our universe: mass, energy, and space-time seem to have eternal qualities. If that is the case, time may be of no consequence on the primordial singularity of our universe prior to expansion. The existence of physical objects or energy is possible without the passage of time; photons are proof of that. Therefore, what happened next?

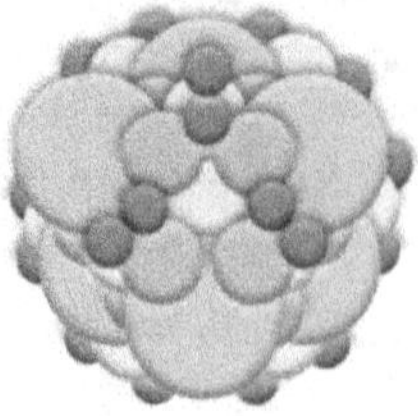

Figure 1

The ontological singularity encompasses the most compact spatiotemporal existence of all its wave probability potential. The singularity has the probability potential for an outward asymmetry that increases its entropy and unfolds the dimensions of its infinite frequency domain. As the spatiotemporal dimensions unfold, each realm has a frequency domain extending to outer infinity. The integral sum of frequencies represents the integral sum of compacted spatiotemporal dimensions of the universe at any scale of expansion. At the mathematical scale of a sentient being, each temporal spherical wavelet or point, emerges and traverses the complex plane of real and imaginary numbers, enabling temporal measurement, mapping of any point on the expanding spherical wave to any point on the plane. Thus, any spatiotemporal state of the universe is derivable and tightly connected to the vanishing singularity through the expanding densely-packed spatiotemporal spherical wavelets of its origin.

All physical dimensions, energy, and mass, spring from the unfolding expansion of the inner infinite of the vanishing singularity. Infinite frequencies unfold onto our reality of space, time, and energy in all its forms. Hence, it is reasonable to ask the following questions: Is mass manifested in high density nodes of low spatiotemporal frequencies along the infinite spectrum? Is energy manifested in high density nodes of high spatiotemporal frequencies? Is space-time manifested in spatiotemporal regions where infinite spatiotemporal frequencies of infinitely small amplitudes cancel in the absence of physical signals? Light is an example of pure energy at a very high realm of spatiotemporal frequency.

Three-dimensional space consists of an infinite number of points, each consisting of an infinite number of frequencies with infinitely small amplitudes which scale up to the isotropic and homogeneous realization of each dimension of the physical world. The integral sum of the discrete frequencies that manifests each dimension in the frequency domain of reality materialize at any observable scale the isotropic and homogeneous smoothness of the existence of space-time, energy, and mass. The smoothness of the time domain of the physical world springs from the discreteness of the frequency domain of its source, as if the physical world were an infinite continuous frequency projection at grand-scale.

During the existence of the ontological singularity, the fundamental frequency is a finite state of minimum entropy that tips over to yield asymmetry. All potential frequencies emerge eventfully from the fundamental. All possible outcomes of entropy begin to emerge as probabilistic wavelets expand and entropy increases. Presently, our observations using the cosmological red-shift technique performed from an earth reference frame, according to an earth reference clock, on far away stars and supernovae, which are on different relative reference frames and clocks, indicate that our universe is expanding and accelerating. However, the observed acceleration may be due to the difference in the rate of change of space and time according to the General Theory of Relativity. If time contracts or dilates in a spatiotemporal region relative to the earth frame of reference, the spatiotemporal wave medium in that spatiotemporal region would appear to accelerate or decelerate relative to the earth reference frame.

Local observers in different inertial frames of reference will still measure the same speed c for light according to the General Theory of Relativity, even if their clocks tic toc differently.

The passage of time in a distant inertial frame of reference where time is more contracted and space is more extended appears to accelerate away during universal expansion when observed from an inertial frame of reference where time is less contracted and space is less extended.

However, the speed of light c is the same in both inertial frames of reference. The illusion of acceleration may come from mixing relative rulers and clocks from different inertial frames of reference.

Nonetheless, in comparisons between inertial frames of reference in spatiotemporal regions with the same or nearly the same temporal rates of change (tic toc of the clocks), but with different spatial rates of change (rulers extend differently), there would be observable spatiotemporal expansion. This implies that to an earth observer of the universe, it is possible that some distant parts of the universe may appear to be expanding at different rates depending on the direction of observation. This is the case when we compare from an earth frame of reference the present era of the universe to the early inflationary phase at a very distant frame of reference from our past.

During the initial spatiotemporal expansion of our universe as shown approximately in figure 2, the rate of change of space over the rate of change of time increases per unit of time, the universe is accelerating. As time contracts, the passage of time is faster as the spatial dimensions extend. Thus, the temporal frequency is higher than the spatial frequency of the spatiotemporal wave during the beginning phase of the universe.

During the inflationary phase, as time contracts even more, the temporal frequency increases considerably more than the spatial frequency. During our present expanding phase, time contracts less than during the inflationary phase from our earth reference frame, spatiotemporal regions of the universe of more contracted time appear accelerating from our earth reference frame.

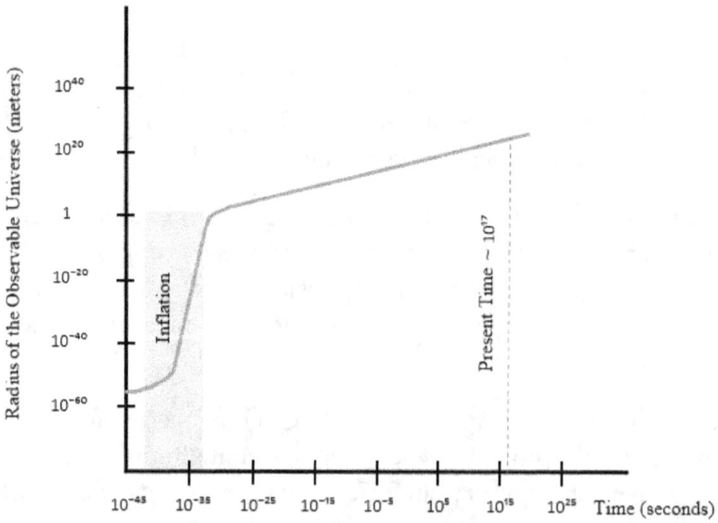

Figure 2

In spatiotemporal regions that are near to the earth frame of reference, as ordinary luminous celestial objects get farther away, they appear fainter and smaller. The surface brightness of the luminous celestial objects per unit area remains constant. The expansion of the universe at every spatiotemporal point may not always cause the inner spatiotemporal distances of a galaxy, like our own Milky Way, the spatiotemporal dimensions between local celestial bodies, or the size of the celestial bodies themselves, to increase in scale over time due to the Huygens-Fresnel principle. It is possible that the universe may have mixed regions of space-time which have static, dynamic, and hyper-dynamic expansion. The expansion of space-time is more readily observable in the vast expanse between distant galaxies or clusters, or radially out toward the outer boundary of the universe. Observations of near or far galaxies have shown that the brightness per unit area of very similar near or far galaxies remains nearly identical. The same geometrical laws apply to near or far celestial objects. It is reasonable to assume that the laws of physics of our universe may originate from the fundamental wave theory of quintessence and the inherent relationship of its physical attributes. (Tolman, 1934)

Energy, matter, frequency, and vibrations are cyclical in nature. The probability potential wave of existence of all material things in our universe exhibits an advanced and a retarded wave at any point that manifests this cyclical nature of space-time-energy. The eternal cycle of motion and transformation of the universe may be one of many in the multiverse of the bulk. There seems to be an eternal universal design of motion, balance, symmetry, and transformation still beyond our understanding, and an evolutionary intrinsic design for self-awareness and self-enlightment.

Any expanding or contracting point connects to its ontological singularity where science predicts retarded spatiotemporal waves and advanced spatiotemporal waves. The singularity event symbolizes the collapse or the rise of the mathematical paradigm of the spatiotemporal expansion or contraction. All possible geometrical shapes, ratios, or dimensional artefacts are expressions of the spatiotemporal expansion or contraction, or the intersection between the expressions. All mathematical or geometrical points are ultimately connected and interdependent. All forms and shapes of matter are entangled to a common source of pure ontological energy.

What is a possible or fanciful shape for a universe?

As the limit of the symmetrical spatiotemporal expansion, $e^{\pm i\pi}$, goes toward either positive or negative infinity-eternity, the spatiotemporal metric goes to a zero point. Let us imagine the spatiotemporal substance and energy of an expanding or contracting toroidal field that may go back into itself. Such a spatiotemporal toroidal field becomes a self-organizing field. At the zero point, there would only be spatiotemporal wavelets, and electromagnetic energy, without any particles in the field of potential.

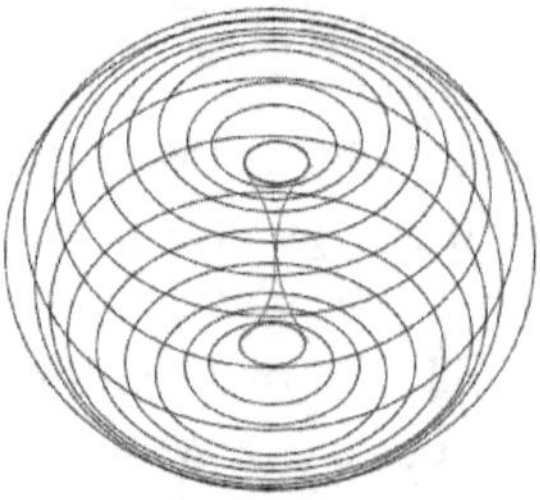

Figure 3

If such a model of a universe were possible, the universe would be able to grow as it expands or contracts at every spatiotemporal point, but it would not collapse on itself. A planetary system from a galaxy located on the side of the source, or sink, of the spatiotemporal dipole, would experience spacetime expanding or contracting at its points since its big bang event, while time passes. Space-time may flow into the side of the sink of the spatiotemporal dipole where space-time may expand, or contract, at every point as time passes. Space-time may flow through matter that exists anywhere in the model of the universe.

The kernel of spatiotemporal growth in such a model of a universe may represent a naked spatiotemporal singularity, an ontological singularity without an event horizon, an eternal singularity. Every point in space-time would expand or contract for every layer of reality in the congruent multiverse as time passes. The six-dimensionality of space-time endows the nonlinear temporal structure for a concentric, congruent, and concurrent dynamic multiverse in the bulk. A multiverse consisting of foliated interrelated universes. A layer of reality, or universe, of the multiverse, may embrace the mathematical framework that supports life. The multiverse as a whole may be considered an infinite and eternal toroidal spatiotemporal field.

Such a multiverse may be able to share all the manifested forms of energy, matter, and particles, with distinct histories. In a multiverse where entropy increases as time passes in all possible directions, each layer of self-contained reality would differentiate according to its history. These very condensed manifestations of space-time-energy would pinch the local layers of reality. Therefore, if the laws of physics are the same, it is possible that the gravitational field may be shared and conserved throughout a multiverse.

Therefore, it is possible to ask the following rhetorical questions. Would it be possible to travel between distinct layers of the multiverse? If travel between distinct layers were possible, would interdimensional travel lead to cultural growth, positive exchanges, or conflict? Are supermassive blackholes shared by layers of the multiverse for a functional purpose? Those layers of self-contained realities that are adjacent would share very striking similarities, cultures, evolution, and histories, that may lead to a commonwealth of kin worlds. Would identical universes be possible within a multiverse or between multiverses? These kin worlds would co-exist at different spatiotemporal coordinates in a multi-dimensional space-time. This would be a Class I multiverse.

The spatiotemporal distance between an arbitrary layer of reality and any superior layer may be considered hyperspace and the spatiotemporal distance between an arbitrary layer of reality and any inferior layer may be considered hypospace. If it were possible to travel between layers of reality, an interdimensional hyperdrive would have to create an interdimensional spatiotemporal bridge. From the perspective of an observer in an arbitrary six-dimensional universe of the multiverse, hyperspace or hypospace, may represent additional dimensions, or the interdimensions between layers of reality. The space-time between space-times.

It is possible to suggest an ensemble of multiverses such as: a Class II Multiverse, a multiverse that is off-centered, or eccentric, to a Class I multiverse, where its axis and overlapping layers are shifted by an arbitrary spatiotemporal distance and/or angle, a Class III multiverse, a multiverse that is separate and distinct in all its layers, but still within the light cone of other classes, and a Class O, a multiverse that is outside of the light cone of other classes, but it may be any of the three classes, O-I, O-II, or O-III. A suffix letter would denote a multiverse that is the same, e.g. Class I-A, or different, e.g. Class II-B through Z, etc. Would the laws of physics be the same between each class of multiverse? Would the initial conditions be the same in each class of multiverse?

If every point in a universe of the multiverse may expand or contract, the wavefunction at any arbitrary point is the resultant spatiotemporal wavelet from the interference of all spatiotemporal wavelets at that point. A single wavefunction may or may not evolve smoothly and deterministically over time and may or may not split or parallel at its point of origin. An evolving wavefunction may endow parallel histories that may split, merge, or manifest quantum phenomena. From the perspective of the point of origin of a single spatiotemporal wavelet in a layer of reality of the multiverse, the behavior of an arbitrary wavelet may not be unitary and would appear as a diminutive randomness that may provide unique initial conditions.

Was Nicolaus Copernicus' heliocentric universe assumption, right? The modified Copernican principle asserts that neither the Sun nor the Earth are in a central, specially favored position in a universe. From a perspective that is on or near the Earth, it is possible to assert, within an estimated Hubble volume of observation, that the Copernican principle seems to hold. Galaxies, interstellar matter, and the cosmological microwave radiation, are observed in a large fraction of the observable universe, in concentric layers around the centrally favored position of the Earth in our universe. Would the nature of space-time in our universe be responsible for the observed Copernican effect?

The Huygens-Fresnel principle of expanding or contracting spatiotemporal wavelets would build concentric spatiotemporal resultant waves within a Hubble volume. Our universe is much larger than a Hubble volume. If a second observer was observing the universe from a perspective that is far from, or at a Hubble volume diameter, away from the Earth, would the Copernican effect be observed? Every observation is based on the history of spatiotemporal waves through our universe, from the perspective of the observer or the recording device. It takes light years for spatiotemporal waves to travel the vast distances of a Hubble volume, so any observation would demonstrate the interference of all waves, in an expanding and accelerating universe, according to the Huygens-Fresnel principle. The Copernican effect reflects the historical lensing of waves toward the center of observation from the very far observable past of the universe to the present of the observer or the observing device.

1.1. A point in space-time

A point is an event in space-time that has unique spatiotemporal coordinates or a unique spatiotemporal address. We may describe a unique point as three-dimensional spatial coordinates at a three-dimensional temporal period that may also be continuously changing with space-time. A point is a dynamic spatiotemporal conceptualization as a function of its locality that is able to represent an event for an observer in a dynamic coordinate system. The spatiotemporal address of the event may change depending on the observer's frame of reference. The intersection of the world-lines of two events may characterize a theoretical point, or the collision of two particles, but a world-line is the complete history of events in space-time of an object or particle, not the spatiotemporal event itself. In the most strict and natural sense, a world-line may pass through an event, or a unique point in space-time, only once in its history, or an event, or a unique point, may be conceptualized existing in a world-line only once, in its uniquely dynamic spatiotemporal address. If this strict condition were transgressed, the space-time of a singular universe would be able to encompass closed time loops.

It is the collective property of physical space-time and its changing geometry that defines all physical events that involve energy and matter. The interstitial spatiotemporal distances between physical events are always present and mediate synchronous or asynchronous events. Events are synchronous whenever and wherever the geometry and collective property of space-time allows it. The state of a vibrating string can be modeled as a spatiotemporal point in a Hilbert space. The fragmenting of a vibrating string into its oscillations in distinct frequencies, that are greater than the fundamental frequency of the string, is given by the projection of the spatiotemporal point onto the coordinate axes in the Hilbert space. A Hilbert space is a complete abstract vector space having the structure of an inner product that enables the measurement of spatial distance and angle. A point is a dynamic concept of space-time which has infinite dimensions in its frequency domain. Each dimension of an expanding wavelet with an infinitely small amplitude manifests a distinct frequency. A static mathematical point is but a snapshot in time of a dynamic point, a slice of now as shown in figure 4. A point embodies energy, space-time, motion, frequency, and infinite information on past, present, and future physical reality, becoming either cause or effect. A spatiotemporal point is complex and intricately encoded in its frequency domain.

Figure 4

The spatiotemporal expansion is compounded from extended spatiotemporal points. Points are the physical manifestation of the eternal motion of an infinite number of frequency domains as the spatiotemporal dimensions unfold. This eternal motion of infinite frequencies compound to emerge as the spatiotemporal wavelets in the observable reality of expanding preemptive imaginary time and real space. Physical signals travel through the spatiotemporal wave medium as all infinite frequencies of infinitely small amplitudes cancel in the absence of the signal. Physical signals such as electromagnetic waves can travel through the vast spatiotemporal distances of the observable universe for billions of years without any loss of energy.

Time creates space, and from that space, more time emerges. The location of points is constantly expanding temporally and spatially, if not obstructed by matter, momentum, pressure or energy. When a point expands unobstructed, there may be a measurable spatiotemporal extension at its locality that drives the overall spatiotemporal expansion of the universe. If the point expands obstructed, dimensional symmetry prevails at the locality of the point. Cosmic expansion in the plenum of matter, the expanse of the solar system, or between galaxies, is dimensionally symmetrical. Cosmic expansion in the great expanse of the observable universe is highly symmetrical, isotropic, and homogeneous.

1.2. On the nature of complex space-time.

The concept of complex space-time broadens the impression of real-valued spatial and temporal coordinates to a dynamic complex wave medium of complex-valued spatial and temporal coordinates.

In quantum mechanics, wave functions are complex-valued functions that describe particles in real-valued spatial and temporal coordinates, and the set of a system's wave functions is an infinite-dimensional complex Hilbert space. Wave functions and fields extend to complex space-time, and this complex space-time may be interpreted as an extended complex spatial and temporal coordinate system. Consequently, complex space-time is the wave medium for the extended fields of complex spatiotemporal wavelets.

The notion of complex space geometry has been previously considered by prominent researchers such as Albert Einstein, with a complex metric tensor, but without complex space-time. Hence, it seems that the notion of complex space-time has not received an equal amount of attention.

Imaginary time has real physical meaning and it allows mathematical analysis to endure the analytic extension of the real temporal variable onto the complex plane, which makes some solutions practicable. Time is dimensional in nature; every spatial dimension has an orthogonal conjugate temporal dimension. A temporal dimension has a dimension of one because only one coordinate is needed to specify any point within the temporal dimension. The temporal coordinate on the measurable extent of a temporal dimension may be specified by an imaginary number. In this conception of complex space-time, every event may be specified by complex coordinates. Each complex coordinate has a spatial part (real part) and a temporal part (an imaginary part), where both parts are associated with an apparent or complex value. A single complex coordinate system may be applied to a spatiotemporal surface with two existing dimensions of space and time. Complex coordinates may specify spatiotemporal distance, surface, or volume. A tesseract may be represented using a single complex coordinate system where the corners of the inner cube, a three-manifold, represents a volume of "now" with complex coordinates that have imaginary parts with zero values, and the outer cube represents a volume of a "future now" with complex coordinates that have non-zero spatial and temporal parts. In a complex coordinate system, a six-dimensional real manifold is a tesseract (a complex three-manifold) with three real spatial dimensions (real number parts) and three real temporal dimensions (imaginary number parts). In differential geometry, any complex n-dimensional-manifold is simultaneously a $2n$-dimensional real manifold. Hence, any complex one-manifold (Riemann surface) is a smooth oriented surface with an associated complex structure. Every closed surface admits complex structures. Thus, a complex two-manifold is a complex surface, and simultaneously a real four-manifold.

In the above example of a tesseract, if all six sides of the tesseract are considered adjacent to other identical tesseracts, then the spatiotemporal wave on each side and adjacent side of the tesseract would interfere with each other. If space-time is homogeneous and isotropic on each side, then the waves would interfere on each side and subtract. The real and imaginary parts of each adjacent pair of complex spatiotemporal coordinates, on each side of the tesseract, would cancel each other, and the resultant expansion of space-time would be null. If the real and imaginary parts of each adjacent pair of complex spatiotemporal coordinates on each side of the tesseract do not cancel each other, then space-time expands into a larger volume of a "future now". Time is a dimensional process. Time and space are the outgrowth of every dimensional extent of space-time. Thus, time, as it relates to every spatial dimension, is both a process and a dimension. The properties of time include the passage of time at different rates where slower rates in gravitational fields are continuously connected to faster rates as a function of distance from the object of mass or gravitational source. Time in all its dimensions is an emergent phenomenon. Time emerges from the expansion of space. The expansion of space is the direct cause of time. The passage of time is a function of the rate of expansion of space-time.

In turn, the rate of expansion of space depends on the spatiotemporal pressure at any point in space-time which is a function of mass and energy. The rate of expansion of the spatiotemporal wave medium, or the passage of time, manifests a gravitational field near an object of mass. Electromagnetic waves propagate at the speed of light through complex space-time because that is the speed of expanding space, or the speed of the passage of time, at any point or event in complex space-time. Time, and space, are emergent properties of the expansion of space-time, endow motion to particles and fields at any point in complex space-time. All motion is a function of the dimensions of space and time, and motion exists due to the expansion of space-time. As an object of mass travels through space, it would travel less through time because the temporal wave would travel at slower rate (lower frequency) through the object moving through less temporal extent of the temporal wavelength, as the temporal wave dilates in the local reference frame of the object in the direction of motion. Hence, it is space-time that endows the passage of time, the motion of objects of mass and fields, and perpetuates motion through its property of inertia.

1.3. The cosmic relative coordinate system.

The properties of the spatiotemporal structure of a volume are conformal as the properties are considered radially inward toward a center point in homogenous and isotropic space-time, preserving the correct angles between directions within small areas, though distorting distances. However, as two arbitrary points on the curved manifold approach each other, the curvature and torsion between those two points may decrease significantly. Consequently, events occurring between two arbitrary points at a quantum scale of a Lorentzian manifold with spatiotemporal properties that are functions of scale, such as curvature, contraction, expansion, and torsion, may differ from identical events with those properties at a much higher celestial scale. Hence, time may not pass between two arbitrary points on a quantum manifold at the same rate that it passes between two arbitrary points on a parent manifold at a celestial scale at the same location. As the scale of a quantum event between two arbitrary points increases, the spatiotemporal manifold at a larger scale may have greater curvature and torsion, where the temporal extent dilates more. As space-time between two arbitrary points becomes smoother and less curved, the passage of time quickens, clocks run faster. The spatiotemporal scale of quantum particles is orders of magnitude above the scale of any possible foaming, strings, D-branes, that may exist. Time is relative according to scale in non-homogenous non-isotropic space-time.

At a quantum scale, a thin slice of the dynamic block of the universe has a very negligible time differential between two arbitrary points, even though at a much higher scale and at a greater thickness of the slice, the General Theory of Relativity would apply to a greater time differential between two higher arbitrary points. The interference of spatial wavelets and temporal wavelets may generate far less curvature at a quantum scale that they do at a celestial scale. Time endows motion at the quantum scale of fundamental particles, potentially creating more space, and vice versa, while the spatiotemporal properties may be far less prominent.

The General Theory of Relativity was conceived to deal with very large-scale spatiotemporal wave theory, pressure, and energy density of matter. Macroscopic objects moving at classical speeds have minute relativistic changes. The General Theory of Relativity has been confirmed on subatomic particles traveling at relativistic speeds, or by measuring minute relativistic changes with very sensitive instruments. Elementary particles traveling at close to the speed of light have larger relativistic effects.

Quantum Mechanics has a very different spatiotemporal background at a quantum scale where very small-scale spatiotemporal wavelet theory is applicable, and quantum effects between particles dominate over less significant spatiotemporal properties. The General Theory of Relativity focuses on the distortion of the smoothness of space-time, and the reciprocal relationship of space and time wave theory, whereas Quantum Mechanics focuses on the local properties of particles embedded in the spatiotemporal medium under the quantum-scale distortion of the smoothness of space-time. The equivalence principle for spatiotemporal wave theory, is that gravitational acceleration is completely equivalent to the contraction and acceleration of the spatiotemporal wave toward the center of mass while the temporal wavelength decelerates and expands and the spatial wavelength accelerates and contracts. A particle entering a gravitational field of a body of mass is guided by the spatiotemporal curvature on its non-linear path, accelerates as the spatiotemporal wave contracts, while the spatial wave contracts spatial distances and the temporal wave dilates temporal extents. The change of gravitational potential energy of the incident particle due to the spatiotemporal wave differential increases the velocity and the kinetic energy of the particle as it travels toward the center of mass or drives the angular momentum of the orbit around the center of mass. The EFEs are coordinate-independent, a property called general covariance, meaning the equations would produce correct, consistent descriptions of the universe regardless of which coordinate system is used.

The General Theory of Relativity and Quantum Theory may be reconcilable through the concept of General Covariance. The EFEs and Quantum Theory are reconcilable when solutions are expressed without any kind of rigid and incongruent coordinate system dependence, or through a single congruent relative coordinate system that is applicable to any configuration of space-time for small or large-scale spatiotemporal regions. In such scenario, the general covariance is not broken when solving every problem in every configuration of space-time. The cosmic relative coordinates are not one privileged choice of coordinates since the coordinates would adjust to every configuration of space-time. Any of the existing coordinate systems may be converted to cosmic relative coordinates. The cosmic relative coordinates would transform according to the local properties of space-time at any specific point.

Let us propose a cosmic relative coordinate system that transforms its coordinates as functions of scale, curvature, contraction, expansion, and torsion, of space-time according to the General Theory of Relativity. Let us start with a relative Cartesian (rectangular) coordinate system where the spatiotemporal distance at any one coordinate point x_0 transforms according to the retarded and advanced wavelets of space and time. The wave interference at any one relative coordinate point, or between any two relative coordinate points, would transform to any configuration and combination of properties of space-time. Regardless of the comparative magnitude of relative spatial and temporal distances, the speed of light remains constant.

$$\Delta x' = \Delta ct' e^{ix_0} = \Delta ct' \langle \cos(x_0) + i\sin(x_0) \rangle \qquad (1.1)$$

$$\frac{1}{c}\frac{\partial x'}{\partial t'} = \frac{\left(e^{ix_0} + e^{-ix_0}\right)}{2} + \frac{\left(e^{ix_0} - e^{-ix_0}\right)}{2i} \qquad (1.2)$$

The relative coordinate magnitude, at each point would represent the relative spatiotemporal distance from an orthogonal plane on a relative frame of reference in six-dimensional space-time. The physical layout of the relative rectangular coordinate system is uniform, but every relative spatiotemporal coordinate point may differ in magnitude and properties. Each spatiotemporal coordinate direction has its corresponding relative length, time, and properties, for the specific point. Each six-dimensional point P_0 may be represented by three relative spatiotemporal coordinates $(\tilde{x}, \tilde{y}, \tilde{z})$, corresponding to three relative coordinate tensors, that completely describe the point in terms of its spatiotemporal attributes.

$$\tilde{x} \equiv \nabla_{\vec{u}} \vec{v}_x \langle x' \rangle = \nabla_{\vec{u}} \vec{v}_x \langle ct'_x e^{ix_0} \rangle \qquad (1.3)$$

$$\tilde{y} \equiv \nabla_{\vec{u}} \vec{v}_y \langle y' \rangle = \nabla_{\vec{u}} \vec{v}_y \langle ct'_y e^{iy_0} \rangle \qquad (1.4)$$

$$\tilde{z} \equiv \nabla_{\vec{u}} \vec{v}_z \langle z' \rangle = \nabla_{\vec{u}} \vec{v}_z \langle ct'_z e^{iz_0} \rangle \qquad (1.5)$$

Where $\nabla_{\vec{u}} \vec{v}$ is the covariant derivative of a vector field $\vec{v}$ in the neighborhood of a point P_0, and $\vec{u}$ is a vector, defined at point P_0. The output is the vector, $\nabla_{\vec{u}} \vec{v}(P_0)$, also at the point P_0. The covariant derivative $\nabla_{\vec{u}} \vec{v}$ is independent of the way it is expressed in a coordinate system. Hence, the coordinate point $(\tilde{3}, \tilde{6}, \tilde{9})$ identifies a specific location where each coordinate number represents a specific tensor in the sense of direction of its respective axis that intersects the directions of the other two coordinate numbers. For example, the coordinate $\tilde{3}$ is the relative distance in the direction of the $\tilde{x}-$axis from the $\tilde{y}-\tilde{z}$ plane, at a relative distance $\tilde{6}$ in the direction of the $\tilde{y}-$axis from the $\tilde{x}-\tilde{z}$ plane, and at a relative distance $\tilde{9}$ in the direction of the $\tilde{z}-$axis from the $\tilde{x}-\tilde{y}$ plane.

As a straightedge travels through space-time, it does not measure distance, it measures the metric along the path of its worldline in six-dimensional space-time. Similarly, as a timepiece travels through space-time, it does not measure time, it measures the metric as the metric is applied on its worldline in six-dimensional space-time. All measuring devices measure the metric. The metric is a distance function for a space-time, whose value is the distance between two points, it quantifies and translates the behavior of spatiotemporal configuration. The metric gives the distance between two spatiotemporal coordinates that may be so infinitesimally close to each other that curvature, or other properties of space, become irrelevant. The metric tensor is the derivative of the distance function between a pair of coordinate points that gives the infinitesimal distance on the manifold or metric space. Therefore, it is possible to propose that the numbering system of each axis of the cosmic relative coordinate system, in each spatiotemporal direction, be based on the metric.

A vector is a geometrical object, i.e. a rank-1 tensor, that retains its own identity independently of how it is described in a basis. If there is a change in coordinates, the covariant derivative transforms in the same way as a basis through a covariant transformation. All cosmic relative coordinate operations are possible between two or more points through the applicable coordinate tensors involved.

The curvature, torsion, and geodesic, between two relative coordinate points, or on the neighborhood of a point P_0, may be described by the covariant derivative through parallel transport. If the spatiotemporal distance from a perpendicular coordinate plane to a point P_0 is torsion-free, the Levi-Civita connection may be used as the covariant derivative. The covariant derivative does not use the metric. In the case of the Levi-Civita connection, the covariant derivative of the metric is zero.

If a particle is moving in the direction of the positive $\tilde{z}-\text{axis}$ were passing through the origin of a cosmic relative coordinate system, any future point on its path would lie within a 45 degree cone centered on the positive $\tilde{z}-\text{axis}$, and any point on its previous path would be on the opposite 45 degree cone centered on the negative $\tilde{z}-\text{axis}$. From each of these relative rectangular coordinates it is possible to derive any other relative coordinates, such as relative spherical, cylindrical, polar, curvilinear coordinates, etc.

The Minkowski principle of complex space-time exemplifies the conjugate nature of space and time in the spatiotemporal continuum. Space and time are but two sides of the same coin, and space-time is the mint. The less there is of space, the more there will be of time, and vice versa, in the wavelength and period of the variable spatiotemporal wave. As space-time expands at every point, there will be more space and time. The framework of the independent reality of space-time is the reciprocal coexistence of the spatial wave and the temporal wave as the spatiotemporal wave changes or preserves its properties. Therefore, time is emergent and creates more space, which in turn allows the emergence of more time.

1.4. On the existence of time

Time exists regardless of what physicists would say; time exists even though our perception of time is a persistent illusion to our senses. The passage of time in our perception utilizes the process of memorization of impressions of experiences, activities, and observations from the real and the imaginary world as entropy increases or decreases from the perspective of the impression. An impression can be stored, memorized, recalled, sequentially, or non-sequentially, to be reconstructed if or when it is accessible. Our memorization process relies on the increasing entropy of real events in the processes of our universe. Our perception of time is biological, biocentric, and based upon a real temporal process. Space and time are fundamental in nature, space-time is an objective reality.

Throughout history, people observed the rising of the sun in the east and the setting of the sun in the west and created all kinds of explanations, before realizing through scientific observation and calculation that it was the earth turning from the west to the east. Do we presently have a similar misunderstanding about the passage of time?

Space endows time, and time emerges from the future of space, to its present, then to the past of space. In fact, the word 'time' may be considered as a metaphor for expanding space-time. Space-time expands from the future, through the present, to the past, opposite to our typical sense of motion through time which is perceived to be from the past, through the present, to the future.

In essence, an object does not move through time, time moves through an object, the faster the object moves through space, the lesser time will move through the moving object in any sense of its direction of motion, as predicted by the General Theory of Relativity. Events may exist in the future of time, tempus incognito, before the present or past. Time is the cause and change is the effect. In such events, the future through the present may constrain the past. The consequences of the future event are confined to the quantum level of the present and past. If an object is illuminated by a source of light in the inertial frame of reference of an observer, the emission of light by the source represents the future event from the perspective of the object. The illumination of the object becomes the unobservable present event, and then the observable past event, from the perspective of the observer.

Times past encompass space, the present emerges inaccessible to our biological senses, and the future exists as purely imaginary (temporal). Our inability to experience the reality of the present biases our perception of time. Our impression of a "now" is the impression of a very recent yore or yesterday. A "now" of our reality is but a fleeting moment to our senses. A fleeting moment, like a perceived "now", is a temporal construct, not a quantum of space-time that is instantaneously rising from the laws of nature. Our memory and perception are based on past events. Hence, our perception of the passage of time is the perception of expanded space-time in the past. Our process of remembrance, reminiscence, or recollection, uses our memory of impressions from our past and most recent past, to construct a sequence of events that follows the increase in entropy of physical processes in our universe. The perception of the passage of time, the temporal duration between events, relies on a sequence of indefinite and unfolding events that may be real or imaginary, conscious awareness of our sensory inputs, biological clocks constructed by our mind depending on the task at hand, and stimuli from the surrounding environment.

Our perception of the passage of time, a moment in time, may be asynchronous to the apparent passage of time at our locality in space-time. A perceived moment in time exists as a construct of a quantum of time, not as an apparent quantum of time that coexists as a separate synchronous natural event. The future of space-time is out of reach from the most recent process of memorization due to spatial and temporal locality. In a physical and spatiotemporal sense, our minds can only memorize the past, the most recent past, and imaginary events. Our senses are incapable of perceiving the passage of time instantaneously. An instant "now" in the passage of time is inaccessible to our biological senses and physical systems. Time emerges and exists independent of our individual perception of time. A "now" is an illusion. What we consider a "now" is a time period consisting of the delay of the arrival of an external signal to any of our senses, the delay of an electrical signal, chemical signal, or bio-structural action through our senses and physical system, the delays of recognition, perception, memorization, and interpretation processes of the external signal, the delay of adding emotional content, etc., to feel or sense the passage of a "now". By the time we sense a "now", the instant of the temporal dimension has become a spatiotemporal distance, or a very recent part of our past. Our "nows" are only yesterdays.

In our description of time in our universe, the future is not fundamentally different from the present and the past. The conceptual perception of the past, present, and future, is continuous and contiguous at a certain scale, as much as space-time is continuous, contiguous, and not demarked, to the benefit of the rationalization. However, all three spatiotemporal concepts share the same context of reality and emergence. The outcomes of events may be perceived as the same, or different, depending on the probability of events, one's observation and judgement. The laws of nature are time-reversal invariant, and space-time is emergent. Hence, to the extent of present knowledge and experience, the laws of nature are time reversal-invariant, omnipresent in our universe, and emergent.

The emergence of space and time is the underlying fundamental theory of our universe. Newton's Law of Universal Gravitation and the General Theory of Relativity are consequences of the Fundamental Theory of Space and Time. Both the Newton's Law of Universal Gravitation and the General Theory of Relativity are not fundamental, the former breaks down at very high speeds for objects approaching the speed of light, and the latter breaks down when gravitational fields become very strong. the Fundamental Theory reproduces both the Newton's Law of Universal Gravitation and the General Theory of Relativity and gives rise to our present notion of space-time.

The equations of the Fundamental Theory of Space and Time are symmetrical. The laws of physics underlying the fundamental theory are time reversal-invariant which means that the equations of processes remain the same when the direction of the expansion of space-time is reversed. Albeit, the processes of very large systems do not remain the same. The processes of very large systems may be reversible, but not invariant. It is possible to run processes of very large systems backwards, even though both the initial conditions and the outcomes will not be the same. Entropy increases over time in very large natural processes.

From the ideas of Hermann Minkowski, we learned that space and time are combined into space-time. Objects move through space-time as a wave medium. There needs to be the emergence of space-time to have past, present and future expanding space. Time and space are two sides of the same coin. Time exists as a different phase of space that is measurable but not yet observable. If time did not exist neither would space. Time exists due to the expansion of space, and space exists due to the emergence of time. Space-time is dimensional in nature as we can assign complex coordinates to locate points on its dimensional extents. Temporal coordinates are relevant to predict observable spatial events under the Fundamental Theory of Space and Time.

Time is real, omnipresent, inexorably passing in its flow, from its source to its sink, in any sense of direction. Our perception of the passage of time assigns order to events, even when the order of events is opposite to a sense of direction of time. Time flows in all directions, time is multidirectional. Space-time is the universal background through which all events exist, where entropy increases for large systems, and where our perception and conceptualization recognize order, sequence, direction, and duration of events. All directions and tenses of time are equally real and emergent, so the future, present, and past are equally real and emergent. As Isaac Newton allegedly stated, “even if absolutely nothing at all happened, time would be passing”, and even though the possibility of absolutely nothing happening in physical reality is improbable, Newton’s intuition was truly commendable.

Space-time is tenuous, measurable, and malleable according to the General Theory of Relativity, while assumed as a non-observable background in Quantum Mechanics. These contrasting views of time involve the scale of physical structures and properties such as the size of large systems at slow speed through space and at high speed through time, the size of very small systems at very high speed through space, near the speed of light, and at very slow speed through time. Time may pass equably in the space-time background of the very large systems and the very small systems, but the effect of time on the scale of physical structures and properties of systems may be correspondingly different. Space-time is fundamentally complex, real and imaginary. Space-time is elemental at the innermost foundation of nature; time is expanding space-time, an irreducible fundamental element that emerges from fundamental space-time to construct reality.

Space-time is the deepest notion to the fundamental nature of objective reality. Space-time is an essential notion that provides the key to understanding the laws of nature. Space-time is a construction of elemental space and time. It is fundamental to nature, and emerges from a deeper level of reality. As time emerges, it consists of elemental wavelets of space and time, that take on physical properties from how those wavelets interfere with each other, building up the structures of space-time at all levels of scale, and endowing the expansion of space-time. The concepts of space and time emerge from a universal reality that, at its root, is utterly dynamic. In a predictive manner, the future gives rise to the present, and the present updates the past. Our intuition conjures the perception that the future is unstructured, open to possibilities and probability, until it becomes the present where actions and decisions render it structured, and consequently the past becomes cast. As time flows, this intuitive perception moves forward in time, becoming part and parcel of human culture, language, way of thinking, and behavior.

A point *P* in space-time, between point *A* and point *B* along an imaginary straight line, experiences opposite spatiotemporal wavelet pairs Ψ^+ / Ψ^- in all directions. Let us imagine one pair of wavelets arriving at point *P*, a forward wavelet from point *A* and a reverse wavelet from point *B*. Both wavelets are the result of the passage of time, or the expansion of space, at each spatial and temporal locality of each point *A* or *B*, that interfere at point *P*, where there may be expansion, contraction, or stasis. Point *P* is also expanding and its spatiotemporal wavelet would equally affect points *A* or *B*. The spatiotemporal wavelets expand from the ontological compacted singularity with the probability potential for an outward asymmetry that increases its entropy and unfolds the spatiotemporal dimensions. All spatiotemporal points are tightly connected to the ontological singularity through the fabric of space-time and the passage of time. Space-time is the medium of entanglement between past, present, and future, of particles, energy, and all phases of matter. Quantum entanglement is a consequence of the medium of space-time and its properties.

Points of reference are useful constructs in the space-time of reality where events may be taking place. The spatial or temporal intervals between points of reference may change or stay the same. The waves of space and time invoke the change of space-time or its stasis. Therefore, there is time even without change in happenstance events, or in the variation in the spatiotemporal intervals, between points of reference. Time is viable and measurable with or without observable change in the spatial distances between two or more points of reference. The measurable temporal distance between two points of reference would be the time it would take a photon to travel at the speed of light between the two points. Change is a function of time. Time is real and viable with or without the perception and acknowledgement of the minds of sentient beings, or with or without the involvement of any life event, at any point of reference. Space and time are one and the same, two sides of the same coin. As mathematical entities, we can replace space and time for each other in the context of space-time. In Minkowski space-time, time is treated exactly as space except with a factor of c, the speed of light in vacuum, and a factor of "i", the imaginary number equal to $\sqrt[2]{-1}$. An interval of space has an equivalent interval of time and vice versa. This view of space-time springs from the emergence of space and time. Time is emergent and creates more space which in turn endows greater emergence of time. As far as we know, the space-time continuum is contiguous, without missing points in space or instants of time at any measurable or observable scale. Both space and time can be subdivided without any observable or measurable limit in spatial or temporal extent. Reality is embedded in the space-time continuum; events, places, instants, and actions are all described in terms of their location in the space-time continuum. Space-time evolves as it exists. The world-line of an object exists due to the expansion of space and the passage of time. Every particle is located along its word-line. The past and future light cones of a particle are located in dynamic space-time continuum with boundaries demarked by the speed of light. Space-time is not static within the boundaries of each light cone with the passage of time. Thus, the world-line of an object is dynamic with the passage of time. When time travels through an object along its world-line, the object may change in some way.

Time may be viewed as a resultant dimension of space-time or a geodesic temporal curve. An object does not travel through time necessitating degrees of freedom, but instead time travels through an object with degrees of freedom in every sense of direction in space-time. Hence, an object does not require being entirely free to move around in its timescape, because the object does not travel through time, time is entirely free to move through and around the object. By moving through space, the object allows less time to move through and around its physical form in the direction of its trajectory. An object in space-time may change its spatial coordinates or it may stay at the same location relative to other objects, but if the object is moving through space at less than the speed of light or standing still, time changes the temporal coordinates as it passes all around and through the object. If it were possible for the object to move at the speed of light, it would not travel through time in the direction of its trajectory, and time would not change the temporal coordinate in the direction of travel. An object traveling at luminal speed may travel on a geodesic between two points on a temporal wave front without the passage of time. The temporal coordinates of the object would change as the object travels luminally in the radial direction of temporal wave propagation, and luminally in a direction between the two points perpendicular to the radial direction. If the object were to travel through an Einstein-Rosen bridge between the two points, then it would appear to an observer in an inertial frame of reference that the object traveled superluminally. In such case, the spatial and temporal coordinates would have changed.

Space and time exist even in vacuum. If we consider space-time in a universe where time passes without entropy and no sentient being is there to experience the passage of time or the expansion of space, will time still exist in that universe? The answer to this question may be speculated as affirmative if time is part and parcel of the fundamental natural laws of that universe without entropy. Even though in the vacuum of that universe, there may not be an arrow of causality, or life as we know it, only the spatiotemporal waves of the medium.

A thought has duration of time. If we consider a temporal interval as an equivalent spatial distance, the time between ideas is a spatial distance that is measurable in space-time. Are thoughts emanating from a source that is corporeal (physical) in nature? Or are thoughts coming from an exotic source or substance that is incorporeal (metaphysical) in nature? These questions are as old as philosophy or religious faith. Nonetheless, thoughts exist in space-time and time is the spatial distance that binds thoughts together in the stream of consciousness. Temporal distances bear the same mathematical relationship to spatial distances as imaginary numbers do to real numbers. Moreover, in Lorentz transformations, temporal distance and spatial distance partly transform into each other as a function of relative velocity. Thoughts are the building blocks of knowledge. Does the accumulation of knowledge follow the nature of thoughts? Are the laws of physics applicable to thoughts or accumulated knowledge in sentient beings? What is the speed of thoughts in space-time? Are thoughts complex like space-time? A physical body and mind (consciousness) are submerged within the passage of time at the speed of light, when they are not translating in any direction of space-time. A sentient being's physical form (body) may travel as fast as its legs or vehicle, but its metaphysical form (consciousness) may travel as fast as its thoughts. Physical locomotion involves force (push or pull) and reaction, thought follows motion, while metaphysical motion may involve the free will of an idea (conceptualization) that may be as fast as the speed of thoughts, motion follows thought. The speed that individual instruments play at in a symphony relate directly to the motion of the music. If the metaphysical form travels at the speed of thoughts through space-time, does space contract and time dilate? After all, space-time is the fabric of thoughts, consciousness, and reality.

1.5. The Dynamic Block Universe

In ancient Greece, Heraclitus allegedly stated that reality is always changing, everything is perpetually flowing and moving, but Heraclitus, as far we presently know, did not expand on the relationship between perpetual changes and time. Decades later, Aristotle concluded that only the present existed, not the past or future, the past has been and now it is not, while the future is going to be, but it is not yet, in the three-dimensional reality of the world. Centuries later, Saint Augustine inferred that the present is an instant without duration in an inaccessible ever-present eternity. The dynamic block universe view emerges from these classical views of reality to propose the existence of a block universe (an ever-present eternity) that is always changing with the flow of time, per our perception, from the future, through the instantaneous present, to the past, through the arrangements of events, endowing motion in our universe. In the alleged words of Heraclitus: "No man ever steps in the same river twice, for it's not the same river and he's not the same man. Nothing endures by change."

As space-time expands in the dynamic block universe, complexity increases and leads to the formation of stable structures of certain arrangements of matter. These arrangements are informational records. Gravitational fields are spatiotemporal fields that maintain the structures of those arrangements of matter which in turn evolve through other processes into the structures of life. These records contain information about earlier states of reality that are embedded in space-time. Space-time is the recording medium for these Akashic records which may evolve into a compendium of objects, thoughts, memories, events, and emotions that are encoded in the spatiotemporal plane of existence of creation.

The higher the complexity of reality, the greater the amount of informational records that are imprinted in the Akashic field of space-time. Memory of past events is built upon this Akashic field. The block universe of the future, present, and past of space-time is constantly emerging and evolving as the indefinite future becomes the definite past. Time emerges and space expands. The block universe extends from the inner infinite of space-time to the outer infinite with a continuous and contiguous flow of time through the inner space and the outer space of all objects in our reality. Changes in the block universe occur as space-time emerges from the future and changes space-time in the past through isotropic, anisotropic, and homogeneous space-time.

The dynamics of these changes are covered by the equations of the General Theory of Relativity during an expansion or contraction of space-time. We may regard the intangible boundary between the future arrangement of events and the past arrangement as the present. The indefinite future may change the definite past. A dynamic block universe is immersed in the current of the passage of time.

The spatiotemporal arrow of time describes the passage of time from the future through the present to the past in a counterintuitive way. The psychological arrow of time of perception is an inexorable flow of time from the past through the present to the future in an intuitive meaningful way for our everydayness. The thermodynamic arrow of time is characterized by the growth of entropy from the past through the present to the future. The cosmological arrow of time is distinguished by the expansion of the universe from the past through the present to the future. The spatiotemporal arrow of time departs from the intuitive flow of the psychological arrow of time, the thermodynamic arrow of time, and the cosmological arrow of time, which are more biocentric views of the passage of time. The indefinite future states determine the definite past states of the dynamic block universe. Quantum uncertainty supports a non-deterministic view of the dynamic block universe. However, relative probabilities of different possible states are still determined by laws of nature. There lies the role of chance in quantum mechanics within the uncertainty of the laws of the block universe. As the Copenhagen interpretation of quantum physics asserts, reality is what one measures it to be, and nothing more. Even though, one always measures the metric of objective reality, not spatial and temporal distances.

In this view of the dynamic block universe, the future of time exists as indefinite future states a priori to the existence of future arrangement of events of the block universe. Contingent statements about future events are neither true nor false. The dynamic block universe becomes a changing picture of continuous happenstance that at any instant of time is an image of change.

The perception of the present or "now" that we experience in the passage of time is a product of our consciousness while time flows inexorably through our physical body and mind. The atomic structure of our physical body is spread out in space, but not in time. The stream of time endows the stream of our consciousness. The "now" of our consciousness can move through space, but the "now" does not move through the stream of time of the tenses of time in the dynamic block universe. All tenses of time of our perception exist simultaneously in space-time. The future, present, and past, are experienced in reference to the "now" of our consciousness.

The illusion of motion through time from the past, through the present, to the future, comes from our order of perception, through the same stream of time. Every instant or moment in time that we experience, every "now", is equally real to our perception in the stream of time.

The whole space and time are laid-out in one changing space-time block. Space and the existing spatiotemporal arrangements of events are modulated by the passage of time. The evolution of three-dimensional existence becomes the physical reality of the four-dimensional existence of present physics, where the tenses of time of our perception may be regarded as an illusion (a figment), but where changes are continuously real.

Time is dimensional and defined within the dynamic block universe. Observers can inspect how the arrangements of events, the configurations of objects, change according to the time-axes of their frames of reference. The apparent changes in the dynamic block universe imply that the universe may have emerged from the ontological singularity through an inflationary process, and expands at all points in space-time.

There is free will to make decisions in the dynamic block universe. For every individual, there is the ability to consider a range of many possible courses of action, and to select only one course of action (one outcome) from that range of possibilities. Hence, there would be one outcome and one related stream of events in the stream of time for every individual.

1.5.(a) The General Theory of Relativity versus The Theory of Quantum Mechanics

The General Theory of Relativity encompasses the dynamic block universe with past, present, and future, in a deterministic spatiotemporal structure that may be accessible, and predictable, to the observer. The entire wavefunction of the world tube of an object simultaneously exists in all dimensions of space and time for infinity and eternity. The simultaneous existence of all there is, and all there is to know, may allow the laws of physics to access anything in any direction of space and time. All possible events exist simultaneously in the dynamic block of the universe, so there is nothing left to chance. If 'all-there-is' is already there, then why play dice with the universe?

The Theory of Quantum Mechanics encompasses a 'now' slice of the dynamic block universe with a dynamic present in a slice of three-dimensional non-expanding space with non-expanding time, in an indeterministic spatial structure that agrees with the uncertainty principle of the observer. The section of the wave function in the 'now' slice collapses when measured by the observer. There is not world tube for an object, and the dimensions of space and time exist now. Hence, if the position of a particle is determined, its velocity is indeterminate, and vice versa. So now, why not play dice in the now section of the universe?

Both deterministic and non-deterministic perspectives play important roles in their respective regions of observation of objective reality, depending on the scale and relative view point of the observer. As a 'now' wave flows through the dynamic block universe, our path through space-time is determined by what our consciousness decides to observe. For observers that consider an expansive large-scale region of the dynamic block of the universe, the General Theory of

Relativity is applicable. For observers of a sliced quantum-scale region of the dynamic block of the universe, the Theory of Quantum Mechanics is applicable. A universal observer has access to all there is.

§ 2. The Cosmological Constant: Einstein's Biggest Blunder

The cosmological constant Λ is a coupling constant in the Einstein's field equations to compensate for the expansion of the universe. Lambda (Λ) represents the energy density of the universe (J/m^3) and is related to the radial distance r_Λ of a spherical observable universe, even though the actual volume has been considered flat, spherical, or elliptical.

$$R_{\mu\upsilon} - \frac{1}{2} R g_{\mu\upsilon} + \Lambda g_{\mu\upsilon} = \frac{8\pi G}{c^4} T_{\mu\upsilon} \tag{2.1}$$

The cosmological constant was added to the EFE's as a compensating term to the General Theory of Relativity because the EFE's did not allow for the static universe Einstein imagined. The gravitational field of a dynamic universe initially at equilibrium may cause the universe to contract. So, the cosmological constant counteracted that possibility. Nonetheless, observations of the universe using the cosmological red-shift technique indicated that the universe appeared to be expanding, which was consistent with the Friedmann's solutions to the original equations of General Theory of Relativity. Initially, Einstein did not accept the validation of his own equations of General Theory of Relativity, which he allegedly later called his biggest blunder. Even though, we now know, with the benefit of hindsight and the progress of physics, that adding the cosmological constant to the original equations of General Theory of Relativity does not yield a static universe at equilibrium because the equilibrium is unstable. (Riess, 1998) If the universe were expanding or contracting slightly, then more energy would be released or absorbed from the spatiotemporal medium, which itself originates more expansion or contraction. (Carroll, 1998, 2000)

Let us express the cosmological constant in terms of the spatiotemporal acceleration as follows:

$$|\Lambda| = \frac{8\pi G\rho_{vac}}{c^2} = \frac{\rho_{vac}}{mc^2}\left(\frac{d^2v}{dt^2}\right) = \frac{\ddot{a}\rho_{vac}}{mc^2} = \frac{\ddot{a}}{ac^2} \qquad \left(\frac{1}{m^2}\right) \tag{2.2}$$

Where is the energy density ρ_{vac} of the spatiotemporal medium or vacuum?

Let us indicate approximate values for the cosmological constant during the present era of the universe from the earth frame of reference.

$$A = \frac{1}{10^{52}m^2} = \frac{1}{10^{35}\sec^2} = \frac{1}{10^{47}GeV} = \frac{1}{10^{25}}\frac{Kg}{m^3} = \frac{1}{10^{122}} \tag{2.3}$$

(dimensionless in Planck units)

The radial distance of the spherical observed universe may be given as

$$r_\Lambda = ct_\Lambda = \frac{\sqrt{3}}{\sqrt{\Lambda}} \tag{2.4}$$

$$\frac{r_\Lambda}{\sqrt{3}} = \frac{1}{\sqrt{\Lambda}} \tag{2.5}$$

Thus, the cosmological constant represents the reciprocal of the square of the one-dimensionally projected radial distance of the observed universe. The cosmological constant is defined as an inverse-square relationship. (Hawking, 1973)

$$\Lambda = \frac{1}{\left(\frac{r_\Lambda}{\sqrt{3}}\right)^2} = \frac{\ddot{a}}{ac^2} \tag{2.6}$$

The cosmological constant is equal to the ratio of the spatiotemporal acceleration to the volume of the observed universe divided by the square of the speed of light.

§ 3. What are the relationships between mass, spatial distance, and temporal distance in our expanding universe?

Let us consider the relationship between spatial and temporal distance across the vast expanse for the expanding universe.

$$10^{35} s^2 = 10^{52} m^2 \tag{3.1}$$

$$\frac{s^2}{m^2} = \frac{10^{52}}{10^{35}} \approx 10^{17} \tag{3.2}$$

$$s^2 = 10^{17} m^2 \tag{3.3}$$

The magnitude of the above approximate spatiotemporal ratio is approximately equal to the magnitude of the square of the speed of light.

$$c^2 = (299792458)^2 \frac{m^2}{s^2} \cong 9 \, x \, 10^{16} \frac{m^2}{s^2} \approx 10^{17} \frac{m^2}{s^2} \tag{3.4}$$

Therefore, at vast distances in the observable universe, the spatiotemporal ratio of the spatiotemporal wave is approximately equivalent to the magnitude of the speed of light at an inertial earth frame of reference.

Thus, the above result affirms the constancy of the measurement of the speed of light throughout the expanding universe according to the General Theory of Relativity, and the symmetrical property of the cosmic expansion.

§ 4. What would the universal gravitational constant G_u be at the outer observable universe?

According to our present estimates we obtain

$$G_u \equiv \frac{m^3/Kg}{s^2} \cong \frac{\left(\sqrt{10^{52}}\,m\right)^3 / 10^{53}\,Kg}{10^{35}\,s^2} \cong 10^{-10} \qquad (4.1)$$

$$G_u = \frac{\ddot{a}_u}{m_u} \approx 10^{-10} \qquad (4.2)$$

In comparison, the gravitational constant at the earth frame of reference is approximately

$$G_{earth} \cong 6.67408\,x\,10^{-11}\,\frac{m^3}{Kg \cdot s^2} \qquad (4.3)$$

Thus, if we compare the approximate values above, we have

$$\frac{G_u}{G_{earth}} \approx 1.5 \qquad (4.4)$$

Would the acceleration of the volume of space-time per unit of mass be the same at any point in the universe? The answer to this question will arrive with more accurate cosmological data from future observations and experiments on the observable universe. If the gravitational constant were the same at any point of the observable universe, then the mass (ordinary matter) of the observable universe would have to be approximately 1.5 x 10^{53} *Kg*, or approximately fifty per cent more than our previous estimate, ceteris paribus.

§ 5. On the weakness of gravity and gravitational attraction

Matter is very porous. An estimate gives a typical atom about 10^{18} parts of space-time to every part of matter. The weakness of gravity is in part the nature of highly porous matter. The wavelength of space-time is nearly 4 x 10^{-17} meters or approximately ten million times smaller than the dimensions of the atomic structure of matter.

Gravity is a force exerted upon the body of an object, a push, not a force of attraction from another body of a second object of mass, or a pull.

Two objects in proximity shield each other from the expansion forces of the spatiotemporal wave medium, but the component forces acting oppositely in the spatiotemporal medium between the masses, subtract, ending in a resultant force in the direction of the largest mass per the Huygens-Fresnel principle as shown in figure 5. The resultant force is in the same direction as the force of space-time acting on the object of lesser mass on its opposite side.

$$m_1 > m_2 \tag{5.1}$$

$$\vec{F}_1 > \vec{F}_2 \tag{5.2}$$

$$\vec{F}_n = \vec{F}_1 - \vec{F}_2 \tag{5.3}$$

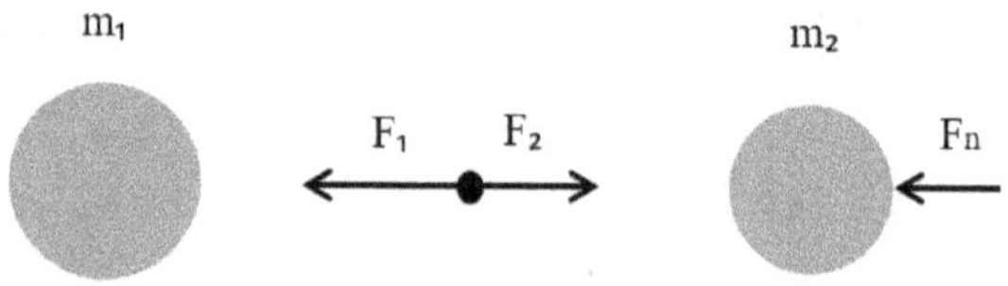

Figure 5

5.1. The Principle of Balance of Energy Flows

Energy is an exchange between an open system and its interactive environment. An open system is a system that has external interactions that can take the form of energy, matter, or information transfers into or out of the system boundary. The interactive environment consists of all other potentially open systems at any given time. *The sum of the energy flows of all open systems and their interactive environments, or the sum of the energy flows of a single open system, and the energy flows of its interactive environment, equal zero.*

Let us consider the energy of a particle with mass *m* traveling through space-time, the force, F_1, that may be exerted by the mass of the particle on space-time in the direction of travel is equal and opposite to the force, F_2, exerted by space-time on the mass.

The energies exerted by forces F_1 and F_2 are

$$E_1 = F_1 \cdot dx_1 \tag{5.4}$$

$$E_2 = -F_2 \cdot dx_2 \tag{5.5}$$

The difference of the energies of the system is given by

$$E_1 - E_2 = F_1 \cdot dx_1 - F_2 \cdot dx_2 = mv_1^{\,2} - mv_2^{\,2} = m\left(v_1^{\,2} - v_2^{\,2}\right) = 0 \tag{5.6}$$

$$v_1 = v_2 \tag{5.7}$$

The sum of the two energies is

$$m\vec{v}_1^{\,2} + m\vec{v}_2^{\,2} = m\left(\vec{v}_1^{\,2} + \vec{v}_2^{\,2}\right) = 2m\vec{v}_1^{\,2} = 2m\vec{v}_2^{\,2} \tag{5.8}$$

$$m\vec{v}_1^{\,2} + m\vec{v}_2^{\,2} = 2m\left(\vec{v}_1 \cdot \vec{v}_2\right) \tag{5.9}$$

$$\frac{1}{2}m\vec{v}_1^{\,2} + \frac{1}{2}m\vec{v}_2^{\,2} = m\left(\vec{v}_1 \cdot \vec{v}_2\right) \tag{5.10}$$

$$E_1 + E_2 = m\left(v_1 \cdot v_2\right) \tag{5.11}$$

By the product rule it follows that

$$d\left(\vec{v}_1 \cdot \vec{v}_2\right) = \left(d\vec{v}_1\right) \cdot \vec{v}_2 + \vec{v}_1 \cdot \left(d\vec{v}_2\right) = 2\left(\vec{v}_1 \cdot d\vec{v}_1\right) = 2\left(\vec{v}_2 \cdot d\vec{v}_2\right) \tag{5.12}$$

Expressing the sum of energies as vectors we have

$$\vec{E}_1 + \vec{E}_2 = m\left(\vec{v}_1 \cdot \vec{v}_2\right) \tag{5.13}$$

$$d\left(\vec{E}_1 + \vec{E}_2\right) = \int_0^t md\left(\vec{v}_1 \cdot \vec{v}_2\right) = \int_0^t 2m\left(\vec{v}_1 \cdot d\vec{v}_1\right) = \int_0^t 2m\left(\vec{v}_2 \cdot d\vec{v}_2\right) \tag{5.14}$$

$$\frac{d\left(\vec{E}_1 + \vec{E}_2\right)}{2} = \int_0^t m\left(\vec{v}_1 \cdot d\vec{v}_1\right) = \int_0^t m\left(\vec{v}_2 \cdot d\vec{v}_2\right) \tag{5.15}$$

The force of the particle on space-time, and the counter force of space-time which is part of the inertia system of the body of mass, are equal and opposite when v^2/c^2 is small.

$$d\vec{E}_1 = \frac{1}{2}\int_0^t md(\vec{v}_1 \cdot \vec{v}_2) = d\vec{E}_2 \tag{5.16}$$

$$E_1 = \int_0^v d\left(\frac{m(v_1 \cdot v_2)}{2}\right) = \frac{1}{2}m(v_1 \cdot v_2) = E_2 \tag{5.17}$$

It is interesting to note that in the General Theory of Relativity we define Kinetic energy to be

$$K_E = mc^2 - m_0c^2 \tag{5.18}$$

Where m is the relativistic mass and m_0 is the rest mass.

$$K_E = \frac{m_0c^2}{\sqrt[2]{1-\frac{v^2}{c^2}}} - m_0c^2 = m_0c^2\left[\left(1-\frac{v^2}{c^2}\right)^{-\frac{1}{2}} - 1\right] \tag{5.19}$$

After computing the Taylor series of the term in square brackets we have

$$K_E = m_0c^2\left[\left(1-\frac{v^2}{c^2}\right)^{-\frac{1}{2}} - 1\right] = m_0c^2\left(\frac{1}{2}\frac{v^2}{c^2} + \frac{3}{8}\frac{v^2}{c^2} + ...\right) = \frac{1}{2}m_0v^2 + m_0c^2\left(\frac{3}{8}\frac{v^2}{c^2} + ...\right) \tag{5.20}$$

and the higher terms can be ignored if v^2/c^2 is small.

$$K_E \approx \frac{1}{2}m_0v^2 \tag{5.21}$$

Therefore, the above kinetic energy equation in the General Theory of Relativity is an approximation. Nevertheless, the sum of the energy flows of a single open relativistic system, and the energy flows of its interactive relativistic environment, equal zero. Thus, the principle of balance of energy flows states that the energy flow equation of a single open system in its interactive environment is

$$E = \frac{1}{2}\int_0^t \vec{F} \cdot d\vec{s} = \frac{W_T}{2} \tag{5.22}$$

Where W_T is the total available work flow between a single open system and its interactive environment.

Some equations of non-linear energy flows for single open systems are:

$$K_E = \frac{1}{2}mv^2 \tag{5.23}$$

$$E_G = \frac{1}{2}mgh \tag{5.24}$$

$$E_{SPRING} = \frac{1}{2}Kl^2 \tag{5.25}$$

$$E_{RESISTOR} = \frac{1}{2}QV \tag{5.26}$$

$$E_{INDUCTOR} = \frac{1}{2}Li_L^{\ 2} \tag{5.27}$$

$$E_{CAPACITOR} = \frac{1}{2}Cv_C^{\ 2} \tag{5.28}$$

$$E = \frac{1}{2}m_0c^2 \tag{5.29}$$

$$E_{PHOTON} = \frac{1}{2}hf \tag{5.30}$$

$$E_{SPACE-TIME} = \frac{1}{2}\mu_0 l^3 \tag{5.31}$$

$$E_{MAGNETIC-FIELD} = \frac{1}{2}\iiint_V \mu_0 |H|^2 \, dv \tag{5.32}$$

All the above non-linear energy flow equations exhibit the same relationship of a single open system with its environment under the principle of balance of energy flows.

5.2 The Holodeck Principle

The holodeck principle is a holographic manifold projection principle applicable to string theory, or M-theory, describing a fundamental property of quantum gravity that allows a time-like volume of space to be encoded on a lower-dimensional time-like boundary, or on a future gravitational horizon, in a region of the volume. Space may be defined as a de Sitter-like space with a positive cosmological constant, simply connected for $n \geq 3$, and may be described as $dS_6 \ x \ S^6$ for expanding space-time with three spatial dimensions and three temporal dimensions in de Sitter space, equivalent to $dS_4 \ x \ S^6$, for string theory, or with an additional dimension equivalent to $dS_4 \ x \ S^7$ for M-theory. In the case of string theory, or M-theory, the three temporal dimensions are folded into a resultant temporal dimension in de Sitter space.

First, let us consider the geometry of space for a Conformal Field Theory. The angles between directed curves through a local point, the orientation of a tangent basis mapped to a basis of the same orientation, and the shapes of infinitesimally small figures, would remain the same during a conformal transformation, but not the curvature or scale. Thus, triangles would remain triangles, and squares would remain squares, but the manifold these shapes are on may be curved, and the scale of the shapes may be changed. Conformal maps can be defined between domains on a Riemannian or a semi-Riemannian manifold.

Secondly, let us illustrate a six-dimensional sphere with constant curvature that is tessellated with triangles and squares. From the frontal perspective shown, the fuzzy edge on the circumference of the sphere, is the boundary of an expanding time-like space with positive curvature that exists all around the sphere. Let us imagine that the scale of the sphere is larger in orders of magnitude than the scale of the observer. So, the observer would see the tessellated figures getting smaller and smaller toward the edges of the circumference of the enormous sphere. The outer edges of the time-like space surrounding the sphere extend to infinity. The outer time-like space around the sphere is infinitely far from any point on the

outer tessellated manifold. Under the outer tessellated manifold, there is an identical tessellated manifold on an inner sphere with a scale that is orders of magnitude smaller than the tessellated manifold of the outer sphere, and in between the inner and outer tessellated manifolds there is expanding time-like space.
The outer time-like boundary, dS_{d-1}, has fewer dimensions than the dimensions of the sphere, dS_d, and it is infinitely far from any point on the tessellated manifold. The time-like space and boundary represent the expansion of space in the wave medium of the bulk, and the passing of non-linear time. Scale transformation in the time-like boundary is holographically dual to radial time-like extension in the time-like space of the bulk.

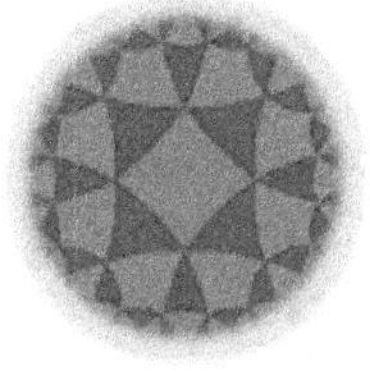

Figure 6.

The above illustration represents a dS_6 / CFT conjecture that describes a vacuum solution of the six-dimensional EFEs with a positive cosmological constant. Each sphere may be interpreted as the embodiment of a spatiotemporal wave. The spatiotemporal waves emerge as time passes from the inner infinity to the outer infinity of space-time for a wave function $\Psi(r,t) \propto e^{idS_d}$. The inner future space becomes the outer past space. Space-time may expand or contract through the interference of spatiotemporal waves. As time-like space expands at an accelerated rate, the spatial wave lengthens and the temporal wave contracts infinitesimally. The temporal wave continues to contract eternally, and the spatial wave extends toward its full extent at infinity, as every point in space-time expands.

The spatiotemporal sphere may be described as consisting of stacked layers of manifolds like the layers of an onion. So, every higher dimension of a manifold encompasses a hologram of the information on space of lower dimension at a specific frequency. This positive holographic property preserves the three-dimensional information of space onto a two-dimensional manifold or holographic plate. The time-like space may represent time-like layers of parallel manifolds that may exist in between the inner and outer tessellated manifolds. These time-like layers in-between may be interpreted as parallel manifolds of specific sibling dS_6 universes that consist of multiple collections of the stacked tessellated spheres, or multiple collections of parallel sibling dS_6 universes at specific frequencies. On a larger scale, hypothetical D-branes may co-exist on multiple layers as a wave function that collapses when observed at a specific layer or manifold. In a sense, the quantification of mass would act like a pinched collection of layers, and perhaps an attractive sink to free strings, gravitational potential, or Higgs bosons. A high energy particle collision at a specific point of a manifold may provide a larger pinch point, or throat, as a gateway between layers.

This conformal field theory is equivalent to a gravitational theory on the bulk of de Sitter space, in the sense that there is a conversion for every interaction in one theory into an interaction in the other. Every holographic entity in the CFT has a counterpart in the gravitational theory. Every point on a manifold of inner space is connected to a collection of points on the outer holographic plate of the conformal boundary. Moreover, when a sphere represents a dS_4 space, the three temporal dimensions would be folded into one resultant dimension, and it would be possible to conjecture compacted spatial dimensions; in the case of string theory, there would be six compacted spatial dimensions, which may result in various string theories, depending on the coordinate system that is used.

A six-dimensional universe with a positive cosmological constant, dS_6, is likely to be asymptotically de Sitter, with a time-like boundary in the far future, but it is unlikely to have an anti-de Sitter boundary. The AdS_6 / CFT correspondence provides a non-perturbative holographic definition of quantum gravity in Anti-de Sitter space with a *CFT* that lives on a time-like conformal boundary. However, the far future dS_6 boundary may share mathematical properties with a dual AdS_6 boundary. Thus, it is possible to define dS_6 quantum gravity as a *CFT* living on the dS_6 conformal boundary of the far future. It is worth noting that non-linear time emerges holographically in a dS_6 / CFT correspondence, while in an AdS_6 / CFT , the radial direction r emerges holographically from the *CFT*. Consequently, dS_6 / CFT correspondence would be expected to deliver microscopically complete examples. The properties of the dS_6 / CFT correspondence of the universe appear to be useful for the analysis of cosmic microwave background data as applicable to the dS_6 era proposed by inflation.

Our current universe has been observed to be a maximally symmetric solution of a de Sitter space with a positive cosmological constant. Every maximally symmetric space, or 'manifold', has constant curvature, or 'constant sectional curvature'. The quadratic equation that describes a six-dimensional universal de Sitter space is

$$-X_r^{\,2} + \left\{ X_{ct}^{\,2} + \sum_{n=1}^{d-3} X_n^{\,2} \right\} = l^2 \tag{5.33}$$

Let us use the spherical parametrization, that is comparable to an Anti-de Sitter space, given by

$$X_r = l\,Sinh\left(\frac{ct}{l}\right) = l\left(\frac{e^{\frac{ct}{l}} - e^{-\left(\frac{ct}{l}\right)}}{2}\right) \tag{5.34}$$

$$X_{ct} = l_{ct}Cosh\left(\frac{ct_{ct}}{l_{ct}}\right)\left|\vec{a}_{ct}\right| = l_{ct}\left(\frac{e^{\frac{ct_{ct}}{l_{ct}}} + e^{-\left(\frac{ct_{ct}}{l_{ct}}\right)}}{2}\right)\left|\vec{a}_{ct}\right| \tag{5.35}$$

$$X_n = l_n Cosh\left(\frac{ct_n}{l_n}\right)\left|\vec{a}_n\right| = l_n\left(\frac{e^{\frac{ct_n}{l_n}} + e^{-\left(\frac{ct_n}{l_n}\right)}}{2}\right)\left|\vec{a}_n\right| \tag{5.36}$$

$$ct = \sqrt[2]{(ct_x)^2 + (ct_y)^2 + (ct_z)^2} = \sqrt[2]{(ct_4)^2 + (ct_5)^2 + (ct_6)^2} \quad (5.37)$$

$$l = \sqrt[2]{l_x^{\,2} + l_y^{\,2} + l_z^{\,2}} = \sqrt[2]{l_1^{\,2} + l_2^{\,2} + l_3^{\,2}} \quad (5.38)$$

In the above parametrization equations, $|\vec{a}_n|$ and $|\vec{a}_{ct}|$, are magnitudes of radial unit vectors, that represent respectively, a $(d-3)$ – dimensional sphere, and a three-dimensional temporal extent of the expanding sphere, of the quadratic equation. Additionally, l is the spatiotemporal radius of the sphere, t is the resultant temporal extent of three-dimensional time, c is the speed of light, where n is the number of dimensions, $n = 1, 2, 3, 4, 5, \ldots d$, where dimensions 1 through 3 are spatial, dimensions 4 through 6 are temporal, and d is the sixth temporal dimension in this case.

The spherical parameter, X_r, representing the contracting or expanding factor of the dS_6 sphere, is equal to zero when t is zero, while the spherical parameter, X_n, representing the spatial $(d-3)$ – sphere, is equal to l_n in the direction of each dimension when t is equal to zero. If the spatial radius, l *or* l_n, of the sphere is zero, then both parameters, X_r and X_n, are zero. If $n = ct$, then $\mathrm{X}_n = \mathrm{X}_{ct}$, which represents the resultant temporal extent of the sphere in the de Sitter space. Thus, in this universal de Sitter space representation, there is a space-time singularity at t is equal to zero.

The expansion or contraction of three-dimensional space and time are functions of the angular frequency, ω, of the wave function of space-time and may be expressed as

$$\frac{\partial^2 \Psi(\omega)}{\partial t^2} = c^2 \frac{\partial^2 \Psi(\omega)}{\partial r^2} \quad (5.39)$$

$$\frac{1}{c^2}\left[\frac{\partial^2 \Psi(\omega)}{\partial t_x^{\,2}} + \frac{\partial^2 \Psi(\omega)}{\partial t_y^{\,2}} + \frac{\partial^2 \Psi(\omega)}{\partial t_z^{\,2}}\right] = \frac{\partial^2 \Psi(\omega)}{\partial x^2} + \frac{\partial^2 \Psi(\omega)}{\partial y^2} + \frac{\partial^2 \Psi(\omega)}{\partial z^2} \quad (5.40)$$

The metric element corresponding to the above parametrization may be expressed as

$$ds^2 = -dt^2 + l^2 Cosh^2\left(\frac{ct}{l}\right)da_n^{\ 2} \qquad (5.41)$$

The quadratic equation is described fully by these global coordinates. The $Cosh^2(ct/l)$ factor expands or contracts the spatial term of the metric, representing each spatial dimension *n* in the six-dimensional de Sitter space, as a function of time. The volume of the dS_6 universe contracts from *t* equal to negative infinity to *t* approaching zero and expands indefinitely when *t* is greater than zero to positive infinity. Furthermore, a positively curved universe is described by elliptic geometry, and can be thought of as a three-dimensional hypersphere. However, the models of a universe may have shapes other than spherical.

The above model of the universe is a six–dimensional sphere with a radius *l* equal to $\left(\sqrt[2]{3}\right)l_n$ at *t* is equal to zero, and with a radius that is a function of *t* and *l* for any constant value of *t*. The spherical space-time may also be mapped into cylindrical space-time with the parameter *t* as the symmetry axis. Let us consider hypothetically the following rhetorical question for the above scenario: is it possible for a parent black hole in one universe of the multiverse, with contracting space within its event horizon leading to a space-time singularity, spring a Big Bang in a daughter universe?

Additionally, for the above model of the universe, the metric element may be expressed with a Poincare coordinate patch to give the half-space coordinatization of dS_6 space as follows:

$$ds^2 = \frac{l^2}{c^2\tau^2}\left(-d\tau^2 + g_{\mu\nu}dx^{\mu}dx^{\nu}\right) \qquad (5.42)$$

The one-boundary background metric, $g_{\mu\nu}$, is Euclidean and describes half of the dS_6 space as τ goes to zero.

The Ricci tensor is proportional to the metric, so a manifold in dS_6 space is a manifold of the six-dimensional EFEs.

$$R_{\mu\nu} = \frac{n-3}{l^2} g_{\mu\nu} \tag{5.43}$$

The Ricci scalar is given by

$$R = \frac{(n-3)}{l^2} \tag{5.44}$$

Spaces of constant curvature can be considered as vacuum solutions to the six-dimensional EFEs with a cosmological constant. Hence, the dS_6 space is a vacuum solution of the six-dimensional EFEs with a positive cosmological constant given by

$$\Lambda = \frac{1}{l^2} = \frac{n-2}{2n} R \tag{5.45}$$

Every point in space-time expands. The existence of Einstein-Rosen bridges, or interdimensional bridges, across the layers of manifolds in a specific dS_6 universe would influence the acceleration of space-time, as high-pressure areas of a dimensional manifold would feed low-pressure areas of another dimensional manifold. Space would be ducted between dimensional manifolds, so there is more space available for expansion per unit of temporal volume, which assists an accelerated expansion of space in the terminal space of the interdimensional wormhole.

This condition would also imply that the spatial wavelength lengthens more than the temporal wavelength during an accelerated expansion in a radial direction. Hypothetically, objects from string theory such as D-branes, closed strings, on lower dimensional manifolds, would be funneled between dimensional manifolds, through interdimensional bridges, trickling away and weakening the gravitational potential on a specific dimensional manifold.

Another characteristic of dS_6 space is entropy, which is described by the entropy of a black hole,

$$S_{BH} = \frac{k_{Boltzmann} A_{horizon}}{4\hbar_{Planck} G_{Newton}} = \frac{k_B}{4c^3} \tag{5.46}$$

Every other variety of entropy that occurs in nature including most varieties of black holes, are presently described in terms of counting some category of quantum microstates. So, it is natural to try to do the same for dS_6 space. But, unlike a black hole with a static event horizon, the horizon of a dS_6 space moves around with an observer. If there was an observer with a world-tube in dS_6 space, the observer would have a changing horizon. Are there observables in dS_6 space? Do particles come in from the past and go out in the future such as in an *S*-matrix? Is the event horizon of a black hole static?

The entropy of the black hole, S_{BH}, is a measure of the multiplicity of microstates that hide behind one particular macrostate. Black hole entropy counts the number of horizon gravitational states, or the number of internal states of matter and gravity. Black hole entropy is the amount of entropy that must be assigned to a black hole in order for it to comply with the laws of thermodynamics as they are interpreted by observers external to that black hole. For instance, if the area of the horizon of a black hole is divided into Plank length square units, $l_p^{\,2}$, then entropy must be assigned to one Plank length square unit out of every four Plank length square units. Since a black hole exists in six-dimensional space-time, how would the entropy of the black hole be ascribed in six-dimensions?

Let us consider an expanding spatiotemporal tesseract that lives on the event horizon of the black hole as the diameter increases. The six-dimensional tesseract has an inner cube in three spatial dimensions and an outer cube in three temporal dimensions. Each dimension for every vertex of each cube is asymptotic to the other two dimensions as they asymptotically meet the radial line from the center of the inner cube that passes through the center of each vertex.

As the dimensions at each vertex asymptotically curve either inward or outward for either the outer temporal vertex or the inner spatial vertex, there is an infinitesimal temporal throat that connects each inner vertex to each corresponding outer vertex of the tesseract. Every tesseract is tangent to an identical tesseract on each of its six sides.

Let us imagine four adjacent tesseracts which are on top of four other identical adjacent tesseracts. Each vertical pair of stacked adjacent tesseracts forms a rectangular cuboid. Each square side at each end has a width and a height of $\sqrt[2]{\pi}$, so that the area at each end is π . The depth of the rectangular cuboid is equal to π , the area of the rectangle on each side is 2π , and the volume of the rectangular cuboid is π^2 . Hence, the distances between vertices are transcendental numbers that represent the expansion of transcendental space-time from inner infinity to outer infinity. A transcendental number is an irrational number that as a decimal representation never ends and never settles into a permanent repeating pattern. So, there are eight cubes forming a hypercube with spatiotemporal volume $\left(s^3/m^3\right)$, or $4\pi^2$, with each vertical pair of stacked cubes forming a rectangular cuboid with ¼ of the spatiotemporal volume of the hypercube, $\frac{1}{4}\left(s^3/m^3\right)$, which is equivalent to $1/\left(4c^3\right)$. Therefore, entropy must be assigned to the six-dimensional spatiotemporal volume of one rectangular cuboid (two tesseracts) out of every four identical six-dimensional rectangular cuboids (eight tesseracts).

It is interesting to point out that the surface area of a single tesseract is 6π and the surface area of the hyper-tesseract is 24π . Thus, the ratio of the surface area of one tesseract to the surface area of the hyper-tesseract is exactly ¼. The surface area of a single tesseract increases by a factor of four to the surface area of a hyper-tesseract. The same would be true for an inner spatial sphere with a radius of $\sqrt[2]{\pi}$ and a volume of $4\pi^2$, compared to an outer temporal hypersphere with radius of $2\left(\sqrt[2]{\pi}\right)$ and a volume of $16\pi^2$. Again, the ratio of the surface area of the inner spatial sphere to the surface area of the hypersphere is ¼. It would be reasonable to suggest that the expansion scale of space-time for these spatiotemporal volumes is one to four. Therefore, all the entropy of an expanding spherical black hole must be ascribed to ¼ of the area of the surface of the event horizon due to the transcendental expansion of space-time and the assignment of the precedent entropy onto the subsequently expanded surface of the event horizon.

The transcendental property of the dimensions of the tesseract, or hyper-tesseract, implies that the transcendental value of a spatiotemporal dimension is not the root of any non-zero polynomial having rational coefficients, and that it is impossible to solve the ancient challenge of squaring the circle with a compass and straightedge. Nevertheless, the transcendental property does allow to hypercube a sphere with a spatial volume of $4\pi^2$ in six-dimensional space-time, and to square a unit circle with a radius equal to $\sqrt[2]{\pi}$ into a square whose sides are equal to π, since both the unit circle and the square have a transcendental area equal to π^2. These transcendental areas and volumes exist in space-time since the dimensions of space-time have transcendental properties. Regrettably, it would be very difficult to find a transcendental compass and straightedge.

There is a straightforward way to calculate entropy of a black hole using temperature, heat, and mass, with natural units $(G = c = \hbar = k_B = 1)$, since the heat that enters serves to increase the mass.

$$dS_{BH} \rightarrow \frac{dQ_{BH}}{T_{BH}} \rightarrow 8\pi M_{BH} dQ_{BH} \rightarrow 8\pi M_{BH} dM_{BH} \rightarrow d\left\{4\pi\left(\sqrt[2]{2}M_{BH}\right)^2\right\} \rightarrow d\left\{4\pi R_{BH}{}^2\right\} \qquad (5.47)$$

$$A_{horizon} = 4\pi R_{BH}{}^2 \qquad (5.48)$$

$$S_{BH} = \pi R_{BH}{}^2 = \frac{A_{horizon}}{4} \qquad (5.49)$$

The area of the circle with the diameter of the black hole is equivalent to entropy using natural units, and the radius of the black hole is equivalent to the effective value of its mass.

A black hole exists on an inner sphere in the past and on an outer sphere in the future. Hence, the time-like space between a collection of points on the outer sphere of a black hole to a point on the inner sphere of the same black hole may be interpreted as a black funnel, from future space to present and past space. Black holes may regulate the pressure, or the energy density, of space-time, in areas of high pressure and curvature like the centers of galaxies. In such scenario, space may be able to enter a supermassive black hole on the outer sphere, contracting through the black funnel into the inner sphere, successfully regulating pressure and maintaining a more uniform energy density in the surrounding space-time.

The pressure of space-time may be regulated through a dynamic change in the diameter of the event horizon. The surface area of the event horizon of a black hole may increase or decrease depending on the transformation process of the black hole. The surface area of the event horizon of a black hole increases when the black hole gains mass or energy, but black holes that do not gain mass or energy are expected to shrink and ultimately vanish due to Hawking radiation. Hence, the diameter of a black hole is dynamic. The pressure, or energy density, of an expanding supermassive black hole would tend to weaken over time. Thus, black holes may be used to illustrate the holodeck principle in a space-time with a positive cosmological constant.

A dynamic change in the diameter of the event horizon of a spherical non-extremal black hole is directly proportional to the acceleration of space-time and indirectly proportional to the gravitational acceleration of the black hole as shown below.

$$\Delta D_{BH} = \sqrt[2]{\frac{\Delta \ddot{a}}{2\pi \Delta g_{BH}}} \tag{5.50}$$

It is worth noting that the holodeck emerges from the property of expansion of space-time as well as the transcendental property of the dimensions of space-time. Both properties of space-time have an effect on the holographic response time of a physical system.

What is the effect of these holographic properties on the time constant of a physical system?

The time constant of a particular physical system in the holodeck represents the temporal extent that the system takes to respond to change, or the temporal extent of a specified parameter to vary by a factor of $1-(1/e)$, which is approximately 63%, where the number e , like π , is a transcendental number.

Let us determine the time constant of a physical system consisting of an inlet pipe with a faucet feeding water through the top of an open tank that has an outlet pipe at the bottom of the same small diameter compared to the radius of the tank. Both the inlet and outlet pipes have identical diameters and instantaneous control valves. Let us start our experiment by having the tank filled to a 100% mark after the inlet control valve shuts off, then the outlet control valve is opened at t equals zero, our timer starts counting from zero seconds to a time τ in seconds when the output valve shuts off, and the volume of the tank is at approximately 37%. Thus, the starting volume of the tank has decreased by approximately 63%. If we repeat the experiment again at the approximately 37% level, with the 37% level being the new 100% volume, opening and later closing the outlet control valve only, the starting volume would decrease again by approximately 63% after the same τ seconds to a new 37% volume.

The physical system of the tank may be represented by the following first-order linear time-invariant differential equation,

$$A\frac{dh}{dt}-\frac{h}{\Omega}=w_{out} \tag{5.51}$$

$$\tau=\Omega A \tag{5.52}$$

Where A is the cross-sectional area of the tank, (m^2) , h is the head or level of water in the tank, (m) , Ω is the resistance of the outlet valve, (s/m^2), w_{out} is the velocity of the volume of water in the outlet pipe, $\left(m^3/s\right)$, t is coordinate time, $\left(s\right)$, and τ is the time constant of the system, $\left(s\right)$. Hence, the expansion of space-time is displacing a volume of water of the capacity of the tank, as space-time presses upon the surface area of the water, in the form of a gravitational field.

$$w_{out}\tau = \dot{a}\tau \tag{5.53}$$

The spatiotemporal displacement after one time constant, τ, is given by

$$\dot{a}\tau = \left(\frac{1}{e^t}\right)\pi r^2 h \tag{5.54}$$

Redefining the equation for the physical system we have

$$\pi r^2 \dot{h} - \frac{h}{\Omega} = \dot{a} \tag{5.55}$$

$$\tau\pi r^2 \dot{h} - \frac{\tau h}{\Omega} = \left(\frac{1}{e^t}\right)\pi r^2 h \tag{5.56}$$

$$\tau\frac{\dot{h}}{h} - \frac{\tau}{\Omega\pi r^2} = \frac{1}{e^t} \tag{5.57}$$

$$\tau\frac{\dot{h}}{h} - 1 = \frac{1}{e^t} \tag{5.58}$$

$$\tau\frac{\dot{h}}{h} = \left(\frac{e^t + 1}{e^t}\right) \tag{5.59}$$

The ratio of the velocity of the volume of space-time to its volume,

$\dot{a}/a$, and the ratio of $\dot{h}/h$, is the same and equal to the reciprocal of coordinate time, $1/t$, in the region of space-time of the physical system.

$$\tau\frac{\dot{a}}{a}=\left(\frac{e^t+1}{e^t}\right) \qquad (5.60)$$

The ratio of the time constant (proper time), τ, to coordinate time, t, is given by

$$\frac{\tau}{t}=\left(\frac{e^t+1}{e^t}\right) \qquad (5.61)$$

$$\tau\frac{d}{dt}\log_e t=\left(\frac{e^t+1}{e^t}\right) \qquad (5.62)$$

When t equals 100% of the total temporal volume of the displacement, and after one time constant, τ, results in 37% of the temporal volume, the temporal volume has decreased by 63% of t.

$$\tau-t=1.37t-t=0.37t \qquad (5.63)$$

It is interesting to note the proper time value of the time constant, τ, of the physical system does not vary in the same holographic region of space-time as time passes. The temporal ratio, τ/t, does not depend on the capacity, or physical dimensions, of the system. The temporal ratio, τ/t, is only time-dependent and transcendental. When the expansion of space displaces the water in the tank, the time constant of the system provides a temporal extent, or proper time, that is 37% longer for the duration of the displacement. The coordinate time provides a shorter temporal extent for the same expansion of space in an empty tank, ceteris paribus. Let us consider a hypothetical D-brane that lives on an infinitesimal flat space that is tangent to the nearly flat manifold around a point of an enormously large sphere, where the ten-dimensional Minkowski metric of dS_6 / CFT with a six-dimensional rotational spherical symmetry is given by

$$ds^2 = H(r)^{-\frac{1}{2}}\left[-dt_x^{\ 2} - dt_y^{\ 2} - dt_z^{\ 2} + dx^2 + dy^2 + dz^2\right] + H(r)^{\frac{1}{2}}\left[dr^2 + r^2\Omega_5^{\ 2}\right] \quad (5.64)$$

$$ds^2 = H(r)^{-\frac{1}{2}}\left[-dt^2 + d\Sigma^2\right] + H(r)^{\frac{1}{2}}\left[dr^2 + r^2\Omega_5^{\ 2}\right] \quad (5.65)$$

$$H(r) = 1 + \frac{R^4}{r^4} \quad (5.66)$$

Where $H(r)$ is a warping or scaling factor and the gravitational potential, R is the radius of curvature, and r is a radial distance. The warping factor depends on the radial distance.

The proportional value of the radius of curvature R to the radial distance r determines the degree of warping. R gives a measure of where gravity is strong if $r << R$, and the deformation of space-time. If $r >> R$, then the warping factor approaches unity and there is less deformation of space-time.

$$R^4 = \frac{4}{\pi^2} G T_3 N \quad (5.67)$$

Where G is the Newton's gravitational constant, T_3 represents the tension on the D-brane, and N is the number of D-branes. The product $T_3 N$ represents mass. The hypothetical D-brane would have a throat that extends radially outward in a time-like space that extends to infinity. R would represent the radius of the curvature of the space-time. A cross section of the throat is an S_5 space and the sides of the cylindrical throat is the conformal boundary. Therefore, it is predictable that the holodeck principle may subsume other holographic principles.

Some rhetorical questions that arise from these concepts are: Do hypothetical D-branes traverse the time-like distance between layers? Do the throats of hypothetical D-branes connect lower dimensional manifolds and higher dimensional manifolds of a specific frequency? Are free strings able to travel through the throats of D-branes from lower dimensional manifolds to higher dimensional manifolds? Moreover, would hypothetical S-branes live on an infinitesimally flat space-like manifold, which is tangent to the nearly flat manifold around a point of an enormously large sphere, that undergoes a time-like scale transformation orthogonal to the manifold of D-branes? Would the wick transformation process of a D-brane result in an S-brane localized in time on the same space-like manifold? If the energy is high enough, would decaying hypothetical S-branes spontaneously produce very long open string particles during a phase transition, at a higher temperature than the theoretical Hagedorn temperature of string theory, which may couple to closed string particles on the same manifold?

§ 6. The Lanczos Tensor

The Lanczos tensor potential, $H_{\mu\nu\varepsilon}$, is the source of the Weyl tensor, the traceless part of the Riemann tensor. The Weyl tensor describes how a cosmological Lorentzian manifold is curved in response to the gravitational field represented by the Lanczos tensor.

Thus, the Lanczos tensor serves as the gauge field of the cosmological gravitational field. The Weyl tensor can be expressed using partial derivatives of the Lanczos tensor and its permutations. Gauge freedom exists in the Lanczos tensor under an affine group which endows multiple solutions.

The Lanczos tensor may be defined through the Weyl-Lanczos equations which generate the Weyl tensor or cosmological tensor. The Lanczos tensor is a third order, antisymmetric tensor in one pair of indices, in a similar way to the second order, antisymmetric electromagnetic torsion tensor $\Psi_{\varepsilon\beta}$. We can express the Weyl tensor, or cosmological tensor, in terms of covariant derivatives of the Lanczos tensor. (Takeno, 1964)

$$C_{abcd} = H_{abc;d} + H_{cda;b} + H_{bad;c} + H_{dcb;a} + \left(H^{e}{}_{(ac);e} + H_{(a|e|}{}^{e}{}_{;c)}\right)g_{bd} + \left(H^{e}{}_{(bd);e} + H_{(b|e|}{}^{e}{}_{;d)}\right)g_{ac} - \tag{6.1}$$

$$\left(H^{e}{}_{(ad);e} + H_{(a|e|}{}^{e}{}_{;d)}\right)g_{bd} - \left(H^{e}{}_{(bc);e} + H_{(b|e|}{}^{e}{}_{;c)}\right)g_{ac} - \frac{2}{3}H^{ef}{}_{f;e}\left(g_{ac}g_{bd} - g_{ad}g_{bc}\right)$$

In the weak gravitational field approximation, we express the Lanczos gravitational potential metric, $h_{ab} \equiv h_{\mu\nu}$, as

$$h_{\mu\nu} = g_{\mu\nu} - \eta_{\mu\nu} \tag{6.2}$$

where, $g_{\mu\nu}$, is the metric of a nearly-flat cosmological Lorentzian manifold, and $\eta_{\mu\nu}$ is the metric tensor of special relativity.

The Lanczos gravitational field acceleration tensor may be expressed as

$$4H_{abc} \approx h_{ac,b} - h_{bc,a} - \frac{1}{6}\left(\eta_{ac}h^{d}{}_{d,b} - \eta_{bc}h^{d}{}_{d,a}\right) \tag{6.3}$$

The Lanczos tensor has been defined and refuted for higher dimensions, even though it is widely accepted for three spatial dimensions and one temporal dimension (a folded resultant temporal dimension) as originally introduced by Cornelius Lanczos. (Lanczos, 1949)

Let us express the Lanczos tensor in terms of ordinary derivatives in a more convenient and recognizable form, $H_{abc} \equiv H_{\mu\nu\varepsilon}$, of the Lanczos gauge in the region of a weak cosmological gravitational field,

$$H_{\mu\nu\varepsilon} \approx \frac{1}{4}\left(h_{\mu\varepsilon,\nu} - h_{\nu\varepsilon,\mu}\right) - \frac{1}{24}\left(\eta_{\mu\varepsilon}h^{\sigma}{}_{\sigma,\nu} - \eta_{\nu\varepsilon}h^{\sigma}{}_{\sigma,\mu}\right) \tag{6.4}$$

Contracting $H_{\mu\nu\varepsilon}$ with the cosmological stress-energy-momentum first order tensor Λ^{ε} in the direction of the basis $\vec{e}_{\varepsilon}$, we obtain

$$H_{\mu\nu} \approx \frac{1}{4}\left(\frac{\partial h_{\mu}}{\partial x^{\nu}} - \frac{\partial h_{\nu}}{\partial x^{\mu}}\right) - \frac{1}{24}\left(\eta_{\mu}\frac{\partial h}{\partial x^{\nu}} - \eta_{\nu}\frac{\partial h}{\partial x^{\mu}}\right) \tag{6.5}$$

$$H_{\mu\nu} \approx \frac{1}{4}\left\{\frac{\partial\left(h_{\mu} - h_{\mu}/6\right)}{\partial x^{\nu}} - \frac{\partial\left(h_{\nu} - h_{\nu}/6\right)}{\partial x^{\mu}}\right\} \approx \frac{5}{24}\left(\frac{\partial h_{\mu}}{\partial x^{\nu}} - \frac{\partial h_{\nu}}{\partial x^{\mu}}\right) \tag{6.6}$$

$$H_{\mu\nu} \approx \frac{1}{5}\left(h_{\mu\nu} - h_{\nu\mu}\right) \tag{6.7}$$

The cosmological gravitational field acceleration tensor, $H_{\mu\nu}$, for four-dimensional space-time with folded temporal dimensions is given by

$$H_{\mu\nu} = \begin{vmatrix} h_{\tau\tau} & h_{\tau x} & h_{\tau y} & h_{\tau z} \\ h_{x\tau} & h_{xx} & h_{xy} & h_{xz} \\ h_{y\tau} & h_{yx} & h_{yy} & h_{yz} \\ h_{z\tau} & h_{zx} & h_{zy} & h_{zz} \end{vmatrix} \tag{6.8}$$

The resultant time-time component, $h_{\tau\tau}$, and the space-space components, h_{ss}, for the antisymmetric cosmological gravitational field acceleration tensor are equal to zero.

$$h_{\tau\tau} = h_{xx} = h_{yy} = h_{zz} = 0 \tag{6.9}$$

The antisymmetric cosmological gravitational field acceleration tensor consists of six components that can describe the four-acceleration of a particle. Three of the components represent the three-dimensional cosmological acceleration field strength vector and the other three components represent the three-dimensional solenoidal cosmological acceleration vector.

In the weak gravitational field of the cosmological curvature about a particle, we have

$$\frac{h_{\mu\nu}}{c}\vec{e}_1 = -\nabla\varphi - \frac{\partial\vec{U}}{\partial t} \tag{6.10}$$

$$\hbar_{\mu\nu}\vec{e}_2 = \nabla\times\vec{U} \tag{6.11}$$

where c is the speed of light, φ is a scalar potential, $\vec{U}$ is the vector potential of the cosmological acceleration field, $\left(\frac{m}{s}\right)$, $\frac{h_{\mu\nu}}{c}\vec{e}_1$ is the cosmological acceleration field strength vector, $\left(\frac{m}{s^2}\right)$, and $\hbar_{\mu\nu}\vec{e}_2$ is the solenoidal cosmological acceleration vector, $\left(\frac{1}{s}\right)$, or cycles per second (Hz). (Fedosin, 2016)

Therefore, for a specific point particle with velocity $\vec{v}$,

$$\frac{h_{\mu\nu}}{c}\vec{e}_1 = -c^2\nabla\gamma - \frac{\partial\gamma\vec{v}}{\partial t} \tag{6.12}$$

$$\hbar_{\mu\nu}\vec{e}_2 = \nabla\times\gamma\vec{v} \tag{6.13}$$

$$\gamma = \frac{1}{\sqrt{1-\frac{v^2}{c^2}}} \tag{6.14}$$

The determinant of the antisymmetric cosmological gravitational field acceleration tensor, $\det(H_{\mu\nu})$, $(1/s^4)$, is useful as a temporal or spatial curvature scaling factor, $\sqrt{\det(H_{\mu\nu})/c^4}$.

The value of the determinant only involves some of the space-time terms and time-space terms of the fully expanded (3 x 3) arrays, which are collapsed in folded three-dimensional time.

$$\det(H_{\mu\nu}) = \frac{3h_{x\tau}}{c} \cdot -\frac{3h_{\tau x}}{c} \cdot -\frac{3h_{\tau x}}{c} \cdot \frac{3h_{x\tau}}{c} = \frac{81}{c^4}\left(h_{x\tau}{}^2 \cdot h_{\tau x}{}^2\right) \qquad (6.15)$$

The gravitational acceleration field tensor in the local curvature of matter, $H^{\alpha\beta}$, is the inverse of the cosmological gravitational field acceleration tensor, $H_{\mu\nu}$, from the perspective of the cosmological gravitational field.

$$H^{\alpha\beta} = g^{\alpha\nu} g^{\mu\beta} H_{\mu\nu} \qquad (6.16)$$

In a region of cosmological curvature, as $gr \to c^2$, the cosmological gravitational field acceleration tensor, $H_{\mu\nu}$, may be expressed as

$$H_{\mu\nu} = \begin{vmatrix} 0 & -\frac{3h_{\tau x}}{c} & -\frac{3h_{\tau y}}{c} & -\frac{3h_{\tau z}}{c} \\ \frac{3h_{x\tau}}{c} & 0 & \hbar_{xy} & -\hbar_{xz} \\ \frac{3h_{y\tau}}{c} & -\hbar_{yx} & 0 & \hbar_{yz} \\ \frac{3h_{z\tau}}{c} & \hbar_{zx} & -\hbar_{zy} & 0 \end{vmatrix} \qquad (6.17)$$

If the antisymmetric cosmological gravitational field acceleration tensor, $H_{\mu\nu}$, is contracted with the metric tensor, it vanishes as an invariant, $g^{\mu\nu} H_{\mu\nu} = 0$, due to its antisymmetry.

The determinant of $H_{\mu\nu}$, or the contraction of the tensor $H_{\mu\nu}$ with itself, is a Lorentz invariant. Therefore, the *n*-dimensional cosmological weak gravitational acceleration field equations may be expressed as

$$\left(\bar{R}_{\mu\nu}+\bar{H}_{\mu\nu}\right)-\frac{1}{(n-1)}\bar{g}_{\mu\nu}\left(\bar{R}+\bar{H}\right)=-\frac{8\pi G}{c^4}\left(\Lambda_{\mu\nu}+B_{\mu\nu}\right) \quad (6.18)$$

where $\bar{R}_{\mu\nu}$ is the cosmological Ricci tensor, $\bar{R}$ is the trace of the cosmological Ricci tensor, $\bar{H}_{\mu\nu}$ is the Lanczos cosmological curvature tensor, $\bar{H}$ is the trace of $\bar{H}_{\mu\nu}$, $\Lambda_{\mu\nu}$ is the cosmological stress-energy-momentum tensor, $B_{\mu\nu}$ is the cosmological acceleration stress-energy-momentum tensor, and $\bar{g}_{\mu\nu}$ is the metric tensor of a cosmological Lorentzian manifold. The antisymmetric cosmological gravitational field acceleration tensor, $H_{\mu\nu}$, may be treated as the source of the cosmological gravitational acceleration stress-energy-momentum tensor, $B_{\mu\nu}$, given by

$$B_{\mu\nu}=\frac{1}{4\pi G}\left(g_{\mu\alpha}H^{\beta\alpha}H_{\beta\nu}-\frac{1}{(n-1)}g_{\mu\nu}H^{\alpha\lambda}H_{\alpha\lambda}\right)^2=\frac{1}{4\pi G}\left(H_{\mu\nu}-\frac{1}{(n-1)}g_{\mu\nu}H\right)^2 \quad (6.19)$$

Thus, we may also express, $H_{\mu\nu}$, in terms of the cosmological gravitational acceleration stress-energy-momentum tensor,

$$H_{\mu\nu}-\frac{1}{(n-1)}g_{\mu\nu}H=\sqrt{4\pi G B_{\mu\nu}} \quad (6.20)$$

The cosmological gravitational acceleration field stress-energy-momentum tensor is defined as a second order symmetric n-dimensional tensor, which describes the density and flux of energy and momentum, or the pressure, of the cosmological gravitational acceleration field in space-time. By the continuity equation, to conserve energy and momentum covariantly, we take the covariant derivative of the Lanczos cosmological curvature tensor, $\bar{H}_{\mu\nu}$, in n dimensions, or of the cosmological gravitational acceleration field stress-energy-momentum tensor, $B_{\mu\nu}$, to have

$$D_\mu\left(\overline{H}^{\mu\nu}\right)=0 \qquad \text{and} \qquad D_\mu\left(B^{\mu\nu}\right)=0 \tag{6.21}$$

The components of the cosmological gravitational acceleration field stress-energy-momentum tensor consist of the cosmological acceleration field strength, $\frac{h_{\mu\nu}}{c}\vec{e}_1$, and the solenoidal cosmological acceleration vector, $\hbar_{\mu\nu}\vec{e}_2$. Hence, for a weak field, we have

$$B_{\mu\nu} = \begin{vmatrix} B_{\tau\tau} & \frac{B_{\tau x}}{c} & \frac{B_{\tau y}}{c} & \frac{B_{\tau z}}{c} \\ cB_{x\tau} & B_{xx} & B_{xy} & B_{xz} \\ cB_{y\tau} & B_{yx} & B_{yy} & B_{yz} \\ cB_{z\tau} & B_{zx} & B_{zy} & B_{zz} \end{vmatrix} \tag{6.22}$$

The components of the cosmological gravitational acceleration field stress-energy-momentum tensor are as follows:

The time-time terms,

$$B_{\tau\tau} = \frac{3}{4\pi G}\left(\frac{h_{\tau\tau}^{\ 2}}{c^2} + c^2\hbar_{\tau\tau}^{\ 2}\right) \tag{6.23}$$

The space-space terms,

$$B_{\chi\chi} = \frac{1}{4\pi G}\left(\frac{h_{\tau\tau}^{\ 2}}{c^2} + c^2\hbar_{\tau\tau}^{\ 2}\right) - \frac{1}{4\pi G}\left(\frac{h_{\chi\chi}^{\ 2}}{c^2} + c^2\hbar_{\chi\chi}^{\ 2}\right) \tag{6.24}$$

The symmetrical space-space terms on each side of the diagonal,

$$B_{xy} = B_{yx} = \frac{1}{4\pi G}\left(\frac{h_{\tau x}}{c}\frac{h_{\tau y}}{c} + c^2\hbar_{yz}\hbar_{xz}\right) \tag{6.25}$$

$$B_{xz} = B_{zx} = \frac{1}{4\pi G}\left(\frac{h_{\tau x}}{c}\frac{h_{\tau z}}{c} + c^2\hbar_{yz}\hbar_{xy}\right) \tag{6.26}$$

$$B_{yz} = B_{zy} = \frac{1}{4\pi G}\left(\frac{h_{\tau y}}{c}\frac{h_{\tau z}}{c} + c^2\hbar_{xz}\hbar_{xy}\right) \tag{6.27}$$

The space-time and time-space terms,

$$cB_{\chi\tau} = \frac{3c}{4\pi G}\left|\frac{\vec{h}_{\chi\tau}}{c} \times \vec{\hbar}_{\chi\tau}\right| \tag{6.28}$$

$$\frac{B_{\tau\chi}}{c} = \frac{3c}{4\pi G}\left|\frac{\vec{h}_{\tau\chi}}{c} \times \vec{\hbar}_{\tau\chi}\right| \tag{6.29}$$

The cosmological gravitational acceleration stress-energy-momentum tensor, $B_{\mu\nu}$, enters the cosmological field equations through the cosmological stress-energy-momentum tensor, $\Lambda_{\mu\nu}$. (Fedosin, 2016)

Celestial bodies of mass may contract or expand, changing their gravitational acceleration potentials and fields which impact the cosmological curvature of a region of space-time. The gravitational acceleration is a direct consequence of the acceleration or deceleration of space-time. Hence, the square of the velocity of light c^2 embodies the relationship between the wavelengths of space and time as they change and manifest gravitation. The metric $g_{\mu\nu}$ encompasses the spatiotemporal manifestation of c^2 that directly relates to the gravitational field.

Let us consider the linearized cosmological field equations for a weak cosmological gravitational field that are theoretically useful for gravitational radiation, where the cosmological gravitational field from cosmological sources may be approximated by these equations. (McDonald, 2016) Let there be a weak cosmological gravitational field, so the cosmological metric perturbation, $h_{\mu\nu}$, is a small correction to flat cosmological space-time. The cosmological curvature tensor becomes

$$\tilde{G}_{\mu\nu} = \frac{1}{2}\left(\partial_\mu \partial^\varepsilon h_{\varepsilon\nu} + \partial_\nu \partial^\varepsilon h_{\varepsilon\mu} - \partial_\mu \partial_\nu h - \left(\frac{\partial^2 h_{\mu\nu}}{\partial t^2} - \nabla^2 h_{\mu\nu}\right) + \eta_{\mu\nu}\left(\frac{\partial^2 h}{\partial t^2} - \nabla^2 h\right) - \eta_{\mu\nu}\partial^\varepsilon \partial^\beta h_{\varepsilon\beta}\right) \quad (6.30)$$

where $h \equiv \eta^{\mu\nu} h_{\mu\nu}$ is the trace of the cosmological metric perturbation, $h_{\mu\nu}$, and the d'Alembertian, $\frac{\partial^2}{\partial t^2} - \nabla^2 \equiv \eta^{\mu\nu}\partial_\mu \partial_\nu$, is the flat cosmological space-time wave operator.

We can linearize $h_{\mu\nu}$ to simplify the cosmological curvature tensor as follows:

$$\bar{h}_{\mu\nu} = h_{\mu\nu} - \frac{1}{2}\eta_{\mu\nu} h \quad (6.31)$$

Using the trace-reversed cosmological metric perturbation, $\bar{h}_{\mu\nu}$, we have

$$\tilde{G}_{\mu\nu} = \frac{1}{2}\left(\partial_\mu \partial^\varepsilon \bar{h}_{\varepsilon\nu} + \partial_\nu \partial^\varepsilon \bar{h}_{\varepsilon\mu} - \left(\frac{\partial^2 \bar{h}_{\mu\nu}}{\partial t^2} - \nabla^2 \bar{h}_{\mu\nu}\right) - \eta_{\mu\nu}\partial^\varepsilon \partial^\beta \bar{h}_{\varepsilon\beta}\right) \quad (6.32)$$

In a linearized cosmological gravitational field, we can take advantage of gauge-freedom by adjusting coordinates to simplify the above equation further. Hence, we change coordinates $x^\mu \to x^\mu + \zeta^\mu$ requiring that $\partial_\varepsilon \zeta^\mu << 1$, so that

$$h_{\varepsilon\beta} \to h_{\varepsilon\beta} - \partial_\varepsilon \zeta_\beta - \partial_\beta \zeta_\varepsilon \quad (6.33)$$

All cosmological curvature tensors can be left unchanged by changing gauge. Thus, we can take advantage of our gauge-freedom to choose ζ^μ so that the Lorenz gauge condition is $\partial^\varepsilon \bar{h}_{\varepsilon\beta} = 0$. The Lorenz gauge makes the components of the gauge seem to be radiative, even when they are only static. Moreover, only a subset of the metric components of all gauges has the radiative degrees of freedom. Fortunately, this change of gauge also allows us to simplify our cosmological curvature tensor significantly. Hence, the cosmological curvature tensor becomes

$$\tilde{G}_{\mu\nu} = -\frac{1}{2}\left(\frac{\partial^2 \bar{h}_{\mu\nu}}{\partial t^2} - \nabla^2 \bar{h}_{\mu\nu}\right) \tag{6.34}$$

The Cosmological Field Equations for the linearized cosmological weak gravitational field simplify to

$$-\frac{1}{2}\left(\frac{\partial^2 \bar{h}_{\mu\nu}}{\partial t^2} - \nabla^2 \bar{h}_{\mu\nu}\right) = -\frac{8\pi G}{c^4} \Lambda_{\mu\nu} \tag{6.35}$$

$$-\frac{\partial^2 \bar{h}_{\mu\nu}}{\partial t^2} + \nabla^2 \bar{h}_{\mu\nu} = -\frac{16\pi G}{c^4} \Lambda_{\mu\nu} \tag{6.36}$$

where $\Lambda_{\mu\nu}$ is the cosmological stress-energy-momentum tensor and $\bar{h}_{\mu\nu}$ is the cosmological metric perturbation.

Thus, we find the solution for two points, r and r', in cosmological nearly-flat space-time, using a radiative Green's function, to obtain

$$\bar{h}_{\mu\nu}(r,t) = \frac{4G}{c^4} \int \frac{\Lambda_{\mu\nu}\{r,(t - |r - r'|/c)\}}{|r - r'|} d^3 r' \tag{6.37}$$

Therefore, we have in the above equation an exact solution to the linearized weak-field Cosmological Field Equations.

6.1 The energy of a gravitational field.

In a self-modifying gravitational field, a quantity of intrinsic curvature, $g_{\mu\nu}R/(n-1)$, is subtracted from the Ricci tensor, $R_{\mu\nu}$, to account for the conservation of mass-energy in the EFEs.

However, gravitational fields are also finite, so any compounding effect of mass-energy of a gravitational field is a convergent effect. The mass-energy density of the stress-energy-momentum tensor, $T_{\mu\nu}$, does not include the energy of a gravitational field. Thus, we can subtract the intrinsic curvature produced from the mass-energy of a gravitational field from the Ricci curvature on the left side of the EFEs. The mass-energy density of a gravitational field at any point in space-time, $U_{\mu\nu}$, excluding any quantum vacuum energy, equals the difference between the energy of the local gravitational field and the energy of the cosmological gravitational field.

$$U_{\mu\nu} = -\frac{1}{(n-1)} g_{\mu\nu} T + \frac{1}{(n-1)} g_{\mu\nu} \Lambda = \frac{1}{(n-1)} g_{\mu\nu} (\Lambda - T) \quad (6.38)$$

The resultant mass-energy density tensor of the gravitational field, $U_{\mu\nu}$, may be added to the EFEs in trace-reversed form, to equal the local Ricci curvature tensor, $R_{\mu\nu}$.

$$R_{\mu\nu} = \kappa\left(T_{\mu\nu} - \frac{1}{(n-2)} g_{\mu\nu} T + \frac{1}{(n-1)} g_{\mu\nu} (\Lambda - T) \right) \quad (6.39)$$

$$R_{\mu\nu} = \kappa\left(T_{\mu\nu} - \frac{1}{(n-2)} g_{\mu\nu} T + U_{\mu\nu} \right) \quad (6.40)$$

§ 7. On the three-dimensionality of time and the movement of particles.

Let us imagine that time is a kind of place, and in that place, we have three temporal dimensions. Each axis has two directions or senses in which a particle can move through time in space-time. A particle moving in space-time moves in a spiral motion through the three dimensions of time. It is possible to propose that three-dimensional time, or timescape, moves around and through matter, or around mass, even if a particle of mass, is not moving through space. Hence, it is also possible to consider a particle as stationary in space and visualize the three-dimensional temporal coordinates as comoving around a particle of mass, and around or through matter. In such scenario, a stationary particle at the origin, and at the center of its past and future cones, may have its worldline moving through its center as three-dimensional time passes. Therefore, two particle-referenced axes rotate about the non-coordinate axial or linear temporal dimension that may be called axial or linear time t_L by convention. The other two non-coordinate dimensions of time may be referred to as radial t_r and apparent t_a . The temporal distance from the origin of motion of the particle to its final destination may be referred to as apparent time t_a or actual time. The apparent time is the shortest direct path or temporal distance between the start position and the final position of the moving particle.

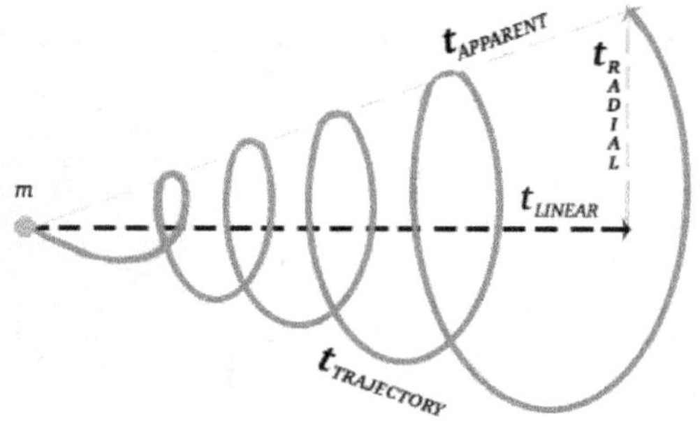

Figure 7

The distance in time of the trajectory is the longest temporal path or temporal distance traveled, longer than the apparent, linear, or radial, as shown in figure 7. Within an atom itself time flows so that particles do not occupy the same position in space to preserve the atomic structure. If the radial and trajectory axes of a moving particle rotated, it would cause the particle to move in an orbital or spiral fashion through time due to the spatiotemporal expansion. Space emerges as a result of the spatiotemporal expansion. Time creates space, space endows more time, motion proceeds from space-time. All matter is submerged in space-time so that matter may move through space-time or be moved through space-time by forces acting on it. Energy and matter emerge through space-time.

In homogeneous and isotropic space-time, the proportionality and distances in time are conserved. The size, form, constitution, atomic structure, or gestalt of matter binds its temporal flow rate or passage of time. An electron shifts through time when it changes orbital distance from its nucleus. Thus, energy is stored in the spatiotemporal field of a particle. Particles may distort space-time to avoid collisions or interchange positions with other particles in space-time and that distortion creates a gravitational effect.

Some particles may travel in longitudinal waves through space-time, with compression and rarefaction cycles, which divides the total energy of particles into, but not limited to, particle energy related to mass, field potential, momentum, and wave energy. The expansion of time separates the moment before from the moment after, allowing the awareness of the human mind on the passage of time.

The electron of the simplest atom may be speculated to move forward in time ahead of its nucleus and appear to be tenuous, large in size, and small in mass in comparison to other fundamental particles. The same tenuous condition of the electron moves backward in time with respect to the nucleus at a lesser time rate. Rest mass is bound and moves forward in time in an expanding universe like ours. A particle moving faster than light in its trajectory within the six senses of three-dimensional time may be able to move backward in time. It is widely accepted in current physics that no particle of mass can move faster than the speed of light according to the General Theory of Relativity, but that applies to motion through linear temporal coordinate distance. It is reasonable to conceptualize that if a particle moves in a helical trajectory through time, the three-dimensional speed of the particle may exceed the magnitude of the linear coordinate speed of light, and may be able to reverse its direction of motion backward in time. Would this explain how particles, virtual or not, may appear and disappear in the observable spatiotemporal medium of the present?

As a particle moves in a spiral trajectory through space-time at a velocity equal or greater that c, it may be able to reverse direction in a spiral trajectory to its original arrow of time (arrow of causality), and follow a new spiral trajectory in the opposite sense arrow of time into recently created space-time in the past. The particle may continue its trajectory, as a tachyon or an antimatter particle, until its velocity decreases below c and starts moving forward in space again, as a tardyon, some-when in the past space-time. Would the antiparticle have the probability to collide with its particle in the past? Would the antiparticle have the probability to coexist or interact with its particle or would it exist in an alternate space-time?

7.1 What are the implications of multidimensional time?

Let us consider the effect on the laws of physics for a universe like ours with multiple spatial and multiple temporal dimensions. Does reality appear like our existing reality to an observer in a universe with three spatial dimensions and three temporal dimensions? How would an observer be able to perceive time as two-dimensional, or three-dimensional, if the observer's realization of time conceptualizes the temporal dimension as linear coordinate time in each sense of every spatial dimension, surface time as the area of a temporal coordinate plane of an extension of an existing spatial coordinate plane, and three-dimensional time as the temporal coordinate volume of an expansion of existing three-dimensional volume of space? (Tegmark, 1997)

An observer in such a six-dimensional universe would be able to have thoughts in a one-dimensional sequence because of the ability of the thought process of the observer to collapse two-dimensional or three-dimensional time into a one-dimensional resultant time. Motion through time in any temporal direction would not necessarily require motion through any of the temporal coordinate directions just like motion through space in any spatial direction would not necessarily require motion through any of the spatial coordinate directions. Even though, any linear motion of an observer, or localized object, through space-time may be conceptualized as a three-dimensional or two-dimensional resultant motion of an observer or localized object frame of reference through the spatiotemporal wave medium.

The observer's ideas that require spatial conceptualizations that are a function of one-dimensional time: velocity, area growth, linear extension, etc., would be very useful in everyday activities. For the observer, a mechanical clock prolongs the illusion of a useful linear concept of time even though it would measure the angular frequency of three-dimensional time at a specific locality. An observer, or localized object, that travels through linear coordinate time may be conceptualized as traveling through a resultant temporal geodesic of three-dimensional time of an alternate rotated inertial coordinate system. Thus, the observer, or localized object, may be conceptualized as traveling in a linear temporal world-line in six-dimensional space-time. The observer's, or localized object's, proper time would be measurable coincident events.

Therefore, energy may be considered a constant or a resultant vector in a certain direction of three-dimensional time whose temporal coordinate dimensions may be mathematically collapsed as a resultant temporal geodesic direction. The direction of the resultant vector, or the direction of the world line, may differ from any of the three temporal coordinate directions. Two non-relativistic observers traveling in different directions of space and time may meet, stay together, not drifting apart, as long as they are not traveling away from each other through spatial directions, because their temporal directions are extensions of the spatial directions. The slower they travel through space, the faster they will travel through time. Each spatial dimension has a conjugate temporal dimension.

There may be backward causation arguments concocted that describe the interactions of tachyons for forward and retro-causality that would explain spooky-action-at-a-distance, entanglement, in six-dimensional reality, which challenge our present four-dimensional view. A tachyon $i(m)$ may be regarded as the quantum of entanglement, a particle with imaginary mass $\sqrt[2]{-1}$, of instantaneous quantum abstract information exchange between entangled quantum states of a system. Thusly, entangled quantum states of particles may share wavelength phase or spin information through the exchange of tachyons.

Quantum mechanical spin describes an intrinsic form of angular momentum for elementary particles, atomic nuclei, and hadrons (composite particles). The Stern-Gerlach experiment was the first experimental evidence for elementary particle spin. Later, spin was described as a two-valuedness that was not describable classically.

Spin is not, as originally conceived, the rotation of a particle around some axis, except as long as spin obeys the mathematical laws as quantized angular momenta. Spin quantum numbers may take half-integer values. The direction of spin may change but an elementary particle cannot be made to spin slower or faster. An elementary particle's spin is associated with a magnetic pole moment with a g-factor differing from 1. The spin angular momentum, *S*, of any physical system is quantized. The allowed values of *S* are calculated by $\hbar\sqrt{s(s+1)}$, where $\hbar$ is the reduced Planck constant and *s* is the spin quantum number.

Tachyons may be conceptualized as the quantum objects of entanglement, emitted by faster-than-light particles through a specific medium, and quantum states information in an electromagnetic bundle of energy with no effective charge. A photon traveling through six-dimensional space-time in a helical motion may be conceptualized to have the probability of reversing the direction of its trajectory, traveling backward in time and appearing to travel faster than light, from the perspective of an inertial frame of reference in the retarded spatiotemporal wave, but slower than light from an inertial frame of reference in the advanced spatiotemporal wave. The realizations, or objections, of faster than light velocity through a specific medium, and the multidimensionality of time, are intimately related. The existence of stable tachyons would be assured as the existence of ordinary particle through the spatial or temporal maximal property in six-dimensional space-time. Furthermore, the photon would be unstable against decay as a tachyon-antitachyon pair.

An electron and a positron orbiting around their common center of mass is a bound quantum state, or exotic atom, known as positronium. The positronium system is unstable, and the electron and positron annihilate each other to predominantly produce two or three gamma-rays, or high-energy photons, depending on the relative spin states. Researchers have long sought experimental evidence for tachyons from a long-standing anomaly in low energy atomic physics, which is the disagreement between theory and experiment for the orthopositronium annihilation decay rate in vacuum. Long-standing, anomalous measurements of the orthopositronium decay rate are interpreted as evidence for two tachyons being occasionally emitted when orthopositronium decays. Thus, a discrepancy between QED theory and an atomic physics experiment is interpreted as indirect evidence for tachyon emission. Besides, there is a hypothesis that tachyons do not interact directly with ordinary matter, but just with photons. (Skalsey et al, 2000)

Tachyons are a way of describing unstable field configurations. Thus, the rate of tachyon emission describes the decay of an unstable state to a stable state. The fluctuations about an unstable equilibrium configuration exemplify the tachyonic properties of imaginary mass and faster than light travel. Imaginary mass particles have the weird property that they speed up as they lose energy, the value of their imaginary mass being defined by the rate at which this occurs. The possibility of particles whose four-momenta are always space-like, and whose velocities are therefore always greater than the speed of light is not in contradiction with the special theory of relativity, and such particles might be created in pairs without any necessity of accelerating ordinary particles faster than the speed of light. (Feinberg, 1967) Tachyons may be regarded as particles with space-like worldlines, or within a space-like worldtube, outside the light cone; entirely different than anti-particles traveling backward in time inside the light cone, along time-like worldlines. (Feynman, 1964)

An observer may opine that our current four-dimensional view of physics may be referred to as three dimensions of space and a resultant dimension of time that suffices our conceptual applications. Thus, that observation leaves open the possibility that the time that we experience is an amalgamation of the three coordinate temporal dimensions of a resultant temporal geodesic. If the physical behavior of particles in multiple dimensions of time appear disturbing to our present way of thinking, perhaps our way of thinking may be less disturbed with greater degrees of freedom. The human body and all its particles are created to exist in multiple dimensions of space-time, so the human mind is up to the challenge. It is known that fundamental particles would still be stable if their kinetic energies were low enough even in a conjugate multidimensional medium of space or time. Observers would be self-aware and able to predict using the laws of physics with information-processing abilities, measurements of fields, in the presence of well-posed causality. (Dorling, 1970) Under the condition of multidimensionality, if the solution to a problem is uniquely determined by boundary conditions then the problem is well-posed, and the dependence of the solution on linear boundary data is bounded, changing by a finite amount if the boundary data changes by a finite amount. Thus, the above predictability is met by a small class of partial differential equations which are hyperbolic. Moreover, the full system of coupled partial differential equations in nature are actually non-linear which allows ill-posed problems in small neighborhood of a hypersurface inside the lightcone, to have well-posed problems in a larger local neighborhood. (Tegmark, 1997)

An ill-posed problem may also be formally solved measuring the initial data with infinite accuracy to be able to place finite error bars on the solution which may presently be beyond the reach of our science, technology, or credibility. It is reasonable to conclude that the impossibility that observers may not exist in multidimensional space or time has not been rigorously demonstrated by any of the above arguments until predictability for multi-temporal dimensions, or stability and complexity for multi-spatial dimensions, have been carefully and precisely analyzed. Causality is the fundamental principle behind the requirement of sufficient predictability, stability, and complexity. Hence, a convincing paradigm shift in the present conceptualization of causality may precede the formalization of multi-temporal dimensions.

A geodesic represents the shortest line between two points that lies in a surface. That shortest line, or segment, is a collection of point with distance values from each of the coordinate axes. Geodesic lines are time-like, space-like, and null. An observer may affirm that the maximal property of a resultant time-like geodesic in ordinary space-time is a necessary condition for the existence of stable particles. Under the conditions of a three-dimensional temporal model, it would be possible to assert that this maximal property of a resultant temporal geodesic would be successful if time is multi-dimensional. It is possible to generalize Minkowski geometry to any number of spatial and temporal dimensions. Thus, let us analyze the maximal property of a resultant temporal or spatial geodesic and a coordinate temporal or spatial geodesic in six-dimensional space-time. (Dorling, 1970)

Disregarding the null, under the three-dimensional temporal model of six-dimensional space-time, the three-coordinate time-like axes originate a resultant time-like geodesic for a particle's trajectory. Each coordinate time-like axis, or line, has the property that if two points on the line are kept fixed, we may bend the line slightly in between the points in a time-like direction. The time-like line has been bent in a time-like direction if every perpendicular dropped from a point on the bent line onto the original line is time-like. The bent time-like line is longer than the original time-like line.

However, we have also curved three-dimensional time by bending one of the coordinate time-like axis, contracting its amplitude, which shortens the resultant time-like geodesic of the particle's trajectory. An observer may regard the resultant time-like geodesic as a particle's trajectory, not necessarily as one of several coordinate temporal dimensions. Likewise, under the three-dimensional spatial model of six-dimensional space-time, the three-coordinate space-like axes originate a resultant space-like geodesic for a particle's trajectory.

Each coordinate space-like axis, or line, has the property that if two points on the line are kept fixed, we may bend the line slightly in between the points in a space-like direction. The space-like line has been bent in a space-like direction if every perpendicular dropped from a point on the bent line onto the original line is space-like. The bent space-like line is longer than the original space-like line. Nevertheless, we have also curved three-dimensional space by bending one of the coordinate space-like axis, contracting its amplitude, which shortens the resultant space-like geodesic of the particle's trajectory. The observed resultant space-like geodesic of the particle's trajectory is not necessarily one of several coordinate spatial dimensions. (Dorling, 1970)

Furthermore, if we were to bend a space-like line in a time-like direction, the bent line is shorter than the original unbent line. If we were to bend a time-like line in a space-like direction, the bent line is shorter than the original unbent line. The resultant space-like or the resultant time-like geodesic is a construct of the three-coordinate space-like or time-like axes. These constructs give the observer the three-dimensional concept of spatial and temporal distance on the

geodesic of a particle's trajectory. A particle may travel through existing space-time in an observed geodesic of its trajectory, not subsequently creating a new dimension of space or time.
Hence, the resultant space-like or resultant time-like geodesics are maximal, but not coordinate space-like axes or coordinate time-like axes. Besides, the relationship between nonhomogeneous and anisotropic space and time is asymmetrical. As particles move in geodesics through six-dimensional space-time, they move through resultant space-like or resultant time-like geodesics, not through each corresponding coordinate space-like or time-like axis. An observer would reasonably expect that the four-dimensional theory of space-time can be reconciled with the six-dimensional theory of space-time as a functional simplification in all its practical purposes.

There is an intimate connection between the maximal property of time in four-dimensional space-time and the existence of stable particles. The energy-momentum conservation of a particle decaying process states that the vector sum of the energy-momentum four-vector of the decay products should equal the energy-momentum four-vector of the original particle.

As a limit of six-dimensional space-time, the temporal vectors of particles, objects, and observers are taken to be parallel, and the limits of the temporal vector angles are taken to zero, to obtain the same results in four-dimensional space-time as the results from six-dimensional space-time. The above vector simplification allows us to fold, in a sense, six-dimensional space-time into four-dimensional space-time, to obtain scalar temporal equations from vector temporal equations, or 4-D Lorentz transformation equations.

Thus, for a particle to decay it is not sufficient that there should exist a set of particles with the same quantum numbers, but it is also well known and necessary that the sum of the masses of the particles should be less than the rest mass of the original particle. The proton and the electron are stable because of the latter restriction of the masses which has been absolutely contingent upon the one-dimensionality of time because of perceived maximal property.
Conclusively, an observer may determine that the property of a geodesic in three-dimensional time in six-dimensional space-time is maximal, and that particles are stable in three-dimensional time.
The eminent physicist Max Tegmark has indicated through his

research that physics has no predictive power for an observer when the partial differential equations of nature are elliptic or ultra-hyperbolic, or that certain combinations of spatial and temporal dimensions are unstable. The multidimensionality of space-time for the observable universe may be illustrated as six-dimensional space-time as shown in figure 8 to illustrate particles, charges, and fields that would be realizable under simpler combinations of spatiotemporal dimensions.

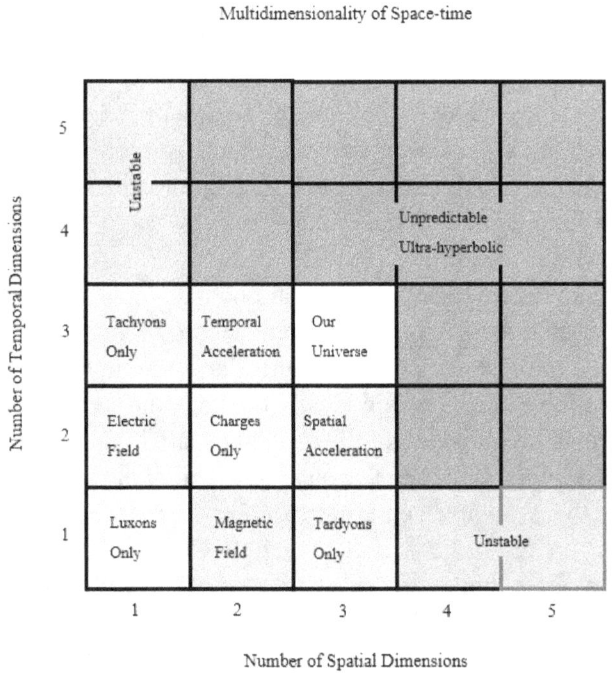

Figure 8

7.2 How do we measure coordinate time in three-dimensional time?

Let us consider an experiment that involves three linear photonic clocks mounted on a non-rotating frame of reference mounted on a jet aircraft that is able to fly at very high speed in several trajectories through the hypothetical six-dimensional space-time. Each linear photonic clock is aligned with each Cartesian coordinate axis (x, y, z), so that depending on the trajectory of the jet aircraft, each linear clock would count time relatively even at slow speeds compared to the speed of light. The three-dimensional linear photonic clock device should maintain the same three-dimensional orientation at all times from takeoff to landing. The aircraft navigation computer would be programmed for linear travel (x-axis), geodesic travel (x-y axes), and resultant travel (x-y-z axes).

In principle, each linear photonic clock measures time by measuring time per cycle between the emission and absorption of a photon in the direction of its axis. The x-axis is the axis of roll (longitudinal), the y-axis is the axis of pitch (lateral), and the z-axis in the axis of yaw (vertical). If our hypothesis of three-dimensional time is supported, each type of travel would involve a measurement of time by each clock according to the General Theory of Relativity. For linear travel, the x-axis linear photonic clock would tick tock slower than the other two axes. Time would dilate in the direction of the contracted longitudinal spatial dimension.

For geodesic travel, the x-y axes linear photonic clocks would tick tock slower than the z-axis linear photonic clock. Time would dilate on the plane of the contracted x-y plane. For resultant travel, all x-y-z linear photonic clocks would ideally tick tock the same. Resultant time dilates in the direction of the resultant spatial trajectory. There are optional x-y-z trajectories, such as a helical or curved sinusoidal trajectory, that would be achievable for a jet aircraft.

The technology to build a linear photonic clock already exists. A single-photon avalanche diode (SPAD), also known as a Geiger-mode avalanche photon detector (G-APD), is a solid-state photodetector in which a photon-generated carrier can trigger an avalanche current due to the impact ionization mechanism as shown in figure 9.

Surprisingly, it is even possible to measure the number of photons absorbed within a certain short time interval in the active region of an avalanche photodiode. For that purpose, it is necessary to precisely measure the rise of photocurrent at the beginning of the avalanche. A SPAD is able to detect low intensity signals, down to a single photon, and to signal the arrival times of the photons within a few tens of picoseconds.

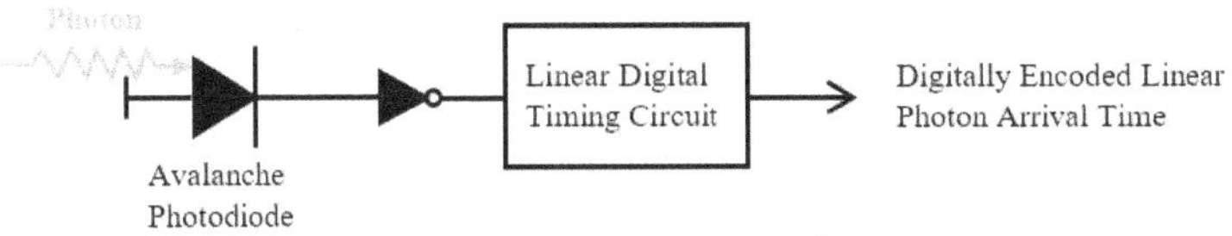

Figure 9

Avalanche photodiodes operated in Geiger-mode can be used even for photon counting with dark count rates well below one kilohertz and with a quantum efficiency of several tens of percent, sometimes even well above fifty percent. The Geiger mode means that the diode is operated slightly above the breakdown threshold voltage, where a single electron–hole pair, generated by absorption of a photon or by a thermal fluctuation, can trigger a strong avalanche. In the case of such an event, an electronic quenching circuit reduces the voltage at the diode below the threshold voltage for a short time, so that the avalanche is stopped and the detector is ready for detection of further photons after some recovery time of typically a hundred nanoseconds. That dead time constitutes a substantial limitation of this technology. It limits the count rate to the order of ten megahertz, whereas an avalanche diode in linear mode, which is operated with lower reverse voltage, may be operated with a bandwidth of many gigahertz. (Renker, 2006)

§ 8. On the positive pressure of space-time inside matter

Spacetime inside matter may be loosely described as a frothy fluid with negative pressure. A positive spatiotemporal energy density resulting from a cosmological constant implies a negative spatiotemporal pressure, and vice versa. The negative spatiotemporal pressure will drive an accelerated expansion of the universe.
In the case of a star being contracted by its own gravitational field, the positive thermal pressure of the star's matter counteracts the collapse of the star. With high enough positive pressure inside the star, the gravitational attraction from pressure will be greater than the repulsion from the pressure gradient, contributing to setting a theoretical upper limit to the mass of a neutron star in our universe.

As the matter of a star compresses, its energy causes greater curvature of space-time, resulting in a stronger gravitational field.

In extreme cases, the gravitational effect of positive pressure may overcome the effect of expansion and pressure and may collapse the star. Matter is negative pressure space-time enclosed by mass, energy, and physical fields. Space-time is the fundamental framework of mass, energy, and physical fields. The dimensional curvature of space and time is the manifestation of the space-time negative pressure gradient. Negative space-time pressure is the mechanism by which a gravitational field is manifested. Negative space-time pressure inside matter varies with the mass density, energy density, momentum, and physical fields. Celestial bodies may have non-homogeneous distribution of mass and other attributes. The gravitational field acceleration of the earth increases slightly from its surface value of approximately 9.8 m/sec^2 through the upper mantel and then increases more rapidly through the lower mantel as it approaches the boundary of the outer core with the lower mantel approximate value of 10.7 m/s^2 according to the Preliminary Reference Earth Model as shown in figure 10. (Dziewonski, 1981)

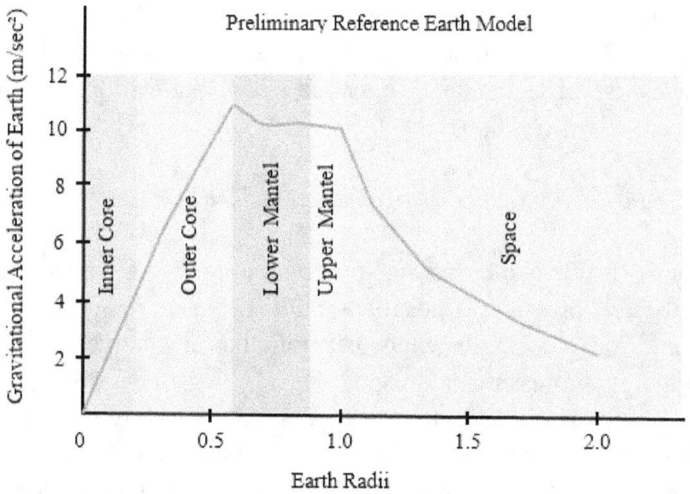

Figure 10

The density of the earth (kg/m^3) increases substantially at the core-mantel boundary, from approximate 5500 Kg/m^3 to 10000 Kg/m^3; it almost doubles! Then, beyond the core-mantel boundary, it continuously increases non-linearly up to approximately 13000 Kg/m^3 at the inner core. Outside of the earth's matter, the buoyancy of the atmosphere or ocean water reduces the apparent strength of the gravitational field, as measured by an object's weight. The magnitude of this effect depends on air pressure (density) or volume of water. The gravitational effect of other celestial bodies may have a small effect on the earth's gravitational field strength. The strength of the gravitational field outside the matter of the earth changes non-linearly from approximately 9.8 m/s² at its surface to approximately 5.7 m/s² at 2000 Km from the surface as shown in figure 11.

Hence, as an object moves away from the surface of a massive celestial body, like the Earth, the gravitational acceleration applied by the massive celestial body on the moving object decreases by the inverse-square law of the distance. The negative pressure of space-time also decreases away from the surface of the large mass as the inverse-square of the distance. The gravitational gradient is the non-linear spatiotemporal negative pressure gradient.

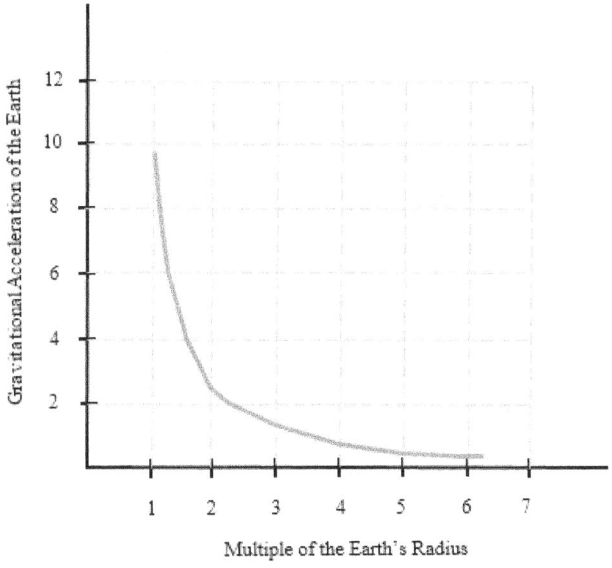

Figure 11

Regrettably, Isaac Newton's law of gravitation does not fully explain the gravitational field in the plenum of matter where there are fundamental particles, molecules, momentum, pressure, energy, and physical fields, which will not obey Newton's law, unless all the above attributes are, or an individual attribute is, considered a single mass of uniform density and pressure at a specific radius from the center of mass to a distance above the surface of the mass. Einstein's field equations in six dimensions describe the curvature even at the atomic level between particles and the relationship of the curvature of space-time to the stress-energy-momentum tensor of those attributes but provide scarce insight on the nature of the resultant gravitational field between numerous attributes of a system in the plenum of matter.

It is possible to theorize that the curvature of space-time is the result of the Huygens-Fresnel principle when spatiotemporal waves interfere constructively or destructively in the spatiotemporal wave medium. The spatiotemporal wave medium acts on matter through the spatiotemporal wave actions. The spatial wave expands giving the space about an object of mass the curvature about its form and contents. Curvature of space does not create gravity.

The gravitational acceleration is the result of the spatiotemporal wave action upon matter, or mass, as the spatiotemporal wave decelerates. However, the spatiotemporal deceleration manifests the effect of the gravitational field internally or externally of matter, not exclusively on the external curvature of space-time of the entire system of mass and attributes.

The Newtonian gravitational field is an external inverse-square law of a system of mass, or matter, as shown in figure 11, but in the plenum of matter, there is a different relationship between the gravitational field, mass, attributes, and distance, as we can see in the preliminary reference earth model as shown in figure 10. Thus, a paradigm shift is needed that improves our understanding of the law of gravitation.

Let us now consider the spatiotemporal acceleration about any individual mass and attributes for a fundamental particle or system of mass. From previous research, the spatiotemporal acceleration at a slow speed, $v << c$, may be expressed as shown below.

$$g = G\frac{m_0}{r_0^2} = \frac{\ddot{a}}{8\pi m_0}\frac{m_0}{r_0^2} = \frac{\ddot{a}}{8\pi r_0^2} \tag{8.1}$$

$$\ddot{a} = 8\pi g r_0^2 = 8\pi G m_0 \tag{8.2}$$

The spatiotemporal acceleration for the Planck mass of a hypothetical particle may be expressed as

$$\ddot{a} = \frac{\hbar c}{m_p} \tag{8.3}$$

The spatiotemporal acceleration for a relativistic mass and length is given by

$$\ddot{a} = 8\pi g r_0^{\,2}\left(1 - v^2/c^2\right) = \frac{8\pi G m_0}{\sqrt{1 - v^2/c^2}} \tag{8.4}$$

The spatiotemporal acceleration is a function of the physical dimensions and contents of particles, objects, and systems of mass. Consequently, the physical dimensions, attributes, and motion through space of mass, energy, and fields, affect the spatiotemporal acceleration to be relativistic for masses at speeds approaching the speed of light.

§ 9. The divergence and the passage of time in the plenum of matter

Let us consider the plenum of matter for a system of mass. Time is emergent in the plenum of matter, but an object, or a system of mass like the earth, does not necessarily have uniform density, uniform attributes, and uniform spatiotemporal pressure in its plenum. As the density of the earth decreases in its outer and inner cores, time is more contracted, space is more extended, toward the center of matter, and the gravitational field decreases. Thus, as the plenum spatiotemporal pressure decreases, the divergence of time decreases proportionally, and the gravitational acceleration decreases. Hence, as time emerges at a locality, the passage of time is a function of the existing spatiotemporal pressure.

$$P_{t-plenum} \propto div\ \vec{T} \propto g \tag{9.1}$$

$$div\,\vec{T} = \frac{\partial T_X}{\partial x} + \frac{\partial T_Y}{\partial y} + \frac{\partial T_Z}{\partial z} \tag{9.2}$$

Therefore, a clock would run faster near the center of the inner core of the earth than in the lower mantel or the surface of the earth. Time dilates away from the center of the earth as the gravitational field increases. The deceleration of the spatiotemporal field is the underlying structure and mechanism for the gravitational field. On the other hand, space extends near the center of the inner core and then contracts toward the lower mantel of the earth. Then, time and space expand or contract in the plenum of matter as a function of the stress-energy-momentum tensor of matter. The greater the effect of the stress-energy tensor, the lesser the temporal pressure, the greater the spatial pressure, and the greater the gravitational acceleration of the spatiotemporal plenum.

The spatiotemporal plenum pressure from the center of mass may be expressed as the force exerted on matter in units of force per unit of area, or as the jolt of the energy of matter from the center of mass of the plenum of matter in units of energy per unit of volume.

$$P_{st-plenum} \equiv \frac{m\frac{\partial^2 r}{\partial t^2}}{\partial s^2} \equiv \frac{\partial^3 E}{\partial r^3} \tag{9.3}$$

Hence, the higher the spatiotemporal pressure of the plenum, the higher the energy density of the spatiotemporal plenum of matter. The spatiotemporal plenum pressure is exerted as the spatiotemporal waves interfere constructively or destructively, according to the Huygens-Fresnel principle in the plenum of matter.

§ 10. On the magnetic fields passing through matter or free space

All space-time inside matter is plenum, the quality of being full of pressurized space-time. Therefore, we may say that matter is equivalent to mass, energy, momentum, pressure, physical fields, and spatiotemporal plenum.

Let us consider what happens when magnetic field lines pass through homogeneous and isotropic free space before they pass through a nearby system of mass (matter) of high permeability as shown in figure 12.

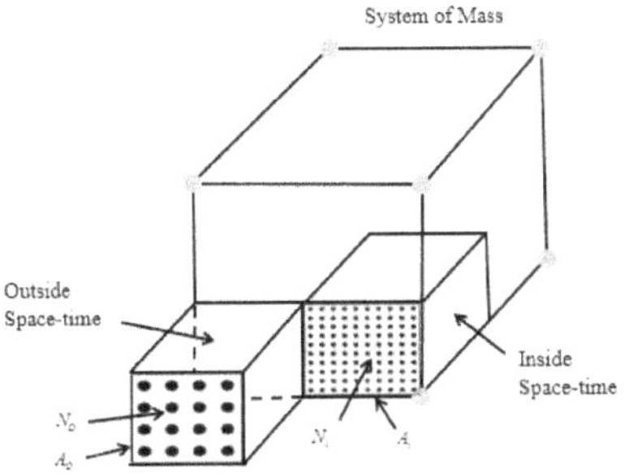

Figure 12

We consider rectangular spatiotemporal volumes of homogeneous and isotropic free space outside of matter and an equal spatiotemporal plenum volume inside of symmetric matter when magnetic lines fully saturate symmetric matter, ceteris paribus. Saturation is a characteristic of ferromagnetic and ferrimagnetic materials. Magnetic saturation is the degree of magnetization that a material obtains in response to an applied magnetic field. Lines of magnetic field N_i represent magnetic lines passing through the spatiotemporal plenum volume of a rectangular prism of the symmetrical system of mass when fully saturated. Lines of magnetic field N_o represent magnetic lines passing through the rectangular prism volume of free space outside of matter.

By comparing the spatial ratio of the two identical spatiotemporal volumes, we have a relative measurement of how compact the spatial volume of space would be inside or outside of symmetric matter at the same inertial frame of reference. The relative permeability of the material at maximum saturation $\mu_{\max}$ is given by

$$\mu_{sat} = \frac{B}{H} \equiv \frac{\phi/m^2}{I/m} \equiv \frac{\phi \cdot s}{q \cdot m} \equiv \frac{\phi \cdot s}{m \cdot s \cdot m} \equiv \frac{\phi}{m^2} \equiv B_{sat} \qquad (10.1)$$

$$\mu_{\max} = \frac{\mu_{sat}}{\mu_o} = \frac{B_{sat}}{B_o} \equiv \frac{\partial N_i}{\partial N_o} \qquad (10.2)$$

Comparing the magnetic flux outside and inside of the above symmetrical matter for the same spatiotemporal volume we have

$$\frac{\partial \phi_o}{\partial \phi_i} = \frac{\dfrac{\partial N_o}{\partial A_o}}{\dfrac{\partial N_i}{\partial A_i}} = \frac{\partial N_o}{\partial N_i} \cdot \frac{\partial A_i}{\partial A_o} = \frac{1}{\mu_{\max}} \cdot \frac{\partial A_i}{\partial A_o} \tag{10.3}$$

$$\frac{\partial A_i}{\partial A_o} = \mu_{\max} \frac{\partial \phi_o}{\partial \phi_i} \tag{10.4}$$

$$\frac{\partial A_i}{\partial A_o} < 1.0 \tag{10.5}$$

The ratio of compactness of the spatiotemporal areas, the spatial compactness ratio, is equal to the relative maximum permeability times the ratio of the outside magnetic flux of space-time to the inside magnetic flux of symmetric matter.

§ 11. Conclusion

The spatiotemporal wave medium of our universe stores infinite energy. The spatiotemporal wave medium is a transmitter and a receiver of this infinite energy density or pressure. Science needs a paradigm shift to include the actions and reactions of the spatiotemporal wave medium as interactive with matter, energy, and other attributes. Every now, every yesterday, and even every tomorrow, becomes part of the space-time of our experience, but it all happens with the motion of space-time where space and time emerge into our reality within the growth of our spatiotemporal medium. We may regard space-time as the illusion that realizes gravity which does not exist. Then, we may realize that gravity comes from the pressure of space-time which is part of the illusion. Motion may be perceived as the cause of change, and change may be considered the effect of the spatiotemporal wave. Therefore, the spatiotemporal wave is ultimately the cause of all changes that result in motion, and changes that result from motion are ultimately the effects of the spatiotemporal wave.

PART II

KINEMATICS

Chapter 4

On the forces of six-dimensional space-time and mass

§ 1. Introduction: the concept of force

Force is defined as the capacity to do work or cause physical change; energy, strength, or active power. Long ago, ancient philosophers conceptualized force in studies of stationary and moving objects and machines, and slowly developed an understanding of the nature of natural motion. Aristotle famously described a force as anything which causes an object to undergo unnatural motion. Force is a quantitative interaction between two physical bodies, such as an object and its background. Force is proportional to acceleration and it is the derivative of momentum with respect to time.

Aristotelian thought prevailed for centuries to explain the reasons why motion occurs and why motion might change. Both Galileo and Newton conceptualized that motion needed no explication, but it is only the change in motion that demands a physical motive. The incredibly lasting ideas of Aristotle that natural motion indicated the tendency of objects to go to their natural place, animalistic motion was voluntarily willed, and forced motion occurred when an object acted on another, formed the base upon which the study of forces in modern physics began, and science and reason started discovering and interpreting the forces of nature. (Newton, 1999)

The concept of force is used to describe an influence which causes a free body to accelerate or change its velocity with respect to time. Force can also be described by insightful concepts such the force of a push or pull that can cause an object with mass to change its velocity to begin moving from a state of rest or to accelerate. An applied force has both a magnitude and a direction, making the applied force into a vector quantity. The SI unit for force is the Newton (N). One Newton of force is equal to 1 $Kg \cdot m/s^2$. The eminent physicist Isaac Newton perfected understanding of forces and motion with mathematical insight that remained unchanged for nearly three hundred years.
Newton explained gravity as an attractive force between the masses of

objects. He explained the relationship between force and motion, and changed the contemporary view of the universe by showing that the same physical laws applied to all matter anywhere in the universe. His mathematical principles were uncontradicted by experiment for nearly two centuries.

1.1. Newton's second law

Newton's laws of motion are only true in inertial frames of reference that are not accelerating. Newton's first law of motion states that if all the forces on an object cancel each other out, then the object continues in the same state of motion.

The first law of motion is a more refined form of the principle of inertia expressed by Galileo Galilei. Newton's second law of motion predicts the acceleration of an object along the direction of motion given the total force acting on it and its mass, by the formula $a = F_T / m$. The formula of the second law states that an object with a constant mass will accelerate in proportion to the net force acting upon the object and in inverse proportion to its mass, an approximation which breaks down near the speed of light. In other words, the net force acting upon an object equals to the change rate of its momentum. (Brackenridge, 1995)

$$F = ma = \frac{\partial p}{\partial t} \tag{1.1}$$

Early in the 20th century, Albert Einstein developed a theory of relativity that correctly predicted the action of forces on objects traveling near the speed of light with increasing momenta, and also provided invaluable insight into the forces produced by gravitational fields and inertia. (Rucker, 1977) Einstein expounded how the curvature of space-time resulted in the gravitational field of a mass. (Einstein, 1952)

1.2. The characteristics of forces

Forces are categorized as vector quantities because they act in a specific direction with a specific magnitude depending on intensity of the push or pull. These characteristics set forces apart from other physical quantities that have no direction such as scalar quantities. In order to determine the resultant force if two or more forces act on an object, it is necessary to know the direction as well as magnitude of each force acting on the object, only then can one calculate the resultant force and its effect. Once all forces are expressed as vectors, all magnitudes and directions are available to decide without uncertainty the action and result of the resultant force using the mathematical operations of vector analysis.

Quantitative forces were first researched for static equilibrium conditions where forces acting opposite to each other canceled out in pairs. Such research showed the additive vector qualities of forces because of the properties of direction and magnitude.

The resultant force followed the parallelogram rule of vector addition and depends on the magnitudes of its components and the angle between the components' line of action.

A useful tool to illustrate and follow the actions of forces acting on an object is the free-body diagram. These diagrams are drawn to depict the directions or lines of action, magnitudes and angles of all forces acting on the body of the object and can be used as a graphical method to calculate and determine the resultant force or resultant.

Forces are often resolved into their independent orthogonal components as a set of basis vectors, so that their components are uniquely determined by the scalar addition of the components of the individual vectors. Orthogonal components are independent of each other because forces acting at ninety degrees have no effect on the direction or magnitude of each other. An orthogonal force vector can be three-dimensional with the third component being at right-angles to the other two. It is often mathematically convenient to choose a set of orthogonal basis vectors that are parallel to a force vector so that the force vector would be resolved to have only one non-zero component.

At present, four fundamental force fields are said to comprise all the forces in the universe. At very short distances, the weak force and the strong force are responsible for the interactions between elementary particles in nature.

The gravitational force acts between masses of objects and the electromagnetic force acts between electric charges. These fundamental interactive forces provide the basis for all other forces to exist.

Examples and manifestation of these fundamental force fields are the force of electromagnetism acting between two surfaces that manifests itself as the friction force, the force of a compressed spring, and the force of the exclusion principle for atoms that does not allow an atom to pass through another, as well as the centrifugal force of acceleration of rotating objects.

Forces, such as the gravitational force and the electromagnetic force, can exert themselves in free space. (Taylor et al, 1966)

§ 2. Nomenclature and definition of six-dimensional forces

The fundamental elements of a force are mass, space and time. Let us define a force with two subscripts for the elements of time and space as follows

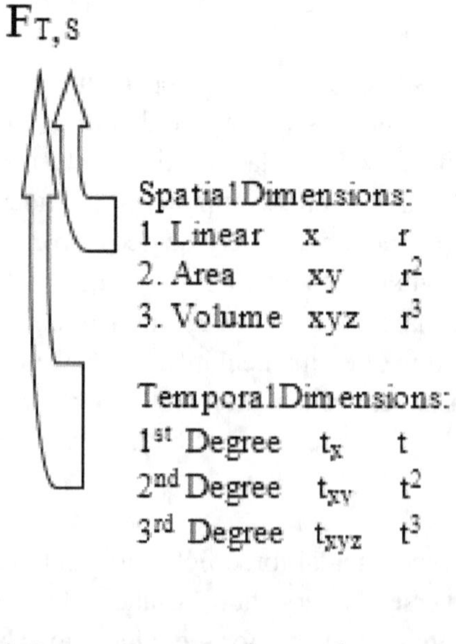

Figure 1. *Representation of a six-dimensional force*

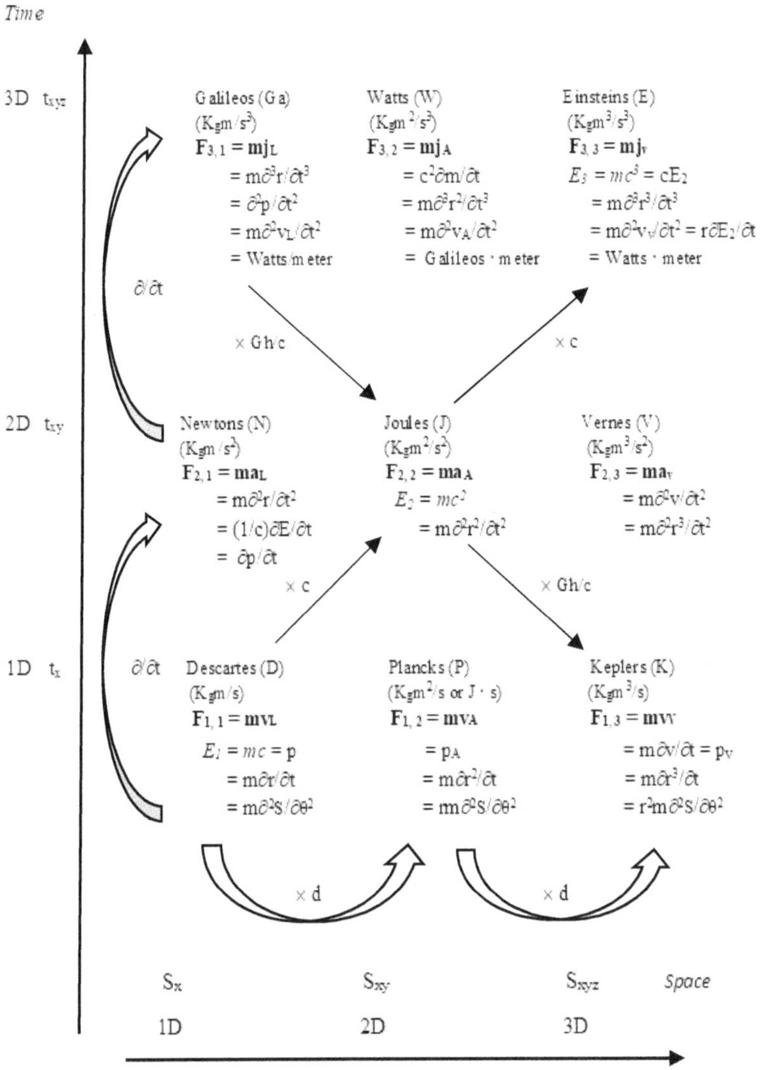

***Figure 2.** The six-dimensional force matrix*

2.1. Defining and deriving six-dimensional forces

Let us define the derivatives of six-dimensional forces such that

$$\text{v} = \text{velocity} = \frac{\partial r}{\partial t} \tag{2.1}$$

$$a = \text{acceleration} = \frac{\partial^2 r}{\partial t^2} \tag{2.2}$$

$$\bar{j} = \text{jolt} = \frac{\partial^3 r}{\partial t^3} \tag{2.3}$$

where a Jolt is defined as the time rate of change of the acceleration for a distance r, an area r^2, or a volume r^3, such that Jolts are found to be

$$\overline{j_L} = \text{linear jolt} = \frac{\partial^3 r}{\partial t^3} \tag{2.4}$$

$$\overline{j_A} = \text{area jolt} = \frac{\partial^3 r^2}{\partial t^3} \tag{2.5}$$

$$\overline{j_V} = \text{volume jolt} = \frac{\partial^3 r^3}{\partial t^3} \tag{2.6}$$

§ 3. On the nine forces of six-dimensional space-time and mass

Let us now describe the forces, units of measurement and their dimensions:

3.1. Force of Descartes (or Force of Linear Momentum)

$$F_{1,1} = mv = p \tag{3.1}$$

$$\text{Descartes} \left(\frac{K_g m}{s} \right) \tag{3.2}$$

$$E_1 = mc \tag{3.3}$$

$$= m\frac{\partial r}{\partial t} = m\frac{\partial^2 S}{\partial \theta^2} \tag{3.4}$$

$$1 \text{ Descartes} = \frac{1 \; Joule}{c} \tag{3.5}$$

$F_{1,1}$ is manifested as linear energy E_1 per unit of velocity in the form of linear momentum p. It is an illustration of how forces in six-dimensional space-time-mass may be conceptualized as being a manifestation of energy and vice versa. Momentum is usually defined as mass times velocity. Here, and more generally, it is a quantity like energy but related to motion through space.

Let us derive the acceleration above from the equation of a sector

$$S = r\theta \tag{3.6}$$

$$\frac{\partial S}{\partial \theta} = r \tag{3.7}$$

$$\frac{\partial r}{\partial \theta} = \frac{\partial^2 S}{\partial \theta^2} \tag{3.8}$$

Thus, we find that if $\partial\theta = \partial t$, then

$$\frac{\partial r}{\partial t} = \frac{\partial^2 S}{\partial \theta^2} \tag{3.9}$$

3.2. Force of Newton

$$F_{2,1} = ma \tag{3.10}$$

$$= m\frac{\partial^2 r}{\partial t^2} = m\frac{\partial p}{\partial t} \tag{3.11}$$

$$\text{Newtons} \left(\frac{K_g m}{s^2} \right) \tag{3.12}$$

$F_{2,1}$ is the legendary force of Newton's second law. It is the force of the falling apple as it hits the ground.

3.3. Force of Galileo

$$F_{3,1} = m\overline{j_L} \tag{3.13}$$

$$= m\frac{\partial^3 r}{\partial t^3} = m\frac{\partial^2 v_L}{\partial t^2} \tag{3.14}$$

$$\text{Galileos} \left(\frac{K_g m}{s^3}\right) \tag{3.15}$$

$$1 \text{ Galileo} = 1 \; \frac{Watt}{meter} \tag{3.16}$$

$F_{3,1}$ is the force of power per unit of distance or watts/meter. The Galilean force equals mass times the linear jolt of the object of mass "*m*", and the time rate of action of the Newtonian force.

3.4. Force of Planck

$$F_{1,2} = mv_A \tag{3.17}$$

$$= p_A = m\frac{\partial r^2}{\partial t} = rm\frac{\partial^2 S}{\partial \theta^2} \tag{3.18}$$

$$\text{Plancks} \left(\frac{K_g m^2}{s}\right) \tag{3.19}$$

$$1 \text{ Planck} = 6.626070954 \times 10^{-34} \; \frac{k_g m^2}{s} \tag{3.20}$$

$F_{1,2}$ is the force of the quantum of action in quantum mechanics. The instantaneous angular momentum (a bivector) of a particle traveling in a helical trajectory through space-time.

3.5. Force of Joules

$$F_{2,2} = ma_A \tag{3.21}$$

$$= m\frac{\partial^2 r^2}{\partial t^2} = m\frac{\partial v_A}{\partial t} \tag{3.22}$$

$$E_2 = mc^2 \tag{3.23}$$

$$\text{Joules} \left(\frac{K_g m^2}{s^2}\right) \tag{3.24}$$

$F_{2,2}$ is the force of the world-famous equation of energy, $E = mc^2$, from The well-known physicist Albert Einstein. A specific fission reaction that conforms to this space-time geometry would release E_2, a well-known vast amount of energy. Energy is related to motion through time.

3.6. Force of Watts

$$F_{3,2} = m\overline{j_A} \tag{3.25}$$

$$= m\frac{\partial^2 v_A}{\partial t^2} = m\frac{\partial^3 r^2}{\partial t^3} = c^2\frac{\partial m}{\partial t} \tag{3.26}$$

$$\frac{\partial m}{\partial t} = \left(\frac{1}{c^2}\right)\frac{\partial E}{\partial t} \tag{3.27}$$

$$\text{Watts} \left(\frac{K_g m^2}{s^3}\right) \tag{3.28}$$

$F_{3,2}$ is the force of power of a mechanical engine or the power dissipated by current passing through a resistive load. The watts force equals the mass times the area jolt of the charges.

3.7. Force of Kepler

$$F_{1,3} = mv_v \tag{3.29}$$

$$= p_v = m\frac{\partial r^3}{\partial t} = r^2 m\frac{\partial^2 S}{\partial \theta^2} \tag{3.30}$$

$$\text{Keplers} \left(\frac{K_g m^3}{s}\right) \tag{3.31}$$

$$1 \text{ Kepler} = 1 \text{ Verne} \times 1 \text{ second} \tag{3.32}$$

$F_{1,3}$ is the force of the velocity of an expanding volume along one spatial dimension with respect to linear time, as in the pneumatic force of a volume of air, during a positive displacement, pushing on a piston within a rigid cylinder. It is the force of the momentum of a changing volume.

3.8. Force of Vernes

$$F_{2,3} = ma_v \tag{3.33}$$

$$= m\frac{\partial^2 r^3}{\partial t^2} = Gm^2 \tag{3.34}$$

$$\text{Vernes} \left(\frac{K_g m^3}{s^2}\right) \tag{3.35}$$

$$G = \frac{F_{2,3}}{(mass)^2} = 3.35673536 \times 10^{14} \; Vernes \tag{3.36}$$

$$1 \text{ Verne} = \text{hc} = 1.987821286 \times 10^{-25} \; \frac{K_g m^3}{s^2} \tag{3.37}$$

$F_{2,3}$ is the force of the acceleration of a changing volume with respect to a temporal plane. The force of Vernes acts in the gravitational constant G for an object of mass "m".

3.9. Force of Einstein

$$F_{3,3} = m\overline{j_V} \tag{3.38}$$

$$= m\frac{\partial^2 v_V}{\partial t^2} = m\frac{\partial^3 r^3}{\partial t^3} \tag{3.39}$$

$$E_3 = mc^3 \tag{3.40}$$

$$\text{Einsteins} \left(\frac{K_g m^3}{s^3} \right) \tag{3.41}$$

$$1 \text{ Einstein} = 1 \text{ Joule} \times \text{c} \tag{3.42}$$

$F_{3,3}$ is the force of the energy of a star going supernova, the mass of the object times the jolt of the volume of the object. or the force of 6-D phonons, a six-dimensional oscillating atomic lattice structure during heat transfer to surrounding atomic structures. The energy released by $E_3 = mc^3$ is three hundred million times greater than the energy that would be released by the equation $E_2 = mc^2$ in a sustained nuclear chain reaction.

§ 4. On the fundamental principle of energy-mass equivalence

Let us consider the property of mass of an object; we can acknowledge that even rest mass translates at some speed through space-time, as a system, or as part of a much larger moving system in the universe. Moreover, rest mass has some of its particles traveling or spinning relativistically within the system of mass itself. Thus, one may argue that rest mass is relativistic in nature. The principle of relativistic mass asserts that the masses of particles or objects in a dynamic motional universe are relativistic in nature, or relativistic as a result of relativistic translation as measured at a universal or local scale.

Let us now consider an energy-mass equivalence formulation for each expression of energy above in six-dimensional space-time, and a fundamental theory of energy.

4.1. Cartesian Energy

$$E_1 = \frac{mc^2}{\sqrt[2]{(c^2 - v^2)}} \tag{4.1}$$

4.2. Einsteinian Energy

$$E_2 = \frac{mc^3}{\sqrt[2]{(c^2 - v^2)}} \tag{4.2}$$

4.3. Hawking-Feynman Energy

$$E_3 = \frac{mc^4}{\sqrt[2]{(c^2 - v^2)}} \tag{4.3}$$

4.4. Energy general form

$$E_n = \frac{mc^{n+1}}{\sqrt[2]{c^{2[(n+1)-n]} - v^{2[(n+1)-n]}}} = \frac{mc^{n+1}}{\sqrt[2]{c^2 - v^2}} \tag{4.4}$$

Therefore, for an object of mass m moving very slowly with respect to the speed of light the equation of E_n simplifies to an equation of energy-mass equivalence of the more familiar form, $E_n = mc^n$, where n is the number of pairs of spatiotemporal dimensions. Every spatial dimension has a conjugate temporal dimension.

4.5. The explosion of a supernova

A supernova is a titanic explosion of a massive star that may shine with the brightness of ten billion suns with an approximate estimated energy of 10^{44} Joules, or 10^{51} ergs, which represents the total output of our sun during its entire lifetime. An erg is a ten-millionth of a

Joule. Supernovae produce heavy elements, occur in stars with at least 8 solar masses, and on average occur once every fifty years in our galaxy. A FOE unit, an acronym for 'ten to the Fifty-One Ergs', is 200 times the mass of the Earth converted completely into energy. The entire energy output of our Sun over its entire lifetime is estimated to be 1.2 FOE.

It is presently believed that about 1% of the energy of a Type I or a Type II supernova is visible as light, 99% of the energy is the kinetic energy of the visible expanding gas cloud of the explosion, that is calculated from the estimated velocity and mass. Not visible but detectable neutrino pulses are emitted that may briefly reach four percent of the visible energy output of the entire observable universe.

Where does the energy come from that causes this titanic explosion?

Currently, it is believed that the energy released in a supernova explosion comes from the gravitational energy released as the star ceases to have enough fusion to support itself and collapses. The iron core of a star may collapse to transform its gravitational energy to heat and motion until it forms a neutron star that sustains its own gravitational potential through fermion pressure, which precedes a rebounding explosion of its collapsing outer layers.

It is theorized that an enormous amount of gravitational potential energy is released during the collapse of the iron core of a star as the iron cannot release energy by fusion because fusion requires a larger input of energy than it releases. The iron core continues to be subjected to the spatiotemporal pressure that exerts a gravitational field, which pushes the electrons closer to the nuclei beyond the allowable quantum limit, causing electrons and protons to combine to form neutrons and neutrinos in the process. As the gravitational collapse continues, an enormous amount of energy is released that blows the outer layers of the star off into interstellar space-time in a titanic explosion. The core collapses to become a tremendously dense neutron star, if the core is massive enough that the gravitational field continues to collapse the core, the remaining core may become a black hole.

The released gravitational energy of the star transforms into heat and motion, but it is not part of the explosion. The loosely bound outer surface of hydrogen and heavier elements, is involved in the explosion. So, where is the rest of the energy of 10^{44} Joules coming from? The answer that has been currently proposed is that the neutral and massless neutrinos, that react weakly with matter, are ejected transferring the energy of gravitational collapse to give about 1% of that energy to the loosely bound elements of the surface to initiate the titanic explosion. However, further studies into white dwarfs indicate that the characteristics of the explosion might be more thermonuclear in nature, not principally or exclusively from the gravitational collapse of the core.

Let us calculate the Hawking-Feynman Energy for a star that is going supernova.

$$E_3 = \frac{mc^4}{\sqrt[2]{\left(c^2 - v^2\right)}} \tag{4.5}$$

If the velocity of the star is a lot slower than the speed of light, $v \ll c$, we have

$$E_3 = mc^3 \tag{4.6}$$

The energy released by $E_3 = mc^3$ is three hundred million times greater than the energy that would be released by the equation $E_2 = mc^2$ in a sustained thermonuclear reaction. The energy of a star going supernova, E_3, is equivalent to the mass of the object times the jolt of the volume of the object.

Let us estimate one FOE in ergs for a mass that is 200 times the mass of the earth,

$$E_3 \approx \left(200 \times 5.972 \times 10^{27} g\right)\left(3 \times 10^{10}\ cm/s\right)^3\ Ergs \cdot c \tag{4.7}$$

$$E_3 \approx 32248.8 \times 10^{57} \approx 3.22488 \times 10^{61}\ Clausius \tag{4.8}$$

Using the Einsteinian equation, we have

$$E_2 \approx \left(200 \times 5.972 \times 10^{27} g\right)\left(3 \times 10^{10}\, cm/s\right)^2 \; Ergs \qquad (4.9)$$

$$E_2 \approx 10749.6 \times 10^{47} \approx 1.07496 \times 10^{51} \; Ergs \qquad (4.10)$$

Let us estimate the energy of a star that is going supernova that has a mass equal to eight solar masses,

$$E_3 \approx \left(8 \times 1.989 \times 10^{30} Kg\right)\left(3 \times 10^{8}\, m/s\right)^3 \; Joules \cdot c \qquad (4.11)$$

$$E_3 \approx 429.624 \times 10^{54} \approx 4.29624 \times 10^{56} \, Einsteins \qquad (4.12)$$

Using the Einsteinian equation, we have

$$E_2 = mc^2 \approx \left(8 \times 1.989 \times 10^{30} Kg\right)\left(3 \times 10^{8}\, m/s\right)^2 \; Joules \qquad (4.13)$$

$$E_2 \approx 143.208 \times 10^{46} \approx 1.43208 \times 10^{48} \; Joules \qquad (4.14)$$

It is interesting to note that the energy calculated using the Einsteinian equation, $E_2 = mc^2$, in a sustained thermonuclear reaction would not produce the necessary energy for the entire process of the smallest possible type of supernova that has been theorized. Moreover, it is estimated that a neutron star has a radius on the order of 10 to 20 Kilometers and a mass between 1.1 and perhaps 3 solar masses. If the visible energy of a supernova has been estimated to be 10^{44} Joules, which is a fraction of the total energy, then it is necessary to estimate the energy of the entire process with the Hawking-Feynman energy equation and a precise quantity of mass for the star.

It is possible to theorize that in order to estimate the total quantity of Einsteinian energy involved in a Supernova process it may be necessary to account for the observable energy of the mass of the exploding outer layers in addition to the energy of the mass of the remaining neutron star. In the ideal case, the particles of the outer layers deform elastically during their collision with the core. Let us calculate the total Einsteinian energy of a star going supernova that has a mass equal to eight solar masses, and a core equal to 1.1 solar masses, as theorized above.

$$E_3 = \left(m_{core} + m_{outer-layers}\right)c^3 \tag{4.15}$$

$$E_3 \approx m_{core}c^3 \approx 1.1 \times 1.989 \times 10^{30}\, Kg \times \left(3 \times 10^8\, m/s\right)^3 \approx 5.90733 \times 10^{55}\ Einsteins \tag{4.16}$$

Let us assume the magnitude of the Einsteinian energy from the outer layers is equal to the magnitude of the observable energy 10^{44} Joules for E_2 times *c*, equal to 10^{52} Einsteins.

$$E_3 \approx m_{outer-layers}c^3 \approx 1.0 \times 10^{52}\ Joules \cdot c \tag{4.17}$$

The total estimated Einsteinian energy of a star going supernova is

$$E_3 \approx 5.90733 \times 10^{55} + 1.0 \times 10^{52} \approx 5.90833 \times 10^{55}\ Einsteins \tag{4.18}$$

The Einsteinian energy of the outer layers represents 0.0169253% of the total estimated Einsteinian energy of the star. The core represents approximately the remaining 99.98% of the total estimated Einsteinian energy of the star.

4.6. The Fundamental Theory of Energy

I. Law of Symmetry of Energy.

Space, time, and phase are homogeneous forms of energy. Phase is the position of a point, during an instant of time, on a wave function. Space and time are two aspects of the same continuous quality, or stable state, of energy. The conservation of momenta of isolated systems, the conservation of charge, or the conservation of energy, underlie the continuous differentiable symmetry of space, time, and phase.

II. Law of Conservation of Energy.

Energy is neither created or destroyed. Energy is transferred, transformed, or transmuted. The conservation of energy underlies temporal symmetry.

III. Law of Stability of Energy.

Unstable states of energy beyond a threshold will settle into stable states of energy. The stable states of energy will prevail. Space-time can sustain an energy density up to a threshold. If a manifestation of energy, such as a particle, transfers energy beyond a threshold upon a locality in space-time, the unstable state of energy at the locality will manifest stable states of energy of various forms.

IV. Law of Transmutation of Energy.

Any forms or properties of energy transmute into any other forms or properties of energy in accordance to the laws of nature. High energy colliding photons can transmute into quarks, leptons, force carriers, and other photons. Quarks and gluons can transmute into other quarks and gluons. When an electron and a positron collide in an annihilation event, gamma ray photons are produced.

V. Law of Conformality of Energy.

Nature is conformal to itself. Laws of nature are scalable in the medium of reality. Space-time expands at every point unless obstructed. The expansion of space is the passage of time. The motion and projection of space is conformal to itself according to the laws of nature. The laws of nature are omnipresent in our universe.

VI. Law of Continuity of Energy.

Energy is continuous and existential. The forms of realizable manifestations of energy of space-time, and their physical processes according to the laws of nature, may be transitory and discrete, but space-time exists continuously. All transitory manifestations of energy in our universe, the reality of our existence, may be described in terms of creation, initial conditions of creation, and probability, according to the law of transmutation of energy, and other applicable laws of nature. Probability implies an unknown number of indeterministic interactions, outcomes of chance, of all there is. All the energy required, to continue all probabilistic interactions, free will, and physical processes, is already there.

VII. Law of Omnipotence of Energy.

Energy is as unlimited as space-time and exists everywhere. The open set of all state functions of manifestations of energy of space-time are endless.

§ 5. Zero Point Energy of a Quantum Mechanical Oscillator in Space-time

The concept of zero-point energy, as developed by the eminent physicist Max Planck in 1911, also called quantum vacuum zero-point energy, is the lowest possible energy, the energy of the ground state, of a quantum mechanical physical system in space-time. The electromagnetic ground state of a single oscillator in space-time contains fluctuations and an associated zero-point energy, if not Heisenberg's uncertainty principle would be violated. Hence, the states of quanta in space-time and their associated electromagnetic fields are fluctuating values.

Let us describe the average energy of a single quantum mechanical oscillator in homogeneous and isotropic space-time using Planck's equation:

$$E = \frac{h\upsilon}{2} + \frac{h\upsilon}{e^{\frac{h\upsilon}{kT}} - 1} \tag{5.1}$$

Where h is Planck's constant, υ is the frequency, k is Boltzmann's constant, and T is the absolute temperature. Zero-point energy refers to random quantum fluctuations of the electromagnetic and other force fields that are present everywhere in space-time wave medium. Thus, space-time is not a true vacuum, it is a plenum, and it is actually a quantum sea of energy and waves. This energy is present even at absolute temperatures near zero, and even in the absence of matter. These zero-point fluctuations may have subtle and gross effects, on the behavior of fundamental particles and on every scale of our reality.

Solving for the energy of zero-point energy of the single quantum mechanical oscillator we have

$$\frac{h\upsilon}{2} = E - \frac{h\upsilon}{e^{\frac{h\upsilon}{kT}} - 1} = E\left[\frac{e^{\frac{h\upsilon}{kT}} - 1}{e^{\frac{h\upsilon}{kT}} + 1}\right] = -iE \cdot \tan\frac{ih\upsilon}{2kT} = E \cdot \tanh\frac{h\upsilon}{2kT} \qquad (5.2)$$

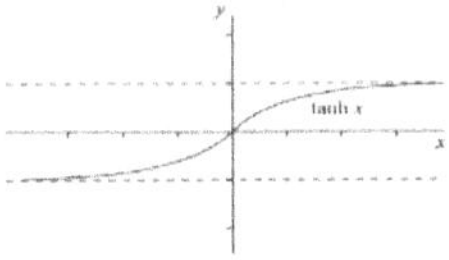

Figure 3

In terms of the hyperbolic functions of the cosh and sinh, we may visualize the graphical relation of both the condition of the Heisenberg's uncertainty principle on the zero-point energy of a single quantum mechanical oscillator, or radiator, and the potential positions of fluctuation in its average energy.

$$\frac{h\upsilon}{2}\cosh\frac{h\upsilon}{2kT} = E \cdot \sinh\frac{h\upsilon}{2kT} \qquad (5.3)$$

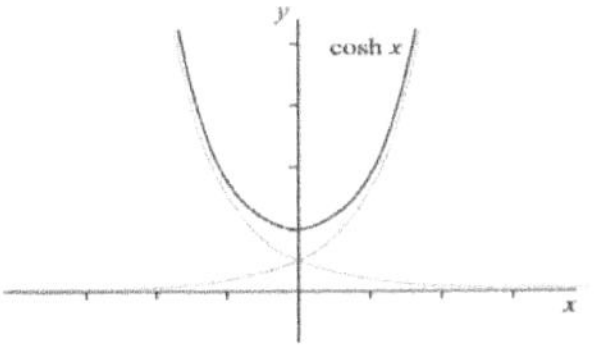

Figure 4

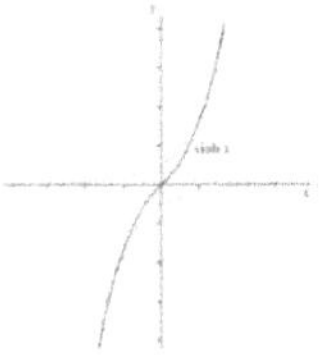

Figure 5

Let us now consider a single quantum mechanical oscillator in homogeneous and isotropic space-time represented by a pendulum of radius *kT* that swings through an angle 2θ with a total average energy $h\upsilon$ as shown below. As pendulum swings half way, the average energy is $h\upsilon/2$ and the angle is θ.

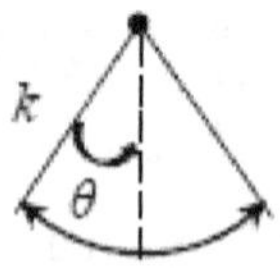

Figure 6

Therefore, we can represent the energy of the sector for half of the arc as follows:

$$\frac{h\upsilon}{2} = \theta \cdot kT \qquad (5.4)$$

$$\theta = \frac{h\upsilon}{2kT} \qquad (5.5)$$

If the angle θ equals 90 degrees or $\pi/2$ radians, we obtain from the above equations:

$$\frac{h\upsilon}{2} = E\left[\frac{e^{\frac{h\upsilon}{kT}} - 1}{e^{\frac{h\upsilon}{kT}} + 1}\right] = E\left[\frac{e^{\pi} - 1}{e^{\pi} + 1}\right] \qquad (5.6)$$

$$\frac{h\upsilon}{2} \cosh\frac{\pi}{2} = E \cdot \sinh\frac{\pi}{2} \qquad (5.7)$$

$$\frac{h\upsilon}{2} = E \cdot \tanh\frac{\pi}{2} = E \qquad (5.8)$$

$$h\upsilon = 2E \qquad (5.9)$$

Where the exponential factor e^{π} is Gelfond's constant, a transcendental number, and the magnitude of the spatiotemporal exponential growth factor at θ equal to 90 degrees when the average energy of a single quantum mechanical oscillator in homogeneous and isotropic space-time varies as a function of the expansion of space-time at every point in the wave medium.

Moreover, the average energy of a single quantum mechanical oscillator in homogeneous and isotropic space-time and the concept of zero-point energy of a single quantum mechanical oscillator are closely related to the Heisenberg's uncertainty principle which is derivable from Planck's equation during one oscillating cycle of 2π radians.

$$E \geq \frac{h\upsilon/2}{2\pi} \tag{5.10}$$

$$\Delta E \geq \frac{h\Delta\upsilon}{4\pi} \tag{5.11}$$

$$\frac{\Delta x \Delta p}{\Delta t} \geq \frac{h}{4\pi\Delta t} \tag{5.12}$$

$$\Delta x \Delta p \geq \frac{h}{4\pi} \tag{5.13}$$

Furthermore, let us consider the energy associated with a single photon oscillator and the frequency υ of its associated electromagnetic wave in homogeneous and isotropic space-time. The energy of a photon oscillator with frequency υ is given by

$$E = h\upsilon \tag{5.14}$$

The average energy of a single photon oscillator in homogeneous and isotropic space-time is given by

$$E = \frac{h\upsilon}{2} + \frac{h\upsilon}{e^{\frac{h\upsilon}{kT}} - 1} \tag{5.15}$$

Thus, half the average energy of a single photon oscillator involves zero-point energy, while the other half, or the rest of the energy, involves the second term of Planck's equation.

Let us consider the ratio of the average energy of a single photon oscillator to its zero-point energy in homogeneous and isotropic space-time.

$$\frac{E}{h\upsilon/2} = \frac{1}{\tanh h\upsilon/2kT} = \frac{e^{h\upsilon/kT}+1}{e^{h\upsilon/kT}-1} \tag{5.16}$$

The single photon oscillator in homogeneous and isotropic space-time may be regarded from the perspective of an observer as in perpetual universal motion such as atoms, molecules, etc. The photonic oscillator obeys the law of conservation of energy and the first law of thermodynamics as an open system, and at the same time it is able to produce an overunity energy effect as a single photon oscillator for small values of the angle θ greater than zero.

$$\frac{e^{h\upsilon/kT}+1}{e^{h\upsilon/kT}-1} > 1 \tag{5.17}$$

Hence, the overunity energy effect of a single photonic oscillator is inversely proportional to the absolute temperature T and directly proportional to the frequency υ. Overunity in this context is a term that indicates that the ratio of initial energy of a single photonic oscillator to the final energy is greater than one as a function of frequency and/or absolute temperature of the medium. Therefore, our universe, where everything is in motion, is no stranger to overunity or perpetual motion, which in the natural world are not impossible. The overunity energy effect has major theoretical implications. Space-time is characterized by a zero-point energy which has the property that its associated pressure is negative and thus makes space-time expand. This expansion of every point in space-time suggests that the cosmological constant and the zero-point energy density of space-time may be one and the same. The amount of zero-point energy in space-time is enormous. There seems to be kinetic energy available at every point in space-time that may be tapped with technology that will power all kinds of devices.

The frequency of the photonic oscillator is given by

$$\upsilon = \frac{2E}{h}\tanh\theta \tag{5.18}$$

Let us denote the angle of the hyperbolic tangent of a single photonic oscillator in isotropic and homogeneous space-time in terms of its relativistic mass to obtain

$$\theta = \frac{h\upsilon}{2kT} = \frac{m_{REL}c^2}{2kT} \tag{5.19}$$

Thus, for any angle θ greater than approximately 11 degrees, we have

$$\upsilon = \frac{2E}{h} \tag{5.20}$$

For any angle θ greater than approximately 11 degrees, ceteris paribus, the ratio of the average energy of a single photon oscillator in homogeneous and isotropic space-time to its zero-point energy is unity. However, for any angle θ equal or smaller than approximately 11 degrees, ceteris paribus, the average energy-to-zero-point energy ratio is over unity. Thus, for lower frequencies of the single photonic oscillator, or higher absolute temperatures, ceteris paribus, the photonic oscillator would obtain the overunity effect.

Let us imagine a theoretical resonating device that is able to receive electromagnetic waves at the frequency υ of hypothetical atomic oscillators at a locality in space-time such that it would connect the vast zero-point energy, the very wheelwork of nature, to existing and future machinery. Thus, the machinery may be driven by a power obtainable at any point in space-time. Accordingly, with the right technology, a directed signal with an appropriate amplitude and a frequency υ might be able to transfer electromagnetic energy to a hypothetical atomic oscillator to induce a lower frequency at higher temperature, or a higher frequency at lower temperature, to obtain the overunity effect at a locality in space-time. The theoretical electromagnetic wave interference, between the signal and the hypothetical atomic oscillator, might produce gravitational effects and perturbations in the local space-time wave medium. Any physical field of an object of mass that changes the energy or waves of space-time such as a dielectric material, an electrical conductor, and a gravitational field, distorts the quantum mechanical state of the space-time wave medium. These changes have already been measured by researchers. The questions that remains is if the zero-point energy can be manipulated, can it be extracted and how?

§ 6. Conclusion

We have seen that some of the forces described are well known and accepted such as the Newton, Watt, and Joule, and that their applications remain unchanged. On the six-dimensional force matrix, as we move up to higher temporal dimensions we differentiate with respect to linear time and as we move to the right of the matrix, we multiply the variable by a linear spatial dimension; nonetheless, each force can still be represented by the mass of the object times an acceleration as in Newton's second law. This implies that energy, motion, momentum, power and other physical notions can be represented by the concept of a force to determine the net result of interactions between such concepts. Energy (E_3) as a six-dimensional property has constructive applications and benefits.

Chapter 5

On relativistic mass, length and time within a Kerr-Newman supermassive black hole

§ 1. Introduction: the attribute of mass

Mass appears to be a characteristic of a body on a macroscale, independent of its environment. Isaac Newton demonstrated that mass determines how much acceleration results when a body is accelerated. Mass is considered to be the charge associated with the force of gravity. The mass of a body resides in its atoms, mostly in its nucleus, which is made of protons and neutrons, which in turn are made of quarks, and other infinitesimal particles. Between these particles or building blocks of matter, there is plenty of space-time residing in the in-between dimensions, for mass at an infinitesimal scale is discrete.

The physicist Albert Einstein described mass as energy without overall motion. Mass is considered the imbalance between energy and momentum, $m_p^{\,2}c^4 = E_p^{\,2} - p^2c^2$. In other words, particles may be thought as having mass because they are throbbing with stored energy even when they are not translating from place to place. (Einstein, 1952) In quantum physics, the Higgs Field is an energy field that exists everywhere in the universe. The field is accompanied by a fundamental particle called the Higgs Boson, which the field uses to continuously interact with other particles, such as quarks. As particles pass through the field they are given mass; luxons passing through the Higgs Field will become slower moving tardyons. If the Higgs field did not exist, luxons would not have the mass required to attract one another and would float around freely at the speed of light. Albert Einstein's principle of special relativity states that luxons, particles without mass, travel at the speed of light, and those with mass, tardyons, travel slower than light speed. Thus, an electromagnetic object, such as a photon, without mass, that moves through space-time relativistically, even at less than light speed, would, for all practical proposes, have mass; as space-time acts as a medium separate from matter. (Larmor, 1900) Hence, if there is an imbalance between a single object's energy and its momentum, even when acting alone, the object would have the attribute of mass. A photon traveling at the speed of light has relativistic mass, but when that photon is at rest, it does not have the attribute of mass. The photon is massless at rest. Thus, *relativistic mass is the property or charge given to a quantum of energy by the curvature of space-time about the physical dimensions or geometry of its physical state*. The higher the mass or its density, the higher the comprehensive effect of curvature of space-time and the more intense the charge of the mass. *Thus, if a physical manifestation of energy is in motion and it compresses space on its trajectory, a similar effect of space-time curvature results and that moving energy can be said to have an attribute of mass.*

Moreover, space-time is the medium and the speed barrier to light since light cannot travel faster than its medium. However, if space is extending faster than light speed, it is entirely possible that an object moving in superluminally extending space may be able to travel as fast as the space medium, in which case the moving object could be traveling faster than our measured speed of light c in a subluminal or luminal space-time frame of reference. If the moving object converts some of its relativistic mass into momentum at the light barrier, its speed will increase in the superluminally extending space. Mass and time dilate in the direction of motion. (Lorentz, 1909, 1920) Thus, as the relativistic mass of the object condenses, the imbalance between its energy and momentum will decrease and the speed of the relativistic mass will increase.

$$m = \frac{E}{c\sqrt{c^2 + v^2}} \tag{1.1}$$

Furthermore, it has been theorized that if we consider frame dragging in a region close to a rotating Kerr-Newman black hole, such as the ergosphere, and send an object at close to the speed of light in an orbit around the ergosphere of the black hole, the object may travel faster than the speed of light with respect to the rest of the universe as observed from a region away from the black hole where the effect of frame dragging is negligible. Moreover, an object can move faster than light relative to another object in homogeneous and isotropic space away from the black hole, as a result of space extension, or contraction, in the region of the black hole where space-time creates the movement.

Other examples for faster than light travel are the hypothetical elementary particles that are called tachyons, imaginary mass particles that have the property to speed up, or slow down, as they lose mass, or energy, the value of their imaginary mass is defined by the rate of loss at which this occurs, or the phase velocity of an electromagnetic wave, when traveling through a medium, can routinely exceed *c*, the free space-time velocity of light. Researchers have suggested that if particles were created initially with faster-than-light speed in particle collisions, no acceleration or infinite energy would be necessary.

$$\left(i\frac{\partial m}{\partial t}\right)^2 = -\left(\frac{\partial m}{\partial t}\right)^2 = -\left(\frac{\partial\left(\frac{E}{c\sqrt{c^2+v^2}}\right)}{\partial t}\right)^2 \tag{1.2}$$

1.1. Sound-in-light travels through space: the music of the stars.

If audible sound to humans cannot travel in the vacuum of space-time due to the lack of enough contiguous particles, or molecules, as in a gas or solid matter, can sound travel through an electromagnetic wave, such as light, considering that light is both a wave and a particle?

Light acts as a wave and a particle in the sea of the Big Bang or cosmic photons. Photons have relativistic mass as they travel at *"c"* after emission from a source of light. Are photons affected by reverberations in the spatiotemporal wave medium? Would the vibrations from a secondary source of sound cause modulation in the spatiotemporal wave medium of light?

Sound is a means of energy transfer through matter, while light is waves, and particles, of pure energy. Thus, sound and light may interact through the spatiotemporal wave medium. The field of acousto-optics relies on this principle to study the interactions between sound waves and light waves. The luminosonic effect may have occurred in the early life of the universe, with x-rays as electromagnetic waves and infrasonic waves as sound. Quantum electrodynamic research has provided an expression for the interaction between cosmic photons, the study of the cosmic photon gas, and the speed of sound in the sea of cosmic photons. (Partovi, 1994)

$$v_{sound} = \left[1 - \frac{88\pi^2\alpha^2}{2025}\left(\frac{T}{T_e}\right)^4\right]\frac{c}{\sqrt[2]{3}} \qquad (1.3)$$

where α is the fine structure constant, "c" is the speed of light in vacuum, and T is the absolute temperature of the photon gas, and T_e is the absolute temperature of the electron mass, $T_e \approx 5.9 \times 10^9$ Kelvin.

An acoustic wave is a longitudinal pressure wave through a wave medium As the acoustic wave propagates through the wave medium, the wave medium experiences localized reverberations. It is possible that if a sound wave propagates through cosmic photon gas, there may be a similar effect.

A photon, also a boson, has no polarity because it has equal opposite spatiotemporal charges, but a photon has an electromagnetic field between its spatial (magnetic) or temporal (electric) poles. Bosons may have identical quantum states, so bosons may occupy very close spatiotemporal localities. The overall charge of a photon is neutral, so it does not have to obey the Pauli exclusion principle. Moreover, the interaction of two virtual photons in high-energy collision experiments is a common occurrence. A photonic wave can interact with another photonic wave constructively or destructively to produce a resultant wave through a wave medium. Because of previous research into the quantum nature of an electromagnetic wave, it is possible to speculate that a photon can decouple into a positive charge and a negative charge, and with its incoming energy above the threshold of the interaction, the photon may undergo monopole and pair production near a nucleus. All conserved quantum numbers (angular momentum, electric charge, lepton number) of the produced particles must sum to zero, so no monopoles remain decoupled. It is also possible that when a photon, with its incoming energy above the threshold of the interaction, is absorbed into an atom the existing electromagnetic forces within the atom, between the positive nucleus and negative electron(s), initiate monopole and pair production.

The absorption and increase of positive and negative charges in the atom from photons may be responsible for the orbital changes of electrons. Similarly, the emission and decrease of positive and negative charges in the atom may initiate photon production and emission. The probability of pair production in the interaction between a photon and an atom increases with the photon energy at approximately the square of the atomic number of the interacting atom. Weak scattering of a photon by another photon may exist, in the pure spatiotemporal vacuum, above some energy threshold of the center of the system of the two photons, where pair production may be initiated.

Can photonic electromagnetic decoupling occur in cosmic photon gas? If light decomposes into charges, e.g. through pair production in photon gas, in the spatiotemporal wave medium, compression and rarefaction waves may also travel through the spatiotemporal wave medium. In such case, the charges would obey the Pauli exclusion principle. Hence, in such possible scenario, a sound wave may travel through light.

Let us now consider the following technological possibility, a sound wave propagating through a particle wave medium onto a material membrane that is able to vibrate and emit photons and electromagnetic waves that propagate through the vacuum. The acousto-optic membrane of such technology emits radiation that encodes the sound wave into an electromagnetic wave that can be decoded by a suitable opto-acoustic receiver. The Pineal gland receives electromagnetic waves from the eardrum which encodes sound waves. If such a bio-acousto-optic mechanism, or similar technology, is combined with a spatiotemporal wormhole as a wave medium, the sound waves may propagate between two distant spatiotemporal localities in our universe at a speed faster than light. Phonons are produced from the vibration of the lattices of atoms or molecules, typically due to heat transfer. Phonons produce electromagnetic waves from modulating spatiotemporal waves through the oscillation, or perturbation, of the spatiotemporal wave medium.

Energy is conserved during the production of phonons $(\hbar\omega)$ and the emission of photons $(\hbar\nu)$. It is possible to envision a technology to perturb the spatiotemporal wave medium at an audible frequency band structure to emulate phonons as virtual sound quanta which emit photons that carry sound waves in the spatiotemporal medium at the speed of light. Such technologies may become the backbone of the intersolar, or intergalactic, infobahn networks, that may interconnect the planetary global internet systems of the future.

The duality of light describes how photons can behave as particles or waves. Let us conceptualize how photons may behave through entanglement between photons themselves, or between a photon, and the mechanism of measurement, that includes recording a quantum state, or an observed behavior or attribute, of the photon. Similarly, the consciousness of a measuring observer may also serve as a mechanism of measurement and recording of events (memory).

A photon can directly couple with another photon through pair production, e.g. leptons or quarks, to either of which the other photon can couple, without violating the uncertainty principle. For instance, the photon may couple directly to a quark inside the target photon, which can be intrinsically described by the photon structure function. An electromagnetic wave may consist of one or more photons of the same frequency that travel through the wave medium after emission from a source.

An electromagnetic wave and its photons may be contiguously emitted by their source as a coherent beam as long as the source is active. It is possible to visualize each wavelength, λ, of the electromagnetic wave as an individual photon representing a Planck quantum of energy at a specific frequency, $E = \hbar\nu$. The energy of the photonic wave consists of the intrinsic energy of its electric field and magnetic field propagated by the spatiotemporal wave. If the wave collapses into the corpus of a photon, the photon propagates as an individual particle, with all its electromagnetic energy condensed within its physical boundary. In such scenario, an entire electromagnetic beam reduces to distinct particles. The mechanism of measurement registers the existence of a photon splitting the photon in the process, causing the decoupling of the photon from the rest of the photons in the electromagnetic beam, as if the photon has reoriented through torsion to entangle with a copy of itself. As the measured photon decouples from the electromagnetic wave beam, all other photons in the electromagnetic wave beam follow suit. The electromagnetic wave of the beam transmutes from a transverse electromagnetic wave to a longitudinal electromagnetic particle wave.

As the electromagnetic wave of the beam transmutes, it is possible that the orbital orientation of the wavelength of the measured photons changes from transverse to orthogonal to the direction of propagation. The wavelength of the measured photon entangles and aligns with its recorded duplicate while still propagating toward its destination. The cross-sectional orbital area of the photon and its quantum of energy are conserved. From an orthogonal perspective to the direction of propagation, each distinct light quantum appears as a distinct particle of a longitudinal electromagnetic wave. Photons and their interactions are timeless, and photonic interactions are timelessly applicable from source to receiving end. Photons can interact timelessly, perceived as instantaneous, or occurring backwards in time, to changes in the mechanism of measurement and recording of temporal events.

Moreover, photons, as bosons, can overlap in the same exact quantum state. An abundance of photons can all exist at the same spatiotemporal locality, propagating in the same direction, with the same polarization, and at the same frequency, as coherent electromagnetic beams. The brightness of the light beam indicates the number of photons and the frequency between the source and the receiving end. If a light source emits a photon in free space, which duplicates itself into an electromagnetic wave in the direction of propagation, the photon becomes as a coherent electromagnetic beam.

Hence, a source, emitting one photon per second in free space, emits an electromagnetic wave that is propagated by the spatiotemporal expansion, like any other electromagnetic wave from any other similar source.

1.2. The Gluonic Interaction

A non-interacting virtual gluon, in the six-dimensional chromodynamic model, consists of one of three color-anticolor pairs of gluons according to the three particle probabilities. Each of the probable color-anticolor gluon pair in the non-interacting virtual gluon may be described as a gluonic dipole with a color field potential that forms the basis of the color singlet state of the electromagnetic force carrier or gauge boson in nature. Gauge bosons mediate the weak, strong, and electromagnetic forces. By the color charge convention, a color gluon represents a positive node and the conjugate anticolor gluon represents a conjugate negative node in the gluonic dipole. Color or anticolor may be real or imaginary, i.e. spatial or temporal.

If a force were to separate any of the three probable positive color gluons from any of the three-corresponding probable negative anticolor gluons that represent each of the three equal particle probabilities of a non-interacting virtual gluon, the result would be a gluonic dipole with a virtual color field potential that emerges as a virtual color field. Color states may combine out of the eight linearly independent interacting color states of virtual gluons, except for the ninth color state or non-interacting virtual gluon state, corresponding to the photon. No available combinations between any of the eight color states produces any other color state, or the color singlet state corresponding to the photon.

Interacting quarks are mediated by color gluons and anticolor gluons. Each interacting quark has either a resultant positive or negative electrical charge. A quark and an antiquark pair have opposite charges. When color charges (red, blue, and green) are combined there is a resultant color charge that is either positive or negative, except in the case of the neutral or non-interacting gluon. Let us suggest that each color charge, or anticolor charge, has the same electrical charge, but each color of gluon (red, blue, or green) may be a real or an imaginary color charge. A positive color charge is attracted to any negative anti-color charge, and vice versa.

The gluon and the photon are considered massless vector bosons with a spin of 1. The gluon has negative intrinsic parity. The intrinsic parity is a phase factor that arises as an eigenvalue of the parity operation, which is a reflection about the origin. The gluon has only two polarizations states. Gauge invariance requires the gluon polarization to be transverse. Let us denote the two polarization states of the gluon to be positive or negative. The gluon with color charge: red, blue or green, is either $\pm$ when the conjugate gluon with anticolor charge is $\mp$.

Thus, any single pair, e.g. $r\bar{r}$, or any double pair, e.g. $2b\bar{b}$, of color-anticolor of a real or imaginary gluon may have real or imaginary electrical charges according to the chromodynamic color charge to electrical charge correspondence principle.

Gluons	Red (r or ir)	Blue (b or ib)	Green (g or ig)	Antired ($\bar{r}$ or $i\bar{r}$)	Antiblue ($\bar{b}$ or $i\bar{b}$)	Antigreen ($\bar{g}$ or $i\bar{g}$)
Real Color	$\pm\frac{2}{3}e$	$\pm\frac{2}{3}e$	$\pm\frac{2}{3}e$	—	—	—
Imaginary Color	$\pm i\frac{2}{3}e$	$\pm i\frac{2}{3}e$	$\pm i\frac{2}{3}e$	—	—	—
Real Anticolor	—	—	—	$\mp\frac{e}{3}$	$\mp\frac{e}{3}$	$\mp\frac{e}{3}$
Imaginary Anticolor	—	—	—	$\mp i\frac{e}{3}$	$\mp i\frac{e}{3}$	$\mp i\frac{e}{3}$

Figure 1. The Chromodynamic Color Charge to Electrical Charge Correspondence of Gluons.

The helicity of a virtual gluon describes a combination of the spin and the instantaneous linear motion that is Lorentz invariant, i.e. the helicity has a value that is the same in all inertial reference frames. If the spin vector of a virtual gluon points in the same direction as the momentum vector, the helicity is positive (right-handed), if the spin and momentum vectors point in opposite directions, the helicity is negative (left-handed). The helicity of a massless virtual gluon, or virtual photon, is always equal to its quantum mechanical chirality. A virtual gluon is chiral if it is indistinguishable from its reflection in a plane mirror. The virtual gluon cannot be superposed onto its reflected image. Human hands are chiral. The combination of spin and instantaneous linear motion of the virtual gluon describes its chirality (laterality) or helicity (polarity). For the massless virtual gluon with a color-anticolor charge, the helicity (polarity) is the same as the chirality (laterality). Chirality or helicity of the massless virtual gluon are either positive (right-handed, counterclockwise) or negative (left-handed, clockwise) corresponding to the electrical charge polarity based on the magnitude of the elementary charge, *e*, carried by a single proton or electron.

By rotational convention, if one imagines a massless virtual round wall clock that is tossed like a frisbee to the right, with its spin vector defined by its hands and its dial facing up, such a clock would have negative chirality and helicity. Opposite color charges of a massless virtual gluon have opposite chirality and helicity. Opposite color charges attract, equal color charges repel. Opposite chirality and helicity complement, equal chirality and helicity separate. The relationship of the color and anticolor charge of a massless virtual gluon, or that of chirality and helicity between gluons, follows a yin-yang swirl principle, in a spatiotemporal realm of scales where electromagnetism emerges from chromodynamics. Chromodynamics is the basis of the strong, weak and magnetic forces.

Chirality and helicity designate the polarity of an electrical charge, or the polarity of a color charge, of a massless virtual gluon. Charge itself is neutral. Charge is spatiotemporal in nature. A Coulomb of charge consists of a unit of length times its orthogonal conjugate unit of time. Charge is an emerging two-dimensional spatiotemporal manifestation. The Planck quantum charge is applicable at the scale of a gluon. Any color charge, or anticolor charge, is a multiple of the Plank quantum charge, $t_p l_p$.

$$r \rightarrow \pm\frac{2}{3}e \rightarrow \pm\frac{2}{3}nt_p l_p \tag{1.4}$$

$$\overline{r} \rightarrow \mp\frac{e}{3} \rightarrow \mp n\frac{t_p l_p}{3} \tag{1.5}$$

$$n = \frac{e}{t_p l_p} \tag{1.6}$$

It is hypothesized that chirality produces a laterality, and spin produces spatiotemporal frame-dragging, about a gluon color charge that is opposite for positive or negative, providing attraction or repulsion. Opposite color charges provide complementary spatiotemporal frame-dragging, decreasing the distance between the charges. Like color charges provide repulsive spatiotemporal frame-dragging, increasing the distance between the charges. The laterality of chirality allows superposition of the color charges.

Color charges of gluons are three characteristics, or three quantum color charge numbers, that do not violate the Pauli exclusion principle. The Pauli exclusion principle of chromodynamics states that two or more identical gluon color charges cannot occupy the same color charge state within the boundary of a gluon simultaneously. It is worthy to hypothesize that there are four quantum color charge numbers of a gluon: polarity of electrical charge, laterality of chirality, polarity of spin, and axial orientation of spin or position of spin. The axial orientation of spin corresponds to any of the senses of the three spatiotemporal directions of the Cartesian coordinates, x, y, and z, where the north magnetic pole, that is centered and perpendicular to the two-dimensional Planck quantum charges, would point to while spinning. The magnetic quantum charge number differentiates the red, green, and blue color charges; consequently, the Pauli exclusion principle is not violated. Color and anticolor, of the same color, have opposite axial orientation of spin that are chiral and can be complementary. Color and anticolor, of different color, have perpendicular axial orientations that mate or mesh at an angle, like Bevel gears, and can be complementary. Color and different color, or anticolor and different anticolor, would be repulsive.

Color-anticolor charge pairs of gluons can be stacked together. The color charge stacks may separate, re-orient into other color-anticolor pairs, before they are absorbed by quarks. The separation and reorientation may occur due to magnetic forces from other nearby color or anticolor charges. A color-anticolor stack, of the same color, may separate and re-orient into other color-anticolor pairs, or it may continue as a stack, with an overall neutral electrical charge that represents a virtual photon, or color singlet state. At low energies, beams of photons crisscross each other through a cloudy or murky medium without the photons interacting or interfering.

Presently, the parameters that distinguish one photon from another are linear frequency, wavelength, color, which are dependent parameters, zero rest mass, polarization (the direction(s) in which the electric field and magnetic field oscillate), spatial length (the spatial extent of the wave packet containing the photon), and temporal length (the temporal extent it takes to pass a fixed point at a speed *"c"*), are independent parameters. Therefore, from this point of view, it is possible to assume all photons do not share equal quantum states of these parameters. Thus, it is possible that there are distinct classes of photons.

It is possible to hypothesize that a photon may be a superglued wave packet of non-interacting strong color-anticolor pairs, i.e. supergluons, of the same hue. A strong color-anticolor superpair may divide through chromomeiosis, into two weaker subpairs, where each subpair may join, through chromosynthesis, to an adjacent subpair from another superpair, to form another distinct color-anticolor superpair. All strong color-anticolor pairs may glue together by this superglue mechanism, into a superglued wave packet. The stability of the superglued wave packet of a photon may depend on the colorness energy and quantity of its gluons. The virtual gluonic mass of a superglue wave packet may be comparatively regarded as zero rest mass.

The superglue mechanism actions may generate oscillations and torsion in the superglued wave packet as superpairs divide into subpairs, the subpairs gravitate toward the center of colorness, the subpairs join into superpairs near the center of colorness, and the superpairs gravitate toward the boundary of the superglued wave packet. The virtual gluonic currents between superpairs and subpairs underlie the superglue mechanism of the virtual wave packet. These gluonic layers are like the layers of a virtual onion that may expand as the spatiotemporal wavelets interfere, and other layers may be manifested.

The superglued wave packet, or virtual photon, may have an orbital motion. As the virtual photon orbits, the spatiotemporal wavelets of the medium may expand or contract in all directions as they interfere. As a spatiotemporal wavelet expand in all directions about the virtual photon, the orbital motion becomes a wave packet in all directions of the emerging wavelet. Let us hypothesize that as the virtual wave packet, or virtual photon, orbits, the orbital plane may rotate about its axis, which would account for the dependent parameters of linear frequency, wavelength, and color, depending on the phase angle of the orbital plane in each sense of coordinate direction, as the virtual photon moves along the direction(s) of the expanding spatiotemporal wavelet. The linear wave motion, packet oscillations, and rotational wave motion about the axis, would account for the dependent parameters of the virtual photon.

The axis of the virtual photon may rotate, not precess, which would account for polarization of the electromagnetic wave. If the axis rotates on a vertical plane, the wave packet in the linear direction is not polarized, but it would be polarized in a perpendicular direction of the same vertical plane. Photons may crisscross each other in a spatiotemporal volume and not interfere. Therefore, the tumbling combination of the orbital, axial, rotating, pulsating, and translating motions, and colorness, of a virtual photon, may endow the wave packet with its dependent and independent parameters.

The electromagnetic characteristic of the superglue wave packet of a virtual photon may emerge from the orientation and alignment of the electrically charged gluonic dipoles that add to an overall zero charge. Hence, the electrical composition of a virtual photon of a distinct class may be illustrated by the concept of a static electrical dipole, where the spatiotemporal medium of the wave packet is the temporal source and spatial sink of the closed electrical circuit, and the electric field is the return path between the opposite charges. *The electric field is the nonvirtual part of the circuit, as it becomes part of the spatiotemporal system, and it may be expressed as electric field potential energy*. An electric point dipole of a virtual photon is the limit obtained by letting the separation between the opposite electrical charges approach zero while keeping the electric dipole moment fixed. A theoretical magnetic point dipole of a virtual photon has a magnetic field of the same form as the electric field of an electric point dipole. The dynamic behavior of the photonic dipole follows the laws of energy which include the conservation of energy. The conservation of energy underlies temporal symmetry.

The electric point dipole of a virtual photon provides us with a theoretical framework that allows us to consider the electrical circuit of the dipole from a different perspective. The space-time of the wave packet is the source of energy of the virtual photonic dipole, the gluonic electrical charges are the terminals of the circuit, and the electric field potential through local space-time is the load. If external energy is used to polarize a virtual photon to create the electric point dipole, the photonic dipole provides a source of electric field potential that can perform work. By the principle of superposition, virtual photonic dipoles may orient and align to increase electric field potential in the spatiotemporal region of the wave packet. If a virtual photonic dipole is sustained, the electric field gushes forth between opposite gluonic charges, becoming nonvirtual energy.

If there are several equally likely outcomes, then spontaneous symmetry breaking can occur. The virtual photonic dipole system would be symmetric with respect to these outcomes. However, if the virtual photonic dipole system is used or interacted with in any way, a specific outcome would occur. Hence, the virtual photonic dipole system is symmetric, even though not encountered with such symmetry, but with a specific asymmetric state. In such case, the symmetry of the virtual photonic dipole is spontaneously broken.

Symmetry breaking represents the breaking of the exact symmetry of the underlying laws of physics of the composition of a photon by the random formation of the structure of a virtual photonic dipole. Thus, symmetry breaking occurs in the pattern formation of a virtual photonic dipole. In spontaneous symmetry breaking, the equations of motion of the electric field are invariant, but the system is not because the spatiotemporal medium of the system is non-invariant. Such a symmetry breaking is parametrized by an order parameter. In the case of the virtual photonic dipole, the order parameter are the polarities of the gluonic charges. The virtual photonic dipole through symmetry breaking becomes nonvirtual electric field potential.

As the gradient of the electric field potential projects through temporal extent from the positive gluonic charge towards the negative gluonic charge in a conventional flow model, the spatiotemporal gradient of the electric field potential may exert a force laterally, or in the direction of its trajectory. As the spatiotemporal gradient of the electric field potential extends onto the negative gluonic charge, it couples with the existing spatiotemporal substance of the negative gluonic charge to complete the circuit. The spatiotemporal gradient points in the direction where the electric field increases or decreases most with the electric field potential.

The electric field potential exists between electrically charged virtual gluons in the composition of the virtual photon. The gradient represents the virtual tubes of stress through the infinitesimal spatiotemporal medium from the forces of attraction between electrically charged virtual colors and anticolors within the realm of each virtual photonic dipole. It is hypothesized that as gluons get pulled apart within the boundary of a virtual photonic dipole, the virtual photonic dipole orients, aligns, and strengthens, as if the varying strength of the collective electric field may be producing the virtual tubes of stress through the spatiotemporal medium, that increase the energy of the electric field potential.

Energy is emitted between opposite charges, the spatiotemporal region of emission is the positive region, and the spatiotemporal region of receival is the negative region, of the conventional dipole model. Space-time is the source, the sink, the medium, and the return path of the closed circuit, in the transfer of energy of the virtual photonic dipole.

Complex Energy, $S\Delta t$, diverges, as reactive energy from the temporal region, $iQ\Delta t$, is transmuted into nonvirtual electric field potential, that converges back, as nonvirtual real energy, $P\Delta t$, onto the spatial region of the virtual photonic dipole.

$$S\Delta t = P\Delta t + iQ\Delta t \qquad (1.7)$$

Dividing the above equation by, Δt, will render the equation for complex power.

Based upon the above hypotheses and previous theoretical research, if a graviton may be represented as either an expanding or a contracting spatiotemporal volume, it is possible for a graviton to hold Planck quantum charges within itself or at its boundary, in the gravitational field of a charged body of mass, having both curvature and polarized charges.

A chiral sphere or tube may illustrate a graviton that is static or dynamic. The chirality and helicity of the charged graviton designates its overall polarity. A homogeneous and isometric spatiotemporal volume with a spin 2, that completes a spin at $c/\sqrt[2]{3}$, may not expand or contract, manifesting a static virtual graviton.

Let us discuss the weak interaction, or weak force, in terms of the electro-weak theory, as the mechanism of interaction between sub-atomic particles that causes radioactive decay and plays an essential role in nuclear fission. The weak interaction is sometimes referred to as flavordynamics. Moreover, the weak interaction is the only fundamental interaction that breaks charge parity symmetry. For instance, the weak interaction occurs within the boundary of a proton as a fundamental interaction of nature.

A neutron may beta-decay into a proton, an electron, and an electron anti-neutrino, through the weak interaction. Carbon-14 (6 protons and 8 neutrons) is an example of a beta-minus decay to Nitrogen-14 (7 protons and 7 neutrons) plus an anti-neutrino and an electron.

Colors and anticolors play an essential role in the weak interaction at the deepest levels of nature. All weak interactions are very near-field gluonic interactions within the boundary of a particle. The exchange of W and Z gauge bosons not only causes the transmutation of a quark, i.e. changing a quark flavor, inside hadrons, but also changes the hadrons themselves. For example, a proton can decay into a neutron, transforming an up quark (*udu*), with an electrical charge of $+\frac{2}{3}e$, into a down quark (*ddu*), with electrical charge of $-\frac{e}{3}$, while emitting a W+ gauge boson, to transmute a positron into an electron neutrino. The W+ gauge boson, with an electrical charge of +1*e*, transfers energy from its gluonic fields, and color charges, to the positron to render an electron neutrino.

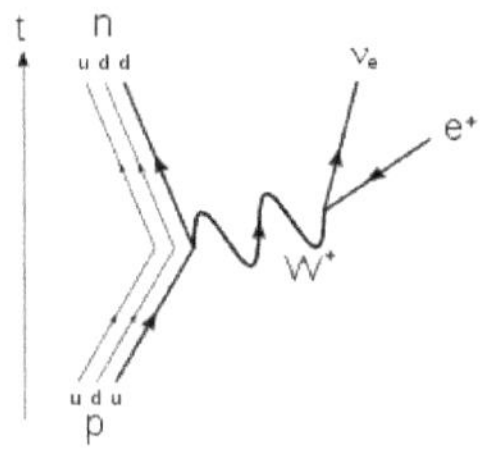

Figure 2. Feynman diagram of electron neutrino emission.

Similarly, during a beta-minus decay of a neutron into a proton, the transformation of a down quark to an up quark occurs, while emitting a W− gauge boson, with an electrical charge of $-1e$, drawing energy from the gluonic fields of the electron anti-neutrino (lepton), and transferring color charges, to transmute the electron anti-neutrino into an electron (lepton).

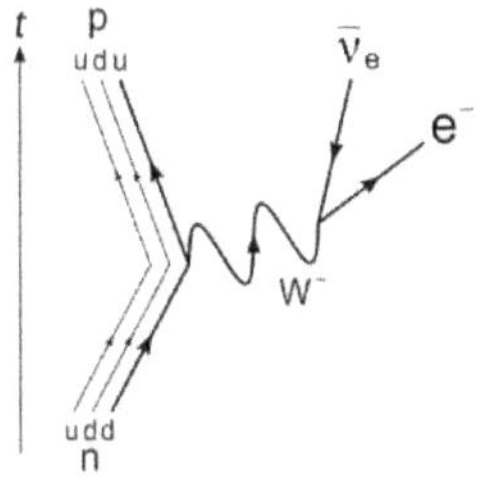

Figure 3. Feynman diagram of electron emission.

The $W\pm$ gauge bosons are gluonic mechanisms to transfer or draw energy from gluonic fields, and to transfer color charges, during lepton transmutations. The Z^0 gauge boson has a neutral charge.

The neutral Z boson may be emitted or absorbed by a quark, or a lepton, during a neutral current interaction, i.e. $e^- + Z^0 \rightarrow e^-$. The neutral Z boson has a fast decay into a color charge and an anticolor charge, $Z^0 \rightarrow r + \overline{r}$.

Thus, the Z^0 gauge boson transforms neither mass nor charge. The gluonic constitution of the Z^0 gauge boson is similar to the photon. It is no coincidence that the electromagnetic force and the weak interaction are two aspects of a single electro-weak force.

Therefore, the strong interaction, the weak interaction, and electromagnetism, are very near-field, intermediate-field, and far-field gluon interactions within and outside the boundary of a particle. All manifestations of particles, anti-particles, and virtual particles involve gluonic interaction, gravitational interaction, and energy, that spring from the spatiotemporal medium. It has been observed that gluons and electrons can interact at the lowest order of the strong interaction to produce or transmute heavy flavors such as the top quark. The following Feynman diagrams are tree level type interactions, without loops, between an electron and a top quark.

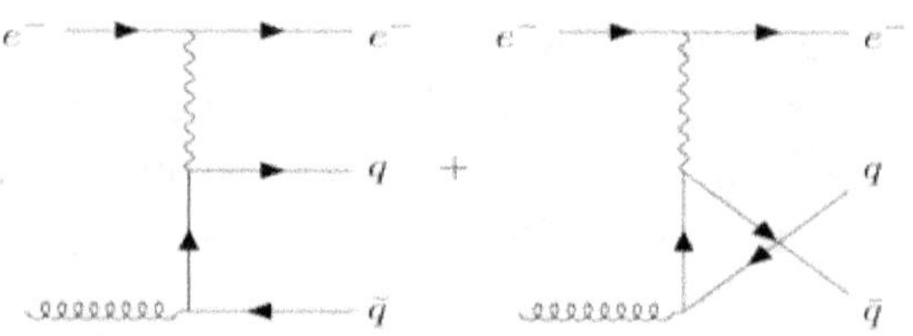

Figure 4. Feynman diagrams for strong interaction between an electron and a quark.

Gluons are virtual particles that, from the scale of an observer, may appear or disappear from what we would call reality. Gluons and their fields make up most of the mass of observable particles. On a very short temporal scale, a gluon may be emitted by a quark, and after a brief existence, it may be absorbed by another quark. A gluon may generate a color change in a quark. A gluon may split into a color-anticolor pair that may be absorbed by another gluon. The properties of a gluon that persist, not those that fluctuate, partake in strong interactions.

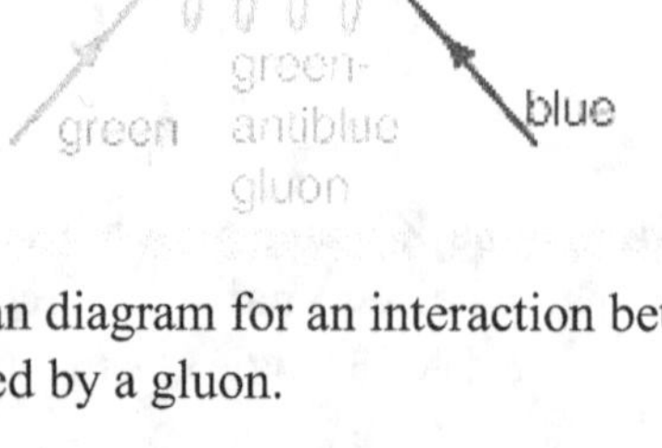

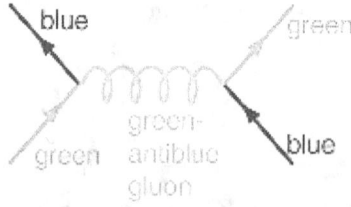

Figure 5. Feynman diagram for an interaction between quarks mediated by a gluon.

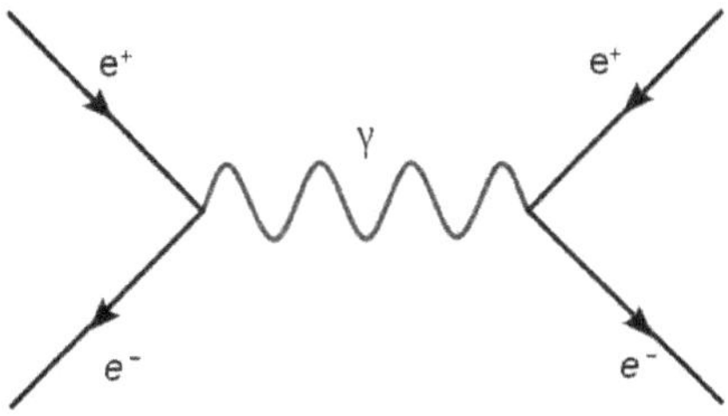

Figure 6. Electromagnetic Interaction of Electron and Positroncreating a Virtual Photon.

A proton, or a neutron, is made of three quarks that are moving around, within the boundary of each particle, at near the speed of light, surrounded by other virtual particles, and a cloud of virtual gluons that produce a gluonic field potential. Most of the mass of the proton, or neutron, comes from the gluonic fields, not from the quarks. Quarks move within the boundary of a proton, or neutron, as if there is no force between them at all, i.e. they are asymptotically free. (Gross, 1973)

Gluons consist of color-anti-color dipoles. As these color-anticolor dipoles surround a quark, they align themselves parallel to the color, or anticolor, of the quark, strengthening the color, or anticolor field. Consequently, the color-anticolor field of the quark is amplified and strengthened.

If a quark were to enter the color-anticolor cloud of another quark, the closer it approaches, the lesser would be the attraction, but the farther it gets the stronger it would feel the attraction or repulsion of the color-anticolor cloud of the other quark. The gluonic cloud acts like a negative gluonic field. Moreover, there may be a runaway process of interaction, where quarks emit gluons, gluons emit other gluons, or gluons may transmute into virtual quark-antiquark pairs that exchange more gluons. Hence, at very small scale within the boundary of a fundamental particle, it all seems to be gluonic field potential of color-anticolor dipoles. The gluonic field potential binds all there is.

The quark-gluon plasma that embodies the manifestation of the quark exists at a dynamic but definite energy density. The energy density at the center of the quark is almost point-like in the middle of the gluonic cloud where the electrical charge gets infinitesimally small. Hence, the quark is distributed energy points in the spatiotemporal volume of its boundary. The quark exists as a concentrated gluonic field potential in a gluonic sea. The quantity and frequency of gluonic interactions play an essential role in the strength of the strong force. A gluon can temporarily become a virtual color-anticolor pair, or a virtual quark-antiquark pair, before becoming a gluon again.

The color-anticolor pair fluctuation is a weaker interaction than the stronger interaction of the quark-antiquark fluctuation. However, the gluon pair, or color-anticolor pair fluctuations, are much more widespread in the boundary of a particle, such as a nucleon, so they represent a more significant contribution to the strong force. (Politzer, 1973)

1.2.(a) The formation of the early universe.

Gluons and quarks swarmed around freely in the spatiotemporal medium of the early universe. The temperature of the early spatiotemporal medium was too high for hadrons to form into the elementary particles.

The early quantum state of the universe was a quark-gluon plasma that has already been replicated in experiments through the collision of atomic nuclei. The evolution of the quark-gluon plasma during cooldown has provided great insight into quark-gluon interaction and the changing conditions of the early universe.

1.2.(b) The manifestation of particles under the law of transmutation of energy.

Unusual combinations of gluons and quarks have been created beyond the usual neutrons and protons. These exotic states of particles may shed more light on particle interactions and the framework of matter. Theoretical simulations of gluons and quarks have predicted that gluons and quarks may combine to create other particles in addition to neutrons and protons. Gluons may combine to produce glueballs, a larger multigluon particle state, or may combine into hybrid particles that consist of bound states of quark-antiquark-gluon triplets. Two quarks bound to two antiquarks to form a tetra quark have already been found. These possibilities suggest an extended gluonic family of particles.

A remarkable thing happens when a neutron, or a proton, is accelerated near the speed of light, the gluons inside the neutron or proton divide and multiply into other gluons. At very high kinetic energy, gluons undergo chromomeiosis into daughter pairs, each pair having a lighter color of less color energy than the parent pair, but the electrical charge of the pair is conserved. The chromomeiosis becomes a runaway process up to the theoretical maximum gluonic occupancy state, or color glass condensate state, of the neutron, or proton, involved.

1.2.(c) The different states of matter.

It is important to examine the familiar configurations of matter, such as nucleons, and their possible constituents, such as quarks, gluons, and gluonic interactions. The existence of different states of a hadron may be unfamiliar from the familiar nucleons, but these different states are still allowed by the current principles and hypotheses of Quantum Chromodynamics. Possible configurations of colorless hadrons have been theoretically predicted such as hybrid states of quark-antiquark-gluon states, molecules of coupled pairs of quark-antiquark, or glueball states consisting only of gluons. Tetraquark molecules have been experimentally predicted, but more of these hybrid states need to be experimentally verified. Expectations for more conclusive results are very high.

The recent discovery of an extreme state of matter indicates that the quark-gluon plasma forms when atomic nuclei traveling close to the speed of light collide with each other. It is theorized that when nucleons collide at close to the speed of light, the confinement of quarks and gluons breaks, releasing all constituents and gluonic energy into an extreme plasmatic state of matter that may have existed in the early universe. The formation of the quark-gluon plasma is a reverse process to the Chromoquantization of matter. The extreme quark-gluon plasma state of matter flows with extremely low friction, and at an extreme temperature of over four trillion degrees Celsius, in comparison, the surface of the sun is, approximately, only fifteen million degrees Celsius.

1.2.(d) The strong or weak color charge bond.

Even though the quantity of quarks inside a neutron, or a proton, may vary significantly, a neutron, or a proton, consists of three quarks, each quark carries a gluon, or color-anticolor pair, each color or anticolor has color charge that may be red, green, or blue. Neither the neutron, nor the proton, has a net color charge. Each color charge may have a different matching intensity and a different hue. It is possible to hypothesize that the strongest color bond is between colors and anticolors of the same hue with the strongest color intensity, i.e. $r\overline{r}$, $g\overline{g}$, or $b\overline{b}$. However, colors, or anticolors, of different hue and color intensity, may bind together to form a gluonic quantum state, i.e. $r\overline{b}$, $r\overline{g}$, $g\overline{b}$, etc. The weakest color bond is between colors and anticolors of different hue, with the weakest color intensity. The color-anticolor pair, or gluonic quantum state, inside each of the three quarks, does not have a net color charge, so it is color neutral or colorless. Similarly, the quark-antiquark pair inside the hadron of a π-meson is color neutral. A color neutral, or colorless state, is a non-interacting state that still consists of color and anticolor charges.

As the color-anticolor pair of a quark exchanges color charge, or anticolor charge, with other quarks in the hadron of a neutron, or a proton, it is possible to hypothesize that there is a very high probability that a color charge of a hue will pair with an anticolor charge of the same hue, to make the strongest possible color-anticolor bond. Once the strongest bond is established, in each quark of the hadron, the quarks may exchange gluons, or exchange non-interacting color-anticolor pairs, always maintaining in this manner the overall colorless quantum state of the neutron, or proton. From previous theoretical research, it was hypothesized that the photon also uses the strongest color bond to maintain its colorless quantum state. The strongest color-anticolor bond principle underlies the force carrier of the electromagnetism, enabling the derivation of nuclear physics from six-dimensional Quantum Chromodynamics. The strongest color-anticolor bond principle complements the concept of confinement for color charged gluons and quarks within the hadrons of nucleons. A strong color-anticolor bond may have a $\pm$ strong color, paired to two corresponding $\mp$ strong anticolors. The same polarity relationship applies to a weak color-anticolor bond.

The nuclear force between nucleons, i.e. protons and neutrons, is attractive at an approximate range between the radius to the diameter of a nucleon. At less than the radius of a nucleon, the nuclear force becomes repulsive, and at greater than the diameter of a nucleon, it decreases exponentially. The nuclear force is mediated by gluons. At the attractive-range, the nuclear force, or strong color force, overcomes the electromagnetic force, enabling protons to bind with other protons, or neutrons, and storing electromagnetic potential energy from the repulsive forces of protons within the nucleus. At the repulsive-range, the repulsive force is equivalent to the Pauli exclusion force for the nearly identical but different nucleons, or between quarks of the same strong color-anticolor charge within different nucleons. The residual attractive effect of the strong color-anticolor interaction force becomes the nuclear force, extending slightly beyond the boundaries of nucleons to produce the attractive nuclear force.

1.2.(e) The nature of the nuclear force as a strong color field potential.

It is hypothesized that the strong color gradually diminishes from the boundary of a quark inward toward the center of color and outward toward the nearby nucleon. As the strong force diminishes toward the nearby nucleon, the strong force weakens from the particle's boundary up to approximately the radius of a nucleon. The center of color charge of a nucleon is approximately four times its radius, or twice its diameter, from the center of color charge of a nearby nucleon. Thus, half-way between nearby nucleons, the strong color force is at its weakest quantum state. The strong color force of a nucleon repulses the strong color force of another nucleon at a short range of less than the distance of the radius of a nucleon from the particle's boundary. However, as the strong color forces between two nucleons repulse, the nucleons move apart to a distance between boundaries of approximately a diameter of a nucleon, where the strong color force weakens, reaching a non-repulsive equipotential point. It is possible to consider the relationship between strong color forces, or the gradient, of nucleons as strong color field potential, so that the equipotential point represents the lowest strong color field potential between nearby nucleons.

From the point of lowest strong color field potential between nearby nucleons, each nucleon's strong color field potential strengthens toward each nucleon's boundary, becoming increasingly repulsive to the opposite strong color field potential of a nearby nucleon.

A nucleon may exchange a pion meson, which is a strong color-anticolor pair, or a quark-antiquark pair, at short-range with another nucleon. The receiving nucleon senses the incident strong color-anticolor pair of a specific hue and responds with a reflective strong color-anticolor pair of the same hue, to provide a corresponding counter-attractive weak gluonic force, or short-range nuclear force, at distances between approximately the radius of a nucleon to the diameter of a nucleon, within the nucleus of an atom. The short-range strong color-anticolor force, or short-range nuclear force, between nucleons, counteracts the long-range electromagnetic repulsion between charged nucleons (protons) and maintains the status in quo of the electromagnetic quantum state of neighboring nucleons. A nucleon may change the colors of the quarks in its hadron during gluonic exchanges with other nucleons, but eventually the nucleon returns to its stable color state of the three colors of charges: red, blue, and green.

It is possible to hypothesize that when protons are interacting at long-range, i.e. at a distance longer than the diameter of a nucleon, each proton can sense an incident proton, emitting a deflective photon, which causes the incident proton, or the emitting proton, to alter its trajectory not to collide. The deflection can occur if the gluonic energy of the strong color-anticolor force is greater than the total kinetic energy of an incident proton. Deflective photons share the same polarity and color.

Hence, a proton may respond to another proton with a strong color-anticolor force that may not exclusively depend on the long-range distance, color charge, momentum, and chirality, of both protons.

1.2.(f) The mass gap hypothesis.

Gluons may divide, but without enough space for their gluonic fields, gluons may recombine; in that manner putting a limit to the runaway process of chromomeiosis within the boundary of the quark, as the process reaches a gluonic saturation state. The strong and weak color force serve as delimiters for color-anticolor pairs, or gluons, in the occupancy quantum state of the quark, or antiquark, since gluons would separate during saturation up to the maximum occupancy quantum state, or color glass condensate. The pixel-antipixel theory and the strong color field potential (nuclear force) theory underlie the delimitation of chromomeiosis, or the confinement of gluons and gluonic fields, inside a quark or antiquark.

It is hypothesized that proton spin results from the spins and orbital motion of quarks, and gluons, within the hadron of a proton, and the angular torsion of the spatiotemporal medium within the boundary of the proton. The existence of a mass gap within nucleons and in the solutions of the Yang-Mills equations may support this hypothesis.

1.2.(g) The Greenberg-Zweig-Gell-Mann Principle of Confinement.

Gluons can interact with each other in ways and with properties that are very different from other force carriers or particles. A hadron consists of a definite number of quarks-antiquark pairs and a dynamic cloud of color-anticolor pairs, or gluons, in equilibrium. Gluons can produce virtual quark-antiquark pairs.

Let us hypothesize that colorness consists of virtual pixes, i.e. pixels and antipixels. A strong dark color, or anticolor, of any hue, divides into weaker lighter colors, or anticolors. A color charge has pixels, and its complementary anticolor charge has antipixels. Pixels, or antipixels, can divide into subpixels, or antisubpixels. Subpixels, or antisubpixels, of any hue, can re-group into pixels, or antipixels, of any hue, that become a lighter color. Dark colors of any hue have a stronger repulsive force between themselves than lighter colors, of any hue, do. Hence, dark colors stay farther apart, and conversely, lighter colors stay closer together. Thus, lighter colors gravitate towards the center of colorness, while darker colors gravitate farther apart from the center of color. This property of colorness generates a colorness field potential, or gluonic field potential, that is weaker at the center, and stronger away from the center. A gluonic field potential behaves like a spring, when stretched it has greater spring potential, when unstretched the spring has less potential. The gluonic field is the gradient of the colorness field potential. The gluonic field potential may be considered a negative field, if an electromagnetic field, or a gravitational field, is considered a positive field. A positive field becomes weaker farther away from the center of electrical charge, or gravitational mass.

As color charges get closer in a gluonic field potential, the spatiotemporal medium toward the center of colorness becomes more compressed by the negative gluonic field potential. The spatiotemporal pressure increases at the center of colorness, but gradually decreases outward in a radial direction as a positive gravitational field. As lighter color charges gather toward the center of colorness, since the electrical charge of the color-anticolor is conserved, the collective electrical charges towards the center strengthen with the continuous generation of lighter color charges.

Consequently, a virtual gravito-gluonic boundary emerges about the periphery of the center of colorness, for the gluon moving outward, that becomes the range of confinement for that gluon. As a color-anticolor charge of a gluon, or that of a quark, moves outward from the center of colorness due to a repulsive or outward force, the gravitational field lessens, the gravitational field potential increases, the electrical field decreases, and the electrical potential increases, as expected for positive fields. Conversely, as the same color-anticolor charge moves outward from the center of colorness, the color charge field of any hue strengthens effectively repulsing the color-anticolor charge inward toward the center of colorness. The combined actions of the positive fields of gravitation and electrical charge, and the action of the negative field of color-anticolor charges, on gluons or quarks, results in a field of confinement that binds quarks together and conserves the integrity of particles such as a proton. Heavier quarks gravitate toward the periphery of the gluonic field potential within a particle, and lighter quarks gravitate toward the center of colorness within a particle. The principle of the confinement field potential evolves from the excellent foundational ideas in

Chromodynamics of eminent physicists such as Oscar Wallace Greenberg, George Zweig, and Murray Gell-Mann.

Chromomeiosis is a specialized color, or anticolor, division that reduces the color and intensity of a pixel, or an antipixel, of any hue, creating a lighter color and a weaker color state, but conserving the polarity and value of the electrical charge of the pixel, or antipixel. Colorness, $\pm k$, i.e. the color state of a gluon, is energy. As pixels, or antipixels, divide through generations, the pixel chromomeiosis tends toward a point-like gluonic field potential, but reduction occurs up to a lower threshold where subpixels, or sub-antipixels, no longer divide to conserve colorness. The lower threshold of chromomeiosis occurs due to the principle of conservation of colorness which underlies the principle of conservation of energy.

The energy, E, of colorness for color charge(s), or anticolor charge(s), may be expressed as

$$E = \pm k\vec{g}d \tag{1.8}$$

where $\vec{g}$ is the gluonic field between colors, and anticolors, in Newtons/$\pm$colorness, and d is the distance between color and anticolor.

Each color, or anticolor, has a temperature of colorness. By definition, there are three colors, or anticolors, with three potential temperature ranges for the charge of colorness, namely, high, medium, or low. The higher the temperature, the higher the energy of the colorness. The temperature of colorness is the temperature of a pixel, or antipixel. Any color may have any of the three temperature ranges.

As an example, let us define the color ‘red’ as a high color charge, ‘blue’ as a medium color charge, and ‘green’ as the low color charge, for either color or anticolor. Next, let us consider some possible color-anticolor pairs of equal or different temperature range: (a) if a higher color charge pairs with a lower color charge, there is an unbalance of color charge, or anticolor charge, which results in a weaker color, or anticolor, bond, (b) if equal colors, or anticolors, of different hue pair up, the result is a median color, or anticolor, bond,

(c) if equal colors, or anticolors, of the same hue pair up, the result is the strongest color, or anticolor, bond, (d) if low colors, or anticolors, of the same hue pair up, the result is a weak color, or anticolor, bond, (e) if low colors, or anticolors, of different hue pair up, the result is the weakest color, or anticolor, bond.

A strong color, or anticolor, bond, may join a weak or a median bond pair. If three strong color or anticolor bonds join in gluonic confinement, they create a triad that is non-interacting and able to escape confinement. However, if the triad of strong color or anticolor bonds, is between the three quarks of a nucleon, they may just exchange color or anticolor charges within their hadron.

Let us hypothesize that the temperature of colorness of a pixel, or antipixel, has an associated angular frequency of vibration, ω_p, in Planck units of angular frequency, from the color, or anticolor strings, or D-branes, that constitute the pixel, or antipixel, according to the principles of String Theory.

The color temperature may be generated by the heat energy, $\pm\Delta Q_p$, produced by the vibration of a pixel, or antipixel, that may be expressed as

$$\pm\Delta Q_p = \pm\bar{\lambda}\omega_p \equiv \pm k \qquad (1.9)$$

where the lambda-bar, $\bar{\lambda}$, is the color temperature quantum of action, or reduced color temperature constant, for a pixel, or antipixel, with colorness, $\pm k$.

1.2.(h) The Gell-Mann-Zweig-Greenberg Postulate.

The Gell-Mann-Zweig-Greenberg Postulate states that: the magnitude of the gluonic force of attraction or repulsion between two arbitrary gluons is directly proportional to the magnitudes of color-anticolor charges and to the square of the distance between them. The force is along the straight line joining them.

Let us suggest some approximate mathematical equations for the gluonic field, gluonic field potential, and forces within the confinement of a colorless hadron, assuming that this postulate and the principle of superposition are valid for a static gluonic field and static gluonic distribution.

Chromostatics is the study of stationary color-anticolor charges or fields as opposed to gluonic currents, or dynamic fields. The following chromostatic equations may represent the color-anticolor pairs of gluons, or the pixel-antipixel pairs of pixes.

Therefore, a chromostatic field within the sphere of influence of a gluon may be expressed as

$$\vec{g} = \pm \frac{b\overline{g} \cdot r^2}{4\pi k e^{(d-r)/m}} \vec{a}_\phi \qquad (1.10)$$

where $\vec{g}$ is the gluonic field from an arbitrary color-anticolor charge, e.g. $\overline{b}g$, antiblue-green, that may exert a force on another color-anticolor charge, e.g. $b\overline{g}$, blue-antigreen, r is the length of the radius in meters from the center of color charge, or center of anticolor charge, to the boundary of the sphere of influence, d is the length of a distance in meters from the center of the color charge, or anticolor charge, to the length of the diameter of the sphere of influence, $\pm k$ is the colorness, m is *"1"* meter, and $\vec{a}_\phi$ is a unit vector in the direction of the field.

The gluonic field increases from the center of colorness to the boundary of the sphere of influence, and then decreases from the boundary, diminishing at a distance approximately equal to the diameter of the sphere of influence.

The polarity of the gluonic field stems from the colorness. The colorness is the square of the color-anticolor charge(s) per unit of gluonic pressure (or energy density). The colorness of an arbitrary gluon, color-anticolor pair, e.g. blue-antigreen pair, in the sphere of influence of another arbitrary gluon, e.g. antiblue-green pair, may be expressed as

$$\pm k = \pm \frac{(b\overline{g})^2 \cdot m^2}{F_{\overline{b}g}} = \pm \frac{(b\overline{g})^2}{\text{Gluonic Pressure } (\overline{b}g)} \tag{1.11}$$

where $\vec{F}_{\overline{b}g}$ is the gluonic field force, or Zweig force, of a single arbitrary gluon, or color-anticolor pair, e.g. antiblue-green, and m^2 is a square meter.

The gluonic field force from a single arbitrary gluon, e.g. antiblue-green, may be expressed as

$$\vec{F}_{\overline{b}g} = \pm \frac{b\overline{g} \cdot r^2}{4\pi k e^{(d-r)/m}} \vec{a}_\phi \tag{1.12}$$

The gluonic force between two arbitrary gluons may be expressed as

$$\vec{F}_{12} = \pm \frac{(g_1 \mp g_2) \cdot r^2}{4\pi k e^{(d-r)/m}} \vec{a}_\phi \tag{1.13}$$

where g_1, or g_2, is a gluon of an arbitrary color-anticolor pair, with a specific polarity of colorness, so that g_1 and g_2 may, or may not be, the same polarity. Opposites color-anticolor polarities attract, like color-anticolor polarities repel, giving the force, $\vec{F}_{12}$, its polarity.

The colorness of two arbitrary gluons, two color-anticolor pairs, in each other's spheres of influence, may be expressed as

$$\pm k = \pm \frac{(g_1 \mp g_2)^2 \cdot m^2}{F_{12}} \tag{1.14}$$

The kolorosity is defined as the colorness per unit of distance, $(\pm k/d)$, or the square of the two color-anticolor charges per unit of gluonic pressure. Per the above definitions, colorness may be express in units of Gell-Manns (G_n), and kolorosity in units of

Gell-Manns per meter.

The gluonic field potential may be expressed as

$$\Delta Z_w = \pm\frac{(b\bar{g})}{k} = \pm\frac{F_{\bar{b}g}}{b\bar{g}\cdot m^2} = \pm\frac{\rho_E}{b\bar{g}} \tag{1.15}$$

where Z_w is the gluonic field potential, or Zweigage, in units of Zweigs. A Zweig may be expressed in units of color-anticolor charge(s) per unit of Gell-Mann, or energy density, *J/m³*, per unit of color-anticolor charge. A gluon unit is a unit of color-anticolor charge, e.g. blue-antigreen pair charge, $g_{b\bar{g}}$.

The gradient of the gluonic field potential points in the direction where the gluonic field increases or decreases most with Zweigage potential.

The gluonic gradient may be expressed in rectangular coordinates as

$$\nabla\vec{g} = g^{\bar{b}g}\frac{\partial g^r}{\partial x_{\bar{b}}}\vec{e}_r \otimes \vec{e}_g \tag{1.16}$$

where the Einstein summation notation is used.

For a smooth gluonic field on a Riemannian manifold, we may express the local field as

$$\nabla\vec{g} = g^{\bar{b}g}\frac{\partial\vec{g}}{\partial x^g}\vec{e}_{\bar{b}} \tag{1.17}$$

The gluonic divergence represents the volume density of the outward flux of a gluonic field from a volume around a given color-anticolor charge.

$$\nabla\cdot\vec{g} = \frac{4\pi\rho_E}{\pm k} = \frac{4\pi\cdot b\bar{g}\cdot\Delta Z_w}{\pm k} = 4\pi\left(\Delta Z_w\right)^2 \tag{1.18}$$

1.2.(i) The Greenberg's gluonic tensors of the General Theory of Relativity.

The maximum magnitude of force exerted by an arbitrary gluon at the boundary of confinement may be expressed as

$$\left(\vec{F}_{\bar{b}g}\right)_{\max} = \pm\frac{F_{\bar{b}g}\cdot r^2}{4\pi\cdot b\bar{g}\cdot m^2\cdot(e^0)} \quad (1.19)$$

$$F_{\bar{b}g} = \pm\frac{4\pi\cdot b\bar{g}\cdot m^2\cdot\left(\vec{F}_{\bar{b}g}\right)_{\max}}{r^2} = \pm 4\pi\cdot b\bar{g}\cdot m^2\cdot\frac{G}{c^4}\cdot G_{\mu\nu} \quad (1.20)$$

$$\frac{G}{c^4}\cdot G_{\mu\nu} = \pm\frac{F_{\bar{b}g}}{4\pi\cdot b\bar{g}\cdot m^2} = \frac{1}{8\pi}T_{\mu\nu} \quad (1.21)$$

$$\pm\frac{F_{\bar{b}g}}{m^2} = \frac{4\pi\cdot b\bar{g}}{8\pi}T_{\mu\nu} \quad (1.22)$$

$$G_{\bar{b}g} = \frac{1}{2}T_{b\bar{g}} \quad (1.23)$$

Thus, half the energy density of $b\bar{g}$ generates the gluonic curvature about $\bar{b}g$, and vice versa. The Greenberg gluonic curvature tensor is $G_{\bar{b}g}$, and the Greenberg gluonic energy density tensor is $T_{b\bar{g}}$, that describe the gluonic field potential. The curvature may be positive or negative, depending on the polarity of the gluonic charges involved.

1.2.(j) The source, curvature, and evolution equation for the confinement volume of a quark.

From previous research it was demonstrated that the cosmological curvature acceleration ratio $\ddot{a}/2ac^2$ is equal to the stress-energy-momentum curvature acceleration ratio $\ddot{a}/2ac^2$ in the manifold within a system of mass. The stress-energy-momentum curvature acceleration ratio does not contribute to the spatial velocity square ratio or to the spatial curvature k ratio. Moreover, the spatiotemporal negative curvature about a system of mass plus the stress-energy-momentum curvature of the system of mass is counteracted by the cosmological-constant positive curvature. The stress-energy-momentum curvature acceleration ratio is equal to one-half of the spatiotemporal curvature acceleration ratio of the Ricci tensor.

$$\mp\frac{3\ddot{a}}{ac^2}\approx\mp\frac{8\pi G}{c^4}\left(3\rho-3p\right) \qquad (1.24)$$

The confinement volume of a quark may be conceptualized as the confinement of gluonic point particles of color energy and pressure that may be expressed as

$$\mp\frac{3\ddot{a}}{ac^2}=\mp\kappa\left(T_{\mu\nu}-\Lambda_{\mu\nu}\right)\mp\frac{2\left(g_1\bar{g}_2\pm\bar{g}_1g_2\right)}{c^2}\mp\frac{2ne}{c^2}\mp\frac{2\sigma^2}{c^2}\pm\frac{2\omega^2}{c^2} \qquad (1.25)$$

where *"n"* is equal to "1" for a down, strange, or bottom quark, and *"n"* is equal to "2" for an up, charm, or top quark, $g_1\bar{g}_2$ and $\bar{g}_1g_2$ are arbitrary color-anticolor pairs of gluons, σ , is the shear between pairs of gluons, and ω is the angular frequency of torsion. (Raychaudhuri, 1955)

The source-curvature-evolution equation represents the curves of congruence of time-like geodesics of the flow of gluonic point particles of color energy and pressure, affecting the spatiotemporal curvature, or geometric property, of the volume of confinement. The source of the flow, or color charge current, is the distribution of electrically charged gluons in the confinement volume.

The gradient of the velocity field of the flow, $\vec{v}$, is a second rank tensor that consists of three parts, i.e. the trace, the symmetric traceless part, and the antisymmetric. (Ellis, 1971)

$$\nabla\vec{v}\rightarrow\nabla_a v^a+\sigma_{ab}+\omega_{ab} \qquad (1.26)$$

The coupled system of source, curvature, and evolution equations conveys geometric property statements about the flow of gluonic point particles of color energy and pressure. Moreover, once the initial conditions of flow expansion, angular frequency of rotation, and shear, are specified for a geodesic flow in a given geometry of space-time, it may be possible to prepare and solve the initial value problem of the coupled system of equations.

The trace is the expansion of the flow, and can be denoted as

$$Flow\ Expansion \equiv \nabla_a v^a \tag{1.27}$$

The shear is the symmetric, traceless part, that may be defined as

$$\sigma^2 = \frac{1}{2}\sigma_{ab}\sigma^{ab} \tag{1.28}$$

$$\sigma_{ab} = \frac{1}{2}\left(\nabla_b v_a + \nabla_a v_b\right) - \left(\frac{1}{n-1}\right) h_{ab} \nabla_a v^a \tag{1.29}$$

The projection tensor is h_{ab}, where the minus sign is for spacelike curves whereas the plus sign is for time-like curves, and *"n"* is the number of spatiotemporal dimensions. (Ciufolini, 1995)

$$h_{ab} = g_{ab} \mp v_a v_b \tag{1.30}$$

The antisymmetric angular frequency of rotation is given as

$$\omega^2 = \frac{1}{2}\omega_{ab}\omega^{ab} \tag{1.31}$$

$$\omega_{ab} = \frac{1}{2}\left(\nabla_b v_a - \nabla_a v_b\right) \tag{1.32}$$

The flow expansion, shear, and angular frequency of rotation, are related to the geometry of the finite number of enclosed geodesics of the cross-sectional area perpendicular to the lines of flow. The shape of the cross-sectional area changes from point to point along the curve of the flow. The bundle of geodesic curves may elongate, shorten, change shape under torsion, or shear. Angular frequency of rotation assists divergence, while shear assists convergence.

As the volume of confinement expands or contracts, the cross-sectional area perpendicular to the lines of flow changes by the same rate. The confinement volume may undergo convergence of the bundle of geodesic curves, or divergence.

The volume of confinement has continuous gluonic field flow between the center of color charge and the dynamic boundary.

The cloud of gluonic points of color energy are very far from the cosmological mass and energy components that contribute to a spatiotemporal positive curvature from the perspective of the local curvature tensor side (left hand side) of the source-curvature-evolution equation.

The energy density is calculated from the gluonic energy, E', and the volume *"a"* of the system, $c^2\rho = E'/a$.

Using the six-dimensional EFE, we obtain

$$\mp\frac{3\ddot{a}}{ac^2} = \mp\frac{8\pi G(3\rho-3p)}{c^4} \mp \frac{2(g_1\bar{g}_2 \pm \bar{g}_1 g_2)}{c^2} \mp \frac{2ne}{c^2} \mp \frac{2\sigma^2}{c^2} \pm \frac{2\omega^2}{c^2} \quad (1.33)$$

$$\mp\frac{3\ddot{a}}{a} = \mp\frac{8\pi G}{c^2}(3\rho-3p) \mp 2(g_1\bar{g}_2 \pm \bar{g}_1 g_2) \mp 2ne \mp 2\sigma^2 \pm 2\omega^2 \quad (1.34)$$

$$\mp\frac{\ddot{a}}{a} = \mp\frac{8\pi G}{3c^2}(3\rho-3p) \mp \frac{2(g_1\bar{g}_2 \pm \bar{g}_1 g_2)}{3} \mp \frac{2ne}{3} \mp \frac{2\sigma^2}{3} \pm \frac{2\omega^2}{3} \quad (1.35)$$

$$\mp\frac{\ddot{a}}{a} = \mp\frac{8\pi G}{c^2}(\rho-p) \mp \frac{2(g_1\bar{g}_2 \pm \bar{g}_1 g_2)}{3} \mp \frac{2ne}{3} \mp \frac{2\sigma^2}{3} \pm \frac{2\omega^2}{3} \quad (1.36)$$

$$\mp\frac{1}{a}\frac{\partial^2 a}{\partial t^2} = \mp\frac{8\pi G}{c^2}(\rho-p) \mp \frac{2(g_1\bar{g}_2 \pm \bar{g}_1 g_2)}{3} \mp \frac{2ne}{3} \mp \frac{2\sigma^2}{3} \pm \frac{2\omega^2}{3} \quad (1.37)$$

Therefore, the energy density, or pressure, of the confinement volume may be expressed as

$$\mp\frac{\ddot{a}c^4}{ac^2G}=\mp 8\pi c^2\left(\rho-p\right)\mp\frac{2c^4\left(g_1\bar{g}_2\pm\bar{g}_1g_2\right)}{3G}\mp\frac{2c^4ne}{3G}\mp\frac{2c^4\sigma^2}{3G}\pm\frac{2c^4\omega^2}{3G} \tag{1.38}$$

$$\rho_E=\mp 8\pi c^2\left(\rho-p\right)\mp\frac{2c^4\left(g_1\bar{g}_2\pm\bar{g}_1g_2\right)}{3G}\mp\frac{2c^4ne}{3G}\mp\frac{2c^4\sigma^2}{3G}\pm\frac{2c^4\omega^2}{3G} \tag{1.39}$$

1.2.(k) The Quantization of Colorness of a Gluon: Pixels and Antipixels.

A strongest color, or anticolor, in a color-anticolor pair of any hue has a corresponding value of colorness equal to +1, or –1. This is the highest possible value of colorness.

A superpixel, or a super-antipixel, has a fractional value of colorness down to the lowest possible value of $\pm(1/n)$, a subpixel, or a sub-antipixel, has a fractional value of colorness down to the lowest possible value of $\pm(1/2n)$, where *"n"* is the quantity of superpixels, or super-antipixels, manifested by gluons in the confinement of a quark, hadron, or particle.

Any two fractional values of colorness of two subpixels, or two sub-antipixels, may add, up to the next upper value of colorness, to a superpixel, or super-antipixel, of stronger color.

Any pixel, or antipixel, may divide, down to the next lower value of colorness, into two subpixels, or two sub-antipixels, of the same, but weaker color, where each subpixel, or sub-antipixel, has a value of colorness that is one half the value of the superpixel, or super-antipixel.

Thus, subpixels, or sub-antipixels, have a weaker color charge than the color charge of the original superpixel, or super-antipixel.

Individual color charge of pixels may not be conserved, but the collective color charge of all pixels, or antipixels, is conserved.

The number of possible ancestral pixes at generation X is equal to 2^{X-1}, where X is an integer and $X \geq 1$.

By convention, the superpixel of a color may be represented with a superscript, e.g. r^p, and a subpixel may be represented by a subscript, r_p.

$$\left(+\frac{1}{n}\right)r^p \leftrightarrow \left(+\frac{1}{2n}\right)r_p + \left(+\frac{1}{2n}\right)r_p \tag{1.40}$$

$$\left(-\frac{1}{n}\right)\bar{r}^p \leftrightarrow \left(-\frac{1}{2n}\right)\bar{r}_p + \left(-\frac{1}{2n}\right)\bar{r}_p \tag{1.41}$$

Superpixels, or super-antipixels, have three different colors and anticolors $\left(r^p, b^p, g^p, \bar{r}^p, \bar{b}^p, \bar{g}^p\right)$, subpixels, or sub-antipixels, have three different colors and anticolors $\left(r_p, b_p, g_p, \bar{r}_p, \bar{b}_p, \bar{g}_p\right)$, that match the colors, anticolors, and electrical charges of the original gluons.

Let us describe the process of chromosynthesis for a subpixel, or sub-antipixel, which may recombine with another subpixel, or sub-antipixel, of the same hue, to form a superpixel, or super-antipixel of the same hue.

Two superpixels, or super-antipixels, may recombine to form a superior pixel, or superior antipixel, up to the strongest possible color charge. The original electrical charge and polarity of a superpixel, super-antipixel, subpixel, or sub-antipixel is conserved.

As two colors, or anticolors, recombine, their pixels, or antipixels, recombine into a stronger color charge, creating a darker color and a stronger color state, conserving the polarity and value of the electrical charge of the pixels, or antipixels.

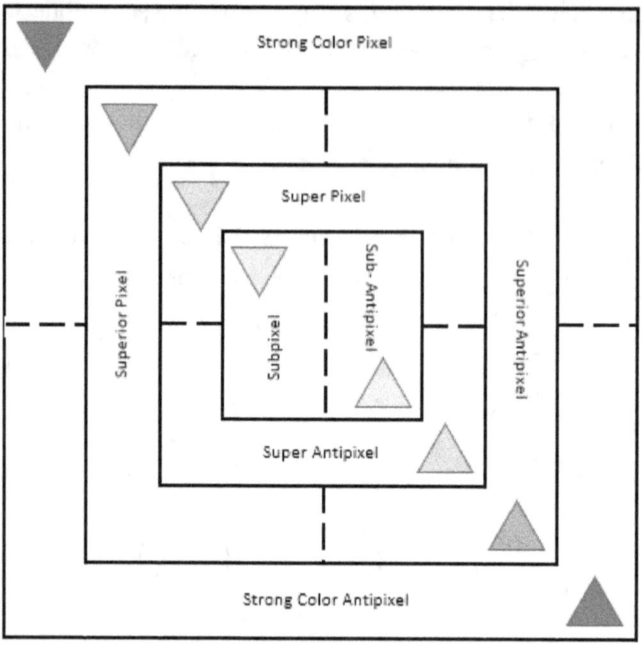

Figure 7. An Illustration of Chromomeiosis or Chromosynthesis.

1.2.(l) The Classical Yang-Mills Theory.

Three of the most important theories in modern physics, quantum electrodynamics, the electroweak theory, and the strong force theory, descend from the classical Yang-Mills Theory. Even the current standard model of particles physics can be traced back to the Yang-Mills Theory. The Yang-Mills equations can generalize Maxwell's equations. (Yang, 1954)

A large class of classical field theories under the Yang-Mills theory, generalizing electromagnetism, also satisfy a generalized type of symmetry. Moreover, a classical Yang-Mills theory may be described as quantized, Lorentz-invariant, with a specified gauge symmetry, that when completely determined leads to the standard model of particle physics.

It is interesting to note that when renowned Physicists Chen Ning Yang and Robert Mills were developing the Yang-Mills theory back in 1954, as Chen described during an interview, they were generalizing Hermann Weyl's previous proposal, of a gauge theory to describe electricity and magnetism, to develop a gauge theory, or a principle of how forces of interaction are governed, including the gravitational force. Not until the late sixties, Chen recognized a similarity between the Yang-Mills equations and the Riemann tensor, related to The General Theory of Relativity, which after checking the details and defining some quantities, turned out to be one and the same. (Yang, 2006) Originally, the equations were meant to represent the vector potential of electromagnetism, not the covariant derivative of curvature of a Riemannian manifold.

The covariant derivative of curvature is a term often used for the Levi-Civita connection in the theory of Riemannian and pseudo-Riemannian manifolds. The connection, in very simple terms, describes parallel transport between two adjacent points on a manifold. The fundamental physical concept of the connection is essential to the Yang-Mills theory. If Maxwell's equations are presented in the language of differential forms as a Yang-Mills theory, then the connection is the electromagnetic vector potential.

The classical Yang-Mills equations over a Riemannian manifold may be expressed in differential form as

$$D\Gamma = 0 \tag{1.42}$$

$$* D * \Gamma = J \tag{1.43}$$

Where D represents the exterior covariant derivative, Γ is the curvature (or the Faraday electromagnetic tensor), $* D$ is the Hodge dual of D, $* \Gamma$ is the Hodge dual of Γ, and J is the source current. Gamma, Γ, is the curvature of the manifold where objects live, and all physical action takes place. The current, *J*, describes the charge distribution and the velocity of charges, $\left(C/m^2/s\right)$.

If the source current, *J*, is regarded as given, then the Yang-Mills equations are just a set of equations that constrain the connection. The source current, *J*, and the connection, completely determine all the physical properties of the system.

The curvature, Γ, may be derived from the connection through commutators of certain differential operators related to the connection. The curvature and the Faraday tensor from electromagnetism may generalize to each other. The Faraday tensor exerts its force on particles. The vector potential of electromagnetism introduces curvature on the manifold. The connection is the physical manifestation of that curvature. Light curves space-time.

The exterior covariant derivative, $D\Gamma$, is a measure of how fast curvature, Γ, is changing as we move around the manifold. The first Yang-Mills equation, $D\Gamma = 0$, tells us that the existing curvature, Γ, is not changing as we move around the manifold, even as the spatiotemporal wavelets expand and interfere. This first equation may be considered source-free for Minkowski space. The second Yang-Mills equation, $* D * \Gamma = J$, tells us that the way the curvature, $* \Gamma$, changes is determined by the source current, *J*, and that the current affects the curvature of the spatiotemporal medium of the connection. The Hodge isomorphism or Hodge star operator $(*)$ is an important linear map defined on the exterior algebra of a finite-dimensional oriented vector space endowed with a nondegenerate symmetric bilinear form. The result when applied to an element is called the element's Hodge dual. The Hodge star operator is a generalization, interchanging certain spatial degrees of freedom, and certain temporal degrees of freedom, on the exterior covariant derivative and on the curvature of the spatiotemporal medium. The Hodge star operator applies to an arbitrary number of dimensions.

Hence, the reason why the notions of the parallel transport of the vector potential of electromagnetism and the description of the covariant derivative of curvature coincide, is related to the spatiotemporal expansion of the medium which is the basic framework and background of all physical fields. The spatiotemporal wavelets expand and interfere in all directions unless obstructed, including along the trajectory of a current on a manifold. The spatiotemporal expansion of the medium is why the notion of a physical field plays a role in the description of parallel transport, or why the mathematical object to describe parallel transport, the exterior covariant derivative, has a physical significance as a field.

Thus, the current may represent moving charged particles through a field from the point of view of electromagnetism, or it may represent moving quanta of mass, real particles, from the point of view of gravitation, affecting the curvature of the manifold, which in turn affects parallel transport, and the connection, between two points.

1.3. The decomposition of a photon

The advent of very high energy electron-positron linear colliders allow the study of the collisions of beams of photons with energies a trillion times higher than those of ordinary light. In simple terms, two photons can directly collide and transform into quarks, gauge bosons, and scalar bosons, according to our present standard model. More explicitly, photon–to–photon collisions may yield W–gauge boson pairs, quark–antiquark pairs, pairs of gluons, pairs of photons, Z^0 gauge bosons, or one or more Higgs bosons, through quarks and W–gauge bosons.

It has long been a rule of classical optics that light cannot be affected by light. Nevertheless, photons can interact with each other through quantum processes. Virtual loops of quarks or leptons may be created when photons travel very near each other. These virtual particles would determine the interactions between photons before transmuting into real photons again. The interaction would be observed as the reflection of a photon by another photon.

Recently, high-energy photon–to–photon scattering experiments have provided new evidence that photons can interact with each other and change direction as previously predicted by quantum electrodynamics. In the high energy collisions of the heavy–ions of lead or gold, photons may be created. Photons created in nearly instantaneous collisions have a chance to collide, scatter off one another, and be observed, even though, the probability of such an event is possible, but minute. These interactions between photons are known as ultra-peripheral collisions. Two photons moving in opposite directions can collide head-on and move off in opposite directions if the photons have equal energies. If the photons have enough energy, an electron-positron pair might be produced. Other final states are allowed at even higher energies by the conservation of energy. The probability, or the cross-section, of a photon–to–photon scattering for a final state is calculated very accurately in the field of quantum electrodynamics.

Photons may also interact when photons transform into quark–antiquark pairs, or virtual mesons, that would interact with each other via the strong force, which binds quarks inside neutrons and protons.

Thus, it is possible that the quantum processes of the strong interaction between photons are also processes of the elementary particles. The research of the elastic collisions of photons, and the quantum processes involved, would lead to greater development of particle physics.

A photon has a charge of 0*e*, where "*e*" is the elementary charge for all particles. From Quantum Chromodynamics, we can define the charges of up and down quarks in terms of "*e*".

$$\uparrow u \equiv +\frac{2}{3}e \tag{1.44}$$

$$(3)\uparrow u \rightarrow +2e \tag{1.45}$$

$$\downarrow d \equiv -\frac{1}{3}e \tag{1.46}$$

$$(3)\downarrow d \rightarrow -e \tag{1.47}$$

Protons and neutrons consists of quarks and gluons. Gluons are the mediating force carriers between quarks. A proton, p^{+}, consists of two up quarks and one down quark. A neutron, n^{0}, consists of one up quark and two down quarks. There are six types of quarks, known as flavors: up, down, strange, charm, top, and bottom, and each flavor has three colors.

A quark of one flavor can transform into a quark of another flavor only through the weak interaction, one of the four fundamental interactions in particle physics.

$$(2)\uparrow u \;+\; (1)\downarrow d \rightarrow p^{+} \tag{1.48}$$

$$(1)\uparrow u \;+\; (2)\downarrow d \rightarrow n^{0} \tag{1.49}$$

From previous research, we know that a photon is neutral; a photon, γ^{0}, consists of the square of oppositely charged spatiotemporal extents, $\left(l_{P}/t_{P}\right)^{2}$, that are equivalent to the energy of a photon. The opposite charges of the photon may be expressed in terms of charged quarks, bosons, and gluons.

$$\gamma^{0} \rightarrow (3)\ \langle +q \rangle \;+\; (6)\ \langle -q \rangle \;+\; (5)\ \langle b^{\pm} \rangle \;+\; (9)\ \langle g^{0} \rangle \tag{1.50}$$

Where the colliding photon consists of three positive quarks, six negative quarks, five bosons, and nine gluons.

Hence, let us propose a table to describe the nine quarks, gauge bosons, scalar boson, and gluons, that may represent the charges emerging from a photon after a high energy collision.

The following chromodynamic table identifies the flavor, charge, colors, approximate mass, and the generation with a roman numeral, of each boson, gluon, or quark.

All quarks have spin-½, bosons and gluons have spin–1, and the scalar boson has spin–0. Quarks tend to decay from the heaviest generation III down to the lightest generation I, to reach greater stability. Moreover, a quark can transmute into another flavor of quark.

The Higgs particle is a boson, but it may be argued to some extent that it is the only boson in the table that does not mediate a fundamental force. W–bosons, Z–bosons, charged leptons, and quarks, may gain mass via the Brout–Englert–Higgs (BEH) mechanism.

Fermions, such as the leptons and quarks, in the Standard Model, can also acquire mass because of their interaction with the Higgs field, but not in the same way as the gauge bosons.

$\frac{2.4MeV}{c^2}$ $+\frac{2}{3}e$ Up Quark I	$\frac{4.8MeV}{c^2}$ $-\frac{e}{3}$ Down Quark I	$\frac{4.8MeV}{c^2}$ $-\frac{e}{3}$ Down Quark I	$<\frac{1.3MeV}{c^2}$ $0e$ Gluon(s) Colors	$\frac{125GeV}{c^2}$ $0e$ Higgs Boson
$\frac{1.275GeV}{c^2}$ $+\frac{2}{3}e$ Charm Quark II	$\frac{95MeV}{c^2}$ $-\frac{e}{3}$ Strange Quark II	$\frac{95MeV}{c^2}$ $-\frac{e}{3}$ Strange Quark II	$\frac{91.19GeV}{c^2}$ $0e$ Z-Boson	$\frac{91.19GeV}{c^2}$ $0e$ Z-Boson
$\frac{172.44GeV}{c^2}$ $+\frac{2}{3}e$ Top Quark III	$\frac{4.18GeV}{c^2}$ $-\frac{e}{3}$ Bottom Quark III	$\frac{4.18GeV}{c^2}$ $-\frac{e}{3}$ Bottom Quark III	$\frac{80.39GeV}{c^2}$ $+1$ W-Boson	$\frac{80.39GeV}{c^2}$ -1 W-Boson

Figure 8. The Chromodynamic Table of a Colliding Photon

Across the chromodynamic table, left to right, the sum of the three quarks represents a neutron. Top to bottom, the sum of the three quarks in the first column represents a positive charge equal to twice the elementary charge *"e"*. Top to bottom, in the second and third column, the sum of the three charges represents a negative elementary charge. It is interesting to note that the decomposition of a photon may render protons, electrons, and neutrons, in numerous permutations of the nine quarks. A proton may consist of two positively charged quarks and a negatively charged quark. A neutron may consist of two negatively charged quarks and one positively charged quark. An electron may consist of three negatively charged quarks. Gluons may be thought of as carrying both color and anticolor charge, having an experimental mass per gluon, up to an upper bound. Each quark has one of the three color charges and each antiquark has one of the three anticolor charges. There are eight independent types of gluon in quantum chromodynamics, since a combination of red, green, and blue color charges is color neutral and non-interacting. Color-charged particles exchange gluons in strong interactions.

When two quarks are close to one another, they exchange gluons and create a very strong color force field that binds the quarks together. The force field gets stronger as the quarks get further apart. Quarks constantly change their color charges as they exchange gluons with other quarks. The sum of the color charges of all nine quarks, five bosons, and gluons, of a photon, equals zero.

The following Gluonic Particle Model illustrates the gluon and the gluonic field as the fundamental building blocks of the photon, quarks, leptons, force carriers, and the Higgs field, which manifest from the spatiotemporal energy.

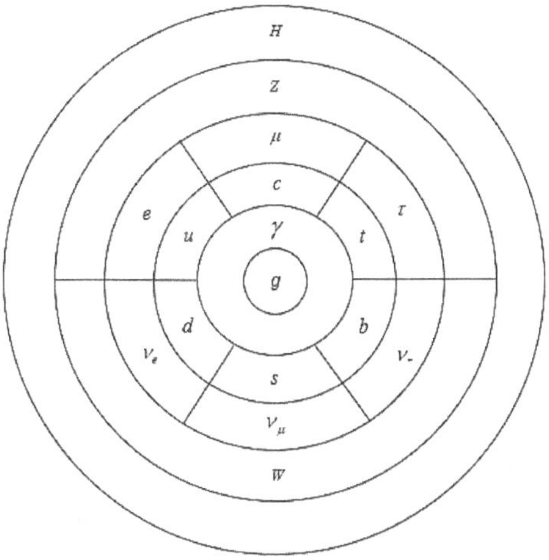

Figure 9. The Gluonic Particle Model

In fact, the Gluonic Particle Model illustrates all the particles that make up the current standard model of particle physics in a way that shows particle composition, massing, decay, and interactions.

Most of the mass, $(\approx 99.97\%)$, on the chromodynamic table of a photon is from the Top Quark, Bottom Quarks, Charm Quark, W-Bosons, Z-Bosons, and the Higgs Boson. Total mass of all quarks, bosons, and gluons, is approximately $650.5 GeV/c^2$. So, where is all this energy coming from? The high energy collisions would require an energy above the sum of the energy thresholds of colliding photons to produce the proposed particles per colliding photon, from the transmutation of the spatiotemporal medium.

If a photon were not massless, there are concerns about losing gauge invariance in the theory of quantum electrodynamics, which would make the theory non-renormalizable, and the conservation of charge would not be absolutely guaranteed. So, experiments have been performed to verify this prediction and place a limit on the rest mass value. The latest experimental limit is $7\times10^{-17} eV/c^2$. It is widely accepted that photons have relativistic mass, energy and momentum.

Thus, the ratio of $650.5GeV/c^2$ to the experimental limit of the rest mass of a photon is approximately 9.29×10^{27}. Hence, the bulk of the energy for the decomposition of a colliding photon into massive and massless particles comes from the kinetic energy impacted upon the spatiotemporal medium by the high energy collider producing the collision of photons.

Moreover, some of the mass of quarks comes from the energy of the interactions of quarks with the Higgs field, but the bulk of the mass of a quark, proton, neutron, or lepton, comes from the energy of the quantum fields of the gluons. It is speculated that as quarks get further apart within the boundary of a particle, the gluon field gets stronger, as if the strength of the gluon field is emerging from the transmutation of the spatiotemporal medium into the quantum energy of the gluon field. Contrary to Epicurus, or Parmenides, something comes into existence out of what is perceived as non-existent, but latent with energy and motion.

The transmutation of the spatiotemporal medium manifests quantum forms, quantum fields, and the innate laws of physics of our reality. Mass is a property of the substance of matter. Mass is the existential quantum form of energy, light, or the behavior of quantum fields. Hence, mass emerges from the actions of quantum fields.

1.4. The Gluon Field Tensor or Electrogravitic Tensor

The interaction of the gluon field between quarks may be represented by a second order tensor, a gluon field strength tensor, $R_{\varepsilon\beta}$, or electrogravitic tensor, in particle physics. The gluon field characterizes the strong interaction between quarks which represents the strong force in the quantum field theory of Quantum Chromodynamics. It is one of the four fundamental interactions of nature mediated by gluons according to their color charge. Quarks consist of three gluons that interact based on their color charge.

Gluons participate in strong interactions that confine color fields to flux tubes which exert force and increase their strength when stretched. The force of the flux tubes constrains quarks to the boundary of a configuration of particles called hadrons. Mesons and baryons are hadrons. Mesons are quark-antiquark pairs, positively charged pions and neutral kaons are mesons. Baryons are triplets of quarks or anti-quarks, protons, anti-protons, and neutrons are baryons. This boundary limits the interaction range to 1×10^{-15} meters, which is approximately the diameter of a proton. The energy of a flux tube increases linearly beyond a certain distance. At greater distances, it becomes more energetically feasible to manifest a quark-antiquark pair from space-time that elongating the flux tube. Quark and gluons may reproduce into more quarks, gluons, and jets of colorless particles, such as photons.

The lack of free quarks of color has verified the confinement of color for quarks. Quarks may be produced in pairs of quark-antiquark, or in a single production of top quark from a pair, to compensate for quantum numbers of color and flavor. The gluon density has been measured inside a proton.

The Higgs field is not responsible for the bulk of the mass acquisition of all particles. The bulk of the mass of baryons, a proton or a neutron, comes from the binding energy of gluons and the kinetic energy of quarks mediating the strong interaction inside the boundary of the baryons. Particles manifest a gain of potential energy when they couple with the Higgs field. (Jammer, 2000)

The gluon field is the source of electrogravitic fields and their manifestations, such as the strong force, charge, most of the mass-energy of a quark, and the spatiotemporal displacement pressure that leads to gravitational fields. The second order gluon field strength tensor, or electrogravitic tensor, is a tensor field on the spatiotemporal medium. The values of the rank-2 tensor are in the adjoint (conjugate) vector bundle, a vector bundle naturally associated to the principal bundle, of the *SU(6)* gauge group, a color symmetry, or gauge symmetry, of Quantum Chromodynamics. Noether's theorem reminds us that any continuous symmetry of the laws of physics gives rise to a conservation law. *SU(6)* underlies the color symmetry conservation of Quantum Chromodynamics. The general special unitary group of degree six, denoted *SU(6)*, is the Lie group of (6×6) unitary matrices with complex determinant and an absolute value of one.

SU(6) corresponds to a general special unitary transformation on complex six-dimensional vectors. The natural representation is that of (6×6) matrices acting on complex six-dimensional vectors. There are (9) parameters, $(n^2 - 1)$ possible generators, from which only (9) generators, $(X_0, X_1, \ldots X_8)$, are applicable, and (8) generators, $(X_1, \ldots X_8)$, are equivalent to the *SU(3)* Gell-Mann generators.

A Hermitian matrix, or self-adjoint matrix, is a complex square matrix that is equal to its own Hermitian conjugate transpose, that is, the element in the *i-th* row and *j-th* column is equal to the complex conjugate of the element in the *j-th* row and *i-th* column, for all indices *i* and *j*, in matrix form. Moreover, the following conditions apply for a matrix *A* that if Unitary, $A^H = A^{-1}$, if Hermitian, $A = A^H$, if Orthogonal, $AA^H = I_n$, where I_n is the identity matrix, and A^{-1} is the inverse matrix of *A*. The traceless and Hermitian generators, $X_i = 1/4\,\lambda_i$, are derived from the six-dimensional Gell-Mann matrices, and they can generate unitary matrix group elements through exponentiation and obey the extra trace orthonormality relation. Thus, the nine-dimensional vectors, corresponding to the matrices, are orthogonal.

These properties were chosen because they naturalize the Gell-Mann's matrices for *SU(3)* to *SU(6)* which forms the basis of the six-dimensional quark model. Similarly, the Pauli matrices naturally generalize to the Gell-Mann matrices, for *SU(2)* to *SU(3)*. This generalization further extends to general *SU(n)*.

If a sum over the index *c* is implied, the (8) interacting infinitesimal generators of the Lie algebra are indexed by *a*, to satisfy the following commutation relation.

$$\left[\frac{\lambda_a}{4},\frac{\lambda_b}{4}\right]=if^{abc}\frac{\lambda_c}{4} \tag{1.51}$$

Orthogonalization involves the process of finding a new set of orthogonal vectors in linear algebra that span a particular subspace of an old linearly independent set of vectors. The new set and the old set have the same linear span. Every vector in the new set is orthogonal to every other vector in the new set. The process of orthonormalization makes the resulting vectors to all be unit vectors.

The six-dimensional Gell-Mann matrices satisfy the orthonormalization condition $trace(\lambda_a\lambda_b)=\delta_{ab}/2$ where δ_{ab} is the Kronecker delta. The trace of an $(n\times n)$ square matrix A in linear algebra is defined as the sum of the elements on the main diagonal. In other words, when two of the six-dimensional Gell-Mann matrices are multiplied, the resultant matrix will have a trace that is one half of the Kronecker delta. Consequently, the six-dimensional Gell-Mann matrices are normalized to ½.

The gluon field strength tensor, or electrogravitic tensor, $R_{\varepsilon\beta}$, may be defined with components that are proportional to the quark covariant derivative, D_ε. (Bilson-Thompson, 2003)

$$R_{\varepsilon\beta}=\pm\frac{1}{ig_s}\left[D_\varepsilon,D_\beta\right] \tag{1.52}$$

$$D_\varepsilon=\vec{\mathfrak{R}}_\varepsilon\pm ig_st_a\mathrm{A}_\varepsilon^a \tag{1.53}$$

Where i is $\sqrt[2]{-1}$, $g_s\left(\approx 1\right)$ is the coupling constant of the strong force, $\vec{\mathfrak{R}}_\varepsilon$ is the six-gradient or the Robertonian operator, t_a are the six-dimensional Gell-Mann matrices $t_a=\lambda_a/4$, "a" is a color index in the adjoint representation of *SU(6)* which takes values 0 through 8 for the nine generators of the group, namely the six-dimensional Gell-Mann matrices, ε is a space-time index,

(t_x, t_y, t_z) for time-like components, and "x", "y", and "z" for space-like components. (Greiner, 1994)

The adjoint representation of a Lie group is a way of representing the group elements as linear transformations of the Lie algebra of the group, in a vector space.

The gluon field, a spin-1 gauge field, or a connection in differential geometry for the *SU(6)* principal bundle, may be expressed by $\mathrm{A}_\varepsilon = t_a \mathrm{A}_\varepsilon^a$. The coordinate-system dependent six components, A_ε, are (6×6) traceless Hermitian matrix-valued functions in a fixed gauge, and A_ε^a, are the (54) real-valued functions of the (6) components for each of the (9) six-vector fields.

The six-dimensional indices $(\mu, \nu, \varepsilon, \beta)$ take temporal values of (t_x, t_y, t_z) and spatial values of (x, y, z) for components of the six-vector and six-dimensional spatiotemporal tensors. Indices (*a, b, c, n*) take values of 0 through 8 for eight gluon color charges and one colorless charge. The Einstein summation convention is used on all tensor and color indices.

The commutator may be expanded as follows,

$$R_{\varepsilon\beta} = \vec{\Re}_\varepsilon \mathrm{A}_\beta - \vec{\Re}_\beta \mathrm{A}_\varepsilon \pm i g_s \left[\mathrm{A}_\varepsilon, \mathrm{A}_\beta\right] \qquad (1.54)$$

We may substitute $\mathrm{A}_\varepsilon = t_a \mathrm{A}_\varepsilon^a$, and $if_{ab}{}^c t_c = [t_a, t_b]$ as a combination relation, for the six-dimensional Gell-Mann matrices with the indices re-labeled, in which the structure constants, or structure coefficients, of *SU(6)* are f^{abc}, and each of the components of the gluon field strength may be expressed as a linear combination of the six-dimensional Gell-Mann matrices.

It is interesting to note that the structure coefficient specifies the product of two basis vectors explicitly, in a Lie algebra over a field, as a linear combination. The Lie bracket product is bilinear, or linear in each of the variables separately, and uniquely extended to all vectors in the vector space.

The basis vectors represent specific directions in the physical vector space, or may represent specific gluons, or other specific particles. The structure coefficients provide specific starting points for the Lie algebra.

Expressing the gluon field strength components as a linear combination of the six-dimensional Gell-Mann matrices we have

$$R_{\varepsilon\beta} = \vec{\mathfrak{R}}_{\varepsilon} t_a \mathrm{A}^a_{\beta} - \vec{\mathfrak{R}}_{\beta} t_a \mathrm{A}^a_{\varepsilon} \pm i g_s [t_b, t_c] \mathrm{A}^b_{\varepsilon} \mathrm{A}^c_{\beta} \tag{1.55}$$

$$R_{\varepsilon\beta} = t_a \left(\vec{\mathfrak{R}}_{\varepsilon} \mathrm{A}^a_{\beta} - \vec{\mathfrak{R}}_{\beta} \mathrm{A}^a_{\varepsilon} \pm i^2 f_{bc}{}^{a} g_s \mathrm{A}^b_{\varepsilon} \mathrm{A}^c_{\beta} \right) = t_a R^a_{\varepsilon\beta} \tag{1.56}$$

$$R^a_{\varepsilon\beta} = \vec{\mathfrak{R}}_{\varepsilon} \mathrm{A}^a_{\beta} - \vec{\mathfrak{R}}_{\beta} \mathrm{A}^a_{\varepsilon} \mp g_s f^{a}{}_{bc} \mathrm{A}^b_{\varepsilon} \mathrm{A}^c_{\beta} \tag{1.57}$$

Where *a*, *b*, and *c* are the color indices with value 0 through 8. $R_{\varepsilon\beta}$ are (6×6) traceless Hermitian matrix-valued functions in a specific coordinate system, with a fixed gauge, and $R^{a}{}_{\varepsilon\beta}$ are real-valued functions, the components of nine six-dimensional second order tensor fields.

There are interactions between gluons and asymptotic freedom, which describes how the bonds of particles asymptotically weaken in some gauge theories as distance decreases and energy increases.

Moreover, it is important to note that quantum chromodynamic group operations are not commutative, or in other words, quantum chromodynamics is a non-abelian gauge theory, so the order of applying the group operation to two group elements does depend on the order in which they are written.

The gluon field strength has extra terms that lead to self-interactions, a complication of the strong force, on the one hand, that makes the gluon field strength inherently non-linear.

On the other hand, the theory of the electromagnetic force is linear. This crucial distinction differentiates quantum chromodynamics from electrodynamics.

The Quantum Chromodynamic Lagrangian density for quarks and their gluon field is given by

$$L = -\frac{1}{2} trace\left(R_{\varepsilon\beta} R^{\varepsilon\beta}\right) + \Psi^*(r,t)(iD_\varepsilon)\gamma^\varepsilon \vec{\Psi}(r,t) \qquad (1.58)$$

Where the trace represents the six-dimensional matrix $\left(R_{\varepsilon\beta} R^{\varepsilon\beta}\right)$,

D_ε is the quark covariant derivative, and γ^ε are the six-dimensional gamma matrices.

The Lagrangian density equation is the equation of motion for the Gluon field, which characterizes the dynamics of the Gluon field strength. (Greiner, 1994) The Gluon field strength tensor, by itself, is not gauge invariant.

The product of two contracted Gluon field strength tensors on all indices is gauge invariant. The six-dimensional Dirac equation is the equation of motion, and the equation that controls and directs the Gluon fields of quarks. (Yagi, 2005)

The equations of motion governing the evolution of the quark fields are:

$$\left(i\hbar c \gamma^\varepsilon \vec{\Re} - \beta\, m' c^2\right)\vec{\Psi}(r,t) = 0 \qquad (1.59)$$

The relativistic six-dimensional quantum mechanical wave equation, including electromagnetic interactions, describes all spin-½ massive particles for fermions (all quarks and leptons), that are symmetric under parity, or symmetric if the sign of one spatial coordinate is flipped in three dimensions.

Six-dimensional parity is also described by the simultaneous flip in the sign of all three spatial coordinates, and all three temporal coordinates, a six-dimensional point reflection.

The Gluon field strength tensor, or electrogravitic tensor, is given by

$$\left[\vec{\Re}_{\varepsilon}, R^{\varepsilon\beta}\right] = g_s j^{\beta} \tag{1.60}$$

The color-charge six-current, j^{β}, is the source of the Gluon field strength tensor, and $\vec{\Re}_{\varepsilon}$ is the six-gradient, or Robertonian operator.

These equations of motion are similar to the Yang–Mills equations for gluons and quarks, or the four-dimensional Maxwell equations in tensor notation.

The color charge six-current is the source of the gluon field strength tensor, similar to the electromagnetic six-current as the source of the electromagnetic tensor.

The color charge six-current is given by

$$j^{\beta} = t^{b} j_{b}^{\beta} \tag{1.61}$$

$$j_{b}^{\beta} = \Psi^{*} \gamma^{\beta} t^{b} \Psi \tag{1.62}$$

which is a conserved current since color charge is conserved. In other words, the color six-current must satisfy the continuity equation:

$$\vec{\Re}_{\beta} j^{\beta} = 0 \tag{1.63}$$

Electrodynamics *(E)*, Color-Charge Current $\left(J^{\beta}\right)$, and Spatiotemporal Curvature (Γ), provide three of the fundamental sources of quantum fields for electrogravitic force fields, energy density, and spatiotemporal pressure.

From the Quantum Triadic Conjunction of $EJ\Gamma$ emerges the standard model of gluonic fields: the strong force, charge, the weak force, electromagnetism, energy, mass, particles, force carriers, matter, and gravitational fields.

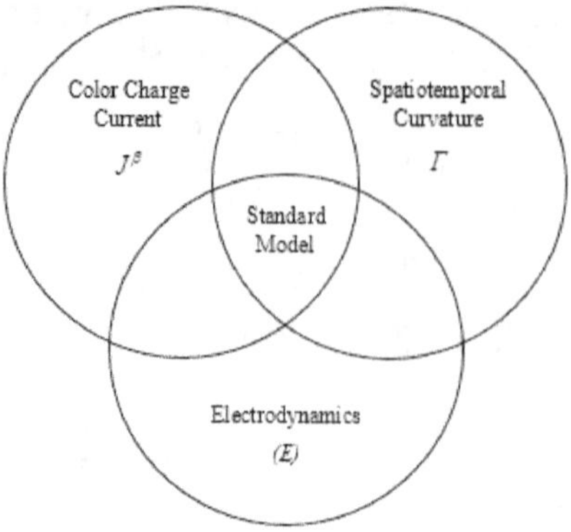

Figure 10. The Quantum Triadic Conjunction of *EJΓ*

The photon is a color singlet state that exists as the non-interacting color state in the color palette of the nine color states of Quantum Chromodynamics. The photon is in a stable colorless state, or color singlet state, a neutral gluon, that does not interact with color states, but it can interact with other color singlet states, or photons, depending on its energy level. The latter high energy gluon interaction of the photon represents the only gluon long-range interaction. A color singlet state is mathematically similar to a spin singlet state.

The decomposition of the photon manifests all elementary particles including a photon as a color singlet state. As this photon emerges, and depending on its energy level, it may collide with another photon in a peripheral collision, and decompose into additional elementary particles, or it may exist as electromagnetic field. The emerging photon does not interact with the strong force but may become the force carrier for electromagnetism. All fundamental particles may originate from light. The photon is a source for the manifestation of all particles, matter, energy, and fields. In a sense, the genesis of our universe emerges from space, time, and light. (Griffiths, 1987)

The color singlet state of the photon is given by

$$\gamma^0 = \frac{r\bar{r} + b\bar{b} + g\bar{g}}{\sqrt[2]{6}} \tag{1.64}$$

Thus, if the color of state of a photon is measured, there would be equal probabilities of the state of the photon being red-antired, blue-antiblue, or green-antigreen.

Both quarks and antiquarks carry three types of color charge: red, green, and blue, and antired, antigreen, and antiblue. A gluon carries both color and anticolor. Thus, the color index gives gluons nine combinations of color charge. These colors are not actual colors, or observed colors, on gluons, but the representations of the effective states that a gluon carries to interact with other gluons. The following color matrix has the names of those combinations in a six-dimensional matrix.

$$\lambda_a = Color\ Matrix = \begin{vmatrix} red & antired & red & antiblue & red & antigreen \\ red & antired & red & antiblue & red & antigreen \\ blue & antired & blue & antiblue & blue & antigreen \\ blue & antired & blue & antiblue & blue & antigreen \\ green & antired & green & antiblue & green & antigreen \\ green & antired & green & antiblue & green & antigreen \end{vmatrix} \quad (1..65)$$

The six-dimensional Gell-Mann matrices, the ninefold way, are given by λ_a divided by 4,

$$t_a = \frac{\lambda_a}{4} \quad (1.66)$$

There are nine gluon colors, but only eight are linearly independent interacting color states, which correspond to the eight colors of gluons, called the color octet. States can be mixed together and presented in many ways.

The color singlet state is the neutral gluon state, or non-interacting gluon state, corresponding to a photon. There are no available combinations between any of the eight color states of the nine states to produce any other color state, or to produce the singlet color state. (Baez, 1996)

The equations and matrices that follow are the six-dimensional Gell-Mann matrices.

$$\lambda_0 = \frac{2\left(r\bar{r} + b\bar{b} + g\bar{g}\right)}{\sqrt[2]{6}} = \frac{1}{\sqrt[2]{3}}\begin{vmatrix} 1 & 1 & 0 & 0 & 0 & 0 \\ 1 & 1 & 0 & 0 & 0 & 0 \\ 0 & 0 & 1 & 1 & 0 & 0 \\ 0 & 0 & 1 & 1 & 0 & 0 \\ 0 & 0 & 0 & 0 & 1 & 1 \\ 0 & 0 & 0 & 0 & 1 & 1 \end{vmatrix} \tag{1.67}$$

$$\lambda_1 = \frac{2\left(r\bar{b} + b\bar{r}\right)}{\sqrt[2]{2}} = \begin{vmatrix} 0 & 0 & 1 & 1 & 0 & 0 \\ 0 & 0 & 1 & 1 & 0 & 0 \\ 1 & 1 & 0 & 0 & 0 & 0 \\ 1 & 1 & 0 & 0 & 0 & 0 \\ 0 & 0 & 0 & 0 & 0 & 0 \\ 0 & 0 & 0 & 0 & 0 & 0 \end{vmatrix} \tag{1.68}$$

$$\lambda_2 = \frac{-2i\left(r\bar{b} - b\bar{r}\right)}{\sqrt[2]{2}} = \begin{vmatrix} 0 & 0 & -i & -i & 0 & 0 \\ 0 & 0 & -i & -i & 0 & 0 \\ i & i & 0 & 0 & 0 & 0 \\ i & i & 0 & 0 & 0 & 0 \\ 0 & 0 & 0 & 0 & 0 & 0 \\ 0 & 0 & 0 & 0 & 0 & 0 \end{vmatrix} \tag{1.69}$$

$$\lambda_3 = \frac{2\left(r\bar{r} - b\bar{b}\right)}{\sqrt[2]{2}} = \begin{vmatrix} 1 & 1 & 0 & 0 & 0 & 0 \\ 1 & 1 & 0 & 0 & 0 & 0 \\ 0 & 0 & i^2 & i^2 & 0 & 0 \\ 0 & 0 & i^2 & i^2 & 0 & 0 \\ 0 & 0 & 0 & 0 & 0 & 0 \\ 0 & 0 & 0 & 0 & 0 & 0 \end{vmatrix} \tag{1.70}$$

$$\lambda_4 = \frac{2\left(r\bar{g} + g\bar{r}\right)}{\sqrt[2]{2}} = \begin{vmatrix} 0 & 0 & 0 & 0 & 1 & 1 \\ 0 & 0 & 0 & 0 & 1 & 1 \\ 0 & 0 & 0 & 0 & 0 & 0 \\ 0 & 0 & 0 & 0 & 0 & 0 \\ 1 & 1 & 0 & 0 & 0 & 0 \\ 1 & 1 & 0 & 0 & 0 & 0 \end{vmatrix} \tag{1.71}$$

$$\lambda_5 = \frac{-2i\left(r\bar{g} - g\bar{r}\right)}{\sqrt[2]{2}} = \begin{vmatrix} 0 & 0 & 0 & 0 & -i & -i \\ 0 & 0 & 0 & 0 & -i & -i \\ 0 & 0 & 0 & 0 & 0 & 0 \\ 0 & 0 & 0 & 0 & 0 & 0 \\ i & i & 0 & 0 & 0 & 0 \\ i & i & 0 & 0 & 0 & 0 \end{vmatrix} \tag{1.72}$$

$$\lambda_6 = \frac{2\left(b\bar{g} + g\bar{b}\right)}{\sqrt[2]{2}} = \begin{vmatrix} 0 & 0 & 0 & 0 & 0 & 0 \\ 0 & 0 & 0 & 0 & 0 & 0 \\ 0 & 0 & 0 & 0 & 1 & 1 \\ 0 & 0 & 0 & 0 & 1 & 1 \\ 0 & 0 & 1 & 1 & 0 & 0 \\ 0 & 0 & 1 & 1 & 0 & 0 \end{vmatrix} \tag{1.73}$$

$$\lambda_7 = \frac{-2i\left(b\bar{g} - g\bar{b}\right)}{\sqrt[2]{2}} = \begin{vmatrix} 0 & 0 & 0 & 0 & 0 & 0 \\ 0 & 0 & 0 & 0 & 0 & 0 \\ 0 & 0 & 0 & 0 & -i & -i \\ 0 & 0 & 0 & 0 & -i & -i \\ 0 & 0 & i & i & 0 & 0 \\ 0 & 0 & i & i & 0 & 0 \end{vmatrix} \tag{1.74}$$

$$\lambda_8 = \frac{2(r\bar{r} + b\bar{b} - 2g\bar{g})}{\sqrt[2]{6}} = \frac{1}{\sqrt[2]{3}} \begin{vmatrix} 1 & 1 & 0 & 0 & 0 & 0 \\ 1 & 1 & 0 & 0 & 0 & 0 \\ 0 & 0 & 1 & 1 & 0 & 0 \\ 0 & 0 & 1 & 1 & 0 & 0 \\ 0 & 0 & 0 & 0 & 2i^2 & 2i^2 \\ 0 & 0 & 0 & 0 & 2i^2 & 2i^2 \end{vmatrix} \quad (1.75)$$

The color state of a gluon describes the color–anticolor pairs that the gluon carries. In the case of quantum particles, or gluons, the particle probability states may be added, or combined, by the principle of superposition, to give several different probability outcomes. If the color singlet state of a gluon, $(r\bar{r} + g\bar{g} + b\bar{b})/\sqrt[2]{6}$, were measured, there would be a 33% chance of it having red–antired, or 33% chance of it having green-antigreen, or 33% chance of it having blue-anti-blue, color charge.

The factor of $\sqrt[2]{2}$, $\sqrt[2]{3}$, or $\sqrt[2]{6}$, is required for normalization. Normalization is the scaling of the six-dimensional Gell–Mann matrices so that all the probabilities add to one. The probabilistic description of the six-dimensional Gell–Mann matrices makes the best sense only when probabilities add to one.

1.5. The transmutation of space-time into mass

The relativistic energy-momentum equation expresses the relationship between the rest mass of a body, m_0, the total energy, E_T, and the magnitude of its momentum, *p,* by

$$E_T^{\ 2} = m_0^{\ 2} c^4 + p^2 c^2 \quad (1.76)$$

$$E_T^{\ 2} - p^2 c^2 = m_0^{\ 2} c^4 \quad (1.77)$$

$$m_0^{\ 2} = \frac{E_T^{\ 2} - p^2 c^2}{c^4} \quad (1.78)$$

The speed of light may be expressed in terms of the permittivity and permeability of free space-time.

$$\varepsilon_0 \mu_0 = \left(\frac{1}{4\pi} \frac{t_p^{\ 2}}{F_q} \right) \left(4\pi \frac{F_q}{l_p^{\ 2}} \right) = \frac{t_p^{\ 2}}{l_p^{\ 2}} = \frac{1}{c^2} \tag{1.79}$$

The photon, or light quantum, may be represented as the linearly polarized complex spatiotemporal unit of area oscillating at an angle. The unit of mass, *Kg*, has been used as a unit of weight and a unit of mass. It is useful to propose a per-unit, or percent, system of mass based on the photon to describe the mass, or energy, of any particle. The mass of any particle may be expressed in terms of a per-unit multiple of the photon mass, m_p , or base mass.

The mass of a particle may be represented by a proportional ratio of the mass, or energy, of a particle, to the relativistic mass, m', or energy, E_p, of a photon.

$$m'c^2 = E_p \tag{1.80}$$

$$\left(\frac{m_p}{m_p} \right) c^2 = E_p \tag{1.81}$$

$$c^2 = \left(\frac{l_p}{t_p} \right)^2 \equiv E_p \tag{1.82}$$

The energy of a particle is an equivalent form of energy that is proportional quanta to the pure energy of a light quantum, or photon. A particle has a rest energy that is orders of magnitude larger than the energy of a photon.

Hence, the ratio of the rest mass of a particle to the mass of a photon may be represented as a proportional ratio between the rest energy of the particle, E_0 , to the energy of a photon.

$$\frac{m_0 c^2}{m_p c^2} = \frac{E_0}{E_p} \tag{1.83}$$

$$\frac{m_0}{m_p} E_p = \frac{m_0}{m_p} c^2 = \frac{m_0}{m_p} \frac{l_p^{\ 2}}{t_p^{\ 2}} = E_0 \tag{1.84}$$

Similarly, the square ratio of rest space, s_0, to rest time, t_0, of a particle in free space-time, may be represented by its rest energy.

$$c_0^{\ 2} = \left(\frac{s_0}{t_0}\right)^2 \equiv E_0 \tag{1.85}$$

The mass of a particle may be expressed as the ratio of the square rest space, s_0, to the square of the rest time, t_0, divided by the square of the speed of light in free space-time, c_0.

$$m_0 = \frac{\sqrt[2]{E_T^{\ 2} - p^2 c^2}}{c^2} = \frac{1}{c^2} \sqrt[2]{\left(\frac{s}{t}\right)^4 - \left(\frac{\partial s}{\partial t}\right)^4} = \frac{1}{c_0^{\ 2}} \left(\frac{s_0}{t_0}\right)^2 \tag{1.86}$$

Where E_T represents the total energy of the particle, $(s/t)^2$, the rest energy of a particle is $(s_0/t_0)^2$, and the energy of momentum of a particle is $(\partial s/\partial t)^2$, in free space-time.

$$m_0 = \left(\frac{t_p}{l_p}\right)^2 \left(\frac{s_0}{t_0}\right)^2 = \frac{\left(\frac{s_0}{l_p}\right)^2}{\left(\frac{t_0}{t_p}\right)^2} = \left(\frac{s_0 \cdot t_p}{t_0 \cdot l_p}\right)^2 = \left(\frac{n_s\, q_s}{n_t\, q_t}\right)^2 = n_{st}^{\ 2} \left(\frac{q_s}{q_t}\right)^2 \tag{1.87}$$

Where q_s is a magnetic charge from the rest space of a particle, q_t

is an electric charge from the rest time of a particle, n_{st} is the ratio of the number of magnetic charges to electric charges that may be manifested from the mass of a particle. At a threshold, as time approaches zero, $t \to 0$, both rest mass and rest energy are manifested, m_0 >> 0 and E_0 >> 0, in the pure spatiotemporal vacuum. Mass may be created from space-time, or vice versa. Mass, Charge, and Energy, are all aspects of space-time.

A volume of highly condensed space-time, transmutes to mass below a threshold as time approaches zero, as the spatiotemporal corpus reaches the speed of light. It is not being suggested that matter particle mechanisms, such as the Higgs boson for the accrual of quark mass, force carriers, or fundamental particles, are not involved in the performance of quantum interactions according to the current standard model of physics, since those matter particles, or their sub-particles, are themselves spatiotemporal manifestations. It is interesting to note that the Higgs boson gives mass to the quarks, and quarks make up the protons and neutrons in the nucleus of an atom, but only approximately two percent of the mass of protons and neutrons is provided by the quarks, and the rest is from the energy in the gluons, which begs the question, where is the mass and energy coming from?

String theory, or M-theory, offer similar conclusions about the nature of matter. Strings, and D-branes, offer an interesting substructure to Planck units of space and time. Dimensions of space may be curled up or condensed. A spatial dimension has a conjugate temporal dimension. If the spatial dimension is curled up, or not extended, the conjugate temporal dimension is not extending, and time has approached zero. We may ask the rhetorical question, has String Theory suggested a similar mechanism for timeless and condensed spatiotemporal volumes in the embodiment of fundamental particles?

Photons are typically described as luxons, or particles that translate at *c*, that are timeless. Tardyons are described as particles that translate slower than *c*. However, spatiotemporal transmutation into mass is speculated to occur as time approaches zero below a threshold of the expansion of space within the boundary of the corporeal manifestation. Hence, any particle or mass must move at *c*, or very close to c, to maintain its corpus. Either, the highly condensed spatiotemporal volume of a particle, or antiparticle, becomes timeless, consequently, the corpus of the particle, or antiparticle, is moving at c, without having to translate through space at the speed of light, or each particle, or antiparticle, is always moving at c with a trembling motion, $\langle v \rangle = \pm c$, due to weaker Coulombic forces near protons at Compton wavelength distances near an atomic nucleus, or in free space-time, which makes the motion appear slower if the orbital translation of the particle is considered perpendicular to the trembling motion, to abide by the General Theory, or the Special Theory of Relativity.

In such scenario, a tardyon may translate at less than c, when acted upon by a force, while still traveling at the speed of light to maintain its physical form as a matter particle in free space-time. If the tardyon were to slow down below c, at a threshold of the expansion of space, its mass would revert to its spatiotemporal volume in the purely spatiotemporal vacuum. All particles would be luxons, but only photons would be luxons without trembling motion.

The fundamental nature of space and time consists of the following guiding principles:

- Space-time is the genuine physical source of all there is.

- Space-time may expand, or contract, in every direction at every point unless obstructed.

- The spatiotemporal wave springs light and relativistic mass.

- Space-time is eternal in its expressions, dimensions, and continuous phases of change, in any of its infinite forms.

Mass is an expression of the infinite forms of space-time. Matter is a permutation of the infinite arrangements of mass. Space-time is the primordial source of energy imbued into light or mass. Energy predates matter in our universe. Light, as an electromagnetic wave, is the primordial and eternal energy of our universe. The infinite forms of space-time are the manifestations of light, mass, and matter, of our universe. Free space-time is the source of the manifestation of light, mass, and matter, that has not metaphorically awakened. Light is omnipresent and existential in our universe. Mass, or matter, may be endowed with other forms of energy from space-time, or when mass, or matter, reverts to the elementary forms of space-time. From space and time, there is light, gravitation, mass, matter, and life in all its aspects.

The three dimensions of space, the three dimensions of time, and the three directions of electromagnetic charge, emanate from the same genuine infinite source. All there is in our universe comes from primordial light. Our senses are compatible expressions of light. Light is the music in the concert of space-time; the first musical of physical creation. Light, mass or matter, are the waves and particles in the symphony of life. Life is an inevitable consequence of the physical properties of space, time, light, and creation.

1.6. Is faster than light speed attainable?

There are factual advantages in considering faster than light speeds in the field of physics.

According to recent estimates our 13.79-billion-year-old observable universe currently has an estimated comoving radial distance of 46.6 billion light years. Astronomers using the Hubble telescope have detected objects moving at four to six times the speed of light relative to the earth reference frame. Thus, it is presently assumed that most of that speed, certainly any speed over c, is due to the expansion of space-time itself. But if a massive celestial object moves away from us, relative to our reference frame, at a speed faster than light, how are we able to see it?

If the light emitted by a massive celestial object was emitted when the object was not moving so fast away from our reference frame, then the light would eventually reach us. As the space-time expands, the incident light travels through different regions of space-time, the red shift occurs, in the trajectory to our reference frame. Consequently, as the massive celestial object is carried away by expanding space-time faster and faster over time from the perspective of our reference frame, the light emitted from that same object will eventually never be able to bridge the gap.

Moreover, it is forecast that in the far distant future more and more celestial objects at the outer regions of our viewing range will eventually be unobservable from our earth reference frame with present technology. Let us consider the Earth from a top view shown below as a reddish spherical object on the circumference of a circle, on a spatiotemporal plane, with three other celestial objects.

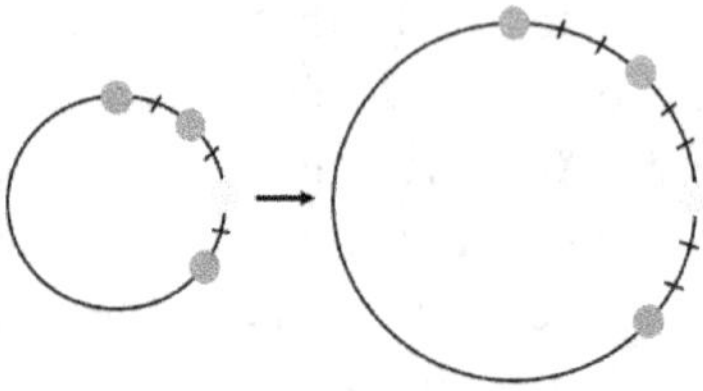

Figure 11.

The distance between the red and the green spherical objects on the circumference of a circle on a plane of space-time are two arc units of distance; after the expansion of the spatiotemporal area of the circle at the speed of light, there are three arc units of distance apart. Let us consider the red and blue spherical objects on the circumference of the circle, they are at six arc units of distance; after the expansion at the speed of light, they are at nine arc units of distance apart. The blue spherical object has comoved three times as fast, or at three times the speed of light, from the red spherical object than the green spherical object. Consequently, the objects are moving apart in space-time at superluminal speed.

However, we may also consider the perspective that each spherical object comoves at the speed of light from a central point on the smaller circle, as all spherical objects sit still at their respective spatiotemporal location, while space-time carries the objects outward at the speed of light, as it expands.

The polarization synchrotron is a technological device, which combines the radio waves with a rapidly spinning magnetic field, to make radio waves travel faster than light. The waves are packed into a very powerful wave the size of a small dot. Some of the potential applications for this technology are: more powerful direct cell phone communications with satellites, faster data transmission in computers, nearly instantaneous interconnecting buses and faster memory devices for microprocessors, faster radio wave-activated drugs for chemotherapy to target specific areas of a patient's body.

A sinusoidal voltage signal is applied to an arc-shaped dielectric material (alumina), with electrodes positioned along its two-meter length at regular intervals to displace the voltage phase very slightly between the electrodes to generate a sinusoidally-varying polarization pattern that moves along the device. The radio waves are made to travel faster than light by carefully adjusting the frequency of the voltage and phase displacement.

Therefore, faster than light technological applications are attainable in certain spatiotemporal plenums of dielectric mediums and across spatiotemporal expanse.

§ 2. On the formation of a supermassive black hole

Let us imagine that space is able to extend or contract faster than light and that the physical laws of the universe remain the same to allow non-homogeneous and anisotropic space to extend or contract at several times the speed of light *c*, through several consecutive light barriers such as *c, 2c*, *3c*, *4c*, *5c,* etc., such as in the case of the formation of a charged rotating supermassive Kerr-Newman black hole. Even though, the spatiotemporal wave contracts from the outer event horizon toward the ring singularity of the Kerr-Newman black hole, the spatial and temporal waves are reciprocal.

Mass and time condense, and length extends at every light barrier reached, seemingly making a very infinitesimally small relativistic mass m' appear to disappear, when the contracted mass minimizes its dimensions from its previous volume before arriving at the light barrier. At the same instant, relativistic time is highly dilated, so when the mass condenses at the light barrier, time is proportionally condensed from its previous temporal magnitude. However, time speeds up as it condenses at each light barrier. As time condenses, the relativistic mass travels faster, increasing its kinetic energy, and time speeds up as length extends. Furthermore, the gravity field of the mass decreases proportionately, as mass and time contract, during every light barrier event.

The electromagnetic forces of the unified electroweak interaction of the mass and charges become stronger than the gravitational field of the mass during every light barrier event. As atomic structures, charge distribution, and energy, collapse into a tighter, more condensed, mass distribution, the increasing pressures and temperatures allow the positively charged nuclei of mass to combine. As the mass coheres, charges condense under the immense tidal forces of the collapsing body, creating an increasingly cohesive charge distribution about the geometry of the forming ring, springing very strong electric fields out of the condensed ring as charges focally direct their electric field lines radially outward. (Melia, 2007 and 2009)

As the density of the mass increases during the collapse of the celestial body, the surface area of its space-time-mass boundary gets smaller and the force of space-time acting on the mass has a decreasing surface area to act upon. Thus, it is theorized that as the mass condenses to a ring, the pressure of space-time on the infinitesimal mass surface exerts a gravitational field, until the mass of the ring collapses into the spatiotemporal dimensions of the wave medium, outside of local space-time, in what becomes the outer event horizon of the Kerr-Newman black hole. Thus, the spatiotemporal wavelength flowing inward contracts, and the spatiotemporal pressure increases and exerts a greater gravitational potential effect on mass as space extends and time condenses inward.

Hence, as the mass speeds up through light barriers, its gravity weakens, and its electromagnetic field strengthens, as the mass condenses. It is theorized that the strength of the electromagnetic field forces of the unified electroweak interaction of the mass and charges acts on the inner area of the outer event horizon of the Kerr-Newman black hole, counteracting the inner gravitational field, to preserve the geometry and boundary of the black hole, creating an effective demarcation for the event horizon between inner and outer space-time. This theoretical action sustains the concentrated electromagnetic field strength of the ring singularity and also establishes a stable gravitational field for the emerging black hole.

As the ring forms, the weak interaction can convert nucleons of one kind to the other, emitting an electron or a positron in the process through beta radioactivity. The strong interaction grows stronger as it binds nucleons together and the nuclei combine under the extreme temperatures and pressures of the extremely dense mass. As the mass continues to collapse further within the spatiotemporal dimensions of the outer event horizon, a spatial distance develops, and its boundary is the inner event horizon of the Kerr-Newman black hole. As the ring reaches a stable existence as an infinitesimal mass, it exists within the inner event horizon space with infinitesimal nonzero dimensions. Thus, it is theorized that the ring has attributes of very condensed mass, spatiotemporal dimensions, and it exists in an infinitesimal localized temporal bubble within the black hole. However, the ring exists in localized time outside of the universal space-time beyond the ergosphere of the black hole. This condition would decouple the gravitational potential effects of the ring's mass within the inner event horizon from the outside universal space-time.

As the geometry of the black hole is sustained by the concentrated electromagnetic field and internal gravity, a gravitational field is exerted to and beyond the outer event horizon and ergosphere of the supermassive black hole, as if the collapsed celestial body was still there in physical form and with its corresponding mass. Hence, according to the above theoretical gravitational model, the external gravitational field of the supermassive black hole results from the spatiotemporal pressure differential sustained by the concentrated electromagnetic field of the unified electroweak interaction of the mass, charges, and internal gravitational field, at the spatiotemporal boundary of the outer event horizon, and not only as a gravitational end result of the finally condensed mass of the ring itself.

The external gravitational field of a Kerr-Newman black hole is theorized to be enough to draw matter in and keep the matter spinning in a stable accretion disk. Material and gas accumulate in an accretion disk around a Kerr-Newman black hole. The spinning material and gas generate their own magnetic fields, and these fields power winds of charged particles that blow away from the black hole. The winds transfer angular momentum from the inner regions of the disk outward. This slows down some of the spinning material and gas, allowing the material and gas to fall onto the black hole. But before matter in the accretion disk can take the final plunge into the ergosphere and then into the outer event horizon of the black hole, the matter must lose some of its rotation speed or angular momentum. If the angular momentum from the disk were not dissipated away, the material and gas in the accretion disk would circle the black hole a very long time in a stable orbit, like planets circling a star. (Newman et al, 1965)

A torsional Alfvén wave is theorized to be generated by the rotational dragging of space near a Kerr-Newman black hole. The wave transports energy outward along the magnetic field lines, causing the total energy of the plasma near the black hole to decrease to negative values. The magnetic field also causes turbulence and friction to build up within the disk. The friction heats up the gas to millions of degrees, causing the plasma to glow brilliantly in the ultraviolet and X-ray bands, when this negative energy plasma enters the outer event horizon of the black hole, the rotational energy of the black hole decreases. Through this process, the energy of the spinning black hole is extracted magnetically.

The magnetic coupling process can transfer energy and angular momentum from a rotating Kerr-Newman black hole to its surrounding disk. Recently, it has been observed through powerful special telescopes that a black hole's powerful magnetic fields create turbulence in surrounding matter that help drive matter inward to be accreted. Tightly coiled magnetic fields close to energetic supermassive black holes expel narrow jets of plasma into space. This mechanism transfers heat to gases attracted to and near the black hole to effectively control the growth of the largest galaxies.

2.1. The radial coordinate of a Kerr-Newman black hole

Furthermore, as a Kerr-Newman black hole forms as previously proposed, the very condensed mass of the celestial body collapses beyond the radial distance *"s"* of space and into the radial distance "τ" of time. Thus, as length extends, time contracts, and the radius of the ring can be measured in terms of its temporal distance. In other words, the ring of the black hole has moved within its temporal bubble as time condenses, outside of external space-time and within the outer event horizon. This physical condition within the outer event horizon allows space-time to flow and curve inward as a river in the direction of the ring into the inner space-time of the supermassive Kerr-Newman black hole.

Hence, from the following integral expression for the spatial and temporal radial coordinates of the collapsing mass of the celestial body we find

$$r = \int_{s_1}^{s_2} \partial s_r + c\int_{t_1}^{t_2} \partial t_r \quad \textit{where } s_2 > s_1 \textit{ and } t_2 < t_1 \tag{2.1}$$

$$r = s_r \Big|_0^S + ct_r \Big|_{t_1}^{t_2} \tag{2.2}$$

$$r = s + c(t_2 - t_1) \tag{2.3}$$

Let the temporal interval $(t_2 - t_1)$ be equal to τ to find the radial distance *"r"*, as the spatial distance *"s"* extends and the mass collapses into a ring singularity, where it is theorized that the internal electromagnetic field and internal gravitational field are repulsive, or negative, from the perspective of the external gravitational field outside of the outer even horizon.

$$r = \lim_{m \to 0} \left[s + c(t_2 - t_1)\right] = c\tau \tag{2.4}$$

Hence, we specify from this physical model that the radius *"r"* of the ring is no longer spatial, but it is temporal. The value "$c\tau$" is the distance of the radius *"r"* in the spatiotemporal dimension, within the outer event horizon of a Kerr-Newman black hole.

2.2. The Schwarzschild factor during the formation of a singularity

Let us consider the Schwarzschild's geometry of a black hole, where $r_s = 2GM/c^2$, about the event horizon with $r > r_S$, described by the following metric:

$$ds^2 = -\left(1-\frac{r_s}{r}\right)dt^2 + \frac{dr^2}{\left(1-\frac{r_s}{r}\right)} + r^2 d\theta^2 + r^2 \sin^2\theta d\phi^2 \qquad (2.5)$$

where we can express the Schwarzschild factor as

$$1-\frac{r_s}{r} = 1-\frac{2GM}{rc^2} \qquad (2.6)$$

As we reconsider the radial coordinate *r*, with space flowing radially inwards at the Newtonian infalling speed of $v^2 = 2GM/r$, in General Relativity, as v reaches the speed of light *"c"*, we have

$$\lim_{v\to c}\left(1-\frac{2GM}{rc^2}\right) = \lim_{v\to c}\left(1-\frac{v^2}{c^2}\right) = \lim_{v\to c}\left(1-\frac{c^2}{c^2}\right) = 0 \qquad (2.7)$$

Therefore, as spatial distance within the black hole extends, it is theorized that the Schwarzschild factor approaches zero as predicted by General Relativity, within the internal region of the theoretical supermassive black hole.

The infalling speed *v* passes the speed of light *c* at the event horizon of the black hole. As $r < r_s$, the speed of contracting of the spatiotemporal wave increases, as space extends and time condenses.

Thus, under these proposed conditions Einstein's law regarding the speed of light still holds because the speed of light applies to the speed of objects moving in space-time as measured locally with respect to an inertial frame of reference.

In this model within the event horizon, it is space-time itself that moves superluminally inward; Hence, theoretically, General Relativity prevails.

2.3. The metric of a Kerr-Newman black hole

The Kerr–Newman metric describes the geometry of space-time in the vicinity of a rotating black hole of mass M, spin α, with a charge Q in standard spherical coordinates.

$$c^2 d\tau^2 = -\left[\frac{dr^2}{\Delta} + d\theta^2\right]\rho^2 + \left[cdt - \alpha \sin^2\theta d\phi\right]^2 \frac{\Delta}{\rho^2} - \left[\left(r^2 + \alpha^2\right)d\phi - \alpha\, cdt\right]^2 \frac{\sin^2\theta}{\rho^2} \quad (2.8)$$

$$\alpha = \frac{J}{Mc} = \frac{a}{c} \quad (2.9)$$

$$\rho^2 = r^2 + \alpha^2 \cos^2\theta = r^2 + \frac{a^2 \cos^2\theta}{c^2} \quad (2.10)$$

$$\Delta = r^2 - r_S r + \alpha^2 + {r_Q}^2 \quad (2.11)$$

$${r_Q}^2 = \frac{Q^2 G}{4\pi\varepsilon_0 c^4} \quad (2.12)$$

Hence, the current capacity at the radius r_Q corresponding to the charge Q of the Kerr-Newman black hole with a mass M can be expressed as

$${I_Q}^2 = \frac{4\pi\varepsilon_0 c^4 {r_Q}^2 {f_Q}^2}{G} = Q^2\left(\frac{\dot{m}}{M}\right) \quad (2.13)$$

$$\left(\frac{I_Q}{Q}\right)^2 = \frac{\ddot{m}}{M} \quad (2.14)$$

where $\dot{m}$ is the matter accretion velocity rate, and $\ddot{m}$ is the matter accretion acceleration rate of the Kerr-Newman black hole at the radius r_Q.

$$\dot{m} = \frac{I_Q M}{Q} \tag{2.15}$$

$$\ddot{m} = \frac{I_Q{}^2 M}{Q^2} \tag{2.16}$$

Let us express the Kerr-Newman metric in terms of the radial coordinate "r", the angular momentum "a" per unit of mass, using the space-time geometry in the vicinity of the rotating Kerr-Newman black hole.

$$c^2 d\tau^2 = -\left[\frac{dr^2}{r^2\left(1-\frac{r_S}{r}\right)+\frac{a^2}{c^2}+r_Q{}^2}+d\theta^2\right]\left[r^2+\frac{a^2\cos^2\theta}{c^2}\right]+\left[cdt-\frac{a\sin^2\theta\, d\phi}{c}\right]^2\left[\frac{r^2\left(1-\frac{r_S}{r}\right)+\frac{a^2}{c^2}+r_Q{}^2}{r^2+\frac{a^2\cos^2\theta}{c^2}}\right]$$

$$-\left[\left(r^2+\frac{a^2}{c^2}\right)d\phi - adt\right]^2\left[\frac{\sin^2\theta}{r^2+\frac{a^2\cos^2\theta}{c^2}}\right] \tag{2.17}$$

As the speed of space-time approaches the speed of light c, the Schwarzschild factor $(1-r_S/r)$ approaches zero as predicted by the General Theory of Relativity and the Kerr-Newman metric for $r \le r_S$, where $r^2 = c^2t^2,\ r_Q{}^2 = c^2 t_Q{}^2$ *and* $dr^2 = c^2dt^2$, can be expressed as

$$c^2 d\tau^2 = -\left[\frac{c^2dt^2}{\frac{a^2}{c^2}+c^2t_Q{}^2}+d\theta^2\right]\left[c^2t^2+\frac{a^2\cos^2\theta}{c^2}\right]+\left[cdt-\frac{a\sin^2\theta\, d\phi}{c}\right]^2\left[\frac{\frac{a^2}{c^2}+c^2t_Q{}^2}{c^2t^2+\frac{a^2\cos^2\theta}{c^2}}\right]$$

$$-\left[\left(c^2t^2+\frac{a^2}{c^2}\right)d\phi - adt\right]^2\left[\frac{\sin^2\theta}{c^2t^2+\frac{a^2\cos^2\theta}{c^2}}\right] \tag{2.18}$$

§ 3. On the Lorentz and Larmor Factors

Thus, for superluminal motion new relativistic factors are needed for mass, length and time transformations in space-time-mass. Let "nc" represent the speed of space-time v_s at which space contracts and v_n is the speed of an object moving through space-time.

Let us denote the Lorentz and Larmor factor.

3.1. The Lorentz factor

$$\Gamma\left(\frac{V_n}{nc}\right) = \frac{1}{\sqrt[2]{1-\left(\frac{v_n}{nc}\right)^2}} \qquad where\ \beta = \frac{v_n}{nc} \qquad (3.1)$$

$\Gamma\left(\frac{V_n}{nc}\right)$ is the Lorentz factor for subluminal speeds of nc where $n < 1$. Thus, by L'Hopital's Rule we have

$$\lim_{v_n \to nc} \Gamma\left(\frac{v_n}{nc}\right) = 0 \qquad (3.2)$$

$$\lim_{v_n \to 0} \Gamma\left(\frac{v_n}{nc}\right) = 1 \qquad (3.3)$$

Thus, a stationary clock in subluminal space-time has a rest mass of "m" and a time of "t" as measured locally. As a moving object approaches the speed of light "c" in subluminal space its mass dilates relativistically, then at "c", the mass of the object collapses to a higher density as spatiotemporal wave contracts, and space extends about the atomic structure of the mass. As the mass condenses, the space pressure manifesting the relativistic mass diminishes and it allows for the Larmor effect when the mass resumes dilation and spatial length resumes contraction as the object speeds up and goes relativistically superluminal.

3.2. The Luminal Larmor factor

$$\Lambda\left(\frac{v_n}{nc}\right) = \frac{1}{\sqrt[2]{2-\left(\frac{v_n}{nc}\right)^2}} \tag{3.4}$$

$\Lambda\left(\frac{v_n}{nc}\right)$ is the Larmor factor for luminal speeds of nc for $n = 1$.

$$\underset{v_n \to nc}{Lim}\ \Lambda\left(\frac{v_n}{nc}\right) = 1 \tag{3.5}$$

$$\underset{v_n \to 0}{Lim}\ \Lambda\left(\frac{v_n}{nc}\right) = \frac{1}{\sqrt[2]{2}} \tag{3.6}$$

Thus, a stationary clock in luminal space-time, v = c, has a mass of $m/\sqrt[2]{2}$ and a time of $t/\sqrt[2]{2}$ as measured locally with respect to an identical clock in an inertial frame in subluminal space-time. Let us define these luminal effects as the effective mass and the effective time of a stationary object in luminal space-time.

3.3. The Superluminal Larmor Factor

$$\mathrm{T}\left(\frac{v_n}{nc}\right) = \frac{\phi}{\sqrt[2]{\left(\frac{v_n}{nc}\right)^2 - 1}} \tag{3.7}$$

$\mathrm{T}\left(\frac{v_n}{nc}\right)$ is the Superluminal Larmor Factor for speeds of nc for $n > 1$.

By L'Hopital's Rule we get

$$\lim_{v_n \to nc} \mathrm{T}\left(\frac{v_n}{nc}\right) = 0 \tag{3.8}$$

$$\lim_{v_n \to 0} \mathrm{T}\left(\frac{v_n}{nc}\right) = \frac{\phi}{\sqrt[2]{-1}} = \frac{\phi}{i} \tag{3.9}$$

Hence, the Superluminal Larmor Factor is useful to calculate the relativistic effects on mass, time and length, at speeds faster than the speed of light c. Thus, a stationary clock in superluminal space-time has a time-like or imaginary mass, $-im$, complex time, $-it$, and time-like length, il.

3.4. Luminal relativistic effect on mass and time

In general, as v → c, for the relativistic effects of either mass or time,

$$\lim_{v \to c} \left(\frac{\phi}{\sqrt[2]{1-\left(\frac{v}{c}\right)^2}} \right) = \infty \quad \textit{or undefined} \tag{3.10}$$

$$\frac{f(v)}{g(v)} = \frac{\phi}{\sqrt[2]{1-\left(\frac{v}{c}\right)^2}} \tag{3.11}$$

By L'Hopital's Rule,

$$\frac{f'(v)}{g'(v)} = \frac{\dfrac{d\phi}{dv}}{\dfrac{-2v}{2c^2\left[\sqrt[2]{1-\left(\dfrac{v}{c}\right)^2}\right]}} \tag{3.12}$$

$$\lim_{v \to c} \frac{f'(v)}{g'(v)} = \frac{0}{-2c} = 0 \tag{3.13}$$

Similarly, as v → c, for the superluminal relativistic effects of either mass or time,

$$\lim_{v \to nc} \left(\frac{\phi}{\sqrt[2]{\left(\dfrac{v}{nc}\right)^2 - 1}} \right) = \infty \quad \textit{or undefined} \tag{3.14}$$

$$\frac{f(v)}{g(v)} = \frac{\phi}{\sqrt[2]{\left(\dfrac{v}{nc}\right)^2 - 1}} \tag{3.15}$$

$$\frac{f'(v)}{g'(v)} = \frac{\dfrac{d\phi}{dv}}{\dfrac{2v}{2(nc)^2\sqrt[2]{\left(\dfrac{v}{nc}\right)^2 - 1}}} = \frac{\left(\dfrac{d\phi}{dv}\right)2(nc)^2\sqrt[2]{\left(\dfrac{v}{nc}\right)^2 - 1}}{2v} \tag{3.16}$$

$$\lim_{v \to cn} \frac{f'(v)}{g'(v)} = \frac{0}{2v} = 0 \tag{3.17}$$

Therefore, for mass

$$\lim_{v \to c} \left(\frac{m}{\sqrt[2]{1 - \left(\frac{v}{c}\right)^2}} \right) = 0 \quad \Rightarrow \quad m' = \frac{m}{\sqrt[2]{1 - \left(\frac{v}{c}\right)^2}} \tag{3.18}$$

$$As \quad v \to c,\ m' \to 0 \tag{3.19}$$

Similarly, for time

$$\lim_{v \to c} \left(\frac{t}{\sqrt[2]{1 - \left(\frac{v}{c}\right)^2}} \right) = 0 \quad \Rightarrow \quad t' = \frac{t}{\sqrt[2]{1 - \left(\frac{v}{c}\right)^2}} \tag{3.20}$$

$$As \quad v \to c,\ t' \to 0 \tag{3.21}$$

For length,

$$\lim_{v \to c} \left(l \sqrt[2]{1 - \left(\frac{v}{c}\right)^2} \right) = 0 \quad \Rightarrow \quad l' = l \sqrt[2]{1 - \left(\frac{v}{c}\right)^2} \tag{3.22}$$

$$As \quad v \to c,\ l' \to 0 \tag{3.23}$$

Therefore, as the speed of the object reaches the speed of light, the mass of the object condenses, length extends, and time condenses, as space extends about the atomic structure of the mass.

§ 4. The Relativistic Mass Cycle

4.1. On the relativistic effects on mass as a function of speed

Hence, for a mass m_n, we can express *the Lorentz-Larmor mass function* as

$$m^n = \frac{m_{n-1}}{\sqrt[2]{1-\left(\frac{v_n}{nc}\right)^2}} - \frac{m_n}{\sqrt[2]{2-\left(\frac{v_n}{nc}\right)^2}} + \frac{m_{n+1}}{\sqrt[2]{\left(\frac{v_n}{nc}\right)^2-1}} \tag{4.1}$$

where m^n is a relativistic mass.

Thus, let us take the limit of m_n as the speed of the mass v_n is less than the speed of light *c*, equal to *c*, or greater than *c*:

a. As $v_n \rightarrow nc$ *and* $nc < c$,

$$\lim_{v_{n\to nc}} m^n = \lim_{v_{n\to nc}} \left[\frac{m_{n-1}}{\sqrt[2]{1-\left(\frac{v_n}{nc}\right)^2}} - \frac{m_n}{\sqrt[2]{2-\left(\frac{v_n}{nc}\right)^2}} + \frac{m_{n+1}}{\sqrt[2]{\left(\frac{v_n}{nc}\right)^2-1}} \right] = 0 \tag{4.2}$$

Thus,

$m_n = 0$, and $m_{n+1} = 0$, when $nc < c$, by the Lorentz-Larmor correspondence principle *b. As* $v_n \rightarrow nc$ *and* $nc = c$,

$$\lim_{v_n \to nc} m^n = \lim_{v_n \to nc} \left[\frac{m_{n-1}}{\sqrt[2]{1-\left(\frac{v_n}{nc}\right)^2}} - \frac{m_n}{\sqrt[2]{2-\left(\frac{v_n}{nc}\right)^2}} + \frac{m_{n+1}}{\sqrt[2]{\left(\frac{v_n}{nc}\right)^2 - 1}} \right] = -m_n \quad (4.3)$$

c. As $v_n \rightarrow nc$ *and* $nc > c$,

$$\lim_{v_n \to nc} m^n = \lim_{v_n \to nc} \left[\frac{m_{n-1}}{\sqrt[2]{1-\left(\frac{v_n}{nc}\right)^2}} - \frac{m_n}{\sqrt[2]{2-\left(\frac{v_n}{nc}\right)^2}} + \frac{m_{n+1}}{\sqrt[2]{\left(\frac{v_n}{nc}\right)^2 - 1}} \right] = m_{n+1} \quad (4.4)$$

Henceforth, we can express m^n as the relativistic mass using *the Lorentz-Larmor mass function* as follows:

$$m^n = m_{n-1}\Gamma\left(\frac{v_n}{nc}\right) - m_n\Lambda\left(\frac{v_n}{nc}\right) + m_{n+1}\mathrm{T}\left(\frac{v_n}{nc}\right) \qquad \textit{for } nc \geq 0 \quad (4.5)$$

Consequently,

$$m^n + m_n\Lambda\left(\frac{v_n}{nc}\right) = m_{n-1}\Gamma\left(\frac{v_n}{nc}\right) + m_{n+1}\mathrm{T}\left(\frac{v_n}{nc}\right) \quad (4.6)$$

Let us define m_n / m_{n-1} *as the light barrier mass contraction ratio*

By applying Maclaurin's series, we obtain

$$\Gamma\left(\frac{v_n}{nc}\right) = 1 + \frac{1}{2}\left(\frac{v_n}{nc}\right)^2 + \frac{3}{8}\left(\frac{v_n}{nc}\right)^4 + \frac{5}{16}\left(\frac{v_n}{nc}\right)^6 + \frac{35}{128}\left(\frac{v_n}{nc}\right)^8 + \ldots. \quad (4.7)$$

$$\Lambda\left(\frac{v_n}{nc}\right) = \frac{1}{\sqrt[2]{2}}\left[1 + \frac{1}{2}(L_r)^2 + \frac{3}{8}(L_r)^4 + \frac{5}{16}(L_r)^6 + \frac{35}{128}(L_r)^8 + \ldots.\right] \quad (4.8)$$

$$where\ L_r = \frac{v_n}{\sqrt[2]{2nc}} \quad and\ L_r\ is\ the\ Larmor\ factor\ ratio \quad (4.9)$$

$$\mathrm{T}\left(\frac{v_n}{nc}\right) = \frac{1}{i} + \frac{1}{2i}\left(\frac{v_n}{nc}\right)^2 + \frac{3}{8i}\left(\frac{v_n}{nc}\right)^4 + \frac{5}{16i}\left(\frac{v_n}{nc}\right)^6 + \frac{35}{128i}\left(\frac{v_n}{nc}\right)^8 + \ldots. \quad (4.10)$$

Thus, the Superluminal Larmor function is a time-like series.

Let us illustrate the concept of relativistic mass on a graph.

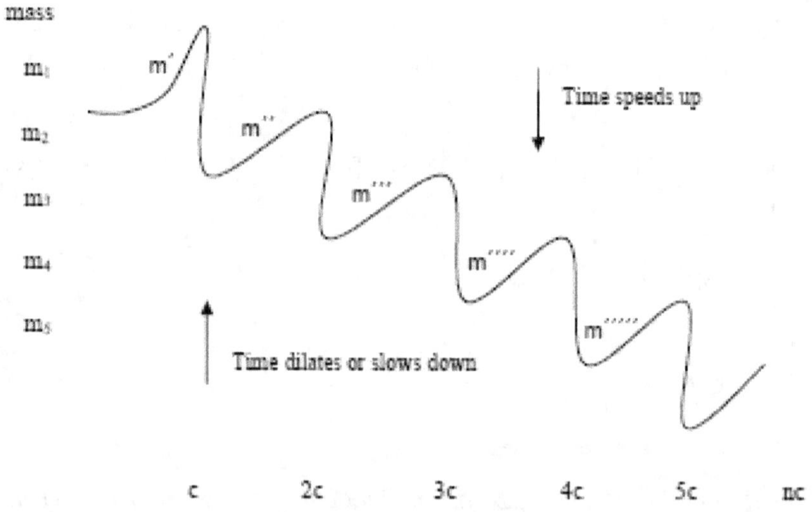

Figure 12. Mass dilation and contraction

4.2. On the mass dilation of a proton near the speed of light

A particle accelerator is a device that uses electric fields to propel ions or charged subatomic particles to high speeds and to contain them in well-defined beams.

Elementary particle physicists tend to use machines creating beams of electrons, positrons, protons, or anti-protons, interacting with each other at the highest possible energies, generally hundreds of GeV or higher, to study the dynamics and structure of matter, space, and time, to seek the simplest kinds of interactions at the highest possible energies.

If a proton is accelerated to near the speed of light *"c"*, let us say to a speed of 0.999957c, or 12,900 *m/s* short of *"c"*, then the proton's ratio of relativistic mass to its rest mass is most nearly 107.83 or about approximately 108 times its rest mass.

Thus, letting that ratio be

$$\frac{m'}{m_1} = \frac{1}{\sqrt[2]{1-\left(\frac{v_m}{c}\right)^2}} = \frac{1}{\sqrt[2]{1-\left(\frac{0.999957c}{c}\right)^2}} \approx 108 \qquad (4.11)$$

Therefore, as the mass of a proton dilates to m', its relativistic mass has dilated about 108 times its rest mass m_1.

Hence, as a mass m_1 travels closer and closer to the speed of light *"c"*, the mass of m_1 dilates to its upper limit m', then as $v = c$, m' goes to zero and m_1 goes to m_2.

$$\lim_{v \to c} m' = \lim_{v \to c} \frac{m}{\sqrt[2]{1-\left(\frac{v}{c}\right)^2}} = 0 \qquad (4.12)$$

$$\lim_{v \to c} \left[m' - \frac{m_2}{\sqrt[2]{2-\left(\frac{v}{c}\right)^2}} \right] = -m_2 \qquad (4.13)$$

$$m_1 \to -m_2 \;\; as \;\; v \to c \qquad (4.14)$$

$$\frac{m_2}{m_1} = \frac{1}{108} \approx 0.93\% \qquad (4.15)$$

Then, m' contracts instantaneously to $-m_2$, to about a very small percentage of its rest mass m_1. The higher the magnitude of m', the lesser the magnitude of $-m_2$. The relativistic magnitude of m' minus the rest mass m_1 is equal to the magnitude of the rest mass m_1 minus m_2.

$$m' - m_1 = m_1 - m_2 \qquad (4.16)$$

After the contraction from m' to m_2, then mass m_2 begins to dilate as it accelerates faster than "c" and the relativistic cycle repeats itself, all other variables being the same, for the next light speed interval $c \le v \le 2c$ if the mass continues to accelerate uniformly towards 2c.

§ 5. The Relativistic Time Cycle

Time undergoes a similar relativistic dilation as mass. Thus, let us imagine a synchronous clock of mass "m" traveling in homogeneous and isotropic space-time at relativistic speed.

Thus, for a time t_n, we can express *the Lorentz-Larmor time function* as

$$t^n = \frac{t_{n-1}}{\sqrt[2]{1-\left(\frac{v_n}{nc}\right)^2}} - \frac{t_n}{\sqrt[2]{2-\left(\frac{v_n}{nc}\right)^2}} + \frac{t_{n+1}}{\sqrt[2]{\left(\frac{v_n}{nc}\right)^2 - 1}} \tag{5.1}$$

where t^n is a relativistic time.

Henceforth, we can express t^n as the relativistic time using *the Lorentz-Larmor time function* as follows:

$$t^n = t_{n-1}\,\Gamma\left(\frac{v_n}{nc}\right) - t_n\,\Lambda\left(\frac{v_n}{nc}\right) + t_{n+1}\,\mathrm{T}\left(\frac{v_n}{nc}\right) \qquad for\ nc \geq 0 \tag{5.2}$$

Consequently,

$$t^n + t_n\,\Lambda\left(\frac{v_n}{nc}\right) = t_{n-1}\,\Gamma\left(\frac{v_n}{nc}\right) + t_{n+1}\,\mathrm{T}\left(\frac{v_n}{nc}\right) \qquad for\ nc \geq 0 \tag{5.3}$$

Let us define t_n / t_{n+1} *as the light barrier time contraction ratio*

a. *As* $v_n \to nc$ *and* $nc < c$,

$$\lim_{v_n \to nc} t^n = \lim_{v_n \to nc}\left[\frac{t_{n-1}}{\sqrt[2]{1-\left(\frac{v_n}{nc}\right)^2}} - \frac{t_n}{\sqrt[2]{2-\left(\frac{v_n}{nc}\right)^2}} + \frac{t_{n+1}}{\sqrt[2]{\left(\frac{v_n}{nc}\right)^2 - 1}}\right] = 0 \tag{5.4}$$

Thus, $t_n = 0$ and $t_{n+1} = 0$ *when* $nc < c$ by the Lorentz-Larmor correspondence principle

b. As $v_n \to nc$ *and* $nc = c$,

$$\lim_{v_n \to nc} t^n = \lim_{v_n \to nc} \left[\frac{t_{n-1}}{\sqrt[2]{1-\left(\frac{v_n}{nc}\right)^2}} - \frac{t_n}{\sqrt[2]{2-\left(\frac{v_n}{nc}\right)^2}} + \frac{t_{n+1}}{\sqrt[2]{\left(\frac{v_n}{nc}\right)^2 - 1}} \right] = -t_n \qquad (5.5)$$

$c.\ As\ v_n \to nc\ and\ nc > c,$

$$\lim_{v_n \to nc} t^n = \lim_{v_n \to nc} \left[\frac{t_{n-1}}{\sqrt[2]{1-\left(\frac{v_n}{nc}\right)^2}} - \frac{t_n}{\sqrt[2]{2-\left(\frac{v_n}{nc}\right)^2}} + \frac{t_{n+1}}{\sqrt[2]{\left(\frac{v_n}{nc}\right)^2 - 1}} \right] = t_{n+1} \qquad (5.6)$$

Henceforth, we can express t^n as the relativistic time using *the Lorentz-Larmor time function* as follows:

$$t^n = t_{n-1}\,\Gamma\left(\frac{v_n}{nc}\right) - t_n\,\Lambda\left(\frac{v_n}{nc}\right) + t_{n+1}\,\mathrm{T}\left(\frac{v_n}{nc}\right) \qquad for\ nc \geq 0 \qquad (5.7)$$

Consequently,

$$t^n + t_n\,\Lambda\left(\frac{v_n}{nc}\right) = t_{n-1}\,\Gamma\left(\frac{v_n}{nc}\right) + t_{n+1}\,\mathrm{T}\left(\frac{v_n}{nc}\right) \qquad (5.8)$$

Let us define t_n / t_{n+1} *as the light barrier time contraction ratio*

Let us illustrate the concept of relativistic time in a graph.

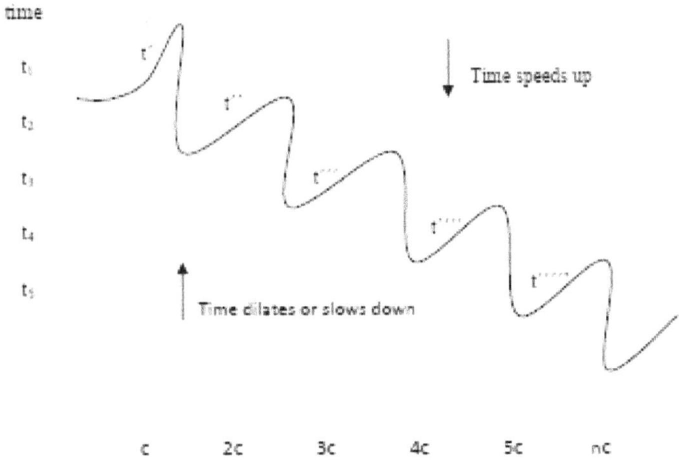

Figure 13. Time dilation and contraction

Hence, as the clock of mass m_1 travels closer and closer to the speed of light "c", time t_1 dilates to its upper limit t', then as $v = c$, t' goes to zero and t_1 goes to t_2. Similarly, t' condenses instantaneously to $-t_2$, to about a very small percentage of its rest time t_1.

The higher the magnitude of t', the lesser the magnitude of $-t_2$. The relativistic magnitude of t' minus the rest time t_1 is equal to the magnitude of the rest time t_1 minus t_2.

$$t' - t_1 = t_1 - t_2 \tag{5.9}$$

After the collapse from t' to t_2, then clock mass m_2 and time t_2 begin to dilate as the mass of the clock accelerates faster than "c" and the relativistic time cycle repeats itself, all other variables being the same, for the next light speed interval $c \leq v \leq 2c$, if the clock mass continues to accelerate uniformly towards $2c$.

§ 6. The Relativistic Length Cycle

6.1. On the relativistic effect on length as a function of speed

For the relativistic effects on length let us imagine a one-dimensional rod of mass m_1 approaching the speed of light c, let us consider the Lorentz-Larmor factor as

$$\lim_{v_n \to nc} r^n = \lim_{v_n \to nc} \left[r_{n-1} \sqrt[2]{1-\left(\frac{v_n}{nc}\right)^2} + r_n \sqrt[2]{2-\left(\frac{v_n}{nc}\right)^2} + r_{n+1} \sqrt[2]{\left(\frac{v_n}{nc}\right)^2 - 1} \right] \quad where\ n \geq 0 \tag{6.1}$$

$$r_n = 0 \text{ and } r_{n+1} = 0 \tag{6.2}$$

and $nc < c$, by the Lorentz-Larmor correspondence principle

$a.\ As\ v_n \to nc\ \ and\ \ nc < c,$

Taking the limit of r^n we obtain

$$\lim_{v_n \to 0} r^n = \lim_{v_n \to 0} \left[r_{n-1} \sqrt[2]{1-\left(\frac{v_n}{nc}\right)^2} + r_n \sqrt[2]{2-\left(\frac{v_n}{nc}\right)^2} + r_{n+1} \sqrt[2]{\left(\frac{v_n}{nc}\right)^2 - 1} \right] = r_{n-1} \quad where\ nc < c \tag{6.3}$$

$$r_n = 0 \text{ and } r_{n+1} = 0 \ \ and\ \ nc < c \tag{6.4}$$

by the Lorentz-Larmor correspondence principle

Thus, let us define r_{n+1}/r_n as the light barrier length contraction ratio

$b.\ As\ v_n \to nc\ \ and\ \ nc = c,$

$$\text{Hence, } r_n > 0\ \ when\ \ nc = c \tag{6.5}$$

$$Lim\ r^n = Lim \left[r_{n-1}\sqrt[2]{1-\left(\frac{v_n}{nc}\right)^2} + r_n\sqrt[2]{2-\left(\frac{v_n}{nc}\right)^2} + r_{n+1}\sqrt[2]{\left(\frac{v_n}{nc}\right)^2-1} \right] = r_n \quad where\ nc = c \tag{6.6}$$

$v_n \to nc \quad v_n \to nc$

$$Lim\ r^n = Lim \left[r_{n-1}\sqrt[2]{1-\left(\frac{v_n}{nc}\right)^2} + r_n\sqrt[2]{2-\left(\frac{v_n}{nc}\right)^2} + r_{n+1}\sqrt[2]{\left(\frac{v_n}{nc}\right)^2-1} \right] = \sqrt[2]{2}\, r_n \quad where\ nc = c \tag{6.7}$$

$v_n \to 0 \quad v_n \to 0$

c. As $v_n \to nc$ and $nc > c$,

Hence, $r_{n+1} > 0$ *when* $nc > c$ (6.8)

$$Lim\ r^n = Lim \left[r_{n-1}\sqrt[2]{1-\left(\frac{v_n}{nc}\right)^2} + r_n\sqrt[2]{2-\left(\frac{v_n}{nc}\right)^2} + r_{n+1}\sqrt[2]{\left(\frac{v_n}{nc}\right)^2-1} \right] = r_{n+1} \quad where\ nc > c \tag{6.9}$$

$v_n \to nc \quad v_n \to nc$

$$Lim\ r^n = Lim \left[r_{n-1}\sqrt[2]{1-\left(\frac{v_n}{nc}\right)^2} + r_n\sqrt[2]{2-\left(\frac{v_n}{nc}\right)^2} + r_{n+1}\sqrt[2]{\left(\frac{v_n}{nc}\right)^2-1} \right] = i\, r_{n+1} \quad where\ nc > c \tag{6.10}$$

$v_n \to 0 \quad v_n \to 0$

Thus, a stationary rod in superluminal space has a time-like length $i\,r$ as measured locally with respect to an identical rod in an inertial frame in luminal space.

Let us define this superluminal effect as the imaginary length of a stationary object in superluminal space-time.

Expressing r^n, r_{n-1}, r_n *and* r_{n+1} in terms of the Lorentz and Larmor factors,

$$r^n = \frac{r_{n-1}}{\Gamma\left(\frac{v_n}{nc}\right)} + \frac{r_n}{\Lambda\left(\frac{v_n}{nc}\right)} + \frac{r_{n+1}}{\mathrm{T}\left(\frac{v_n}{nc}\right)} \tag{6.11}$$

$$r^{n} = r_{n-1}\sqrt[2]{1-\left(\frac{v_n}{nc}\right)^2} + r_n\sqrt[2]{2-\left(\frac{v_n}{nc}\right)^2} + r_{n+1}\sqrt[2]{\left(\frac{v_n}{nc}\right)^2 - 1} \quad (6.12)$$

$$r^{n} - r_{n-1}\sqrt[2]{1-\left(\frac{v_n}{nc}\right)^2} = r_n\sqrt[2]{2-\left(\frac{v_n}{nc}\right)^2} + r_{n+1}\sqrt[2]{\left(\frac{v_n}{nc}\right)^2 - 1} \quad (6.13)$$

Let us illustrate length dilation and contraction in a graph.

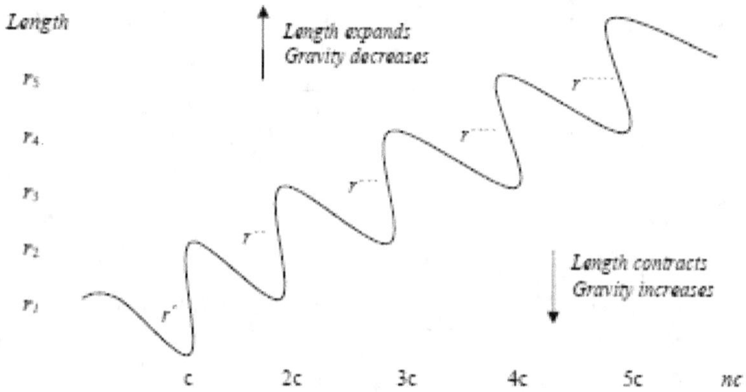

Figure 14. Length dilation and contraction

6.2. On the elasticity of space at luminal speed

Consequently, let us imagine that a meterstick of space is accelerated to near the speed of light c, let us say to a speed of 0.999957c, or 12,900 m/s short of c, then the meterstick's ratio of rest length to its relativistic length in the direction of motion is most nearly 1/107.83 or about approximately 1/108.

All other variables being equal, this ratio is the reciprocal of the ratio for mass or time. (Taylor et al, 1966)

Thus, letting that ratio be

$$\frac{l'}{l_1} = \frac{\sqrt[2]{1-\left(\frac{v}{c}\right)^2}}{1} = \frac{\sqrt[2]{1-\left(\frac{0.999957c}{c}\right)^2}}{1} \approx \frac{1}{108} \tag{6.14}$$

Therefore, as the length of our meterstick contracts to l', its relativistic length l' has contracted to about 1/108 times its rest length l_1. As l' expands to l_2, our meterstick expands to its relativistic length l_2 about 108 times its rest length l_1.

$$\frac{l_2}{l'} = \frac{108}{1} \tag{6.15}$$

Hence, as our meterstick of space l_1 travels closer and closer to the speed of light "c", the length of l_1 contracts to its lowest limit l', then as v = c, l' goes to l_2.

$$\lim_{v \to c} l' = \lim_{v \to c} l \sqrt[2]{1-\left(\frac{v}{c}\right)^2} = 0 \tag{6.16}$$

$$\lim_{v \to c} \left[\frac{l_2 - l'}{\sqrt[2]{2-\left(\frac{v}{c}\right)^2}} \right] = l_2 \tag{6.17}$$

$$l' \to l_2 \;\; as \;\; v = c \tag{6.18}$$

Then, l' dilates instantaneously to l_2, to a magnitude above its rest length l_1. The lesser the magnitude of l', the higher the magnitude of l_2 over the rest length. The magnitude of the rest length l_1 minus the relativistic length of l' is equal to the magnitude of relativistic length l_2 minus the rest length l_1.

$$l_1 - l' = l_2 - l_1 \qquad (6.19)$$

After the dilation from l' to l_2, then our meterstick of space of length l_2 begins to contract as it travels faster than "c" and the relativistic cycle repeats itself, all other variables being the same, for the next light speed interval $c \leq v \leq 2c$ if the imaginary meterstick continues to travel uniformly towards the second light barrier $2c$.

§ 7. The metric expansion of space-time-mass

The metric expansion of the medium of space-time is an intrinsic property of the entire universe both locally and at great distances. The expansion is the increase of the distance between two coordinate points where the scale of space-time itself changes with the passage of time. The expansion is modeled mathematically by the Friedmann-Lemaître-Robertson-Walker metric as an exact solution of Einstein Field Equations of General Relativity. The FLRW metric describes a homogeneous, isotropic expanding or contracting universe.

Let us denote the simplest example for the metric of expanding or contracting space-time in homogeneous and isotropic Minkowski space-time, or flat space-time, with six-dimensional coordinates to be

$$e^{i\sigma} ds^2 = -e^{i\omega t} c^2 dt^2 + e^{i\theta} dr^2 \qquad (7.1)$$

$$e^{i\sigma} ds^2 = -e^{i\omega t} c^2 \left(dt_x^{\ 2} + dt_y^{\ 2} + dt_z^{\ 2}\right) + e^{i\theta}\left(dx^2 + dy^2 + dz^2\right) \qquad (7.2)$$

$$e^{i\sigma} ds^2 = e^{i(\omega t + \theta)} \eta_{\mu\nu} dt^{\mu} dr^{\nu} \qquad (7.3)$$

The FLRW metric defines how a distance can be measured between two nearby points in space-time, in terms of a coordinate system. Thus, the metric is useful as a formula which describes how displacement, scale growth, and curvature in space-time may be expressed mathematically. In expanding space-time, length, mass, and time may change regardless of scale or locality. The dimensional changes over time due to the expansion of space-time are supplementary to the dimensional changes that occur when a mass translates at relativistic speed.

The effect of dilation or contraction is a function of the amplitude of the spatiotemporal wave. The expansion or contraction of space-time is a function of the amplitude and growth (scale) factor of the spatiotemporal wave. As time expands and space expands, space-time expands as a wave medium. Nevertheless, space-time-mass may expand at different rates at different localities of space-time where there may be nodes of mass, energy, or warped space-time regions. *Every point in space-time expands freely unless obstructed.*

Let us imagine that space-time expands as modeled by the FLRW metric by some scale factor, $Ae^{i(\omega t+\theta)}$, that affects time, space, and space-time proportionally so that time, length, and mass expand or contract according to scale. Consequently, if the scale factor of expansion is the same for length, time, and mass, then a yardstick, clock, or mass of a measuring instrument, will remain proportional and imperceptible to the observer's senses and instruments of measurement.

Moreover, if the driving force of change is space-time, and space-time is not changing substantially between its cycles or periods, then the growth (scale) factor is infinitesimal, even at an expanding locality of space-time.

$$\frac{d\Sigma^2}{dt^2} = c^2\frac{dt^2}{dt^2} - c^2\frac{d\tau^2}{dt^2} \tag{7.4}$$

$$Ae^{i\sigma}\frac{d\Sigma^2}{dt^2} = c^2 Ae^{i\omega t} - c^2 Ae^{i(\omega t-\omega\tau)}\frac{d\tau^2}{dt^2} \tag{7.5}$$

$$Ae^{i\sigma}\frac{d\Sigma^2}{dt^2}=c^2Ae^{i\omega t}\left(1-e^{-\omega\tau}\frac{d\tau^2}{dt^2}\right) \tag{7.6}$$

Where T is the period of time, t is coordinate time, $d\Sigma/dt$ is the speed of space, dt/dt is the speed of coordinate time, and $d\tau/dt$ is the speed of proper time.

Mass is very porous with great expanses of space-time between atoms, molecules, or within the atomic structure.

Space-time is a medium for the waves of objects of mass, physical fields, and for the waves of space and time, which interfere within the structure of mass, constructively or destructively, realizing contraction, expansion, or a standing wave, within the boundary of mass.

$$m'Ae^{i\sigma}=\frac{m_0}{\sqrt[2]{1-\frac{v^2}{c^2}}}Ae^{i\sigma} \tag{7.7}$$

Thus, for expanding (+) or contracting (−) space-time we can express the growth (scale) factor as static or dynamic as follows:

If $e^{\pm i\sigma}=+1$, then $\pm\sigma=0$ for static spatial expansion. (7.8)

If $e^{\pm i\sigma}=-1$, then $\pm\sigma=\pm\pi$ for static temporal expansion. (7.9)

If $e^{\pm i\sigma}\neq\pm1$ and $0<+\sigma<\pi$ or $0>-\sigma>-\pi$, then it is a dynamic spatiotemporal expansion. (7.10)

It is possible that the universe may have mixed regions of space-time which have some static and dynamic expansion. The EFEs of 1915 are for a static spatial expansion of the universe where spatial and temporal dimensions overlap.

Let us consider the Lorentz factor of General Theory of Relativity.

$$\gamma = \frac{1}{\sqrt[2]{1-\frac{v^2}{c^2}}} = \frac{dt}{d\tau} \tag{7.11}$$

An alternative representation is

$$\gamma = \cosh\sigma = \frac{1}{\sqrt[2]{1-\tanh^2\sigma}} \tag{7.12}$$

$$\tanh^2\sigma = \frac{v^2}{c^2} \tag{7.13}$$

The ratio of the velocities may be expressed as a function of the rapidity σ, also known as the parameter of velocity:

$$\frac{v}{c} = \tanh(\sigma) = \frac{e^{\sigma} - e^{-\sigma}}{e^{\sigma} + e^{-\sigma}} \tag{7.14}$$

Therefore, we obtain

$$e^{\sigma} = \sqrt[2]{\frac{c+v}{c-v}} \tag{7.15}$$

$$e^{-\sigma} = \sqrt[2]{\frac{c-v}{c+v}} \tag{7.16}$$

Thus, relativistic momentum and energy are given by

$$p = \frac{mv}{\sqrt[2]{1-\frac{v^2}{c^2}}} = mc\sinh(\sigma) \tag{7.17}$$

$$E = \frac{mc^2}{\sqrt[2]{1-\frac{v^2}{c^2}}} = mc^2\cosh(\sigma) \tag{7.18}$$

A very important argument made in the General Theory of Relativity is that the universe expands continuously. The farthest observable sides of the universe are expanding faster than the speed of light.

The parameter of velocity σ is the exponential power of growth of a point in the complex spatiotemporal medium. Hence, the parameter of velocity, $\pm i\sigma = i(\pm\omega t \pm \theta)$, is a complex number that represents the phase angle difference of the sum of the advanced spatiotemporal wave and the retarded spatiotemporal wave as space-time expands in all directions at every spatiotemporal point unless obstructed.

Let us reconsider the Lorentz factor using the magnitude of the parameter of velocity,

$$\frac{dt}{d\tau} = \frac{1}{\sqrt[2]{1-\frac{v^2}{c^2}}} = \frac{1}{\sqrt[2]{1-\left(\frac{e^{\sigma}-e^{-\sigma}}{e^{\sigma}+e^{-\sigma}}\right)^2}} = \frac{1}{\sqrt[2]{\left(\frac{e^{\sigma}+e^{-\sigma}}{e^{\sigma}+e^{-\sigma}}\right)^2-\left(\frac{e^{\sigma}-e^{-\sigma}}{e^{\sigma}+e^{-\sigma}}\right)^2}} \tag{7.19}$$

$$\frac{dt^2}{d\tau^2} = \frac{1}{\left(\frac{e^{\sigma}+e^{-\sigma}}{e^{\sigma}+e^{-\sigma}}\right)^2-\left(\frac{e^{\sigma}-e^{-\sigma}}{e^{\sigma}+e^{-\sigma}}\right)^2} \tag{7.20}$$

$$\frac{d\tau^2}{dt^2} = \left(\frac{e^{\sigma}+e^{-\sigma}}{e^{\sigma}+e^{-\sigma}}\right)^2-\left(\frac{e^{\sigma}-e^{-\sigma}}{e^{\sigma}+e^{-\sigma}}\right)^2 \tag{7.21}$$

$$d\tau^2 = \left[\left(\frac{e^{\sigma}+e^{-\sigma}}{e^{\sigma}+e^{-\sigma}}\right)^2-\left(\frac{e^{\sigma}-e^{-\sigma}}{e^{\sigma}+e^{-\sigma}}\right)^2\right]dt^2 \tag{7.22}$$

It is interesting to note that the first term that multiplies dt^2 indicates the condition during spatiotemporal expansion that equals one, when the advanced wave is offset by the retarded wave. The second term indicates the condition during spatiotemporal expansion that equals a fraction, when there is a remainder wave after interference of the advanced and retarded waves. The difference between the two terms is also a fraction that when multiplied by the squared coordinate time, dt^2 , is equal to the squared proper time, $d\tau^2$. Therefore, the above equation for the interference of spatiotemporal waves demonstrates a possible explanation for spatiotemporal relativity when an object of mass moves in a spatiotemporal direction at a speed less than the speed of light.

7.1. The law of inertia: Newton's first law of motion.

"An object at rest stays at rest and an object in motion stays in motion with the same speed and in the same direction unless acted upon by an unbalanced force."

Let us imagine an isolated object of mass at rest in isotropic and homogeneous space-time that experiences the constant acceleration of space-time around its physical state of mass where the temporal and spatial waves of space-time are in equilibrium around the space-time-mass boundary. As temporal waves emerge and the spatial waves contract near an object of mass, it realizes a gravitational field around the object of mass through its spatiotemporal wave velocity and acceleration. In reality, a gravitational field may be conceptualized as a spatiotemporal field.

As long as the temporal and spatial waves about the object of mass are not disturbed or acted upon by an unbalanced force, the temporal and spatial acceleration and pressure of space-time will remain at equilibrium. The object of mass will stay at rest. Similarly, let us imagine an object of mass that experiences the constant acceleration of space-time around its physical state of mass where the temporal and spatial waves of space-time are in equilibrium around the space-time-mass boundary. All around the object of mass there is a volume of equal space-time acceleration and pressure. The object of mass has a rectangular shape.

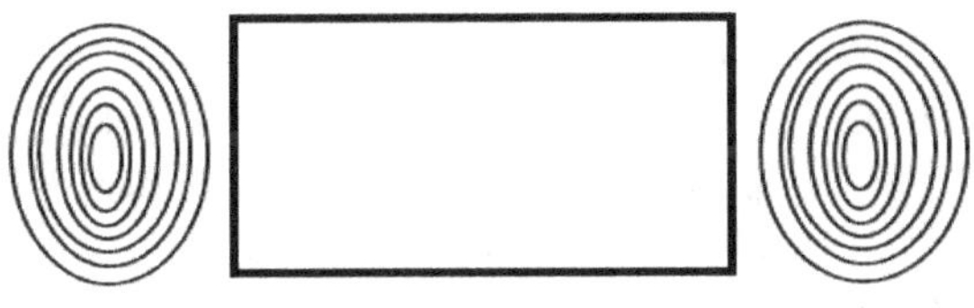

Figure 15.

An unbalanced force acts on one of its smaller end sides that disturbs the equilibrium of the temporal and spatial acceleration and pressure of space-time on the end sides as the object moves, creating a local volume of higher space-time pressure on the leading side and a volume of lower space-time pressure on the tailing side where the unbalanced force acted, or a spatiotemporal differential.

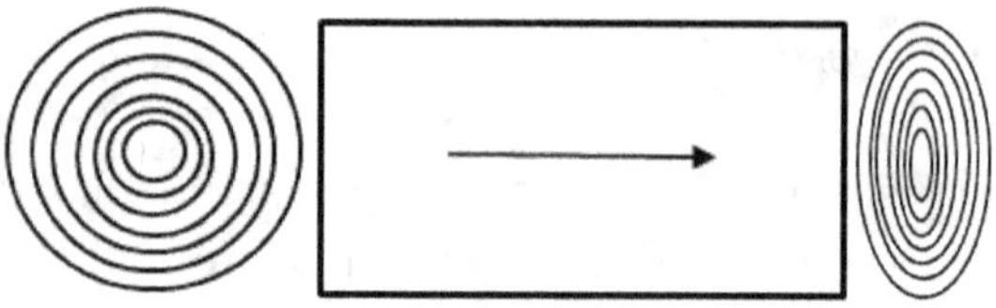

Figure 16.

The pressure of the acting unbalanced force contracts the spatial waves on the leading side of the object. Consequently, the temporal waves expand, the temporal period extends and decreases frequency, which causes amplitude to increase, as a result space contracts and time expands on the leading side of the moving object of mass. Accordingly, space-time expands on the tailing side of the object of mass, creating a spatiotemporal differential between the end sides where it causes an acceleration and a spatial displacement in the direction of the applied unbalanced force. As the leading side of the object of mass moves in the direction of the unbalanced force, it curves or contracts space-time as it applies pressure relative to the speed of the object, at relativistic speed it causes greater and more noticeable curvature.

The object of mass is now in motion through space-time and will remain in motion with the same speed and in the same direction, ceteris paribus, until acted upon by an unbalanced force. The initial spatiotemporal differential is analogous to a warp drive that propels and perpetuates the motion of the object of mass through free space-time if unobstructed or not deviated from its course by an unbalanced force. Hence, the spatiotemporal medium propels the object of mass toward the region of higher spatiotemporal pressure. (Alcubierre, 1994) *The emergent nature of the temporal waves and the reciprocal spatial waves of space-time about an isolated object of mass provides and maintains the inertial equilibrium, or perpetuates the motion, ceteris paribus, regardless of path, state of rest, or motion, in homogeneous and isotropic space-time.*

Hence, the inertia of an isolated object of mass in isotropic and homogeneous space-time is determined by the interaction of the temporal wave and the spatial wave in the acceleration and pressure of space-time with the physical state of the object of mass at its locality, regardless of the inertia of all other external bodies of mass. Mach's principle manifests itself universally through the equivalence principle of General Relativity for the gravitational field potential of a gravitational mass and an inertial mass that upholds the constancy of the speed of light in a frame of reference. The gravitational field potential of the gravitational mass, or the inertial mass of an object, depends on the physical state of the object of mass and may be equal as long as the gravitational mass and the inertial mass of the object are equivalent in a spatiotemporal region.

The effect of inertia is the basis for the relativity of space-time in the General Theory of Relativity for the motion of an object of mass through space-time. The spatiotemporal warp in front and behind the object, a spatiotemporal differential, propels the object continuously. If the object were accelerated linearly by the greater force of a warp drive, the greater the velocity, the greater the effect of inertia, and the greater the length contraction and time dilation. As the speed increases exponentially, so does the relativistic mass of the moving object, to the point that the object would need an extensive amount of kinetic energy to approach the speed of light. A hypothetical warp drive would warp space-time to create a hypothetical topological feature of a shortcut between a departure point and a very distant destination point.

The non-linearity of the General Theory of Relativity for space-time-mass emerges as a consequence of the exponential nonlinear nature of the expansion or contraction of space-time throughout the universe.

7.2. The forcing function

A time-dependent process in a system of linear differential equations may be described by a forcing function that is only a function of time *f(t)*, excluding any of the other variables. The solution of the forcing function equals a constant for every value of its temporal variable "t".

Generally, the solution or result of a non-homogeneous forcing function may be solved employing a superposition of linear combinations of the homogeneous solution and the forcing term.

Let us define a forcing function $f(t)$ representing an external driving function of time, $Ae^{i\omega t}$, which may be regarded as the input or driving function to the system of volume *a*, which *a(t)* is the system output or response as a function of time.

The waves of space-time during the expansion of time and space provide the pre-emptive wave medium as a system response to particle waves and physical fields when the spatiotemporal field as a system input drives the expansion of space and time for mass, energy and physical fields.

$$\frac{\partial a}{\partial t}+\frac{a}{\tau}=A_s e^{i\omega t} \tag{7.23}$$

$$\tau\dot{a}+a=\tau A_s e^{i\omega t} \tag{7.24}$$

$$a=a_0 e^{-\omega t}+\frac{A_s}{\frac{1}{\tau}+i\omega}\left(e^{i\omega t}-e^{-\omega t}\right) \text{ for t} \geq 0 \tag{7.25}$$

Taking the derivative of the volume "*a*" to find the velocity of the volume, we have

$$\dot{a}=-\frac{a_0}{\tau}e^{-\omega t}+\frac{A_s}{\frac{1}{\tau}+i\omega}\left(i\omega e^{i\omega t}+\frac{e^{-\omega t}}{\tau}\right) \tag{7.26}$$

Substituting all terms into the forcing function we obtain

$$\left[\frac{A_s}{\frac{1}{\tau}+i\omega}\right]\left(\frac{e^{i\omega t}}{\tau}+i\omega e^{i\omega t}\right)=A_s e^{i\omega t} \tag{7.27}$$

The left side of the above equation simplifies to the forcing function on the right.

Hence, the magnitude of the volume *"a"* per τ , as the variable *"t"* goes to infinity, is given by

$$\left|\frac{a(t=\infty)}{\tau}\right|=\frac{A_S}{\tau\sqrt{\omega^2+\left(\frac{1}{\tau}\right)^2}}=\frac{A_S}{\sqrt{1+\omega^2\tau^2}} \tag{7.28}$$

The long-term magnitude of the general solution to the forcing function may be described as the amplitude divided by a function of the angular frequency and the time constant. If the angular frequency ω increases, the amplitude A_S of the spatial wave and the magnitude of the volume *"a"* would decrease. The linear frequency increases the angular frequency if the wavelength or period of the spatial wave contracts when acted upon by an external unbalanced force of a mass, energy, or a physical field.

The spatial wave contracts near an object of mass as the temporal wave extends in the gravitational field. As the distance from the object of mass increases, the temporal wave contracts as the spatial wave expands in the weaker gravitational field of the object of mass.

For the general solution of the volume *"a"* given above, the initial volume of space a_0 may be defined equal to zero for an emergent spatial wave, or it may be defined to be non-zero for an existing volume of space in homogeneous and isotropic space-time. The expanding or contracting volume of space is defined as the magnitude of the amplitude times the growth factor or scale factor of resultant time for any value of the temporal variable *"t"* equal or greater than zero.

Let us define a forcing function $f(t)$ representing an external driving function of time, $A_t e^{i\omega t}$, which may be regarded as the input or driving function to the system of volume θ, which $\theta(t)$ is the system output or response as a function of resultant time.

The volume θ is a temporal volume of the three coordinate dimensions of time: amplitude, linear frequency, and angular frequency.

$$\frac{d\theta}{dt} + \frac{\theta}{\tau} = A_t e^{i\omega t} \tag{7.29}$$

$$\tau\dot{\theta} + \theta = A_t e^{i\omega t} \tag{7.30}$$

$$\theta = \theta_0 e^{-\omega t} + \frac{A_t}{\frac{1}{\tau} + i\omega}\left(\mathrm{e}^{i\omega t} - e^{-\omega t}\right) \quad \text{for } \mathrm{t} \geq 0 \tag{7.31}$$

Therefore, the magnitude of the volume θ per τ, as the variable "t" goes to infinity, is given by

$$\left|\frac{\theta(t=\infty)}{\tau}\right| = \frac{A_t}{\tau\sqrt{\omega^2 + \left(\frac{1}{\tau}\right)^2}} = \frac{A_t}{\sqrt{1+\omega^2\tau^2}} \tag{7.32}$$

Taking the derivative of the volume θ to find the velocity of the volume, we have

$$\dot{\theta} = -\frac{\theta_0}{\tau} e^{-\omega t} + \frac{A_t}{\frac{1}{\tau} + i\omega}\left(i\omega e^{i\omega t} + \frac{e^{-\omega t}}{\tau}\right) \tag{7.33}$$

Substituting all terms into the forcing function we obtain

$$\left[\frac{A_t}{\frac{1}{\tau}+i\omega}\right]\left(\frac{e^{i\omega t}}{\tau}+i\omega e^{i\omega t}\right)=A_t e^{i\omega t} \tag{7.34}$$

Similarly, the long-term magnitude of the general solution to the temporal forcing function may be described as the amplitude A_t divided by a function of the angular frequency and the time constant. If the angular frequency ω decreases, the amplitude A_t of the temporal wave and the magnitude of the volume θ would increase.

The linear frequency decreases the angular frequency if the wavelength or period of the temporal wave extends when acted upon by an external unbalanced force of a mass, energy or a physical field. The temporal wave extends near an object of mass as the spatial wave contracts in a gravitational field. As the distance from the object of mass increases, the spatial wave extends as the temporal wave contracts in the weaker gravitational field of the object of mass.

For the general solution of the volume θ given above, the initial volume of time θ_0 may be defined equal to zero for an emergent temporal wave, or it may be defined to be non-zero for an existing temporal volume in homogeneous and isotropic space-time. The expanding or contracting temporal volume is defined as the magnitude of the amplitude times the growth factor or scale factor of time for any value of the temporal variable *"t"*.

Let us illustrate the reciprocal relationship between conjugate space and time as wave attributes such as: amplitude, linear frequency, and angular frequency, change during the expansion or contraction of spatiotemporal waves in space-time.

$$\omega_S=\frac{1}{\omega_t} \tag{7.35}$$

$$A_S = \frac{1}{A_t} \tag{7.36}$$

The angular frequency of the temporal or spatial wave is inversely proportional to its amplitude $\omega \propto 1/A$. The units of the amplitude depend on the type of wave, spatial or temporal, but the units are always in the same units as the oscillating variable of the wave. The amplitude of the waves of space-time, which relate to the spatial or temporal volume, refer to the pressure of space-time upon objects of mass. Pressure is force per unit of area (N/m^2) or density of energy (J/m^3). Thus, the amplitude of a temporal wave refers to the energy per unit volume.

The energy of a temporal wave is proportional to the square of its amplitude, $E \propto A^2$. Energy, or Work, is defined as the product of mass, acceleration, and spatial distance. Let us imagine a temporal wave near an object of mass doing work, as the spatial wave contracts near the mass of the object. The amplitude A_t of the temporal wave is the radius of the angular frequency at its peak value t_p during a period of time *T*.

$$E_t = mc^2 = F \cdot d = m\left(\frac{\partial^2 c^2 A_t^2}{\partial t^2}\right) \cdot \left(\frac{ct_p}{cT}\right) = mc^2\left(\frac{\partial^2 A_t^2}{\partial t^2}\right) \cdot \left(\frac{t_p}{T}\right) \tag{7.37}$$

$$E_t = mc^2\left(\frac{\partial^2 A_t^2}{\partial t^2}\right)\left(\frac{t_p}{T}\right) = m\left(\frac{\lambda^2}{T^2}\right)(2\pi)\left(\frac{t_p}{T}\right) = m\left(\frac{2\pi}{T}\right)\left(\frac{\lambda^2}{T^2}\right)(t_p) = m\omega c^2 t_p = \hbar\omega \tag{7.38}$$

What is the acceleration of the probability of the temporal wave?

$$\frac{\partial^2 A_t^2}{\partial t^2} = \frac{E_t}{mc^2}\left(\frac{T}{t_p}\right) = \frac{T}{t_p} \tag{7.39}$$

Let us illustrate how the amplitude of the temporal wave, A_t, changes, during a period *T*.

$$A_t = e^{-i\omega T}\left(\tau\dot{\theta} + \theta\right) \tag{7.40}$$

Therefore, the amplitude of the temporal wave may be defined as the product of the temporal growth factor times the sum of the product of the time constant and the volume velocity plus the initial volume of the temporal wave. As space-time expands, the amplitude dimension of the temporal wave decreases proportionally. This conceptualization is applicable to the three-dimensional spatiotemporal wave or any other three-dimensional representation of a wave of a physical field.

7.3. The spatiotemporal constant: π

The number π is a mathematical constant that describes the ratio of a circle's circumference to its diameter in Euclidean geometry. It is both irrational and transcendental because it cannot be expressed exactly as a common fraction and it is not the root of any non-zero polynomial having rational coefficients. The number π is commonly approximated to the value 3.141592654 for engineering calculations.

Throughout the history of civilizations, there has been an incredible effort to calculate the accuracy, or number of digits, of the number π for very practical reasons. Nevertheless, it seems that the geometrical origin of the number π has not received an equal amount of attention.

What is π ? What is it about the symmetry of a circle in Euclidean geometry that allows for the number π to describe the same geometrical relationship between circumference and diameter regardless of scale?

$$\pi = \frac{C}{d} e^{i\omega t} \tag{7.41}$$

If we consider the equation of an expanding spiral as $R = e^{\omega t}$ in space-time, where $\omega = 1/k = 1/\text{Tan}\,\beta = Cot\beta$, and ω is the rate of growth of the spiral, and the smaller the angle β the greater the rate of growth. When the angle β is 90 degrees, ω goes to zero, $\omega = Cot\ 90^0 = 0$, and the non-expanding spiral becomes a unit circle at $R = 0$. Therefore, it is possible to consider the unit circle as a special logarithmic spiral with a rate of growth of zero in a space-time that is not expanding. Since a straightedge measures the metric of space-time, the measurement of the circumference, or the diameter of a circle, even in a space-time that was expanding or contracting, would be unchanging.

Let us describe the following mathematical expressions that involve the constant:

$$2\pi = \frac{C}{r} = \frac{\omega}{f} \tag{7.42}$$

The above mathematical expressions involve variables associated with a circle. Thus, let us start our investigation with the equation for the area of a temporal circle.

$$A = \pi t^2 \tag{7.43}$$

$$\frac{\partial A}{\partial t} = 2\pi t \tag{7.44}$$

$$\frac{\partial^2 A}{\partial t^2} = 2\pi \tag{7.45}$$

$$\pi = \frac{1}{2}\frac{\partial^2 A}{\partial t^2} \tag{7.46}$$

Thus, half of the acceleration of the area of a spatiotemporal circle is equal to the number π. Then, the number π is not just a ratio, but also acceleration in expanding space and time. What if the area was the area of a different two-dimensional shape such as a rectangle or a square, what would π be?

The answer lies on a spatiotemporal surface. Let us consider the three-dimensionality of time as a representation of the axes that represent the aspects and attributes of the temporal wave as it emerges. The temporal coordinates may be described as Linear Frequency (Period) (f_T) for $\left(t_{x_2} - t_{x_1}\right)$, Angular Frequency (Rotation or Spin) $\left(\omega_t\right)$ for $\left(t_{y_2} - t_{y_1}\right)$ and Amplitude $\left(A_t\right)$ for $\left(t_{z_2} - t_{z_1}\right)$. Let us imagine a round mechanical clock moving at relativistic speed with an axis of propagation normal to the clock face or dial. The axis of propagation would represent the Linear Frequency-axis, the radius of the dial would represent the Amplitude-axis, and the rotating hands of the clock would represent the Angular Frequency-axis. Thus, time is conceptualized above as a temporal volume, but it may also be conceptualized as a temporal surface.

On the temporal surface, let us imagine a temporal linear-angular frequency plane where there is a diagonal line with a slope equal to ω/f. The temporal frequency plane is rectangular and is bounded by the linear frequency axis and the angular frequency axis. The slope of the diagonal line is 2π. If the area of the rectangle is decreased by halving the magnitude of the angular frequency only, then the number π is the slope of the diagonal line of the halved rectangle. Consequently, the rectangle of area $A_1 = \omega f$ is twice as large as the rectangle of area $A_2 = \omega f/2$. Thus, the acceleration of the area A_2 of the smaller rectangle is half of the acceleration of the area A_1 of the larger rectangle. The number π is equal to the acceleration of the area A_2.

$$\pi = \frac{\partial^2 A_2}{\partial t^2} = \frac{1}{2}\frac{\partial^2 A_1}{\partial t^2} \qquad (7.47)$$

It is interesting to point out that if the diagonal line with slope π, of the rectangle with an area A_1, is rotated to create a circle with the origin at mid-point, the distance that extends from the corner of the rectangle to the circle is *"b"* and the parallel side of the rectangle, to this distance, is *"a"*, the ratio $a/b = (a+b)/a$ equals to the Golden Ratio (1.6180339887...). Moreover, if the distances *"a"* and *"b"* are used as the circumference of a circle, the angle of the *b*–sector equals the Golden Angle $\left(137.5077^0...\right)$.

The equations below were derived by the eminent Swiss mathematician and physicist Leonhard Euler which further illustrate the significance of the number π in Euler's identity and his exact and rigorous solution, using an infinite series, to the Basel problem.

$$e^{i\pi} = -1 \qquad (7.48)$$

Let us imagine a temporal square with a side equal to *"i"* that is represented in Euler's identity as the scale factor of growth whose exponent we may consider to be a rectangle of area $A_2 = \omega f/2$. The rectangle represents a temporal area in space-time that expands with the passage of time with acceleration equal to π.

$$e^{i\pi} = e^{i\left(\frac{\omega}{2f}\right)} = i^2 \qquad (7.49)$$

Similarly, the solution to the Basel problem is given by

$$\pi^2 = \lim_{n\to\infty} 6\left(1 + \frac{1}{2^2} + \frac{1}{3^2} + \frac{1}{4^2} + \ldots + \frac{1}{n^2}\right) \qquad (7.50)$$

Let us imagine a circle A with a half-area that has an acceleration equal to π, if two smaller circles (B, C) with quarter-areas are enclosed by circle A, and within each circle B and C there are two other smaller circles (D, E) and (F, G) with quarter-areas with respect to B and C, all enclosed by circle A. Half the area of each circle has an acceleration equal to π with respect to its total area. The six circles B, C, D, E, F and G would contribute the total half-area of circle A. If the acceleration of half the area of circle A is represented as a square area with a side equal to $\sqrt[2]{\pi}$, then the square would have an area equal to π. Accordingly, the square would have an area equivalent to the sum of the accelerations of the six circles B, C, D, E, F and G.

$$\frac{1}{2}\frac{\partial^2 A}{\partial t^2}=\left(\frac{1}{8}+\frac{1}{8}+\frac{1}{16}+\frac{1}{16}+\frac{1}{16}+\frac{1}{16}\right)\frac{\partial^2 A}{\partial t^2} \tag{7.51}$$

The circles B, C, D, E, F and G are shown below:

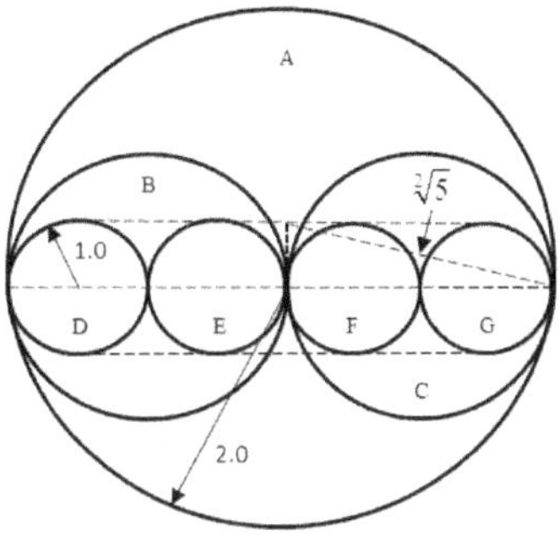

Figure 17.

In mathematics, the golden ratio (secto divino) has fascinated mathematicians and physicists through the ages, including Pythagoras, Euclid, Kepler, and most recently Penrose. Artists have found the golden ratio aesthetically pleasing, while other researchers and scientists have found the golden ratio in the biological proportions of nature and in the atomic scale of matter. It is interesting that it is also found in the proportion of the expansion of space and time between circles F and G, D and E, or B and C, in the dimensions of a triangle, in relation to π.

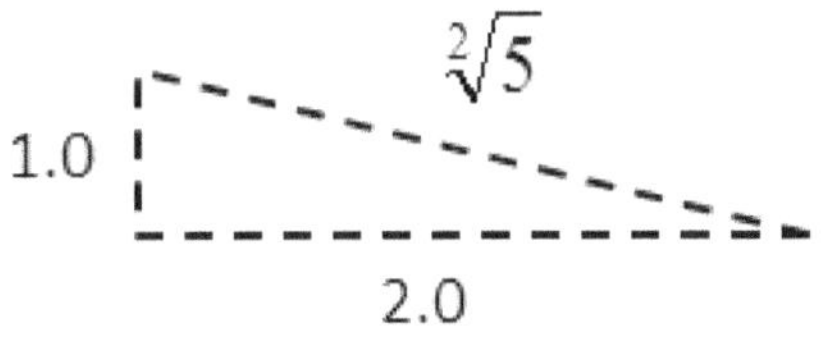

Figure 18.

$$\Phi = \frac{Opposite + Hypothenuse}{Adjacent} = \frac{1+\sqrt[2]{5}}{2} = 1.6180339887... \quad (7.52)$$

If Euler's infinite series for the solution of the Basel problem is equal to a volume *"a"*, we obtain

$$a = \lim_{n \to \infty} \left(1 + \frac{1}{2^2} + \frac{1}{3^2} + \frac{1}{4^2} + \ldots + \frac{1}{n^2} \right) \quad (7.53)$$

$$\pi^2 = \lim_{n \to \infty} 6(\mathrm{a}) \quad (7.54)$$

If the half-area of circle A, with acceleration equal to an area of π, is multiplied by a distance π it makes a rectangular prism with volume π^2. The volume of the rectangular prism is made up of six pyramids with their apexes connecting at the center point of the prism. The base of each pyramid is a side area of the rectangular prism. The sum of the volumes of the six pyramids is equal to Euler's solution to the Basel problem which equals π^2.

The solution of Euler illustrates how a point in space-time may expand six-dimensionally in every sense of direction.

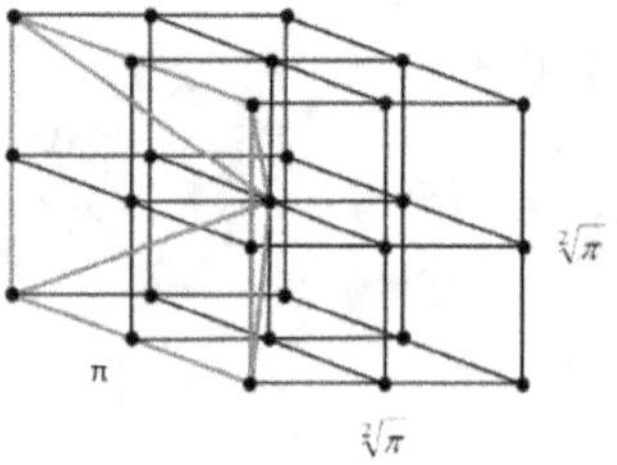

Figure 19.

Therefore, the number π may be defined as, but not necessarily limited to, the acceleration of a spatiotemporal area, a ratio of spatial distances, or the slope of the diagonal line of a temporal frequency plane, depending on the purpose of its mathematical and physical relationship in space-time-mass.

In Euclidean trigonometry, the number π is the length of the hypotenuse of a triangle whose rise is only half, but run is the same, as those of a triangle whose hypotenuse equals 2π .

The insight into the number π reveals the transcendental acceleration and natural symmetry of the surfaces and volumes of space and time which expand or contract around and through objects in the universe. The spatiotemporal acceleration π acts geometrically upon the three-dimensional aspect of a volume.

If a spherical temporal surface of space-time were to contract to half its size, π would be equal to one-eighth of the acceleration of the surface area of the previous surface, and if the latter contracts to one-half of its surface area, then π would again equal to one-eighth the acceleration of the surface area of the previous surface, and so on.

Hence, the number π may also be defined as one-eighth the acceleration of the surface area of a spherical spatiotemporal surface, or a spherical spatiotemporal wave, as it contracts or expands.

7.4. The Casimir Force

The Casimir effect describes a small attractive force between two close, perfectly conducting and parallel, uncharged metal plates in the spatiotemporal wave medium. It was the Dutch physicist Hendrik B. G. Casimir who first predicted this phenomenon in 1948 which described the phenomenon as the difference in pressure on both sides of each metal plate caused by the difference in the spatiotemporal fluctuations of the zero-point energy outside and between the plates. (Casimir, 1948)

The Casimir effect has been surmised as the presence of dielectrics and metal conductors altering the spatiotemporal expectation value of the energy of the second quantized electromagnetic field.

The Casimir effect manifests itself as a force between such close, parallel, and uncharged metal conductors, because the expectation value of the energy depends on the physical shapes and positions of the dielectrics and metal conductors.

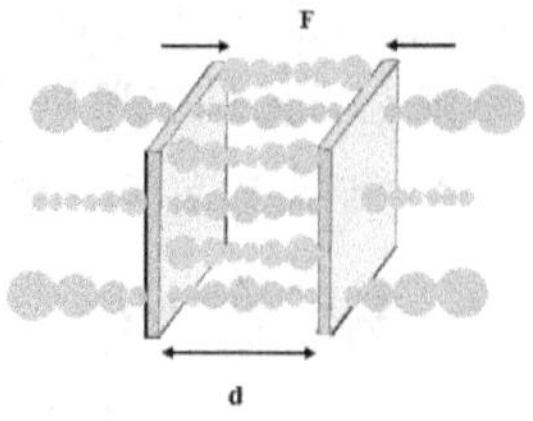

Figure 20.

Let us imagine that we have two close, perfectly conducting and parallel, uncharged square metal plates in the spatiotemporal wave medium, far from other objects or boundaries, where the medium expands at every point amidst zero-point energy and in the presence or absence of mass. As space-time expands between the close square metal plates, the temporal waves diverge (green), and the spatial waves converge (blue), toward each plate, as space contracts and time dilates according to the General Theory of Relativity.

Outside the metal plates, a similar process occurs, but in this external wave medium the space-time encounters a boundary on only one external side of each plate, so the temporal waves (green) and spatial waves (blue) can diverge or converge for a longer distance, or volume, if unobstructed, producing a greater divergence or greater convergence between boundaries, or between boundary and a region of homogeneous and isotropic spatiotemporal wave medium. Consequently, it is reasonable to assume that the pressure of the scalar wave medium outside of both plates is greater than the pressure between the plates.

Let us hypothesize that the pressure difference between the outside and inside scalar forces may account for the Casimir effect as the expansion of space-time, or scalar wave medium, acts on the square metal plates in conjunction with zero-point energy and electromagnetic fields that may exist in the medium. Let us consider a large cube, such as Figure 21, which consists of eight smaller square cubes, so that two of the square cubes represent a square prism, such as Figure 24. One of the faces of an individual square cube represents the inside or outside area of the Casimir metal plates, such as the shaded area of the square cube in Figure 22. The individual square cube may be regarded as an expanding square cube inside the expanding sphere of an expanding point in space-time, such as in Figure 23. The surface of the Casimir metal plate is one eighth of the total surface of the two opposite sides of the large cube, such as in Figure 21.

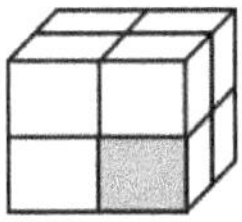

Figure 21.

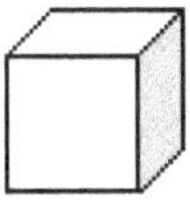

Figure 22.

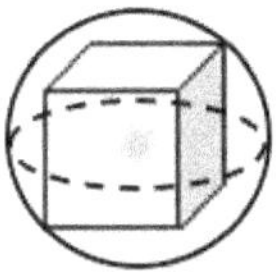

Figure 23.

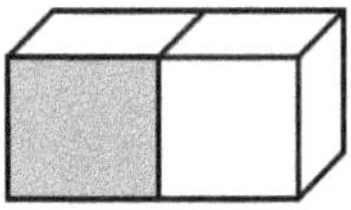

Figure 24.

From previous research, half of the acceleration of the area of a temporal circle is equal to the number π . Then, the number π is not just a ratio, but also an acceleration in expanding space and time. Thus, for the area of a different two-dimensional shape such as a rectangle, such as in Figure 24, or one of the largest sides of a square prism, π would be:

$$\pi = \frac{1}{2}\frac{\partial^2 A}{\partial t^2} \tag{7.55}$$

$$\pi^2 = \frac{1}{4}\left(\frac{\partial^2 A}{\partial t^2}\right)^2 \tag{7.56}$$

Hence, π may represent the acceleration of the shaded area of an individual square cube, such as in Figure 22, or half the area of one of the largest sides of a square prism, such as in Figure 24. If a point expands at the geometric center of a spatiotemporal volume which has the shape of a square prism, such as in Figure 25, the volume of the square prism is made up of six pyramids with their apexes connecting at the point of expansion. The base of each pyramid is a side area of the square prism, such as in Figure 26. The sum of the volumes of the six pyramids is equal to Euler's solution to the Basel problem which equals π^2.

The solution of Euler illustrates how a point in space-time may expand six-dimensionally in every sense of direction to cover the volume of a square prism. Thus, if Euler's infinite series for the solution of the Basel problem is equal to a volume *"a"*, we obtain:

$$\pi^2 = \frac{1}{4}\left(\frac{\partial^2 A}{\partial t^2}\right)^2 = \lim_{n\to\infty} 6(\mathrm{a}) = \lim_{n\to\infty} 6\left(1+\frac{1}{2^2}+\frac{1}{3^2}+\frac{1}{4^2}+\ldots+\frac{1}{n^2}\right) \tag{7.57}$$

$$\frac{\pi^2}{6} = \lim_{n\to\infty}\left(1+\frac{1}{2^2}+\frac{1}{3^2}+\frac{1}{4^2}+\ldots+\frac{1}{n^2}\right) \tag{7.58}$$

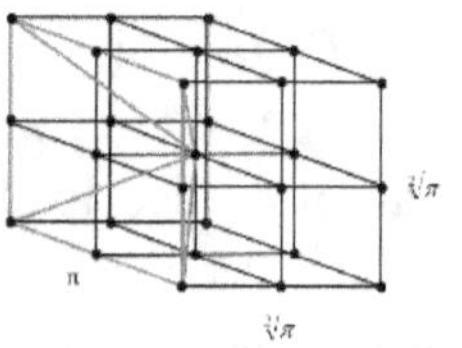

Figure 25.

Figure 26 illustrates how one of the six spatiotemporal pyramids expands six-dimensionally, with the shaded base of the pyramid representing one of five surfaces involved in the volume of expansion for every pyramid. The shaded side is where the pressure of the expanding scalar wave medium exerts pressure, or an attractive force on the area of the Casimir metal plate. Half the shaded area of the base of the pyramid, in Figure 25 and Figure 26, is equivalent to one fifth of the total surface area of the pyramid. The geometrical relationship of π , Euler's solution, and the physical shapes above are physically scalable.

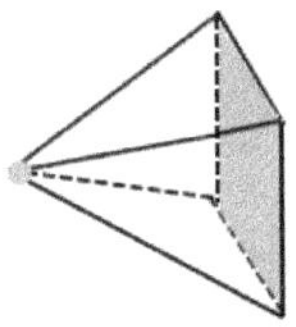

Figure 26.

Let us express the geometrical relationship of the above concepts in a mathematical way for the force per unit area acting on the Casimir metal plates. Let us imagine that we are able through very advanced technology to recreate the Casimir effect experiment at spatial dimensions that are multiples of the Planck scale units.

Let the area of the Casimir metal plate be a common multiple *"n"* of the Planck area, l_p^2 , or A_p, the separation distance between the metal plates be the same common multiple of the Planck length, l_p, and the mass of each metal plate be the same common multiple of the Planck mass, m_p , in homogeneous and isotropic space-time. From previous research, the spatiotemporal acceleration may be expressed as

$$\ddot{a} = \frac{\hbar c}{m_p} \quad (7.59)$$

Then, the Planck-Casimir unit force acting on the surface of the hypothetical Casimir metal plates may be denoted as

$$F = m_p \ddot{a} = \hbar c \quad (7.60)$$

And the spatiotemporal Planck-Casimir unit pressure of the force is given by

$$\frac{F_{PC}}{nA_p} = \frac{\ddot{a}(nm_p)}{nl_p^{\,4}} = \frac{\ddot{a}m_p}{l_p^{\,4}} \tag{7.61}$$

$$\frac{F_{PC}}{A_p} = n\frac{\ddot{a}m_p}{l_p^{\,4}} \tag{7.62}$$

Let us consider the spatiotemporal volume between or outside the hypothetical Casimir metal plates for a point in space-time expanding six-dimensionally in every sense of direction at the center of the volume of a rectangular prism, such as in Figure 24 and Figure 25.

The force exerted by one half of one sixth of the spatiotemporal volume of the rectangular prism, π^2, on each area of the Casimir metal plates, is exerted by one of four forces, F_4, exerted externally by each side of the half-pyramid that remains, such as in Figure 27, and only the area of the base of the five surfaces of a half-pyramid expands toward the Casimir metal plate, which is equivalent to one fifth of the total surface area of the complete pyramid.

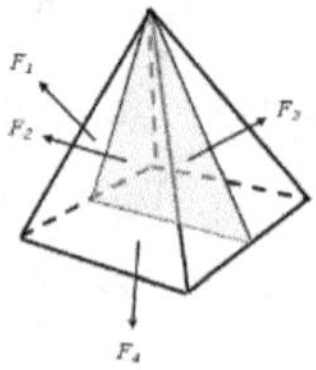

Figure 27.

Thus, the Casimir geometric expansion factor for this hypothetical example is given by

$$n = \frac{1}{2}\cdot\frac{1}{6}\cdot\pi^2\cdot\frac{1}{4}\cdot\frac{1}{5} \tag{7.63}$$

The Casimir geometric expansion factor is a function of the geometry of the spatiotemporal volume, and the physical dimensions, spacing distance, the masses of the Casimir metal plates, the content of zero-point energy, and electromagnetic fields, in the local spatiotemporal volume, the expansion and acceleration of space-time at every point, and the relativistic relations of the spatiotemporal wave medium according to the General Theory of Relativity.

Hence, the equation for the Planck-Casimir unit pressure may be expressed as

$$\frac{F_{PC}}{A_p}=\frac{1}{2}\cdot\frac{1}{6}\cdot\pi^2\cdot\frac{1}{4}\cdot\frac{1}{5}\cdot\frac{\ddot{a}m_p}{l_p^{\,4}}=\frac{\pi^2}{240}\frac{\ddot{a}m_p}{l_p^{\,4}} \tag{7.64}$$

Then, we are able to derive the Planck-Casimir equation for the attractive force per unit area, F_{PC}, to obtain an equation for the pressure on the Casimir metal plates.

$$F_{PC}=\frac{\pi^2}{240}\frac{\hbar c}{l_p^{\,4}}A_p \tag{7.65}$$

Therefore, it is reasonable to assume that the pressure on the Casimir plates may be related to the expansion and acceleration of the spatiotemporal wave medium, and its content of zero-point energy and electromagnetic fields, in the vicinity of the plates.

Hence, we may express the pressure differential between the inside spatiotemporal wave medium of the Casimir metal plates and the outside spatiotemporal wave medium using two Casimir geometric expansion factors and the difference between the external force and the internal force acting on the Casimir metal plates.

$$F_{EXT}-F_{INT}=\left(n_1-n_2\right)\frac{\hbar c}{l_p^{\,4}}A_p \tag{7.66}$$

$$F_{PC}=F_{EXT}-F_{INT} \tag{7.67}$$

$$n_1 - n_2 = n = \frac{\pi^2}{240} \tag{7.68}$$

$$\frac{n_1}{n_2} = e^{i(\omega t - \omega \tau)} \tag{7.69}$$

The external Casimir geometric expansion factor is n_1, the internal Casimir geometric expansion factor is n_2, and the ratio between them is equal to the relativistic spatiotemporal expansion kernel, the natural base rate of growth, raised to the exponent $i(\omega t - \omega \tau)$, representing the growth factor differential of the expanding spatiotemporal volume applying the Casimir force at the location of the hypothetical Casimir metal plates.

$$e = \lim_{\tau \to \infty} \left(1 + \frac{1}{\tau}\right)^{\tau} \tag{7.70}$$

The natural base rate of growth describes the spatiotemporal wave medium as the spatial waves diverge and the temporal waves converge away from a boundary on smaller and smaller spatiotemporal dilations, finer time periods. A boundary represents either an obstruction to a scalar wave by an object or by a region with isotropic and homogeneous space-time of constant pressure. The transcendental "e" is the base rate of growth shared by all continually growing natural processes in the spatiotemporal wave medium.

§ 8. Conclusion

Physics is like a comedy, where particles are the actors and actresses, fields are the scripts, a world line is a stage, space and time are the elemental muses of laughter and tragedy, the divine creator is the genuine magnificent playwright, and physicists are the participant audience. Space is able to move superluminally in some regions of our universe. Therefore, matter as a physical state of energy existing in space-time is also able to move superluminally within those regions of space, as theoretically measured locally with respect to an inertial frame of reference at a subluminal region of homogeneous and isotropic space-time. Hence, theoretically, the speed of light remains constant in either subluminal or superluminal space-time, and General Relativity prevails. As theoretically proposed throughout this document, the relativistic effects on mass, time, and length, act on moving objects approaching, reaching, and passing the light barrier.

Theoretical applications for the Lorentz-Larmor transformation function may be on faster than light travel through hypothetically stable wormholes, space-time manipulation through the generation of a hypothetical warp bubble, or on hypothetical elementary tachyonic particles which always move faster than light and are not excluded by the General Theory of Relativity. Such hypothetical tachyonic particles would travel back in time exhibiting imaginary mass and temporal condensation, and length extension, as theorized in this document for a Kerr-Newman supermassive black hole.

The careful understanding and theoretical modeling of the physical processes within a black hole lead to the principles and interactions of gravitation, matter, electromagnetic fields, gluonic interaction, and the underlying structure of space-time itself. The expansion of space-time is emergent and constitutes the spatiotemporal geometry and physical symmetry in the underlying structure of the universe.

Chapter 6

On the kinematics of motion in six-dimensional space-time

§ 1. Introduction: the kinematics of an object or particle

The kinematics of a particle or the geometry of motion is directly related to the energy of a particle through its movement in six-dimensional space-time or its intrinsic energy of mass formation which is equivalent to the square of its velocity. Therefore, the study of kinematics is the study of the concept of motion, energy, forces, and transformation.

1.1. On the nature of motion

Energy may be originated through the geometry of motion of a particle or an object. Momentum is a prime example as it may originate energy in linear, non-linear, or helical form, as the mass of the object travels through space-time with one-dimensional, two-dimensional, or three-dimensional motion. Velocity is a function of time, and time is also dimensional; as an object travels through space-time it also travels through a temporal path as well as a spatial path. Hence, the structure of the geometry of motion of an object engenders the form of energy that originates in the object of movement. Geometry of movement invokes and transforms energy. (Dugas, 1988)

David Hume was a Scottish philosopher, historian, and economist. An empiricist that supported the theory that all knowledge is based on experience derived from the senses. Hume was an empiricist and a skeptic, believing that scientific concepts must be based on experience and evidence, not reason alone. He also held that time did not exist separately from the movement of objects. He was a big influence on Einstein's thinking about space and time.

According to Hume, time is the effect of succession; time encompasses the perceived impressions of the mind at a fixed and finite motion. The senses of sight and touch provide the human mind with concepts such as space and time. Thus, for Philosopher David Hume, time and motion were intrinsically related. (Hume, 1738)

Space-time expands in all directions at the speed of light. Thus, as a particle moves through space-time, it lags the speed of space-time due to its mass. The acquisition of mass causes the energy of a system, or object, to travel slower than light. If the object was moving linearly on a two-dimensional plane of space-time, its momentum would equal its mass times the speed of light, then its acquired mass would be equal to its energy divided by the speed light. However, if the object was moving in a nonlinear manner through a two-dimensional plane of space-time, its angular momentum may equal its mass times the speed of light times the radius, but its mass would be equal to its energy divided by the square of the speed of light.

The principle of superposition prevents objects of mass embedded in isotropic and homogeneous space-time from moving away from each other due to the expansion of every point in isotropic and homogeneous space-time if there is destructive interference between adjacent space-time wavelets. Consequently, objects retain their proportional scale and distances with respect to one another in isotropic and homogeneous space-time, while time passes as every point in space-time expands in all directions unless obstructed. Hence, motion is possible because of the expansion of space-time and the passage of time.

Furthermore, as the object travels through extra dimensions of space-time, its velocity may change from linear to non-linear, or from non-linear, to helical. In effect, the momentum of the object has a velocity that is directly proportional to a distance, or an area, or a volume, of space-time. The geometry of space-time is invoking the realization of energy. *Any motion in six-dimensional space-time is an expression of a force, and every force in six-dimensional space-time is an expression of a form of energy*. For any motion in six-dimensional space-time may be described by the actions of nine forces or the conceptualization of nine energies.

Therefore, the conceptualization of six-dimensional space-time impacts the underlying framework of the present thermodynamic laws and the formulation of thermodynamic energy since all laws and energy are functions of time and motion. Translation, pulsation, expansion, or contraction, is a form of motion through time. Time is the agent of change, and change is the driver of motion. There is no motion without change, or change without time. (Bottema, 1990)

§ 2. The pulsation of the mass of a particle

For the pulsating volume of the mass

$$V_m^{\ 2} = \left(\frac{4}{3}\right)^2 \pi^2 \left(r^3\right)^2 \left(Sin^2\theta \, dt^2 + 1\right)^2 \tag{2.1}$$

$$V_m^{\ 2} = \frac{16\pi^2 r^6}{9} \left(Sin^2\theta \, dt^2 + 1\right)^2 \tag{2.2}$$

First, let us consider the hypothetical case of a stationary pulsating particle system that encompasses six-dimensional space-time within its boundary of matter where the curvature of space or time is not flat, or zero. The relativistic mass of the particle system consists of photons that are pulsating at the speed of light, *c*. The mass is timeless and represents pure energy in space-time.

During every pulsating cycle, time extends creating space, then space is contracted before the next extension of time, and the pressure of space increases within the boundary of the volume. If the pressure of space-time increases to a level that causes time to extend faster than the speed of light *c*, more space would be created, and expansion would occur. As the pressure within the pulsating particle system decreases at an accelerated rate, more space is created, and expansion accelerates. The particle system may rotate, translate, and pulsate in expanding six-dimensional space-time.

For the pulsating mass with its geometrical center at point $C\left(t_X, t_Y, t_Z, r_X, r_Y, r_Z\right)$, and with a radius of r_m from point C to a point $P\left(t_{X_P}, t_{Y_P}, t_{Z_P}, x_P, y_P, z_P\right)$ on its boundary, density ρ_m , and volume V_m , we have

$$r_m^{\ 6} = \left(r_m^{\ 3}\right)^2 = \left(r_m^{\ 2}\right)^3 \tag{2.3}$$

$$r_m^{\ 2} = c^2\left(t_{X_P} - t_X\right)^2 + c^2\left(t_{Y_P} - t_Y\right)^2 + c^2\left(t_{Z_P} - t_Z\right)^2 \tag{2.4}$$

$$+ a(t)^2\left\{\left(x_P - r_X\right)^2 + \left(y_P - r_Y\right)^2 + \left(z_P - r_Z\right)^2\right\} \tag{2.5}$$

Combining the mass density and volume

$$V_m^{\ 2}\rho_m^{\ 2} = \frac{16\pi^2 r_m^{\ 6}\rho_m^{\ 2}\left\{Sin^2\theta\left(1 - k_t t^2\right)dt^2 + 1\right\}^2}{9} = m^2 \tag{2.6}$$

Let us represent m^2 as a six-dimensional vector $d\vec{M}$ such that

$$d\vec{M} \equiv -\left(V_m\rho_m\right)\vec{a}_{t_X} - \left(V_m\rho_m\right)\vec{a}_{t_Y} - \left(V_m\rho_m\right)\vec{a}_{t_Z} + \left(V_m\rho_m\right)\vec{a}_X + \left(V_m\rho_m\right)\vec{a}_Y + \left(V_m\rho_m\right)\vec{a}_Z \tag{2.7}$$

$$d\vec{M} \equiv -(m)\vec{a}_{t_X} - (m)\vec{a}_{t_Y} - (m)\vec{a}_{t_Z} + (m)\vec{a}_X + (m)\vec{a}_Y + (m)\vec{a}_Z \tag{2.8}$$

§ 3. The six-dimensional vector of a rotating and pulsating particle

Nature shows us examples of how energy is used in pulsating organisms such as the Pulsating Xenid, or Heteroxenia Fuscescens, is a soft coral with large feathery pinnate tentacles, opening and closing its beautiful flower-shaped tentacles in popular places such as the Gulf of Eilat, to spend energy to increase its ratio of photosynthesis-to-respiration to higher levels, or a colorful Jellyfish, a free-swimming marine animal that travels from the surface to the deep sea, in the vastness of the mighty Atlantic Ocean, and consists of a gelatinous umbrella-shaped bell and trailing tentacles. The bell can pulsate to allow the Jellyfish to travel from place to place. Let us consider a similar event involving a particle that rotates, pulsates, and translates through space-time as it transforms its potential energy to kinetic energy. (Reuleaux, 2012)

For the angular velocity of the radius of the mass of the particle

$$\omega = \frac{d\theta}{dt} \tag{3.1}$$

For the counterclockwise speed of the rotating radius "r" of the mass of the particle

$$\frac{ds}{dt} = r\frac{d\theta}{dt} \tag{3.2}$$

$$\frac{ds^2}{dt^2} = r^2\frac{d\theta^2}{dt^2} \tag{3.3}$$

$$\frac{d\theta^2}{dt^2} = \frac{1}{r^2}\frac{ds^2}{dt^2} \tag{3.4}$$

Let us describe a six-dimensional vector, in homogeneous and isotropic space-time, from the origin of an inertial coordinate system to the center point of the pulsating particle of mass m.

$$\vec{r}_{O-C} \equiv -\left(ct_X\vec{a}_{t_X} + ct_Y\vec{a}_{t_Y} + ct_Z\vec{a}_{t_Z}\right) + a(t)\left(r_X\vec{a}_X + r_Y\vec{a}_Y + r_Z\vec{a}_Z\right) \tag{3.5}$$

From the center point C of the particle to a point $P\left(t_{X_P}, t_{Y_P}, t_{Z_P}, x_P, y_P, z_P\right)$ on its boundary, on a plane that is always parallel to the x-y plane passing through the circumference of the circle of its equator, let us denote a six-dimensional vector that rotates and always points to point P, such as

$$\vec{r}_{C-P} \equiv -\left(ct_{X_P}Cos2\theta - ct_{Y_P}Sin2\theta\right)\vec{a}_{t_X} - \left(ct_{X_P}Sin2\theta + ct_{Y_P}Cos2\theta\right)\vec{a}_{t_Y} \tag{3.6}$$

$$-\left(ct_{Z_P}\right)\vec{a}_{t_Z}$$

$$+a(t)\left(x_P Cos2\theta - y_P Sin2\theta\right)\vec{a}_X + a(t)\left(x_P Sin2\theta + y_P Cos2\theta\right)\vec{a}_Y + a(t)\left(z_P\right)\vec{a}_Z$$

Expressing the vector metric for the worldspiral in space-time-mass of the event at point P for a rotating, translating, and pulsating particle, of radius r and mass m, we obtain

$$-c^2d\tau^2 + dM^2 = -c^2d\pi^2 + a(t)^2d\Sigma^2 + dM^2 \tag{3.7}$$

Therefore, the space-time-mass metric equation for the worldspiral of the event at point *P*, for the rotating, pulsating, and translating particle of mass *m* can be expressed as

$$dw^2 = -c^2 d\pi^2 + a(t)^2 d\Sigma^2 + dM^2 \tag{3.8}$$

Where the three-dimensional temporal vector is

$$-c^2 d\pi^2 = -\left(ct_X + ct_{X_P} Cos2\theta - ct_{Y_P} Sin2\theta\right)^2 - \tag{3.9}$$

$$\left(ct_Y + ct_{X_P} Sin2\theta + ct_{Y_P} Cos2\theta\right)^2 - \left(ct_Z + ct_{Z_P}\right)^2 \tag{3.10}$$

and the three-dimensional spatial vector is

$$a(t)^2 d\Sigma^2 = \left(a(t)r_X + a(t)x_P Cos2\theta - a(t)y_P Sin2\theta\right)^2 + \tag{3.11}$$

$$\left(a(t)r_Y + a(t)x_P Sin2\theta + a(t)y_P Cos2\theta\right)^2 + \left(a(t)r_Z + a(t)z_P\right)^2 \tag{3.12}$$

Let us express dw^2 as a six-dimensional worldspiral metric in space-time-mass of the event at point *P* as follows:

$$-c^2 d\tau^2 + dM^2 = -c^2 \pi^2 + a(t)^2 d\Sigma^2 + dM^2 \tag{3.13}$$

The six-dimensional space-time-mass worldspiral vector may be expressed as

$$d\vec{w} = -U_{t_X}\vec{a}_{t_X} - U_{t_Y}\vec{a}_{t_Y} - U_{t_Z}\vec{a}_{t_Z} + U_X\vec{a}_X + U_Y\vec{a}_Y + U_Z\vec{a}_Z \tag{3.14}$$

Let us introduce curvature in space-time-mass, so that space has a curvature of $\left(1 - k_s r^2\right)$ and time has a curvature of $\left(1 - k_t t^2\right)$ where applicable.

Therefore, the magnitudes $U_{t_X}, U_{t_Y}, U_{t_Z}, U_X, U_Y, U_Z$, include spatiotemporal curvature, and mass, as shown below:

$$U_{t_X} = -\{ct_X - ct_{Y_P} Sin2\theta + ct_{X_P} Cos2\theta + m\} \tag{3.15}$$

$$U_{t_Y} = -\{ct_Y + ct_{X_P} Sin2\theta + ct_{Y_P} Cos2\theta + m\} \tag{3.16}$$

$$U_{t_Z} = -\{ct_Z + ct_{Z_P} + m\} \tag{3.17}$$

$$U_X = \{a(t)r_X + a(t)x_P Cos2\theta - a(t)y_P Sin2\theta + m\} \tag{3.18}$$

$$U_Y = \{a(t)r_Y + a(t)x_P Sin2\theta + a(t)y_P Cos2\theta + m\} \tag{3.19}$$

$$U_Z = \{a(t)r_Z + a(t)z_P + m\}\vec{a}_Z \tag{3.20}$$

§ 4. The velocity and acceleration of a rotating and pulsating particle in six-dimensional space-time

We can express the velocity and acceleration of the point P on the particle, as it goes through its worldspiral by applying six-dimensional Einsteinian operators.

The Cartesian Robertonian operator $\vec{\Re}$ of point P on the particle for a six-dimensional space-time-mass vector metric $d\vec{w}^2$ is given by

$$\vec{\Re}\cdot\left(d\vec{w}^2\right) = -\frac{\partial U_{t_x}}{\partial t_X}\vec{a}_{t_x} - \frac{\partial U_{t_y}}{\partial t_Y}\vec{a}_{t_y} - \frac{\partial U_{t_z}}{\partial t_Z}\vec{a}_{t_z} + \frac{\partial U_X}{\partial x}\vec{a}_x + \frac{\partial U_Y}{\partial y}\vec{a}_y + \frac{\partial U_Z}{\partial z}\vec{a}_z \tag{4.1}$$

$$\frac{\partial U_{t_x}}{\partial t_X} = -c - cCos2\theta = -c(1 + Cos2\theta) = -2cCos^2\theta \tag{4.2}$$

$$\frac{\partial U_{t_y}}{\partial t_Y} = -c - cCos2\theta = -2cCos^2\theta \tag{4.3}$$

$$\frac{\partial U_{t_z}}{\partial t_Z} = -2c \tag{4.4}$$

$$\frac{\partial U_X}{\partial x} = a(t) + a(t)Cos2\theta = 2a(t)Cos^2\theta \quad (4.5)$$

$$\frac{\partial U_Y}{\partial y} = a(t) + a(t)Cos2\theta = 2a(t)Cos^2\theta \quad (4.6)$$

$$\frac{\partial U_Z}{\partial z} = 2a(t) \quad (4.7)$$

$$\vec{\mathfrak{R}} \cdot \left(d\vec{w}^2\right) = -2cCos^2\theta\vec{a}_{t_x} - 2cCos^2\theta\vec{a}_{t_y} - 2c\vec{a}_{t_z} + \quad (4.8)$$

$$2a(t)Cos^2\theta\vec{a}_x + 2a(t)Cos^2\theta\vec{a}_y + 2a(t)\vec{a}_z$$

Hence, if the massless particle system travels relativistically through its worldspiral, it may travel near or at the speed of light. However, what would happen if we hypothesize that the particle system travels faster than light through its worldspiral? From the perspective of a temporally forward moving inertial observer, the particle system may seem to disappear as the particle system travels back on an advanced temporal wave.

§ 5. The nature of temperature

5.1. What is temperature?

A temperature is a numerical measurement of the magnitudes of coldness or hotness in substances of mass as perceived by the human mind and its senses. Temperature is a measure of a quality of a state of a material.

Temperature is the effect of oscillations of atomic particles and structures as they vibrate to absorb or release energy through the processes of interaction between atomic particles, natural fields of energy, and atomic structures at the microscale level. (Cengel, 2001) Temperature as measured in Kelvin, degrees Celsius, and degrees

Fahrenheit, reflects the actions of heat radiation, particle velocity, kinetic energy, or the behavior of thermometric properties of matter. The average kinetic energy of atomic particles and structures is proportional to their absolute temperature.

The most common instrument to measure temperature is the thermometer, or a heat radiation detector, which can demonstrate that temperature is a dynamic variable, that changes as a function of time and/or as a function of location, within a volume or surface of mass.

Furthermore, the measurement of temperature is empirical, and is not infinitesimal in time or space (volume of mass), for even though it may be calculated at infinitesimal scales, it may take very advanced technology to measure it accurately at the microscale level.

In a closed system, temperature tends to become uniform over time throughout the system. Heat is transferred through its paths from regions of higher temperature to regions of lower temperature, regardless of the levels of temperatures between regions, but not between two regions of equal temperature, ceteris paribus.

The lowest limit of temperature in a physical system that may be approached but not reached is absolute zero as measured in Kelvin. The motion of quanta of mass and its structures is minimal at absolute zero Kelvin.

5.2. The energy of temperature

In the case of the non-linear energy, Einsteinian energy, of an ideal gas due to mass, absolute pressure, volume, and temperature, with all other forms of systemic energy excluded, for a closed system, we have

$$E_2(p,v) = \frac{mc^3}{\sqrt{c^2 - v^2}} = \Delta TS \tag{5.1}$$

Where "s" is the specific entropy, and the absolute temperature delta, $\Delta T = c^3 / s\sqrt{c^2 - v^2}$, is the change in temperature during an interval of time, $\Delta T = T_2 - T_1$.

Let us express the three-dimensional energy, Hawking-Feynman energy, for an ideal gas as follows:

$$E_3 = \frac{mc^4}{\sqrt{c^2 - v^2}} \tag{5.2}$$

For the three-dimensional energy of an ideal gas in terms of the mass and the absolute temperature we have

$$E_3 = c\Delta TS \tag{5.3}$$

Where the three-dimensional energy of "the ideal gas" is equal to the change in absolute temperature of the ideal gas system for non-linear energy times the product of the entropy with the speed of light. The three-dimensional energy includes, but it is not limited to, rest mass energy, kinetic energy, potential energy, thermal energy, radiation energy, etc.

5.3. The energy of a Phonon

A phonon, a quasiparticle, represents a quantum-mechanical state of excitation of vibration modes of the elastic structures of interacting particles in the microscale. In the elastic arrangement of condensed matter, the phonon is a periodic collective excitation of the atomic particles and structures of the substance.

A phonon is a definite discrete unit or quantum of vibrational mechanical energy, or thermal energy which comes from the kinetic energy of atomic motions and the potential energy of distortion of interatomic bonds.

Phonons are major players, essential elements, in thermal conductivity and electrical conductivity of condensed matter. Phonons vibrating at higher frequencies can produce heat. Phonons are elementary vibrational motions of lattices, or three-dimensional spatial patterns of atomic particles and structures, that oscillate at a single frequency or normal mode, and as quasiparticles share the quantum mechanical principle of wave-particle duality.

$$E_{phonon} = \hbar c \upsilon = \frac{c^4 S}{s\sqrt{c^2 - v^2}} \quad (5.4)$$

$$E_{phonon} = c\Delta TS \quad (5.5)$$

$$\Delta T = \frac{\hbar \upsilon}{S} \quad (5.6)$$

§ 6. The Boltzmann constant and the gas laws

6.1. The nature of k_B

The Boltzmann's constant, k_B , defines the relation between absolute temperature and the kinetic energy contained in each molecule of an ideal gas. The eminent and magnanimous Physicist Max Planck was instrumental in naming this constant after the Austrian Physicist Ludwig Boltzmann, whose work provided the data for Planck to actually calculate the constant and is equal to the ratio of the ideal gas constant *R*, 8.3144621 (75) $(\text{Joules/Kelvin} \cdot \text{Mol})$, to the Avogadro's constant 6.02214129 (27) x 10^{23} (entities per mole), where the term "entities" usually represents atoms or molecules. A mole is the amount of any substance that contains as many elementary entities as there are atoms in 12 grams of pure carbon-12. (Bohr, 1958)

The value of Boltzmann's constant is approximately 1.380488 (13) x 10^{-23} Joule per Kelvin. In general, the energy in a gas molecule is directly proportional to the absolute temperature. As the temperature increases, the kinetic energy per molecule increases. As a gas is heated, its molecules move more rapidly. This produces increased pressure if the gas is confined in a space of constant volume, or increased volume if the pressure remains constant.

6.2. The Ideal Gas Law

An ideal gas is defined as one in which all collisions between atoms or molecules are perfectly elastic and in which there are no intermolecular attractive forces, and the volume of the atoms and molecules is inconsequential.

The ideal gas law is the equation of state of a hypothetical ideal gas. It is a good approximation to the behavior of many gases under many conditions, although it has several limitations. It was first stated as a combination of Boyle's Law and Charles's Law. The ideal gas law, or Clapeyron's Law, is often introduced in its common form as follows:

$$PV = n_m RT \tag{6.1}$$

Where in SI units, P is the absolute pressure of the gas in Pascals, the volume V of the gas is in cubic meters, n_m is the measured amount of substance in moles of the gas, R is the ideal gas constant, and T is for the absolute temperature measured in Kelvin.

The molecular equation is derived from the principles of statistical mechanics is given by

$$PV = nk_B T \tag{6.2}$$

Where P is the absolute pressure of the gas measured in Pascals; n is the number density of the ideal gas (unit-less as used in the above formula). The number density of an ideal gas at $0^0 C$ and 1 atmosphere (atm) as a yardstick: 1 amagat unit (amg) is equal to $2.6867774 \times 10^{25} m^{-3}$ is often introduced as a unit of number density. The Boltzmann's constant, k_B, relates temperature and energy in Joules/Kelvin, and T is the absolute temperature measured in Kelvin. The consistent results of this statistical mechanics formula with experimental results evaluate the principles of statistical mechanics.

The combined gas law is a gas law that combines Boyle's Law, Charles's Law, and Gay-Lussac's Law. The combined gas law is the unification of the previous three ideal gas laws. Each law relates mathematically one thermodynamic variable to another while holding other thermodynamic variables constant.

The combined gas law is given by

$$\frac{P_1 V_1}{T_1} = \frac{P_2 V_2}{T_2} \tag{6.3}$$

The combined gas law can be used to expound the mechanics that act upon absolute values of pressure, volume, and temperature.

6.3. The Universal Gas Law

The form of the Universal Gas Law for an ideal gas is very useful because it links absolute values of pressure, mass density, and temperature in a unique formula independent of the quantity of the considered gas.

Let us first consider the ideal gas law as follows:

$$\frac{P_1 V_1}{T_1} = \frac{P_2 V_2}{T_2} \tag{6.4}$$

$$\frac{P_1}{T_1} = nk_B \tag{6.5}$$

Where P is the absolute pressure of the ideal gas in Pascals, V is the absolute volume of the ideal gas in cubic meters, T is the absolute temperature of the ideal gas measured in Kelvin, n is the number density of the ideal gas, and k_B is the Boltzmann's constant.

$$\frac{P_1}{T_1} = \frac{E_1}{V_1 T_1} \tag{6.6}$$

$$P_1 = \frac{E_1}{V_1} \tag{6.7}$$

For the Universal Ideal Gas Law, we have in terms of the energy

$$\frac{P_1V_1}{T_1} = \frac{P_2V_2}{T_2} \tag{6.8}$$

$$\frac{E_1V_1}{V_1T_1} = \frac{E_2V_2}{V_2T_2} \tag{6.9}$$

$$\left(\frac{E_1}{T_1}\right)^2 = \left(\frac{E_2}{T_2}\right)^2 \tag{6.10}$$

Let us express the mass, $E = Mc^2 = PV$, as mass times the square of the speed of light equal to absolute pressure times absolute volume to define the Universal Gas Law, or ABCD Law, in terms of absolute values of SI units of pressure (*Pa*), volume (m^3), temperature (*K*), mass (*Kg*), and "*c*" the constant speed of light, such that

$$\frac{P_1V_1M_1c^2}{(T_1)^2} = \frac{P_2V_2M_2c^2}{(T_2)^2} \tag{6.11}$$

$$\frac{\sqrt{P_1V_1M_1}}{T_1} = \frac{\sqrt{P_2V_2M_2}}{T_2} \tag{6.12}$$

Let us introduce an absolute volume (*v*) multiplier, for a fixed cubic meter of volume, with the condition that $v_1 = v_2$, and the mass *M* is divided by *v*, to express the mass density, $\rho = M/v$.

$$\frac{\sqrt{\dfrac{P_1V_1M_1}{v_1}}}{T_1} = \frac{\sqrt{\dfrac{P_2V_2M_2}{v_2}}}{T_2} \tag{6.13}$$

$$\frac{\sqrt{P_1V_1\rho_1}}{T_1} = \frac{\sqrt{P_2V_2\rho_2}}{T_2} \tag{6.14}$$

This formulation of the ideal gas law takes into account the variable of mass for an independent system where energy or mass is an extensive property. In this system, absolute variables of volume, pressure, temperature and mass are dynamic variables.

§ 7. The bridge between the macroscale and microscale of particles and classical space-time

Boltzmann's constant, k_B, is considered to be a bridge between the macroscale and the microscale of particle physics given by the equation:

$$PV = nk_BT \tag{7.1}$$

The product of pressure and volume is a macroscopic amount of pressure-volume energy representing the state of the bulk gas as expressed on the left-hand side of the equation.

On the macroscopic scale, the ideal gas law aforementioned states that, for an ideal gas, the product of pressure (*P*) and volume (*V*) is proportional to the product of the amount of substance (*n*), in density units of the ideal gas, the Boltzmann constant (k_B), and the absolute temperature (*T*) in Kelvin.

On the microscopic scale, the product of the "*n*" density units of ideal gas, with each unit of gas having an average kinetic energy equal to k_BT, is a quantum amount of energy for the particles of gas, on the right-hand side of the equation.

Let us consider the ideal gas law when absolute values of pressure, volume, temperature, and density units of the ideal gas are dynamic variables, and functions of time, such that

$$P(t)V(t) = n(t)k_BT(t) \tag{7.2}$$

To find the velocity of the absolute dynamic variables of pressure (*P* or *p*), volume (*V* or *a*), temperature (*T* or τ), and density units (*n*) we get

$$\frac{\partial P}{\partial t}V + P\frac{\partial V}{\partial t} = k_B \frac{\partial (nT)}{\partial t} \tag{7.3}$$

$$\frac{\partial P}{\partial t}V + P\frac{\partial V}{\partial t} = k_B \left\{ \frac{\partial n}{\partial t}T + n\frac{\partial T}{\partial t} \right\} \tag{7.4}$$

$$V\frac{\partial P}{\partial t} + P\frac{\partial V}{\partial t} = k_B T\frac{\partial n}{\partial t} + k_B n\frac{\partial T}{\partial t} \tag{7.5}$$

$$a\dot{p} + p\dot{a} = \frac{\hbar \upsilon}{\tau}(\tau\dot{n} + n\dot{\tau}) \tag{7.6}$$

Where $\hbar$ is the reduced Planck constant and v is the linear frequency.

To find the acceleration of the absolute variables of pressure, volume, temperature, and density units, we get

$$\frac{\partial V}{\partial t}\frac{\partial P}{\partial t} + V\frac{\partial^2 P}{\partial t^2} + \frac{\partial P}{\partial t}\frac{\partial V}{\partial t} + P\frac{\partial^2 V}{\partial t^2} = k_B\left\{\frac{\partial T}{\partial t}\frac{\partial n}{\partial t} + T\frac{\partial^2 n}{\partial t^2}\right\} + k_B\left\{\frac{\partial n}{\partial t}\frac{\partial T}{\partial t} + n\frac{\partial^2 T}{\partial t^2}\right\} \tag{7.7}$$

$$V\frac{\partial^2 P}{\partial t^2} + P\frac{\partial^2 V}{\partial t^2} + 2\frac{\partial V}{\partial t}\frac{\partial P}{\partial t} = k_B\left\{T\frac{\partial^2 n}{\partial t^2} + n\frac{\partial^2 T}{\partial t^2} + 2\frac{\partial n}{\partial t}\frac{\partial T}{\partial t}\right\} \tag{7.8}$$

$$a\ddot{p} + p\ddot{a} + 2\dot{a}\dot{p} = k_B(\tau\ddot{n} + n\ddot{\tau} + 2\dot{n}\dot{\tau}) \tag{7.9}$$

$$a\ddot{p} + p\ddot{a} + 2\dot{a}\dot{p} = \frac{\hbar \upsilon}{\tau}(\tau\ddot{n} + n\ddot{\tau} + 2\dot{n}\dot{\tau}) \tag{7.10}$$

$$k_B = \frac{a\ddot{p} + p\ddot{a} + 2\dot{a}\dot{p}}{\tau\ddot{n} + n\ddot{\tau} + 2\dot{n}\dot{\tau}} \tag{7.11}$$

§ 8. The Laws of Thermodynamics

8.1. Zeroth Law

If two closed thermodynamic systems are each at thermal equilibrium with a third closed thermodynamic system through two distinct diathermal walls, then all three are in thermal equilibrium with each other, and both diathermal walls are thermodynamically equivalent and interchangeable, without changing the thermal equilibrium of all three thermodynamic systems.

All heat consists of phonons or quantum mechanical oscillations. The temperature, T, is constant throughout a thermodynamic system that is in phononic equilibrium. The phononic energy of each of three thermodynamic systems in thermal equilibrium is equivalent. (Atkins, 2010)

$$T_1 - T_0 = T_2 - T_0 = T_3 - T_0 \tag{8.1}$$

$$c\Delta T_1 S_1 = c\Delta T_2 S_2 = c\Delta T_3 S_3 \tag{8.2}$$

8.2. First Law

The total energy of a closed thermodynamic system is conserved, not created or destroyed, but may be converted from one form of energy to another.

In a closed thermodynamic system where there is no transfer of matter, the fundamental transfer of heat, and its related change in absolute temperature at any point in the system, is the transfer of quantum mechanical oscillations or phonons at any point in the system from an atomic particle or structure to another.

In a closed thermodynamic system, any change in the internal energy of the system due to the transfer of heat is the result of the absorption or emission of phonons and the performance of work by the system on its surroundings or vice versa.

$$\Delta U_i = \pm c\Delta T \Delta S \pm W \tag{8.3}$$

8.3. Second Law

An isolated naturally spontaneous thermodynamic system follows the arrow of time toward thermal equilibrium or maximum entropy as every existing atomic particle or structure of the substance transfers or absorbs phononic energy, if not at thermal equilibrium, until all phonons are at an equivalent and uniform state of energy arrangement per unit of temperature that represents the correct informational specification of the distinct isolated system.

The entropy of a closed thermodynamic system is the amount of phononic energy of the system, when the system is at thermal equilibrium with its surroundings, which does not perform work. Entropy is the thermodynamic property of a substance at phononic equilibrium, or a measure of the number of possible phononic states of a thermodynamic system in thermal equilibrium.

$$E_{phonon} = cTS \tag{8.4}$$

For a closed system at thermal equilibrium containing an amount of entropy energy E_S,

$$S = \frac{E_S k_B}{E_{phonon}} \tag{8.5}$$

Where E_S / E_{phonon} is the number of phonons ω, or the coefficient of growth factor for the number of quantum states $\Omega(E_S)$, for the system at thermal equilibrium during an infinitesimal energy interval δE.

$$e^{\omega} = \Omega(E_S) \tag{8.6}$$

$$\omega = \frac{E_S}{E_{phonon}} = \frac{E_S}{cTS} = ln\ \Omega(E_S) \tag{8.7}$$

Let us express the entropy for the closed system at thermal equilibrium as follows:

$$S = \omega k_B = \frac{E_S k_B}{cTS} \tag{8.8}$$

A phononic process of a closed system that is reversible conserves its entropy, but a phononic process that is irreversible does not. Thermodynamic processes follow the arrow of entropy, and entropy follows the arrow of time. Thus, *entropy is a consequence of the passage of time and the expansion of space-time.*

The exchange of phononic energy between two distinct closed thermodynamic systems that are not at thermal equilibrium is a potentially reversible process if entropy is conserved. The phononic energy naturally flows from a region of substance of higher frequency phonons ω_H to a region of substance of lower frequency phonons ω_L through a conductive medium for that energy and frequency. Hence, the quantum number of phonons in a closed system increases by $\Delta\omega$.

$$\Delta\omega = \omega_H - \omega_L \tag{8.9}$$

The internal phononic energy of a closed thermodynamic system with constant volume increases by $\Delta E_{phononic}$ when external phononic energy is transferred to the system.

$$\Delta E_{phononic} = c\Delta T \Delta S \Delta\omega \tag{8.10}$$

The phononic energy difference between two distinct closed thermodynamic systems that are not at thermal equilibrium is available to be converted to mechanical energy. Work is available to be performed from the phononic energy flow, between two distinct regions of substance of two closed thermodynamic systems that are not at thermal equilibrium through a conductive medium for that energy and frequency, in the direction of flow of phononic energy, by natural systems or devices.

$$\Delta W = (\omega_H - \omega_L) k_B \Delta T \tag{8.11}$$

Universal gravitational systems are conversely related to the second law of thermodynamics. This may justify the name for the Inverse Second Law of Thermodynamics. Objects in gravitational systems may have negative specific heat capacity (*J/K*) as they absorb external phonons. Furthermore, universal gravitational systems decrease entropy. Thus, for a gravitationally bound object or device losing gravitational potential faster than gaining kinetic energy we have,

$$\frac{i^2 \omega E_{phonon}}{T} = \frac{\Delta W - \Delta U_g}{\Delta T} \tag{8.12}$$

Therefore, gravitationally bound systems or devices in our universe tend toward non-uniform states of energy arrangement per unit of temperature. Consequently, the inverse second law of thermodynamics may apply to gravitationally bound systems or devices. This is a consequence of the reversal of the arrow of entropy due to the reversal of the arrow of time in the space-time of gravitationally bound systems or objects with negative specific heat capacity.

8.4. Third Law

A closed thermodynamic system at one phononic energy state that is stable, unique, and minimal, at an absolute temperature of zero Kelvin, has entropy equal to zero. Thus, entropy is zero at one ground energy state of the closed thermodynamic system.

$$S_0 = S - k_B ln\, \Omega(E) \tag{8.13}$$

When $\Omega(E) = 1$, the closed thermodynamic system is at one ground energy state and the entropy *S* of the system is equal to the one ground state entropy S_0 of the system.

If a closed thermodynamic system is brought to very low absolute temperatures at one energy state that is not well-defined, where the system is fixed in a finite phononic energy arrangement state above the system's one ground phononic energy state, then that fixed energy state is residual entropy.

As the temperature of a closed thermodynamic system approaches zero Kelvin, the entropy change for a reversible phononic process also approaches zero. The entropy of a closed thermodynamic system with a one ground energy state may be reduced by a finite number of operations equal to the number of phonons " ω " in the closed system, from beginning to end of the entropy reduction process, decremented by one phonon for every operation, reducing the energy of the system by k_B per operation, to reach the one ground phononic energy state. However, this ideal process would require very advanced technology.

$$\sum_{n=1}^{\omega}(\omega - n) \qquad (8.14)$$

Hence, the entropy of a closed thermodynamic system at absolute zero Kelvin is a well-defined constant. This is because a system at zero temperature exists in its ground energy state, so that its entropy is determined only by the degeneracy of the ground energy state and the surge of the phononic energy of the system.

8.5. Fourth Theorem: Enthalpy

The enthalpy of a closed thermodynamic system, a state function of the system, consists of the internal energy U_i of the system plus the product of absolute pressure (*P*) and volume (*V*).

$$H = U_i + PV \quad (Joules) \qquad (8.15)$$

Energy transfer into or out of a closed thermodynamic system at constant pressure causes a change in the enthalpy of the system through the expansion or contraction of the absolute volume or through phononic energy transfer.

Enthalpy is a change in the combined magnitude of phononic energy and internal energy of a closed thermodynamic system, and is not directly measurable. Enthalpy is thermodynamic potential energy (H) from thermodynamic processes under constant pressure. The enthalpy of an ideal gas is an extensive property which does not depend on variable absolute pressure. Hence for a homogeneous closed thermodynamic system, the enthalpy is proportional to the absolute volume of the system. Henceforth, the enthalpy of a closed thermodynamic system is proportional to the product of the entropy of the system times its absolute temperature.

$$H = U_i \pm S(T) \tag{8.16}$$

$$H = U_i \pm \omega k_B(T) \tag{8.17}$$

Where S is the entropy of the closed thermodynamic system, T is absolute temperature, k_B is Boltzmann's constant, and ω is the number of phonons in the system.

The internal energy of the closed thermodynamic system is the amount of energy needed to create the system, and the phononic energy is the amount of energy needed to create the absolute volume of space-time of the closed system under constant absolute pressure at an absolute temperature. The internal energy of a closed thermodynamic system is the combination of all energies absorbed by the system to undergo all thermodynamic processes to create the system.

At one ground internal energy state, the entropy of the system is zero, so we have

$$H_{1G} = U_i \pm 0(T) \tag{8.18}$$

$$\partial H_{1G} = \partial U_i \tag{8.19}$$

Therefore, for a closed thermodynamic system that is brought to an absolute temperature of zero Kelvin, by the third law of thermodynamics, the enthalpy of the system is equal to its one ground internal energy state. The one-ground internal energy state of a closed thermodynamic system is a stable, unique, and minimal thermodynamic potential energy.

The enthalpy of a closed thermodynamic system is stable, unique, minimal, and equal to its one ground internal energy state when the entropy of the system is zero at an absolute temperature of zero Kelvin.

8.6. Fifth Theorem: Time

All thermodynamic processes for any type of physical system are a function of time. Time is primordial and of the essence for any thermodynamic system to change any of its properties or qualities. For without time, there is no change, no motion, and no expansion of space-time, to drive thermodynamic processes. Thus, all forms of energy are reliant on time.

$$\textit{The speed of time} = \frac{\partial \theta}{\partial t} \tag{8.20}$$

Arrows of Entropy and Enthalpy follow the arrow of time. Time is the prime mover of all kinds and classes of thermodynamic processes in the universe. Time is the engine of change and the driving force of thermodynamic change.

All the laws of thermodynamics are dependent on time and the direction of the arrow of time. The arrow of time of thermodynamic processes is reversible, but not all thermodynamic processes are reversible. Time and space are two sides of the same coin and space-time is their mint. For a closed thermodynamic system, enthalpy is proportional to time, entropy is proportional to enthalpy. Entropy, enthalpy, and time are proportional. It is proposed in this document that Time, as well as its conjugate Space, is three- dimensional, and the laws of thermodynamics are observed in six-dimensional space-time. All thermodynamic variables, states, and processes exist in the framework of six-dimensional space-time. The speed of time is the speed of change, and that speed is relative. The speed of time is inversely proportional to the strength of a gravitational field or directly proportional to the dilation of time for a moving object as the object approaches the speed of light. For the speed of proper time we have

$$v_\tau = \sqrt{1 - \frac{v^2}{c^2}} = \frac{\lambda_\tau}{T_\tau} \tag{8.21}$$

Does the rate of the passage of time affect absolute temperature?

Time dilation is the elapsed time difference in the General Theory of Relativity between two events as measured by observers either moving relative to each other or differently situated from gravitational masses. Time dilation describes the relation between time and movement in space-time for tardyons as follows:

$$t' = \frac{ct_0}{\sqrt[2]{c^2 - v^2}} \tag{8.22}$$

Where the proper time is t_0 and t' is dilated time.

There are ways in which time, temperature, and motion, may change:

a. As the motion of a tardyon or object changes, the rate of time changes.

b. As space-time curves near a particle or particle system of mass, the rate of time changes.

c. As temperature changes, the motion of a tardyon, or a particle system, changes.

d. As a tardyon or particle system moves in space-time, the faster the tardyon or particle system moves spatially, the slower the tardyon or particle system moves temporally. In that case, are time and temperature independent or are they related through motion?

The passage of time is inversely proportional to the motion of particles or their masses, and the motion of particles is a function of temperature which in turn is inversely proportional to the passage of time.

$$f(t) \propto \frac{1}{m} \propto \frac{1}{T} \tag{8.23}$$

How would the rate of proper time change for each of two distinct particles or two distinct particle systems moving near a very massive celestial body or particle system when the temperature difference between the two particles or the two particle systems is considerable? Would the particle or the particle system with the lowest temperature experience slower, equal, or faster change in its rate of proper time? Since temperature is a consequential function of time, it would be expected that a considerable temperature difference between distinct particles or distinct particle systems would experience a change on its rate of proper time, not contrary to the spirit of the acclaimed General Theory of Relativity. Thus, is temperature relative to the inertial frame of reference or the free-falling frame of reference of an observer? Would a distinct particle, such as a luxon, experience a change in its rate of time when the ambient temperature of its surroundings changes?

The motion of atomic particles increases as temperature rises as shown empirically and in early research on Brownian motion. As the temperature decreases, motion slows down proportionally up to some distinct ground energy state where the distinct particle system temperature approaches absolute zero as an energy minimum for the system. At such a ground state of energy, there are nonzero molecular oscillations. The rate of time may be zero for a luxon at nonzero temperatures when the distinct particle moves through space at luminal speed, but not temporally. Consequently, zero temporal motion does not mean a particle or particle system is at or approaching zero absolute temperature, or vice versa.

Time and temperature are highly correlated through the relative expansion and contraction of space-time from the perspective of an observer in an inertial frame of reference or free-falling frame of reference. Temperature is a measure of particle density, and any change in particle density will be correlated to a reciprocal change in the mass or energy of a distinct particle, or particle system, that affects the curvature of space-time. Therefore, the proper time of a distinct particle system in a region of higher temperature and higher particle density, will change at a slower rate of time than the rate of time at a region of lower temperature and lower particle density, and even at a slower rate of time in a region of minimum particle density or near absolute zero temperature. *Hence, the proper time of a distinct particle or distinct particle system is inversely proportional, or highly correlated, to the absolute temperature of the particle or particle system.* Temperature is relative because time is relative from the perspective of an observer in an inertial frame of reference or a free-falling frame of reference.

Recent experiments in slowing the speed of light through an Einstein-Bose condensate are examples of the direct proportionality between absolute temperature and particle density or matter condensation, and the correlation of time and absolute temperature in a distinct particle system with high particle density at very low absolute temperature, in a superfluid. Photons may be brought to a state of lowest motion as the absolute temperature approaches zero in a Bose-Einstein condensate which implies that the rate of time of the condensate has been increased considerably. From the perspective of an inertial frame of reference or a free-falling frame of reference, the photons in the Bose-Einstein condensate appear to travel through the temporal volume more, and less so, through the spatial volume of space-time in their surroundings. In other words, the temporal wave function of the photon speeds up as the spatial wave function of light slows down in the superfluid.

The critical temperature for the transition to a Bose-Einstein condensate occurs below a critical temperature T_C , which for a uniform gas in space-time consisting of non-interacting particles with no apparent internal degrees of freedom is given by

$$T_C = \left(\frac{n}{\xi(3/2)}\right)^{2/3} \frac{2\pi\hbar^2}{mk_B} \approx 3.3125 \frac{\hbar^2 n^{2/3}}{mk_B} \tag{8.24}$$

The critical temperature T_C may also be expressed as a vector as $\vec{T}_C = \vec{\nabla} T$ to represent the three-dimensional nature of absolute temperature as a vector.

Thus, the absolute temperature volume in magnitude form may be expressed as T^3 because every spatiotemporal surface of uniform temperature magnitude (isophononic) through a distance has an associated absolute temperature. Hence, we may express the average acceleration of time with respect to space during the proper temporal period for the transition to a Bose-Einstein condensate as follows:

$$\left(\frac{\Delta^2 t}{\Delta r^2}\right)^2 \approx \left(\frac{3.3125}{k_B}\right) \frac{m \cdot n^{2/3}}{T_C} \tag{8.25}$$

$$\frac{\Delta^2 t}{\Delta r^2} \approx \left(\sqrt[2]{\frac{3.3125}{k_B}}\right)\left(\sqrt[2]{\frac{m}{T_C}}\right)\left(\sqrt[3]{n}\right) \tag{8.26}$$

Where m is the mass, n is the number of particles per volume of the particle system, and k_B is the Boltzmann constant.

Let us define the Bose Critical Constant as

$$B_C = \sqrt[2]{\frac{3.3125}{k_B}} \approx 4.898197943 \ \text{x}\, 10^{11} \left(\frac{s \cdot K^{1/2}}{Kg^{1/2} \cdot m^2 \cdot n^{1/3}}\right) \tag{8.27}$$

Hence, the simplified average acceleration of time with respect to space is given by

$$\frac{\Delta^2 t}{\Delta r^2} = B_C \cdot \left(\sqrt[2]{\frac{m}{T_C}} \right) \cdot \left(\sqrt[3]{n} \right) \tag{8.28}$$

Thus, the average acceleration of time with respect to space is equal to the product of the Bose Critical Constant B_C , the square root of the mass-critical temperature ratio, and the cubic root of the particle density of the particle system.

The infinitesimal acceleration of time with respect to space is defined as

$$\frac{\partial^2 t}{\partial r^2} = B_C \cdot \left(\sqrt[2]{\frac{\partial m}{\partial T_C}} \right) \cdot \left(\sqrt[3]{n} \right) \tag{8.29}$$

Therefore, as the absolute temperature of the superfluid drops the average acceleration of time with respect to space increases. If the mass of particle density increases so does the average acceleration of time. As time accelerates with respect to space, the photons in the BEC will travel more through time and less so through space. As a result, light appears to slow down in the Bose-Einstein condensate.

As the rate of time may speed up for something, keeping something at very low or very high absolute temperatures may change the effects of the passage of time. Chemical reactions occur more slowly at lower absolute temperatures. Faster chemical reactions take place as atomic particles move more through the spatial volume at a faster rate of space, and less through the temporal volume or at a slower rate of time. *The speed of chemical reactions is directly proportional to the spatial speed of its atomic elements, and inversely proportional to the temporal speed of its atomic elements.*

Let us imagine an analogy of a two-headed vehicle, similar to a train with two-headed locomotives separated by a single railcar, so that the vehicle may be driven in either direction without breaking apart. As the driver in the spatial end drives her vehicle head faster in her direction, the driver situated in the opposite temporal end starts slowing down his vehicle head in the temporal direction, and the entire two-headed vehicle moves more toward the spatial direction than toward the temporal direction. If both vehicle heads were driven equally fast, the inertial reference frame of the railcar would be at an equilibrium point. *By the principle of spatiotemporal translation, the more a particle or particle system travels in the temporal direction, the less the particle or particle system travels in the spatial direction, or vice versa.*

From the perspective of a luxon in a state of relativistic mass, time stands still, and the speed of the luxon is constant for any observer in an inertial frame of reference or a free-falling frame of reference. Thus, the luxon is at the equilibrium point of its temporal wave function. As temperature changes in the region of the luxon, the spatial wave function and the temporal wave function of the luxon are affected, as the equilibrium point of the luxon appears to change toward the spatial wave function from the perspective of an observer in an inertial frame of reference, or a free-falling frame of reference, in a near region of considerable higher temperature. As the region of lower temperature of the luxon approaches zero absolute temperature, the equilibrium point of the luxon in its temporal wave function appears to shift toward the direction of the spatial wave function, and the luxon appears to travel slower through space to an observer in an inertial frame of reference, or a free-falling frame of reference, of a near region of considerable higher temperature. The relativity of time causes temperature to be relative from the perspective of an observer in an inertial frame of reference, or a free-falling frame of reference, in a region of considerable higher temperature.

Temperature is motion, while motion is relative and dependent on the relativity of space and time, which in turn is affected by the motion of mass. Motion incorporates relativistic mass in photons. Mass, motion, and temperature interact with time. Consequently, mass, motion, or temperature is dependent on the properties of time. They are all functions of time.

$$m \propto T \propto v \propto \frac{1}{f(t)} \tag{8.30}$$

Would the uncertainty or accuracy of a present-day laser-cooled atomic clock be a function of the acceleration of time and the absolute temperature of its parts and surroundings in its inertial frame of reference or its free-falling frame of reference?

Raman cooling is a technique for sub-recoil cooling to allow the cooling of atoms using optical methods below the limitations of Doppler cooling, limited by the recoil energy of a photon given to an atom. Two laser beams may be used to trigger the transition between two hyperfine states of the atom. The first laser beam excites the atom to a virtual excited state because its frequency is lower than the real transition frequency, and the second laser beam de-excites the atom to the other hyperfine state. The transition frequency between the two hyperfine states is exactly equal to the frequency difference of the two laser beams.

Similarly, Raman side-band cooling uses atoms in a magneto-optical trap to obtain a high density of atoms at a low temperature. This optical cooling technique is still not sufficient to reach the Bose–Einstein condensation of Cesium atoms, but the Bose-Einstein condensation can be reached using Raman side-band cooling as a first step in the process.

The proper time of a Cesium atom is inversely proportional, or highly correlated, to the absolute temperature of the atom. Absolute temperature affects the period, and the frequency, of the radiation corresponding to the transition between the two hyperfine states of the ground state of the Cesium 133 atom which constitutes the current definition of time measurement. Cesium 133 is an isotope of Cesium used especially in atomic clocks and one of whose atomic transitions is used as a scientific time standard. An atomic second is defined by the interval of time taken to complete 9,192,631,770 oscillations of the Cesium 133 atom exposed to a suitable excitation.

An energy amount of approximately 0.000038 eV in the microwave region separates the two hyperfine states involved in a Cesium clock, which is about a thousand times smaller than the random thermal energy of about 0.04 eV associated with the 100 °C temperature at which the Cesium clock is typically operated.

The uncertainty of the atomic clock has gone from approximately 10,000 ns/day in the late 1940's to approximately 0.01 ns/day of laser-cooled atomic clocks in the late aught's of the 21st century. Hence, the accuracy of a laser-cooled atomic clock is also a function of temperature, and the rate of time slows noticeably when the absolute temperature of the atomic clock and its surroundings drops considerably to a critical temperature.

The proper time of a Cesium 133 atomic clock is inversely proportional, or highly correlated, to the absolute temperature of its Cesium 133 atoms. Thus, the period and frequency of the radiation associated with the transition between the two hyperfine states of the ground state of the Cesium 133 atoms will be affected as time accelerates or decelerates with the change in absolute temperature. *The least uncertainty or greatest accuracy of a present-day laser-cooled atomic clock is a function of the infinitesimal acceleration of time and the absolute temperature that affect its parts and surroundings at the inertial frame of reference, or the free-falling frame of reference, of the atomic clock in space-time.*

§ 9. The Laws of Black Hole Mechanics and Dynamics

Let us express the laws of black hole mechanics and dynamics as follows:

Zeroth Law:

The surface gravity of a dynamic black hole is directly proportional to its absolute temperature and inversely proportional to its radius.

$$\frac{c^2}{R_{BH}} = g_{BH} \tag{9.1}$$

$$\frac{\partial g}{\partial S} = -\frac{T}{8\pi ma} \tag{9.2}$$

The entropy of a dynamic black hole is directly proportional to product of the mass or energy, and its volume a.

First Law:

The energy in the entropy of a dynamic black hole changes if the black hole accretes or radiates mass over time, or if the temperature of the black hole increases or decreases over time.

$$S_{BH} \leq \frac{m_{BH}c^2}{4T_{BH}} \tag{9.3}$$

$$S_{BH} \leq \frac{\ddot{a}c^2}{32\pi GT_{BH}} \tag{9.4}$$

Second Law:

For perturbations of dynamic black holes, the change of energy is related to a change in the acceleration of space-time, angular momentum, or electric charge, as shown by

$$dE = \frac{\ddot{a}c^2}{8\pi G} + \Omega dJ + \Phi dQ \tag{9.5}$$

Where $\ddot{a}$ is the acceleration of space-time about the surface of the outer event horizon of the black hole, Ω is the angular velocity, J is the angular momentum, Φ is the electrostatic potential, and Q is the electric charge, of the black hole.

Third Law:

The area of the event horizon of a dynamic black hole, assuming the weak energy condition, may change over time as a function of the radius of the black hole. The event horizon surface area of an existing non-extremal dynamic black hole is geometrically or space-time related.

$$\frac{\partial A_{EH}}{\partial t} = f(R_{BH}) \tag{9.6}$$

Fourth Law:

Event horizon surface gravity vanishes in a theoretical extremal black hole with a minimal mass or minimal temperature, and compatible charge and angular momentum, while the entropy of an existing non-extremal black hole, at absolute zero Kelvin, or ground state, is a well-defined constant.

9.1. The surface area of the outer event horizon of a non-extremal black hole

For the surface area of the event horizon of a spherical non-extremal black hole we get

$$A_{BH} \geq \frac{\ddot{a}}{2g_{BH}} \tag{9.7}$$

$$A_{BH} \geq \frac{8\pi G m_{BH}}{\frac{2c^2}{R_{BH}}} \tag{9.8}$$

Where the acceleration of space-time about the surface of the event horizon is expressed in terms of the mass, gravity, and radius of the black hole as follows:

$$\ddot{a} = 8\pi G m_{BH} = 8\pi g_{BH} R_{BH}^{\ 2} \tag{9.9}$$

In the strong-field regime about the outer event horizon of a black hole we find

$$\frac{\ddot{a}}{8\pi} \leq R_{BH}c^2 \qquad (9.10)$$

Aside, $\ddot{a}/4\pi r^2$ is the convergence modulus of expanding space-time about a spherical body of mass.

Accordingly, let us imagine that a mass m_p falls into a spherical non-extremal black hole, and then let us ask the question: how would the area of the outer event horizon grow?

$$\Delta A_{BH} \geq \frac{\ddot{a}}{2(g_{BH} + g_{m_p})} \qquad (9.11)$$

The area of the outer event horizon would grow directly proportional to the acceleration of space-time about the event horizon, and inversely proportional to the sum of the gravitational acceleration of the black hole mass m_{BH} and the infalling mass m_p.

9.2. The space-time-mass bound on information storage capacity

The amount of information or entropy that can be stored or reproduced in the space-time volume of a minimal mass is less than or equal to the limit specified by the Benkenstein bound. If the mass were a singularity, then its saturated boundary or bound, within and about the space-time volume of its entropy, is that of a theoretical black hole.

Nature is able to hide away complex historical information behind the simplicity and about the event horizon of a black hole. Entropy is a measure of storage capacity for complex historical information that exists about and hides inside of the event horizon of a black hole.

The amount of information that perfectly describes a physical system down to the quantum level is also limited by the bound. An information storage device with finite physical dimensions has limited or bounded memory storage capacity. The Bekenstein Bound has been defined to be

$$I \le \frac{2\pi R_{BH} E_{BH}}{\hbar c\,(\ln 2)} \qquad (9.12)$$

Simplifying the above expression, we have

$$\frac{2\pi R_{BH} m_{BH} c^2}{\ddot{a} m_p (\ln 2)} = \frac{4\left(2\pi R_{BH}{}^2 m_{BH} c^2\right)}{4\ddot{a} m_p (\ln 2) R_{BH}} = \frac{8\pi g_{BH} R_{BH}{}^2 m_{BH}}{4\ddot{a} m_p (\ln 2)} = \frac{\ddot{a} m_{BH}}{4\ddot{a} m_p (\ln 2)} = \frac{m_{BH}}{4 m_p \ln 2} \qquad (9.13)$$

$$= \frac{m_{BH}}{m_p \ln 16} = \frac{\frac{m_{BH}}{m_P}}{\ln 16} \approx 83{,}085{,}408.94\, m_{BH} \qquad (9.14)$$

Consequently, the limit on information storage is directly proportional to the ratio of the mass of a non-extremal black hole to the Planck mass divided by the growth factor constant.

If $e^{2.772588722} = 16$, then $\ln 16 = 2.772588722$ is the growth factor "x" for defining the information as 4 times the logarithm to the base 2 of the number of quantum states (e.g. binary 1 or 0).

As a result, the maximum space-time-mass bound on information storage capacity within a virtual sphere that encompasses the entropic system of a non-extremal black hole may be formulated as

$$I \le \frac{\frac{m_{BH}}{m_P}}{\ln 16} \qquad (9.15)$$

$$I \le \frac{\frac{m_{BH}}{m_P}}{\ln e^{X}} \qquad (9.16)$$

Where the variable "*I*" is the number of bits in the space-time volume of the entropic system of the non-extremal black hole with a mass m_{BH}. Then, it is possible to calculate the space-time-mass bound by knowing only the mass of the black hole.

§10. Entropy and Enthalpy for Open Thermodynamic Systems

10.1. The internal energy of an open thermodynamic system

Enthalpy is defined as the internal energy of a system plus the constant pressure times its volume as given by

$$H = U_{INTERNAL} + PV \quad (Joules) \tag{10.1}$$

The internal energy comprises the energy of separation from its surrounding with pressure being a constant, the energy of activation and the energy of breaking the molecular bonds of compounds to create and maintain the system. Thus, it is also the phononic energy received by the system plus the non-mechanical work that has been done. The term *PV* is the amount of work done to displace the surrounding medium (space-time, atmosphere, mass, etc.) in order to vacate the space-time-mass, to be occupied by the system. Thus, *H is made up of the energy of formation and the energy of medium displacement.*

10.2. The specific enthalpy of an open thermodynamic system

The specific enthalpy, *h*, of an open thermodynamic system is the intrinsic energy per unit of mass that accounts for space-time displacement and the energy of the formation of matter.

$$h = \frac{E}{m} = c^2 \quad \left(\frac{Joules}{Kg}\right) \tag{10.2}$$

Where *E* is energy, *m* is mass, and *c* is the speed of light. In terms of the specific enthalpy we have

$$E = mh = H \tag{10.3}$$

$$\delta H = T\delta S \tag{10.4}$$

$$h = sT \tag{10.5}$$

$$h = \frac{U_i}{m} + \frac{PV}{m} \tag{10.6}$$

10.3. The entropy of an open thermodynamic system

For an open thermodynamic system, the relationship between specific entropy, entropy, energy, mass, temperature, and the speed of light, may be expressed as follows:

$$S = ms \tag{10.7}$$

$$E = msT = H \tag{10.8}$$

$$s = \frac{E}{mT} = \frac{c^2}{T} \quad \left(\frac{J}{Kg \cdot K}\right) \tag{10.9}$$

$$E = ST \tag{10.10}$$

$$T = \frac{E}{S} = \frac{h}{s} = \frac{c^2}{s} \tag{10.11}$$

Entropy (S) is a measure of the equilibrium of energy in a thermodynamic system. Entropy is a measure of unexpected changes that tend to average out, or smooth out, differences in temperature, pressure, density, and chemical potential that may exist in a thermodynamic system.

Entropy is a function of the amount of phononic energy in a thermodynamic system that is capable of doing work. Thus, the higher the entropy, the lesser the available energy to do work. Conversely, the lower the entropy, the higher the available energy of a system to do work.

The change in entropy of the thermodynamic system is equivalent to the change in phononic energy of the system per unit of absolute temperature.

$$\partial S = \frac{\partial E_{phononic}}{T} \tag{10.12}$$

$$S = \frac{E_{phononic}}{T} = \frac{H}{T} \tag{10.13}$$

$$\partial S = \frac{\partial H}{T} \tag{10.14}$$

Hence, the change of entropy of the thermodynamic system is equivalent to the change of enthalpy of the system per unit of absolute temperature. Thus, the enthalpy changes as the phononic energy changes, $\partial H = \partial E_{phononic}$. Enthalpy can be converted into phononic energy or vice versa, and consequently, entropy increases or decreases.

10.4. The proportionality of entropy, enthalpy, and time

Entropy and enthalpy are directly proportional to time, follow the arrow of time, and tend to increase or remain the same. Enthalpy, or Entropy, is a time keeper for a thermodynamic system since it is a discretely marked event in the passage of time.

Entropy and enthalpy are proportional to each other. A change in enthalpy is proportional to a change in entropy per unit of absolute temperature. Entropy is a derived construct of enthalpy. Therefore, enthalpy is really the true physical attribute of the system; that is, a change in the energy state of the system. Enthalpy is intrinsic energy.

The work that the thermodynamic system is capable of doing is directly related to the enthalpy of the system.

10.5. The specific enthalpy of internal energy and *G*

Let us consider an object with mass, the amount of energy that has been used for the formation of that mass is equivalent to the amount of energy that was used to displace the space-time medium to be occupied by the mass of the object. Hence, we may consider, *h*, to be the intrinsic energy per unit of mass, to displace space-time and to form the mass.

The acceleration of space-time volume occupied by the mass is given by

$$\ddot{a} = \frac{\partial^2 V}{\partial t^2} = -8\pi G m_p = -\frac{\hbar c}{8\pi m_p} \qquad (10.15)$$

Where m_p is the Planck mass of the object, $\hbar$ is the reduced Planck constant, and *G* is the gravitational constant. It is theorized that neither *G* nor the mass needs to be a true constant.

Let us consider a celestial body, like a planet. The specific enthalpy of a celestial body may be expressed as a function of the gravitational acceleration *g* of the celestial body such that

$$h = \frac{U_i}{m} + \frac{PV}{m} = \frac{U_i}{m} + \frac{U_g}{m} \qquad (10.16)$$

$$h = \frac{U_i}{m} + G\frac{m}{r} = u_i + gr \qquad (10.17)$$

Where *m* is the mass of the celestial body, *r* is the radius, *h* is the specific enthalpy, and *g* is the gravitational acceleration. Therefore, we can express the specific intrinsic energy of the space-time-mass as

$$u_i = h - gr \qquad (10.18)$$

Where $gr = c^2$ if the speed of gravity is theorized to equal the speed of light.

10.6. The proportionality of entropy, mass, spatiotemporal volume, gravity, and temperature

Let us define space-time pressure as the normal force per unit of area "a" applied to the surface of an object of mass or energy. Space-time pressure is applied to the boundaries of objects of mass at every point. The density of energy in a volume, or Joules/Volume, is equivalent to pressure, Force/Area.

Let us express the pressure in an equation of state for an ideal system in terms of absolute temperature (Kelvin), entropy, and volume of space-time "a" as follows:

$$p = T\frac{\partial S}{\partial a} \tag{10.19}$$

$$p = \frac{\partial F}{\partial A} = \sqrt[3]{a}\,\frac{\partial F}{\partial a} = m\left(\sqrt[3]{a}\right)\frac{\partial \ddot{a}}{\partial a} \tag{10.20}$$

$$m\left(\sqrt[3]{a}\right)\frac{\partial \ddot{a}}{\partial a} = T\frac{\partial S}{\partial a} \tag{10.21}$$

$$\frac{\dfrac{\partial \ddot{a}}{\partial a}}{T\dfrac{\partial S}{\partial a}} = \frac{1}{m\left(\sqrt[3]{a}\right)} \tag{10.22}$$

$$\frac{\partial \ddot{a}}{\partial S} = \frac{T}{m\left(\sqrt[3]{\mathrm{a}}\right)} = \frac{c^2 T\left(\sqrt[3]{a}\right)}{E} \tag{10.23}$$

Thus, the acceleration of space-time is directly proportional to the absolute temperature, and entropy is directly proportional to mass and the radius of the volume of space-time.

Let us express gravity as a function of the acceleration of space-time:

$$g = -\frac{\ddot{a}}{8\pi\left(\sqrt[3]{a}\right)^2} \tag{10.24}$$

$$\frac{\partial g}{\partial S}=-\frac{\frac{\partial \ddot{a}}{\partial S}}{8\pi\left(\sqrt[3]{a}\right)^{2}}=-\frac{T}{8\pi m\left(\sqrt[3]{a}\right)^{3}}=-\frac{T}{8\pi ma}=-\frac{c^{2}T}{8\pi E\left(\sqrt[3]{a}\right)} \tag{10.25}$$

Similarly, gravity is directly proportional to the absolute temperature, and entropy is directly proportional to the mass or energy and the volume of space-time. Constant temperature for a normal system in thermal equilibrium is analogous to constant surface gravity on the space-time-mass boundary of the system.

Consequently, the absolute temperature is given by

$$|T|=8\pi ma\frac{\partial g}{\partial S}=m\left(\sqrt[3]{a}\right)\frac{\partial \ddot{a}}{\partial S} \tag{10.26}$$

§ 11. Conclusion

Energy is a consequence of the type of motion of an object or particle. Motion is change that originates energy. Time is cause, and change is effect, in the causality of energy. Kinematic energy is a consequence of the passage of time and the expansion of space-time. The geometric evolution of space-time invokes the realization of energy. Time is the conjugate of space in space-time, and time may be curved as space may be curved at a location where space-time is curved. Absolute temperature is a manifestation of phononic energy that interrelates pressure, volume, and mass in ideal gases. Present thermodynamic laws are direct consequences of phononic energy due to the passage of time and the expansion of space. Entropy, enthalpy, and time are proportional.

PART III

SPACE-TIME AND GRAVITATION

Chapter 7

On the multidimensionality of space-time and motion

§ 1. Introduction: the concept of time

It has been over a hundred years since the theory of relativity was expounded by Albert Einstein and yet we still refer to space-time as a four-dimensional medium in which events happen and the forces of nature are prevalent. It is no surprise that after great thinkers like Galileo Galilei, Isaac Newton and Albert Einstein et alia considered, utilized or conceptualized space-time as a four-dimensional medium or background to be interpreted as three dimensions of space and one of time, present day thinkers still follow the footsteps of such illustrious scientists. All dimensions separately accounted for and at ninety degrees from each other as in a four-dimensional Cartesian coordinate system (x, y, z, t), even though there are currently elaborate theories that consider other spatial dimensions still to be found in the physical world. All forces of nature involve the spatial and temporal dimensions to enact their actions or reactions in all physical processes that occur in space-time as well as in the presence of mass or energy. Thus, it seems almost natural to refer to this medium as space-time-mass, or space-time-energy, since these physical manifestations are to be commonly found in the processes of most physical systems and such features exist in the observed reality of nature.

Long ago, even before the times of Galileo and Newton, scientists and researchers studied the properties of objects and their motion to describe the trajectory of an object with a mass *"m"* through space and time. In our everyday experience space is commonly characterized as having three dimensions: width, depth, and height, where each spatial dimension is a separate dimension with its own direction and sense of motion. The unit of distance of each dimension has a property of its own independent of the other two. Thus, motion has been described in one dimension, or one-dimensionally, when it takes place in the direction of depth, or let us say the x-axis, regardless of its sense of motion, two-dimensionally when in the x-y plane, and three-dimensionally when the motion is in the volume of space *(x-y-z)* such as the wave-like motion of a particle in a curvilinear trajectory or a particle moving in a helical trajectory in space-time.

Since the time of Pythagoras, geometers, and even present-day physicists, use the distance of the hypotenuse of a right triangle in Euclidean space, where the potential curvature, expansion or contraction of space-time has been shadowed, to find the distance between two points in space. In the case of a Cartesian coordinate system, a distance from a point in space to the origin of the coordinate system is a property related to the three distinct properties of space often associated with the x, y, and z coordinate axes of the coordinate system. Also, the distance between two points in space is associated with the three-dimensional trajectory of an object of mass "*m*" or a particle involved in a helical trajectory or wave-like motion in a curvilinear trajectory through the volume of space being observed. Thus, the three dimensions of space have long been conceptually accepted; they are part of everyday experience and observation as distinct properties of space that even when taken as a whole as in three-dimensional motion, or in distance measurement, are visible and familiar to the observer.

Time is also a component of space-time and it is frequently measured to gain a sense of temporal interval magnitude, if any, between two events that are either occurring simultaneously or at different instants of time. Those instants of time are usually seconds, minutes, hours of a day in everyday life, as measured by a clock, that are utilized to commence or end activities, or set up appointments to be kept, among many other applications. The first measurements of time were done using natural cycles to observe the passage of time, then other devices such as, but not limited to, sundials, sand clocks, water clocks, mechanical clocks, electronic clocks, and even atomic clocks, have been invented and built.

Clocks have been very useful to measure the magnitude of intervals of time in a repeating cycle and the magnitude of time measured has been visualized and accepted as one-dimensional with a sense of motion. This interpretation of time has been intuitively sufficient for everyday time-keeping and measurement purposes, so it is no surprise that even eminent physicists continue the tradition of conceptualizing time as a one-dimensional property of space-time. A synchronous clock is a time keeping device that continuously measures equal intervals of time, but it also measures the speed of time at a locality in space-time. In the case of a mechanical clock, the speed of time is measured in radians or degrees per second. The speed of time may also be measured in cycles per second.

$$\text{Speed of Time} = \frac{\partial \theta}{\partial \tau} = \omega \tag{1.1}$$

It is widely accepted in modern physics that a synchronous clock moving fast through space-time experiences time dilation due to its relativistic speed of translation as it approaches *"c"*, the speed of light, but this concept was not always so widely accepted.

Furthermore, if an experimenter has two stationary synchronous clocks that are sitting on the surface of a large mass *M*, such as the earth, moving very slowly compared to *c*, and the experimenter travels, very slowly compared to *c*, with one of the two clocks orthogonally from the surface of the mass *M* to a significant distance *"r"* from the geometrical center of *M*, for a sufficient interval of time, the experimenter would be able to observe the effects of relativistic time.

The effects of relativistic time can be observed when the clock is returned to the surface of the earth and it is compared with the clock that stayed on the surface. These effects are accepted to be the result of the curvature of space-time about the earth, the gravitational field of the earth and the Sagnac effect, if applicable, during the motion of the experimenter and clock, from their starting location on earth, to and from their orbital destination. (Hafele, 1971)

Global Positioning Systems consisting of satellites orbiting the earth are present examples of the relativity of time due to the curvature property of near-earth space-time. The orbiting GPS clocks have to be corrected or compensated to coincide with their counterparts near to or on the surface of the earth, in order for the positioning system to work accurately and avoid large errors on the calculated distances between spatial coordinates of the widely-used system.

The constancy of the speed of light, or constant speed of light *"c"* *(celeritas)*, refers to the constancy of the measurement of the speed of light by an observer in any inertial frame of reference in vacuum, or in any free-falling reference frame, due to the relativity of space-time. Light itself does not have to always travel at the same value (magnitude) of speed from the perspective of a frame of reference of a different medium, but every measurement by every observer in an inertial frame of reference in vacuum, or a free-falling reference frame in vacuum, would result in the same value of "*c*". Space and time would contract or expand relatively in an observer's inertial frame of reference, or free-falling frame of reference, to maintain the perception of constancy of the speed of light for the observer.

The constancy of the measurement of the speed of light is the ratio of the equivalent spatial radial distance traveled by the temporal wave to its duration.

1.1. Is time linear or multidimensional?

Therefore, *sitting clocks tick tock the same, but very fast-moving clocks may tick faster or slower than tock.* Thus, for a specific direction and sense of motion as in the direction of the x-axis in a Cartesian coordinate system it is widely accepted that

$$t' = \frac{t}{\sqrt[2]{1 - \frac{v^2}{c^2}}} \tag{1.2}$$

Where t' is relativistic time, "t" is proper time, "c" is the speed of light, and "v" is the speed of the moving object or particle.

Clearly, clocks are not measuring a dimension of time. A synchronous clock is a tachometer for the spatial interval traveled by an object, or particle, during a temporal interval, between specific locations $[x_a - x_e, y_a - y_e, z_a - z_e]$ in a Cartesian coordinate system, according to the properties of space-time and to the translational speed of the clock. Clocks are cycle-counters or tachometers. *The concept of linear absolute time at very slow velocities is an illusion with very useful purposes in everydayness.*

Perhaps, now would be a good time to ask rhetorical questions about time: "if we think that time is one-dimensional, why do we have to square time as dt^2, as we do space as dr^2, in order to formulate space-time metric equations?

If there are three spatial dimensions, dr^2 is mathematically derived from the dimensions of length, width, and height, in units of meters ^{2}, for its distance squared. If time is one-dimensional, would its temporal distance not be linear seconds as in dt? For it is conceivably factual that if we square time as dt^2 for very practical mathematical purposes and applications in physics, we may have already assumed mathematically, the three-dimensional attributes of time.

Therefore, time, like space, has directions and senses of motion, and linear motion is affected by the properties of linear space-time. In linear motion, the linear property of space or the property of one-dimensional space is accompanied by its corresponding linear property of time. The linearly moving object or particle experiences relativistic effects from both the spatial and temporal properties of space-time as it approaches "c".

Thus, the distance of time $(t_a - t_e)$ traveled by a synchronous clock in space-time between specific temporal locations along the t_X – axis of a stationary six-dimensional Cartesian-type coordinate system is associated with the spatial distance traveled along the *x*-axis $(x_a - x_e)$ according to the properties of space-time for that path.

Each temporal dimensional axis is parallel to its directional spatial dimensional axis and orthogonal to the other two temporal axes. As a result, the angle between two space dimensions or waves, or two temporal dimensions or waves, in homogeneous and isotropic space-time, may not exceed ninety degrees, as the space-time wave propagates onwards; at ninety degrees, the two dimensions or waves are parallel to each other.

Thus, the spatial coordinates may be illustrated as having a potential clockwise torsion as seen from space and the temporal coordinates may be illustrated as having a potential counterclockwise torsion as seen from time.

Hence, space-time may be conceptualized as being very pliable to conditions such as contraction, dilation, curving, torsion, inflating (scale growth), and twisting as it propagates. If there is torsion, then the stress-energy-momentum tensor of general relativity is asymmetric. This corresponds to the case with a nonzero torsion tensor.

The top view of a tripodal superposition of a spatial and a temporal Cartesian-type coordinate system involving the potential torsion of six-dimensional space-time is illustrated below

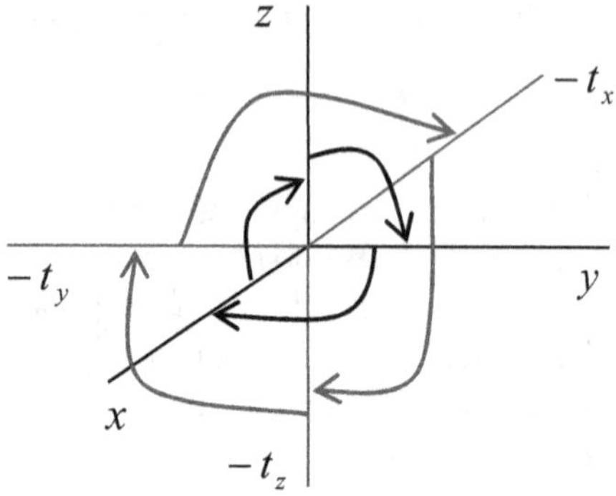

Figure 1. The potential torsion of six-dimensional space-time

If the angle between two space dimensions, or waves, or between two temporal dimensions or waves, were less than ninety degrees, then the spatial dimensions, or the temporal dimensions, would no longer be orthogonal, contrary to the physical properties of homogenous and isotropic flat six-dimensional space-time.

1.2. The equations of six-dimensional space-time

Let us now consider the Friedman-Lemaitre-Robertson-Walker metric in spherical polar coordinates for homogeneous and isotropic six-dimensional flat space-time, where $a(t)^2 = 1$, such that we have

$$d\Sigma^2 = dr^2 + r^2 d\theta^2 + r^2 Sin^2\theta d\varphi^2 \tag{1.3}$$

$$d\pi^2 = dt^2 + t^2 d\theta^2 + t^2 Sin^2\theta d\varphi^2 \tag{1.4}$$

$$d\Omega^2 = d\theta^2 + Sin^2\theta d\varphi^2 \tag{1.5}$$

$$-c^2 d\tau^2 = -c^2 d\pi^2 + a(t)^2 d\Sigma^2 \tag{1.6}$$

$$-c^2 d\tau^2 = -c^2 dt^2 - c^2 t^2 d\theta^2 - c^2 t^2 Sin^2\theta d\varphi^2 + dr^2 + r^2 d\theta^2 + r^2 Sin^2\theta d\varphi^2 \tag{1.7}$$

where $d\Sigma^2$ equals the spatial metric, $d\pi^2$ equals the time metric, and $c^2 d\tau^2$ equals the space-time metric in six-dimensional flat space-time.

Let us illustrate a point T in three-dimensional time $(-t,-\theta,-\varphi)$ where the angle of inclination from the $t_Z - axis$ is $-\theta$, the angle from the t_X – axis on the $t_X - t_Y$ plane is $-\varphi$, and $-t$ is the time distance from the origin to a point "t" in three-dimensional time.

Three-dimensional temporal coordinates or three-dimensional spatial coordinates may be represented by three intersecting complex planes, where a point in space-time is given as a complex number on every complex plane. The three spatial planes are superimposed on the three temporal planes to obtain three complex planes. Every spatial dimension in every sense on a complex plane has an opposite conjugate temporal dimension or vice versa.

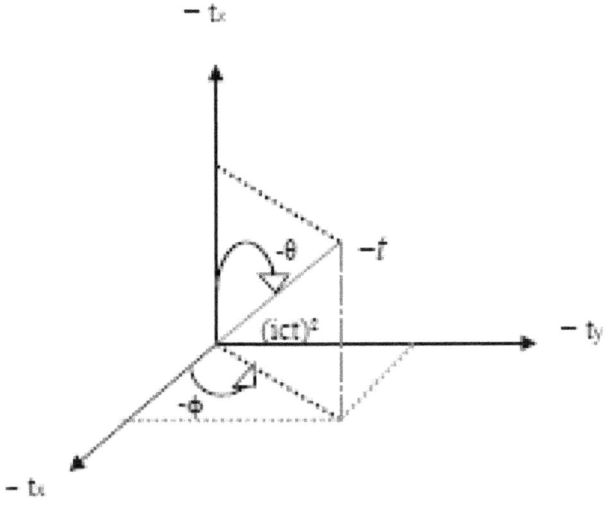

Figure 2. Three-dimensional temporal coordinates

$$-c^2 d\tau^2 = -c^2 dt^2 - c^2 t^2 d\Omega^2 + dr^2 + r^2 d\Omega^2 \tag{1.8}$$

$$-c^2 d\tau^2 = -c^2 dt^2 + dr^2 \tag{1.9}$$

$$-c^2d\tau^2 = -c^2dt_X{}^2 - c^2dt_Y{}^2 - c^2dt_Z{}^2 + dx^2 + dy^2 + dz^2 \quad (1.10)$$

Let us examine the equation of coordinate time $d\pi^2$, proper time $d\tau^2$ and spatial coordinate time $d\Sigma^2 / c^2$ in six-dimensional space-time.

$$-c^2(d\tau^2 - d\pi^2) = d\Sigma^2 \quad (1.11)$$

$$c^2(d\pi^2 - d\tau^2) = d\Sigma^2 \quad (1.12)$$

$$d\tau^2 = d\pi^2 - \frac{d\Sigma^2}{c^2} \quad (1.13)$$

$$\frac{d\Sigma^2}{c^2} = d\pi^2(1-\gamma^2) = -d\pi^2\left(\frac{v^2}{c^2 - v^2}\right) \quad (1.14)$$

We find that as the object slows down, v << c, the proper time distance approaches the coordinate time distance, $d\tau^2 \to d\pi^2$, and at $v = 0$, $d\Sigma^2 / c^2 = 0$, as predicted from the six-dimensional metric and general relativity. The result agrees with the current four-dimensional metric because the four-dimensional metric is a collapsed representation of the six-dimensional metric in six-dimensional flat space-time. The temporal distance element dt^2 is not a linear temporal element component; it is a three-dimensional temporal element component. Thus, the four-dimensional metric is a contracted or resultant six-dimensional metric of space-time that leads to the same results. If there were three spatial dimensions and one dimension of time, then the metric for flat space-time would be $-cd\tau = -cdt + dr^2$, and this is not how it has been defined. Furthermore, there is a correspondence of curvature in space-time, time curves where space curves. Proper time is a physical manifestation of curved time.

In general relativity, space curves and time curves about a slowly moving object of mass. Thus, for curved space-time, the temporal distance element dt^2 should be multiplied by a temporal factor of curvature $(1 - k_t t^2)$ as the spatial distance element dr^2 is divided by a spatial factor of curvature $(1-k_S r^2)$ in the universal metric equation for space-time.

It is proposed that the temporal factor may be denoted as $(1-k_t t^2)\equiv\left(1-v^2/c^2\right)$ where the temporal constant of curvature k_t may be expressed as a dynamic temporal variable of curvature v^2/c^2t^2 and the spatial factor may be denoted as $(1-k_S r^2)\equiv\left(1-v^2/c^2\right)$ where the spatial constant of curvature k_S may be expressed as a dynamic spatial variable of curvature v^2/c^2r^2. The velocity variable "v" would be the velocity of a mass, or particle, moving through the infinitesimal spatiotemporal element of the metric being portrayed.

Therefore, as time and space curve, the curvature of space-time, or its manifestation of a gravitational field, about a celestial body or an object of mass moving slowly relative to the speed of light, or about an object traveling through space-time at relativistic speed, is one and the same effect.

Thus, proper time equals coordinate time minus the spatial coordinate time distance. Let us now find the proper time in terms of the speed of proper time v_τ.

$$d\tau^2 = d\pi^2 - \frac{d\Sigma^2}{c^2} \tag{1.15}$$

$$\frac{d\tau^2}{d\pi^2} = 1 - \frac{d\Sigma^2}{c^2 d\pi^2} \tag{1.16}$$

$$v^2 = \frac{d\Sigma^2}{d\pi^2} \tag{1.17}$$

$$d\tau^2 = d\pi^2\left(1 - \frac{v^2}{c^2}\right) \tag{1.18}$$

$$-\frac{d\Sigma^2}{c^2} = -\frac{v^2 d\pi^2}{c^2} \tag{1.19}$$

$$v_\tau^{\ 2} = \frac{d\tau^2}{d\pi^2} = \left(1 - \frac{v^2}{c^2}\right) \tag{1.20}$$

Therefore, we find proper time as $v \rightarrow c$ to be

$$d\tau^2 = d\pi^2\left(1 - \frac{v^2}{c^2}\right) = \frac{d\pi^2}{\gamma^2} = d\pi^2 v_\tau^{\ 2} \tag{1.21}$$

We observe that when $v = 0$, then $d\tau^2 = d\pi^2$, as expected in general relativity. Moreover, we find that for relativistic speeds at $v < c$, coordinate time distance can be converted to proper time distance or vice versa.

§ 2. On the simultaneity of events and the synchronicity of clocks

The advent of the General Theory of Relativity put an end forever to our cozy and familiar conceptualization of absolute time for any location in space-time regardless of motion. The new way of thinking about space-time put forth by the physicist Albert Einstein eventually changed the mind-sets of physicists about space and time to a far more complex and fluid notion about how time passes according to the curvature of space-time and the space-time effect on gravity at the locality of curvature. The General Theory of Relativity of Albert Einstein has been called the greatest scientific insight in human history, but the thorough interpretation of space-time and its interactions must be its most significant accomplishment for that derivation will have an indelible effect on history and science again. (Einstein, 1952)

In relativistic terms, proper time is a function of motion as well as a function of the curvature properties of space-time, for space-time is curved in the presence of rest mass or by a mass moving at a speed that approaches "*c*", the constant speed of light, as measured by an observer in a stationary or inertial coordinate system. *The interaction of mass and space-time is one and the same, the presence of mass curves the wavelength of space-time, and as a result the wavelength of time contracts, proper time dilates, to conserve the energy of space-time. Therefore, for each dimension of space-time, the reciprocal relationship of the temporal vector and its corresponding spatial vector preserves the physical state of space-time's vector space.*

The times of two or more events may be given simultaneously by specified stationary synchronous clocks located at the places of the events, for either a stationary or an inertial coordinate system, for all time determinations of each specified synchronous clock. Any absolute signification of simultaneity of a coordinate system could not be attached to two or more events, or their respective synchronous clocks, when viewed from a system of coordinates that is in motion relative to that system. However, an observer on a stationary coordinate system would be able to declare the clocks of the events to be synchronous and two or more events to be simultaneous. Thus, it is possible for a stationary observer at the origin of a stationary coordinate system to measure time on specific stationary synchronous clocks for two or more events taking place on inertial coordinate systems with their specific inertial synchronous clocks in the space-time of the stationary coordinate system.

2.1. Synchronicity and simultaneity experimental conditions for temporal events

Hence, with the help of an imaginary physical experiment it is possible to set up the right conditions for the synchronicity of clocks and the simultaneity of events in space-time:

1. In a stationary Cartesian coordinate system (x, y, z), four identical clocks, that are at rest, are synchronized at the origin of the system.

2. Clock A is synchronized to Clock B, Clock B is synchronized to Clock C, Clock C is synchronized to Clock D, which results in Clock A being synchronized to Clocks C and D and Clock B being synchronized to Clock D.

3. All four ideal clocks are synchronous to each other, keeping time exactly at the same rate, and with an ideal accuracy among the measured time of each clock that causes no discrepancies between clocks during the measured time of the experiment for all time determinations. As determined by an observer at the origin of a stationary coordinate system

each time event at each clock coincides or takes place simultaneously with the event of each of the other three clocks.

4. As each of the four clocks is used to measure the constancy of the speed of light, c, each clock agrees with the other three clocks.

The conditions that are defined above by means of the simultaneity of events and the synchronicity of clocks provide the basis for time measurement in a stationary Cartesian coordinate system.

§ 3. On the relativity of time and the types of spatial motions

Two key principles that are instrumental in measuring the relativity of time and the association of time and space in every direction and sense of motion are the universal constancy of the speed of light in space-time and the principle of relativity. Hence, the following fundamental principles are postulated:

1. Any ray of light emitted from a moving object of mass "m", in a stationary coordinate system in space-time, moves with the measurable constant speed of light "c."

2. The laws of physics and the physical states of the objects of mass "m" remain the same during translatory motion for any type of motion in the stationary coordinate system and at a speed less than the speed of light "c."

3. Any time interval at any location in the stationary coordinate system can be measured by a synchronous clock at that location or by dividing the distance traveled by a ray of light between two points in space-time by the constant speed of light, c.

$$\text{Time of Interval} = \frac{\text{distance traveled by light ray}}{c} \tag{3.1}$$

4. In the presence of homogeneous and isotropic space-time in a stationary coordinate system, all objects of mass "m" with identical physical dimensions and physical states, translating relativistically at the same speed, or at the same acceleration, at near the speed of light, experience relativistic effects of

time dilation, length contraction, and mass dilation, in any direction and sense of motion for any type of motion.

3.1. Constructing the framework and performing a temporal thought experiment

Thus, with the help of an imaginary physical experiment, let there be three identical spheres that are made of a clear and uniform material with a smooth surface that are each able to emit a ray of light from the geometrical center of the interior of the sphere without distorting or altering the ray of light in any way or refracting its intended direction or trajectory. All spheres have equal uniform mass *"m"* and the same physical dimensions and physical states.

Let each sphere have a synchronous clock as defined in "§ 2", that will travel with the sphere, and let an observer travel with the sphere to measure time when the sphere has reached its assigned destination in the stationary Cartesian coordinate system according to the fundamental principles postulated above. Then, let the instruments and observers of the experiment be positioned as follows:

a. Let Clock A be positioned at the origin (0, 0, 0) of the stationary Cartesian coordinate system with Observer A.

b. Let the three spheres be positioned on the z-axis above Clock A and Observer A at a spatial distance *"k"* from each other where there will be enough clearance and space between them in such a way that no sphere, clock, or observer, will encroach on one another or interfere in any way with one another before or during the experiment. Let the mass *"m"* of each sphere be small enough, the mass of each observer, the mass of each clock, and let the spatial distance *"k"* be large enough, so that any curvature of space-time that may be produced by any one mass be so small as to produce a negligible effect on the other spheres' surrounding space-time during the experiment.

c. Let Sphere B have Clock B and be positioned at the spatial point (0, 0, z + k) with Observer B.

d. Let Sphere C have Clock C and be positioned above Sphere B at the spatial point (0, 0, z + 2k) with Observer C.

e. Let Sphere D have Clock D and be positioned above Sphere C at the spatial point (0, 0, z + 3k) with Observer D.

In accordance with the principles of the simultaneity of events and the synchronicity of clocks defined in "§ 2", let us imagine the sequence of events of the physical experiment as follows:

1. The Observer A at the origin (0, 0, 0) of the stationary Cartesian coordinate system signals the start to all observers and spheres simultaneously. Let Observer A have an innovative sensing device that can detect instantaneously rays of light arriving at any level or coordinate point at z + k, z + 2k or z + 3k on the z-axis. The sensing device can work together with Clock A instantaneously to detect and measure simultaneously, or at different instants of time, the instants of times of arrival of rays of light.

2. Spheres B, C and D and their Observers B, C and D depart their initial position simultaneously at t_0 and at an equally predetermined acceleration and speed at less than the speed of light, nc, where 0 < n < 1.

3. Sphere B and Observer B travel in a linear trajectory parallel to the positive x-axis in the first quadrant of the stationary coordinate system from (0, 0, z + k) to $(x_X, 0, z+k)$ approaching the speed of light but reaching $(x_X, 0, z+k)$ at a relativistic speed of nc, where 0 < n < 1. At the instant of time that Sphere B arrives at $(x_X, 0, z+k)$, it emits a ray of light from its geometrical center towards the z-axis start coordinates (0, 0, z + k) of the stationary coordinate system and simultaneously Observer B measures the time taken to travel from (0, 0, z + k) to $(x_X, 0, z+k)$ to be $(t_X - t_0)$ on Clock B. The distance traveled by Sphere B along its path is measured to be $(x_X - 0)$.

Let that distance equal "r", so that

$$r = (x_X - 0) = x_X \tag{3.2}$$

and the time interval measured by Observer B is

$$\Delta t_{XO} = (t_X - t_0) = t_X - t_0 \tag{3.3}$$

4. Sphere C and Observer C travel on a curvilinear trajectory always at a distance (z + 2k) above the x-y plane in the first quadrant of the stationary coordinate system from (0, 0, z + 2k) to $(x_a, y_a, z+2k)$ approaching the speed of light but reaching $(x_a, y_a, z+2k)$ at a relativistic speed of nc, where $0 < n < 1$. At the instant of time that Sphere C arrives at $(x_a, y_a, z+2k)$, it emits a ray of light from its geometrical center towards the z-axis start coordinates (0, 0, z + 2k) of the stationary coordinate system and simultaneously Observer C measures the time taken to travel from (0, 0, z + 2k) to $(x_a, y_a, z+2k)$ to be $(t_a - t_0)$ on Clock C. The distance "*r*" traveled by Sphere C along its path in its curvilinear trajectory is measured to be $r = x_X$, the same distance measured in step 3.

If the curvilinear trajectory of Sphere C is followed beyond $(x_a, y_a, z+2k)$, it would smoothly merge and coincide in trajectory with a circle of equation $r^2 = x^2 + y^2$ where the radius of the circle is $r = (x_X - 0)$, equal to "*r*" as measured in step 3.

Thus, an object traveling the curvilinear trajectory of Sphere C would eventually travel in a circular path with a radius "*r*" about the z-axis start coordinate (0, 0, z + 2k) on a plane parallel to and above the x-y plane by a constant distance of (z + 2k) in the stationary coordinate system.

Hence, the distance traveled by Sphere C along its path is measured to be

$$r = x_X \tag{3.4}$$

and the time interval measured by Observer C is

$$\Delta t_{ao} = (t_a - t_o) = t_a - t_o \tag{3.5}$$

5. Sphere D and Observer D travel in a helical trajectory in the first quadrant of the stationary coordinate system from (0, 0, z + 3k) to $(x_e, y_e, z+3k+z_e)$ approaching the speed of light but reaching $(x_e, y_e, z+3k+z_e)$ at a relativistic speed of nc, where 0 < n < 1. At the instant of time that Sphere D arrives at $(x_e, y_e, z+3k+z_e)$, it emits a ray of light from its geometrical center towards the z-axis coordinates (0, 0, z + 3k) of the stationary coordinate system and simultaneously Observer D measures the time taken to travel from (0, 0, z + 3k) to $(x_e, y_e, z+3k+z_e)$ to be $(t_e - t_o)$ on Clock D. The distance "r" traveled by Sphere D along its path in its helical trajectory is measured to be $r = x_X$, the same distance measured in step 3 or 4.

If the helical trajectory of Sphere D is followed beyond $(x_e, y_e, z+3k+z_e)$, it would coincide with the path of a spatial helix or cylindrical spiral with a radius "r" about the z-axis, where $r = x_X$, and $x_e = \cos(\tau), y_e = \sin(\tau)$, and $z_e = \tau + z + 3k$, where $0 \le \tau \le 2\pi$. Thus, an object traveling the helical trajectory of Sphere D would eventually travel in the path of a cylindrical spiral or helix with a radius "r", where $r = x_X$ as in step 3, about the z-axis starting from a constant distance of (z + 3k) above the x-y plane in the stationary coordinate system.

Thus, the distance traveled by Sphere D along its path is measured to be

$$r = x_x \tag{3.6}$$

and the time interval measured by Observer D is

$$\Delta t_{eo} = (t_e - t_o) = t_e - t_o \tag{3.7}$$

In accordance with the principles of the simultaneity of clocks and the universal constancy of the speed of light:

a. At the origin (0, 0, 0) of the stationary coordinate system, Observer A measures the time interval between the departure of each sphere from the start coordinates on the z-axis and the

instant of time at the space-time coordinates where each sphere emits a light ray towards the start coordinates on the z-axis. Thus, let us call this time interval "the time interval of the sphere." Hence, the time intervals of spheres B, C, and D are measured by Observer A to be the same. Furthermore, Clocks B, C, and D are determined by Observer A to be synchronous during the experiment. Next, Observer A measures the time interval for each light ray, from the instant of time of emission at the space-time coordinates to the instant of time of arrival at the z-axis coordinates. Observer A finds that the time intervals differ in magnitude due to the difference in spatial distance between paths *XO*, *XYO* and *XYZO* when divided by the constancy of the speed of light in space-time. Thus, the spatial intervals of the light rays as measured by Observer A are

$$\frac{path\ XO}{c} > \frac{path\ XYO}{c} > \frac{path\ XYZO}{c} \tag{3.8}$$

and the time intervals of the spheres as measured by Observer A are

$$(t_{xo} - t_{oo}) = (t_{ao} - t_{oo}) = (t_{eo} - t_{oo}) \tag{3.9}$$

b. For Sphere B, the time interval measured by Observer B on Clock B does not agree with the time interval measured by Observer A on Clock A at the origin, for the time of Sphere B to travel a distance "*l*" from its point of origin (0, 0, z + k) on the z-axis to coordinates $(x_X, 0, z + k)$ of the stationary coordinate system. Let the time interval of Sphere B be

$$Time\ Interval = \Delta t_{XO} = t_X - t_O \tag{3.10}$$

$$nc = \frac{l}{(t_x - t_o)} \quad \text{where } 0 < n < 1 \tag{3.11}$$

c. For Sphere C, the time interval measured by Observer C on

Clock C does not agree with the time interval measured by Observer A on Clock A at the origin, for the time of Sphere C to travel a distance *"l"* from its point of origin (0, 0, z + 2k) on the z-axis to coordinates $(x_a, y_a, z+2k)$ of the stationary coordinate system. Let the time interval of Sphere C be

$$Time\ Interval = \Delta t_{ao} = t_a - t_o \tag{3.12}$$

$$nc = \frac{l}{(t_a - t_o)} \quad where\ 0 < n < 1 \tag{3.13}$$

d. For Sphere D, the time interval measured by Observer D on Clock D does not agree with the time interval measured by Observer A on Clock A at the origin, for the time of Sphere D to travel a distance *"l"* from its point of origin (0, 0, z + 3k) on the z-axis to coordinates $(x_e, y_e, z+3k+z_e)$ of the stationary coordinate system. Let the time interval of Sphere D be

$$Time\ Interval = \Delta t_{eo} = t_e - t_o \tag{3.14}$$

$$nc = \frac{l}{(t_e - t_o)} \quad \text{where } 0 < n < 1 \tag{3.15}$$

3.2. Conducting a second temporal thought experiment

Let us consider a second physical experiment done by the same observers with the same instrumentation and the same specified conditions, but now let the spheres B, C, and D leave their start coordinates on the z-axis simultaneously and travel parallel to each of the three spatial axes moving relativistically at the same speed and the same acceleration at less than the speed of light for a distance *"l"* in homogeneous and isotropic space-time. Sphere B travels parallel to the x-axis a distance *"l"*, Sphere C travels parallel to the y-axis a distance *"l"* and Sphere D travels parallel to the z-axis a distance *"l"*. At the instant of time that each sphere B, C, and D arrives at the end coordinates, each emits a ray of light from its geometrical center towards the z-axis start coordinates of the stationary Cartesian coordinate system and simultaneously each sphere's observer measures the time taken to travel from start to finish on each Clock B, C, and D.

The observers find that Clocks B, C, and D undergo the same time dilation, even though each clock is moving on a different spatial dimension. Clocks B, C, and D do not agree with Clock A, as measured by Observer A at the origin of the stationary coordinate system, on the time interval for each sphere from start to finish. Observer A measures each sphere to have traveled the same distance *"l"* during an equal time interval.

3.3. Conclusions on temporal thought experiments

Therefore, each of the three spheres could travel in a different dimension because space is three dimensional. Otherwise, some of the spheres may have ended up in the same location, or if space was one-dimensional all spheres would end up at the same end coordinates. Hence, if we consider time to be one-dimensional, then only one clock of one sphere would have measured the passage of time and its dilation. Every spatial dimension is known to be at ninety degrees to the other two spatial dimensions, and each temporal dimension would be at ninety degrees to the other two temporal dimensions. In a Cartesian-type coordinate system for either space or time, the resultant of the time axes and the resultant of the spatial axes coincide on a line that is at forty-five degrees to each axis of space or time along the axis of propagation of space-time. *The underlying structure of space-time forms a dimensional resultant correspondence between space and time along the axis of propagation.* Thus, these dimensional properties and relationships are possible because of the three-dimensional properties of both space and time.

Current time measurement or determination practice affirms that the interval of time for a moving object or particle traveling with a clock for an interval τ , where $\tau > t_p$ (the Planck time is coordinate time), along a path l_p (the Planck length), at a relativistic speed, is the same for any path on any type of translatory motion. If time is considered a one-dimensional time-axis in the direction of the spatial resultant along the axis of propagation of space-time in a Cartesian coordinate system in homogeneous and isotropic space-time, then, an object A traveling relativistically at a specified speed and acceleration along the axis of propagation for a distance l_p would have a certain space-time correspondence $(ic\tau)$, where τ is proper time. Three identical objects B, C, and D traveling at the same relativistic speed and acceleration as object A, and with synchronous clocks with object A, in the same homogeneous and isotropic space-time along the x, y, and z axes, for the distance $l_p / \sqrt[2]{3}$, would experience a proper time passage of only $\tau / \sqrt[2]{3}$ due to the correspondence between each spatial dimension along the x, y, and z axes and the resultant time-axis of propagation for object A. Thus, time behaves as a three-dimensional property of space-time. (Coan et al, 2005)

In 1916 through 1917, the eminent German mathematician David Hilbert lectured on The Foundations of Physics, in which he outlined an extraordinary side effect of Einstein's Theory of Relativity. Hilbert was studying the interaction between a relativistic particle moving in a circular path on a plane toward or away from a stationary mass. Hilbert observed that if the relativistic particle had a velocity greater than $c / \sqrt[2]{3}$ where *c* is the speed of light, a stationary mass would repel it, if less than $c / \sqrt[2]{3}$, it would attract it. At least, that's how it would appear to a distant inertial observer. (Sauer, 2009)

Hilbert's side effect is due to the three-dimensionality of spacc and time. The particle travels on a two-dimensional plane where space-time expands at the speed of light *c* through its resultant axis (*r*), but at $c / \sqrt[2]{3}$ through its coordinate axes (x, y, z). As the particle travels around its circular path, it arrives at its arbitrary point of departure earlier than the radially expanding space-time from the stationary mass. As the space-time wave reaches the moving particle it displaces it in the direction of the curvature of the wave. Every point in space-time expands uniformly in isotropic and homogeneous space-time, so the moving particle is less affected by destructive wave interference as it moves faster than the wavelets of space-time through the direction of coordinate time. Thus, the observed displacement in Hilbert's model is due to the curvature of space-time as it expands without destructive wave interference (Huygens Principle).

Hence, Hilbert's side effect showed that the Schwarzschild solution to the Einstein Field Equations allows a particle that is moving in a circular motion on a plane around a mass at a speed greater than $c / \sqrt[2]{3}$ (0.577*c*) to undergo a space-time displacement that confirms spatial and temporal three-dimensionality. Furthermore, Hilbert's side effect is the manifestation of Einstein's own Mach's principle.

The inertia of an isolated body or mass in isotropic and homogeneous space-time is determined by the interaction of the spatiotemporal wave acceleration and pressure with the physical state of the body of mass at the locality of the isolated body or mass, regardless of the inertia of all other external bodies or masses. Mach's principle manifests itself universally through the equivalence principle, and upholds the constancy of the speed of light.

In accordance with the principles of general relativity involving length contraction and time dilation, for an equal path length *"l"* traveled by all spheres in the imaginary physical experiments above, according to time observations, all time intervals measured by Observers B, C, and D are the same for Clocks B, C, and D during the experiment. Thus, all three clocks are synchronous. Therefore, there are temporal and spatial symmetries for all three types of motion: linear, curvilinear and helical. The symmetry is maintained for any of the three types of motion, or the three spatial dimensions above, by the equality and the three-dimensional existence of the temporal and spatial dimensions along the axes of the coordinate system during the motion through homogeneous and isotropic space-time. As a result, it is essential to see that the symmetry of relativistic effects is due to the three-dimensionality of space as well as to the three-dimensionality of time. Otherwise, the three moving synchronous clocks in the above imaginary physical experiments would not be able to agree for any type of motion in space-time, in accordance to the principles of the general theory of relativity and the constancy of the speed of light.

Thus, the universal constancy of the speed of light holds for all moving spheres and observers such that

$$\frac{l}{(t_x - t_o)} = \frac{l}{(t_a - t_o)} = \frac{l}{(t_e - t_o)} \qquad (3.16)$$

and the three-dimensionally symmetrical relativistic effects affirm that

$$(t_x - t_o) = (t_a - t_o) = (t_e - t_o) \qquad (3.17)$$

Moreover, the magnitudes of the spatial distances from the point of origin on the z-axis to the end coordinates of the trajectory during the experiment in the stationary coordinate system can be calculated for each sphere if we subtract *k*, *2k*, and *3k*, from the z-coordinates of the linear, curvilinear, and helical trajectories, to level the spatial determination such that

$$\{(x_x)^2\} > \{(x_a)^2 + (y_a)^2\} > \{(x_e)^2 + (y_e)^2 + (z_e)^2\} \qquad (3.18)$$

and for the three-dimensional relativistic temporal intervals of the spheres we find

$$\{(t_x)^2\} > \{(t_{x_a})^2 + (t_{y_a})^2\} > \{(t_{x_e})^2 + (t_{y_e})^2 + (t_{z_e})^2\} \qquad (3.19)$$

Thus, we see that we cannot attach any absolute linear signification to the concept of time, but that two or more simultaneous events in six-dimensional space-time, viewed from a stationary coordinate system, can be measured to be simultaneous events regardless of trajectory, or direction and sense of motion, when viewed as events of moving objects or particles in that coordinate system. Furthermore, *time is space-dependent, and space is time-dependent, but they are not type-of-motion dependent.* The path of a moving object, regardless of trajectory, taken at relativistic speed, dilates proper time and contracts space, in homogeneous and isotropic space-time, in all directions and senses of motion. *Thus, time, dilates or contracts three-dimensionally, proportionally and continuously, regardless of the direction or sense of motion, so that the path chosen results in the same proper time dilation, with all other physical variables being equal.*

§ 4. On the six-dimensionality of space-time and the relativistic effects on moving bodies

Consequently, time is three-dimensional, in a similar way to space being three-dimensional, and the passage of time in each direction and sense of motion can be different just as spatial translation of a moving object in a coordinate system can differ in any one direction of space depending on trajectory. So, for an object or particle traveling in space-time a proper time interval between events of the object as seen from an inertial coordinate system will depend on the object's speed with respect to the speed of light. If a stationary clock is compared to a clock traveling at relativistic speed, the temporal interval difference between the two clocks would increase as a function of the length of the path and the speed of the moving clock with respect to the speed of light. (Taylor et al, 1966)

In homogenous and isotropic space-time of a stationary Cartesian time coordinate system, we find that

$$(\Delta t)^2 = (t_{x_2} - t_{x_1})^2 + (t_{y_2} - t_{y_1})^2 + (t_{z_2} - t_{z_1})^2 \qquad (4.1)$$

Where Δt is the three-dimensional temporal coordinate distance, or coordinate time interval between two events, at temporal coordinates $\left(t_{X_1}, t_{Y_1}, t_{Z_1}\right)$ and $\left(t_{X_2}, t_{Y_2}, t_{Z_2}\right)$, in six-dimensional space-time. The temporal coordinates may be described as Linear Frequency or Period (f_T) for $\left(t_{x_2} - t_{x_1}\right)$, Angular Frequency (Rotation or Spin) (ω_t) for $\left(t_{y_2} - t_{y_1}\right)$ and Amplitude (A_t) for $\left(t_{z_2} - t_{z_1}\right)$. The axes of the three-dimensionality of time represent the aspects and attributes of the temporal wave as it emerges. Let us imagine a round mechanical clock moving at relativistic speed with an axis of propagation normal to the clock face or dial. The axis of propagation would represent the Linear Frequency-axis, the radius of the dial would represent the Amplitude-axis, and the rotating hands of the clock would represent the Angular Frequency-axis.

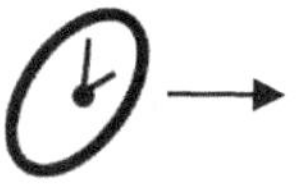

Figure 3.

As the temporal wave emerges and expands, it may be obstructed in its path by nodes of mass or energy in space-time, and its period (or wavelength) would contract increasing linear frequency, its amplitude would increase, and the passage of time would slow down. Using the analogy of the moving clock, the temporal distance of the period would contract increasing frequency (Linear Frequency) in the direction of propagation of the clock, the extending radius (Amplitude) would enlarge all aspects of the dial of the clock, where the radian distance (minute of arc) between minute marks would extend, and the passage of time would slow down (Angular Frequency), when the tip of the minute hand travels longer between minute marks. The temporal wave of space-time has amplitude, linear frequency, and angular frequency, which are inherent properties of three-dimensional time. Particles and physical fields share the attributes of the wave medium.

Space is often represented three-dimensionally by a coordinate system such as the widely used Cartesian coordinate system where the resultant distance of the units of the x, y, and z axes is represented as a line segment or ray extending from the origin to a point in space. If the first quadrant of the coordinate system is used where the x, y, and z axes are all positive, the ray extending from the origin to a point in space would be at forty-five degrees to each of the coordinate axes. If time is represented analogous to space using a Cartesian-type coordinate system, the resultant time ray from the origin to a point in time would be at forty-five degrees to each of the time axes t_X, t_Y, and t_Z. In such representation of space-time, space and time would have opposite line segments, or rays, along the same line for the units of measurement of their respective coordinates, in homogeneous and isotropic space-time.

Let us consider the equations for the magnitudes of relativistic time intervals and speeds in the x, y and z directions of a Cartesian coordinate system for a clock moving three-dimensionally in space-time such that

$$t'_x = \frac{t_x}{\sqrt[2]{1 - \frac{v_x^2}{c^2}}} \tag{4.2}$$

$$t'_y = \frac{t_y}{\sqrt[2]{1 - \frac{v_y^2}{c^2}}} \tag{4.3}$$

$$t'_z = \frac{t_z}{\sqrt[2]{1 - \frac{v_z^2}{c^2}}} \tag{4.4}$$

With the help of these equations, we can visualize that if time is to be consider linear, let us imagine so in the direction of the x-axis, time dilation would only occur in the direction of the x-axis.

Thus, for curvilinear relativistic motion or helical relativistic motion, that is, two-dimensional motion or three-dimensional motion, only the velocity $\vec{v}_x$ would have a relativistic effect on the temporal interval of the trajectory. The velocity $\vec{v}_x$ has magnitude (speed v_x) and a direction of motion along the x-axis. Then, in our imaginary physical experiment of "§ 3", the linear relativistic motion would have had the greatest time dilation ascribable to $\vec{v}_x$, followed by the curvilinear relativistic motion with time dilation attributed to $\vec{v}_x$, but not velocity $\vec{v}_y$, and lastly, the helical relativistic motion would have had the smallest time dilation, again ascribable to $\vec{v}_x$, but not to velocities $\vec{v}_y$ or $\vec{v}_z$. The velocities $\vec{v}_y$ and $\vec{v}_z$ have magnitudes (speeds v_y and v_z) and directions along the y-axis and along the z-axis respectively.

Hence, Clocks B, C, and D would not have agreed on the temporal interval of travel since each type of relativistic motion involved a different magnitude of velocity $\vec{v}_x$ to reach the distance *"l"* at the same speed and acceleration.

Therefore, we conclude that if time dilation is a relativistic effect regardless of the type of relativistic motion, then it is essential to recognize that all velocities $\vec{v}_x$, $\vec{v}_y$ and $\vec{v}_z$ participate in helical relativistic motion, and velocities $\vec{v}_x$ and $\vec{v}_y$ participate in curvilinear relativistic motion. As a result, we can iteratively conclude that relativistic effects are acting in each dimension of six-dimensional space-time and time is unequivocally three-dimensional as its conjugate space.

4.1. Clock speed: what time-speed is it?

We may find it useful to reiterate that clocks are cycle-counters or tachometers. *Thus, a synchronous clock does not measure a dimension of time. A synchronous clock is a tachometer measuring the cycles of time in space-time between six-dimensional events.*

Consequently, we can derive the following equations for the

relativistic temporal speeds in the t_x, t_y, and t_z directions of a temporal Cartesian-type coordinate system for a moving clock to be

$$v'_{t_x}{}^2 = \frac{c^2 v_{t_x}{}^2}{c^2 - v_x{}^2} \tag{4.5}$$

$$v'_{t_y}{}^2 = \frac{c^2 v_{t_y}{}^2}{c^2 - v_y{}^2} \tag{4.6}$$

$$v'_{t_z}{}^2 = \frac{c^2 v_{t_z}{}^2}{c^2 - v_z{}^2} \tag{4.7}$$

$$v'_t = \sqrt[2]{v'_{t_x}{}^2 + v'_{t_y}{}^2 + v'_{t_z}{}^2} \tag{4.8}$$

Clock speeds v'_{t_x}, v'_{t_y}, and v'_{t_z} are relativistic for a fast-moving synchronous clock with respect to the speed of light. Clock speeds v_{t_x}, v_{t_y}, and v_{t_z} are the clock speeds of a very slow-moving synchronous clock.

Similarly, we can derive the spatial relativistic speeds in the x, y and z directions of a Cartesian coordinate system of a moving synchronous clock of mass "m" to be

$$v_x{}^2 = \frac{c^2\left(v'_{t_x}{}^2 - v_{t_x}{}^2\right)}{v'_{t_x}{}^2} \tag{4.9}$$

$$v_y{}^2 = \frac{c^2\left(v'_{t_y}{}^2 - v_{t_y}{}^2\right)}{v'_{t_y}{}^2} \tag{4.10}$$

$$v_z{}^2 = \frac{c^2\left(v'_{t_z}{}^2 - v_{t_z}{}^2\right)}{v'_{t_z}{}^2} \tag{4.11}$$

$$v_r = \sqrt[2]{{v_x}^2 + {v_y}^2 + {v_z}^2} \qquad (4.12)$$

4.2. The speed of space-time

Let us reconsider the Friedman-Lemaitre-Robertson-Walker metric for homogeneous and isotropic space-time of uniform curvature for elliptical space, such that

$$-c^2 d\tau^2 = -c^2 dt^2 + a(t)^2 d\Sigma^2 \qquad (4.13)$$

where the current value of the scalar factor equals one, $a(t) = l_{proper} / l_{t_0} = 1$, where $a(t)$ is a function that relates the proper distance, l_{proper}, between a pair of objects, to the distance, l_{t_0}, at some reference time, such that we have

$$d\Sigma^2 = c^2 dt^2 - c^2 d\tau^2 \qquad (4.14)$$

$$\frac{d\Sigma^2}{dt^2} = c^2 - c^2 \frac{d\tau^2}{dt^2} \qquad (4.15)$$

Thus, we find the speed of space to be

$$v_s^2 = v_t^2 - c^2 v_\tau^2 \qquad (4.16)$$

Fundamentally, time exists for itself and for the purposes of space-time. It does not need or ask for our acknowledgement, observation, measurement or interaction. Time flows amidst the masses of objects or the changes that take place between objects or particles.

It is eventful and always attentive and dependent of its relationship with space. Time and space are but two sides of the same coin and the mint is space-time.

If light travels as fast as time, its wavelength λ_c equals the wavelength of time λ_t , in homogeneous and isotropic space-time. Light's energy, amplitude and frequency, will be at its maximum and its measured speed will be *"c"*. For any luminal speed of space-time, light will match the speed of time and its speed will always be measured by any observer to be *"c"*. Let us call this limitation on the speed of light *"the light barrier."* As time speeds up and accelerates, then a slow clock runs faster and ticks slower than tocks, as space-time expands.

If the wave of space-time propagates with variable acceleration in space-time, as time expands, the qualities of space-time impart that the spatial and temporal speeds are:

$$v_s = \left(\frac{\lambda_s}{T_s}\right) \tag{4.17}$$

$$v_t = -\left(\frac{c\lambda_t}{T_t}\right) \tag{4.18}$$

where λ_s is the wavelength of space and T_s is the spatial period of the wavelength of space. Similarly, λ_t is the wavelength of coordinate time and T_t is the temporal period of the wavelength of coordinate time.

Thus, we derive the speed of space as follows

$$v_s^{\ 2} = v_t^{\ 2} - c^2 v_\tau^{\ 2} \tag{4.19}$$

Consequently, for the squared wavelengths of space we find

$$\frac{\lambda_s^{\ 2}}{T_s^2} = \frac{c^2\lambda_t^{\ 2}}{T_t^2} - \frac{c^2\lambda_\tau^{\ 2}}{T_\tau^2} \tag{4.20}$$

§ 5. On the special relativity and principles of space-time

Let us imagine that space-time expands indefinitely and that we are tasked to conjecture principles of special relativity for homogeneous and isotropic space-time and relativistic motion in vacuum where gravity is not a significant factor. Our careful deliberation leads us to conjecture the following principles:

1. The speed of light is the same for all observers, independent of the uniform relative motion of the observer, the speed of light or the speed of the light source.

2. The squared speed of space is equal to the squared speed of time minus the product of the squared speed of light and the squared speed of proper time (the speed of space-time) relative to any observer.

3. In the absence of mass and energy, space-time expands in all directions at an equal and uniform relative speed in homogeneous and isotropic space-time.

4. The laws of physics are the same in any frame of space-time independent of the motion of the frame of reference.

5. The speed of light cannot exceed the speed of time.

It is no surprise that each principle has a familiar ring for all known forms of energy and mass are framed in the fabric of space-time. Even light travels in the medium of space-time and is bound by the properties of space-time. Indeed, *the speed of space-time is the fastest thing in the universe.*

The first principle is the principle of the special relativity of space-time. Any observer traveling in uniform relative motion would measure the same speed for the speed of light as a result of the relativistic properties of space-time. Thus, the length contraction and time dilation effects would lead an observer to measure the same magnitude for the speed of light regardless of the observer's uniform relative speed and frame of reference. The faster the speed of the observer, with respect to space-time, the shorter the measuring rod and the slower the time interval of the determining clock. Consequently, the speed of proper time for any observer is found to be

$$v_\tau^2 = \frac{v_t^2 - v_s^2}{c^2} \qquad (5.1)$$

The experimental measurement of the universal constancy of light is proof that the principle of the special relativity of space-time is valid and factual. Without the relativistic effects underlay by this principle, it would not be possible for the speed of light to have universal constancy. (Rucker, 1977)

The second principle states that the speed of space-time squared is equal to the squared speed of time minus the squared speed of space relative to any observer. In the absence of mass and energy, the speeds of space and time are consistent and determinable. This principle infers a very important quality of space-time, the reciprocal relationship of space and time as conjugates in space-time. This reciprocal relationship is reflected in the simultaneous and contradictory way that space contracts and the amplitude of time dilates. This effective relationship keeps the speed of space-time on even keel. Thus, the speed of light remains constant.

Forthrightly, we acknowledge the speed of space-time to be

$$c^2 v_\tau = v_t^2 - v_s^2 \qquad (5.2)$$

Hence, this last assertion about the constant speed of light from the second principle leads to the grist of the third principle where in the absence of mass and energy in homogeneous and isotropic space-time, space-time expands in all directions at an equal and uniform relative speed. The speed of space-time remains uniform.

The duality of space and time is expressed mathematically by the fact that the behavior of the spatiotemporal wave does not depend on the independent position of "x" and of time "t", but rather on the combination of position and time, $x - vt$. The waves of space-time between two adjacent coordinate points would destructively interfere at any location along the medium where the two interfering space-time waves have a displacement in the opposite direction.
The waves of space-time between two adjacent coordinate points would constructively interfere at any location along the medium where the two interfering space-time waves have a displacement in the same direction. *By the principle of the superposition of space-time, when two spatiotemporal waves interfere, the resulting*

displacement of the medium at any location is the algebraic sum of the displacements of the individual waves at the same location.

Thus, let us denote the magnitude of the relativistic space-time interval between two events for an observer in uniform relative motion to be

$$(s')^2 = (\Delta l')^2 + (v_s \Delta t')^2 = \frac{\Delta l^2 \left(c^2 - v^2\right)}{c^2} + \frac{c^2 v_s^{\,2} \Delta t^2}{c^2 - v^2} \tag{5.3}$$

$$where \;\; \Delta l^2 = \Delta x^2 + \Delta y^2 + \Delta z^2 \text{ and } \Delta t^2 = \Delta t_x^{\,2} + \Delta t_y^{\,2} + \Delta t_z^{\,2} \tag{5.4}$$

Moreover, for any moving or stationary observer, let us state the space-time interval between two events as

$$s^2 = \Delta l^2 + v_s^{\,2} \Delta t^2 = (s')^2 = (\Delta l')^2 + (v_s \Delta t')^2 \tag{5.5}$$

The fourth principle declares that the laws of physics in any frame of space-time observed by a hypothetical observer traveling at relativistic speed are the same laws of physics as those observed by a stationary hypothetical observer. This principle is the universal principle of conservation of space-time. This fundamentally important principle acknowledges the laws of physics to exist with any observer regardless of the observer's relative uniform motion or frame of reference wherever the observer wishes to go in the known universe. Thus far, wherever scientists have experimented at, or observed in the known universe, the laws of physics have been found to be the same.

The fifth principle states succinctly that light can only travel as fast as time travels. Thus, *the speed of time is a limit to the speed of light, light is not able to travel faster than its spatiotemporal medium in any direction.* If light travels at the speed of time, then the speed of time is *"c"*, and the speed of light is a valid indication or measurement of the speed of time in isotropic and homogenous space-time.

Astrophysicists have recently declared that with the help of the latest most powerful telescopes they have been able to peer into the farthest objects of the observable universe estimated to be about 13 billion light years away from earth. At the present time, it is believed that the universe at large is expanding, the spaces within which everything in the universe resides, the spaces between objects, are stretching, causing galaxies to lie farther apart than they did previously or initially. Powerful telescopes have not been able to peer beyond about 13 billion light years where no light sources are detected, but the darkness of space-time.

These recent astronomical observations validate the fifth principle that "the speed of light cannot exceed the speed of time in any direction", but most importantly the astrophysicists' observations confirm that space-time expands. We are able to visually observe from earth through powerful telescopes near as well as distant galaxies. It is a factual indication of the speed of space-time with respect to the observable and verifiable speed of light in the visible universe.

§ 6. On the fundamentals of the metric of space-time

The Incas regarded space and time as a single concept long ago. The word *Pacha* in the Quechua language, still spoken in South America, means world, universe, space, time, date, or place. The peoples of the Andes have kept this understanding until the present day. In expanding space-time, distance is a dynamical quantity that changes over time. The metric captures all the geometric and causal structure of space-time, being used to define notions such as distance, volume, curvature, angle, future and past. Explicitly, the metric has been expressed in a symmetric bilinear form on each tangent space of a space-time differentiable manifold M which varies in a smooth, or differentiable, manner from point to point.

Given two tangent vectors, the metric has also been expressed as a generalization of the dot product in ordinary Euclidean space. This analogy is not exact, however. Unlike Euclidean space where the dot product is positive definite, the metric gives each tangent space the structure of Minkowski space.

6.1. The Riemann Curvature Tensor

The eminent physicist Georg Friedrich Bernhard Riemann discussed the possibilities of curvature of the universe and suggested that the geometry of space may be related to physical forces. Riemann's doctoral lecture, published in 1866 as 'On the hypotheses which lie at the foundations of geometry', became the cornerstone of differential geometry; it extended the ideas of the physicist Carl Friedrich Gauss from two-dimensional surfaces to higher dimensional surfaces. The ideas of Riemann, and Hermann Minkowski, are the foundation for curved four-dimensional space-time in the General Theory of Relativity. The pseudo-Riemannian and Lorentzian character of spatiotemporal manifolds is the framework of the General Theory of Relativity.

The Riemann Curvature tensor is composed of a Scalar Part, Semi-traceless Part, and a Trace-free or Traceless Part (Weyl Tensor).

Therefore, we have

$$R_{abcd} = S_{abcd} + E_{abcd} + C_{abcd} \tag{6.1}$$

The Scalar Part S_{abcd} originates from the Ricci tensor scalar *"R"*, and it represents information on how the ratio of a spherical volume *Vol (M)* of curvature, tangent at a point on a Riemannian *n*-manifold, to the same spherical volume, tangent at a point on a Euclidean manifold $Vol\ (R^n)$, changes through a geodesic, and assigns a real number with respect to the metrics and the small radius of the spherical volumes. The metric provides a real number associated to curvature at any point of a Riemannian *n*-manifold. The scalar curvature is the average of the Ricci tensor and the Ricci tensor scalar *"R"* is independent of the coordinate system. Riemannian manifolds have the property that they are locally flat.

$$S_{abcd} = \frac{R}{n(n-1)} H_{abcd} = \frac{R}{n(n-1)} \left(g_{ac} g_{db} - g_{ad} g_{cb} \right) \tag{6.2}$$

The Semi-traceless Part has the Ricci tensor that represents the local curvature originated by the presence of matter and energy, tidal waves, frame dragging, and local gravitational waves in a region of space-time. For a celestial body, the curvature produced by the local matter and energy is counteracted by the cosmological curvature of matter and energy everywhere else in the universe. The Ricci tensor describes how a volume element changes as it moves through the curvature of space-time.

$$E_{abcd} = \frac{1}{n-2}\left(g_{ac}S_{bd} - g_{ad}S_{bc} + g_{bd}S_{ac} - g_{bc}S_{ad}\right) \quad (6.3)$$

Where for any $S_{\chi\chi}$ term, we find

$$S_{\alpha\beta} = R_{\alpha\beta} - \frac{1}{n}Rg_{\alpha\beta} \quad (6.4)$$

The Trace-free Part is the Weyl tensor that includes curvature originated by gravitational waves. The gravitational waves are also present in regions of space-time without matter, tidal waves, or non-gravitational fields. Cosmological gravitational waves, a cosmological scalar part, and a cosmological semi-traceless part are included in the Weyl tensor. The cosmological curvature represented by the Weyl tensor counteracts the local curvature included in the local Riemann curvature tensor, the local semi-traceless part (Ricci tensor) and the local scalar part. The cosmological curvature of space-time counteracts the local curvature of space-time for a celestial body of mass.

$$C_{abcd} = \bar{R}_{abcd} + \frac{1}{n-2}\left(\bar{R}_{ad}\bar{g}_{bc} - \bar{R}_{ac}\bar{g}_{bd} + \bar{R}_{bc}\bar{g}_{ad} - \bar{R}_{bd}\bar{g}_{ac}\right) + \frac{1}{(n-1)(n-2)}\bar{R}\left(\bar{g}_{ac}\bar{g}_{bd} - \bar{g}_{ad}\bar{g}_{bc}\right) \quad (6.5)$$

Let us express the Riemannian curvature tensor equation in the following way:

$$R_{abcd} - S_{abcd} - E_{abcd} = C_{abcd} \quad (6.6)$$

The left side of the above equation is the local curvature and the right side is the cosmological curvature. The stress-energy-momentum tensor produces the local curvature for a celestial body of mass.

$$R_{abcd} - S_{abcd} - E_{abcd} = kT_{abcd} \quad (6.7)$$

Let us consider the vacuum solution case for the above equation as the stress-energy-momentum tensor vanishes, $T_{abcd} \to 0$.

$$R_{abcd} - S_{abcd} - E_{abcd} = 0 \quad (6.8)$$

$$R_{abcd} - \frac{R}{n(n-1)}\left(g_{ac}g_{db} - g_{ad}g_{cb}\right) - \quad (6.9)$$

$$\frac{1}{n-2}\left(g_{ac}S_{bd} - g_{ad}S_{bc} + g_{bd}S_{ac} - g_{bc}S_{ad}\right) = 0$$

Multiplying by $n(n-1)(n-2)$ we find

$$n(n-1)(n-2)R_{abcd} - (n-2)R\left(g_{ac}g_{db} - g_{ad}g_{cb}\right) - \quad (6.10)$$

$$n(n-1)\left(g_{ac}S_{bd} - g_{ad}S_{bc} + g_{bd}S_{ac} - g_{bc}S_{ad}\right) = 0$$

Substituting for each $S_{\chi\chi}$ term,

$$n(n-1)(n-2)R_{abcd} - (n-2)R\left(g_{ac}g_{db} - g_{ad}g_{cb}\right) - \quad (6.11)$$

$$n(n-1)\left[g_{ac}\left(R_{bd} - \frac{1}{n}Rg_{bd}\right) - g_{ad}\left(R_{bc} - \frac{1}{n}Rg_{bc}\right) + g_{bd}\left(R_{ac} - \frac{1}{n}Rg_{ac}\right) - g_{bc}\left(R_{ad} - \frac{1}{n}Rg_{ad}\right)\right] = 0$$

Simplifying the last term of the previous equation,

$$n(n-1)(n-2)R_{abcd} - (n-2)R\left(g_{ac}g_{db} - g_{ad}g_{cb}\right) - \quad (6.12)$$

$$n(n-1)\left[g_{ac}R_{bd} - \frac{1}{n}Rg_{ac}g_{bd} - g_{ad}R_{bc} + \frac{1}{n}Rg_{ad}g_{bc} + g_{bd}R_{ac} - \frac{1}{n}Rg_{bd}g_{ac} - g_{bc}R_{ad} + \frac{1}{n}Rg_{bc}g_{ad}\right] = 0$$

$$n(n-1)(n-2)R_{abcd} - (n-2)R\left(g_{ac}g_{db} - g_{ad}g_{cb}\right) - \quad (6.13)$$

$$n(n-1)\left[g_{ac}R_{bd} - \frac{2}{n}Rg_{ac}g_{bd} - g_{ad}R_{bc} + \frac{2}{n}Rg_{ad}g_{bc} + g_{bd}R_{ac} - g_{bc}R_{ad}\right] = 0$$

Factoring terms to obtain a simpler expression,

$$n(n-1)(n-2)R_{abcd} + \{-(n-2)+2(n-1)\}Rg_{ac}g_{bd} + \quad (6.14)$$

$$\{(n-2)-2(n-1)\}Rg_{ad}g_{bc} - n(n-1)\{g_{ac}R_{bd} - g_{ad}R_{bc} + g_{bd}R_{ac} - g_{bc}R_{ad}\} = 0$$

By the symmetry properties of the Riemann tensor, it is possible to contract indices using the first and third symmetry relations to produce a new second-order tensor that is symmetric like the Ricci tensor. Contracting any pair of indices produces a symmetric second-order tensor.

Applying g^{ac} to contract indices,

$$n(n-1)(n-2)g^{ac}R_{abcd} + \{-(n-2)+2(n-1)\}Rg^{ac}g_{ac}g_{bd} + \quad (6.15)$$

$$\{(n-2)-2(n-1)\}Rg^{ac}g_{ad}g_{bc} - n(n-1)\{g^{ac}g_{ac}R_{bd} - g^{ac}g_{ad}R_{bc} + g^{ac}g_{bd}R_{ac} - g^{ac}g_{bc}R_{ad}\} = 0$$

Contracting indices and simplifying coefficients,

$$n(n-1)(n-2)R_{bd} + \{-(n-2)+2(n-1)\}Rg_{bd} + \quad (6.16)$$

$$\{(n-2)-2(n-1)\}Rg_{bd} - n(n-1)\{R_{bd} - R_{bd} + Rg_{bd} - R_{bd}\} = 0$$

Subtracting like terms we find

$$n(n-1)(n-2)R_{bd} - n(n-1)(-R_{bd} + Rg_{bd}) = 0 \quad (6.17)$$

$$(n-2)R_{bd} + R_{bd} - Rg_{bd} = 0 \quad (6.18)$$

$$(n-1)R_{bd} - Rg_{bd} = 0 \quad (6.19)$$

or in a more recognizable form, for $n > 1$, we have

$$R_{\mu\nu} = \frac{1}{(n-1)} g_{\mu\nu} R \tag{6.20}$$

$$R_{\mu\nu} - \frac{1}{(n-1)} g_{\mu\nu} R = 0 \tag{6.21}$$

During the above contraction of the Riemann curvature tensor equation some information was lost. The scalar part cancels with equal and opposite terms from the semi-traceless part (Ricci tensor). The semi-traceless part already contains information about local scalar curvature, local gravitational waves and related gravitational field potential, that counteract or cancel information in both the scalar part and the Weyl tensor. The remaining curvature of the symmetric second-order tensor, or Ricci tensor, is equal to zero to represent a vacuum solution for the resulting second-order tensor. The right side of the Riemann tensor equation is the cosmological curvature. The cosmological stress-energy-momentum tensor may produce cosmological curvature at a region of space-time with or without local curvature.

$$C_{abcd} = -k\Lambda_{abcd} \tag{6.22}$$

Let us consider the vacuum solution case for the above equation as the cosmological stress-energy-momentum tensor vanishes, $\Lambda_{abcd} \rightarrow 0$.

$$C_{abcd} = 0 \tag{6.23}$$

$$\bar{R}_{abcd} + \frac{1}{n-2}\left(\bar{R}_{ad}\bar{g}_{bc} - \bar{R}_{ac}\bar{g}_{bd} + \bar{R}_{bc}\bar{g}_{ad} - \bar{R}_{bd}\bar{g}_{ac}\right) + \frac{1}{(n-1)(n-2)}\bar{R}\left(\bar{g}_{ac}\bar{g}_{bd} - \bar{g}_{ad}\bar{g}_{bc}\right) = 0 \tag{6.24}$$

Multiplying by *(n−1)(n−2)* we find

$$(n-1)(n-2)\bar{R}_{abcd} + (n-1)\left(\bar{R}_{ad}\bar{g}_{bc} - \bar{R}_{ac}\bar{g}_{bd} + \bar{R}_{bc}\bar{g}_{ad} - \bar{R}_{bd}\bar{g}_{ac}\right) + \bar{R}\left(\bar{g}_{ac}\bar{g}_{bd} - \bar{g}_{ad}\bar{g}_{bc}\right) = 0 \tag{6.25}$$

Applying $\overline{g}^{ac}$ to contract indices,

$$(n-1)(n-2)\overline{g}^{ac}\overline{R}_{abcd}+(n-1)\left(\overline{g}^{ac}\overline{g}_{bc}\overline{R}_{ad}-\overline{g}^{ac}\overline{R}_{ac}\overline{g}_{bd}+\overline{g}^{ac}\overline{g}_{ad}\overline{R}_{bc}-\overline{R}_{bd}\overline{g}^{ac}\overline{g}_{ac}\right)+ \quad (6.26)$$

$$\overline{R}\left(\overline{g}^{ac}\overline{g}_{ac}\overline{g}_{bd}-\overline{g}^{ac}\overline{g}_{ad}\overline{g}_{bc}\right)=0$$

$$(n-1)(n-2)\overline{R}_{bd}+(n-1)\left(\overline{R}_{bd}-\overline{R}\overline{g}_{bd}+\overline{R}_{bd}-\overline{R}_{bd}\right)+\overline{R}\left(\overline{g}_{bd}-\overline{g}_{bd}\right)=0 \quad (6.27)$$

$$(n-1)(n-2)\overline{R}_{bd}+(n-1)\left(\overline{R}_{bd}-\overline{R}\overline{g}_{bd}\right)=0 \quad (6.28)$$

$$(n-2)\overline{R}_{bd}+\left(\overline{R}_{bd}-\overline{R}\overline{g}_{bd}\right)=0 \quad (6.29)$$

$$n\overline{R}_{bd}-2\overline{R}_{bd}+\overline{R}_{bd}-\overline{R}\overline{g}_{bd}=0 \quad (6.30)$$

$$(n-1)\overline{R}_{bd}-\overline{R}\overline{g}_{bd}=0 \quad (6.31)$$

$$\overline{R}_{bd}-\frac{1}{(n-1)}\overline{R}\overline{g}_{bd}=0 \quad (6.32)$$

The above result agrees with our previous result for the local Riemann curvature tensor equation and validates the Riemann curvature tensor equation.

The Riemann curvature at a point *"P"* in space-time is the difference between the two Riemann curvatures.

$$R_{abcd}-S_{abcd}-E_{abcd}\leq or \geq C_{abcd} \quad (6.33)$$

If the local Riemann curvature and the cosmological Riemann curvature are equal, the resulting curvature is Riemann-flat. If the local Riemann curvature is less, the resulting curvature is cosmologically curved. If the cosmological Riemann curvature is less, the resulting curvature is Ricci-curved locally. By the continuity equation, to conserve energy and momentum covariantly, we take the covariant derivative of the Einstein tensor in *"n"* dimensions, or of the stress-energy-momentum tensor, to have

$$D_\mu\left(G^{\mu\nu}\right)=0 \qquad \text{and} \qquad D_\mu\left(T^{\mu\nu}\right)=0 \tag{6.34}$$

$$R^{\mu\nu}-\frac{1}{(n-1)}g^{\mu\nu}R=0 \qquad R\neq 0 \ \textit{and} \ \therefore \ D_\mu R\neq 0 \tag{6.35}$$

$$D_\mu\left(R^{\mu\nu}-\frac{1}{(n-1)}g^{\mu\nu}R\right)=D_\mu R^{\mu\nu}-\frac{1}{(n-1)}g^{\mu\nu}\left(D_\mu R\right) \tag{6.36}$$

Calculating with Christoffel symbols we can obtain

$$D_\mu R^{\mu\nu}=\frac{1}{(n-1)}g^{\mu\nu}D_\mu R \tag{6.37}$$

$$D_\mu\left(R^{\mu\nu}-\frac{1}{(n-1)}g^{\mu\nu}R\right)=\frac{1}{(n-1)}g^{\mu\nu}D_\mu R-\frac{1}{(n-1)}g^{\mu\nu}\left(D_\mu R\right)=0 \tag{6.38}$$

$$D_\mu\left(R^{\mu\nu}-\frac{1}{(n-1)}g^{\mu\nu}R\right)=0 \tag{6.39}$$

Thus, the energy and momentum are covariantly conserved,

$$D_\mu\left(R^{\mu\nu}-\frac{1}{(n-1)}g^{\mu\nu}R\right)=0 \tag{6.40}$$

$$D_\mu\left(G^{\mu\nu}\right)=0 \tag{6.41}$$

For $n=3$ spatial dimensions, the Einstein field equations are

$$R_{\mu\nu}-\frac{1}{2}g_{\mu\nu}R=\frac{8\pi G}{c^4}\left(T_{\mu\nu}-\Lambda_{\mu\nu}\right) \tag{6.42}$$

For $n=4$, three spatial dimensions and one temporal dimension, we have

$$R_{\mu\nu} - \frac{1}{3} g_{\mu\nu} R = \frac{8\pi G}{c^4}\left(T_{\mu\nu} - \Lambda_{\mu\nu}\right) \tag{6.43}$$

For $n = 6$, three spatial dimensions and three temporal dimensions, we have

$$R_{\mu\nu} - \frac{1}{5} g_{\mu\nu} R = \frac{8\pi G}{c^4}\left(T_{\mu\nu} - \Lambda_{\mu\nu}\right) \tag{6.44}$$

Since the original Einstein Field Equations are four-dimensional, and they can predict Newtonian gravity for weak-field, time-independent, slow-moving bodies of mass equations, let us reconsider the semi-traceless Part (the Ricci tensor) that represents the local curvature originated by the presence of matter and energy, tidal waves, frame dragging, and local gravitational waves in a region of space-time.

$$E_{abcd} = \frac{1}{n-2}\left(g_{ac} S_{bd} - g_{ad} S_{bc} + g_{bd} S_{ac} - g_{bc} S_{ad}\right) \tag{6.45}$$

$$E_{abcd} = \frac{1}{(n-2)}\left[g_{ac} E_{bd} - \frac{2}{n} E g_{ac} g_{bd} - g_{ad} E_{bc} + \frac{2}{n} E g_{ad} g_{bc} + g_{bd} E_{ac} - g_{bc} E_{ad}\right] \tag{6.46}$$

Applying g^{ac} to contract indices,

$$E_{bd} = \frac{1}{(n-2)}\left\{g^{ac} g_{ac} E_{bd} - \frac{2}{n} E g^{ac} g_{ac} g_{bd} - g^{ac} g_{ad} E_{bc} + \frac{2}{n} E g^{ac} g_{ad} g_{bc} + g^{ac} g_{bd} E_{ac} - g^{ac} g_{bc} E_{ad}\right\} \tag{6.47}$$

$$E_{bd} = \frac{1}{(n-2)}\left(E_{bd} - E_{bd} + g_{bd} E - E_{bd}\right) = \frac{1}{(n-2)}\left(-E_{bd} + g_{bd} E\right) \tag{6.48}$$

$$E_{bd} + \frac{1}{(n-2)} E_{bd} = \frac{1}{(n-2)} g_{bd} E \tag{6.49}$$

$$\frac{(n-1)}{(n-2)} E_{bd} = \frac{1}{(n-2)} g_{bd} E \tag{6.50}$$

$$E_{bd} = \frac{1}{(n-1)} g_{bd} E \tag{6.51}$$

$$E_{bd} - \frac{1}{(n-1)} g_{bd} E = 0 \tag{6.52}$$

or in a more recognizable form, for $n > 1$, we have

$$R_{\mu\nu} - \frac{1}{(n-1)} g_{\mu\nu} R = 0 \tag{6.53}$$

If $n = 4$, we obtain the same left-side of the four-dimensional field equation with three spatial dimensions and one resultant temporal dimension.

$$R_{\mu\nu} - \frac{1}{3} g_{\mu\nu} R = 0 \tag{6.54}$$

Thus, the four-dimensional EFE for only local curvature becomes

$$R_{\mu\nu} - \frac{1}{3} g_{\mu\nu} R = kT_{\mu\nu} \tag{6.55}$$

By contracting the semi-traceless Part (the Ricci tensor) for only the local curvature, we have verified that the Ricci-side of the four-dimensional Einstein field equations is structured to include a fraction of ⅓ to represent the curvature of the three spatial dimensions and one resultant temporal dimension of space-time. If the field equations include a fraction of ½ on the Ricci-side, the equations are structured to represent curvature of temporal manifolds layered on spatial manifolds for three space-time dimensions, so the equations are not structured for space and time to be independent dimensions.

Let us reconsider the four-dimensional field equation,

$$G_{\mu\nu} = \frac{8\pi G}{c^4}\left(T_{\mu\nu} - \Lambda_{\mu\nu}\right) \tag{6.56}$$

$$R_{\mu\nu} - \frac{1}{3} g_{\mu\nu} R = \frac{8\pi G}{c^4}\left(T_{\mu\nu} - \Lambda_{\mu\nu}\right) \tag{6.57}$$

As the stress-energy-momentum tensor vanishes, a region of space-time may have gravitational radiation while being singularity-free, Ricci-flat, but not Riemann-flat. The Ricci tensor suffices to describe curvature, but space is inherently Riemannian.

Substituting for the vacuum solution Ricci tensor when the stress-energy-momentum tensor is zero $T_{\mu\nu} = 0$ in the region of space-time under consideration, with nonzero cosmological matter and energy and cosmological gravitational waves, we have

$$R_{\mu\nu} - \frac{1}{3} g_{\mu\nu} R = -\frac{8\pi G}{c^4}\left(\Lambda_{\mu\nu}\right) \tag{6.58}$$

The metric tensor, $g_{\mu\nu}$, may be obtained from the curvature of cosmological matter and energy, and cosmological gravitational waves from the gravitational field present through a vacuum region of space-time whose limit is the Newtonian potential.

On a nearly-flat local six-dimensional Riemann manifold we have

$$g^{\mu\nu} g_{\mu\nu} = g = 6 \tag{6.59}$$

Using the four-dimensional equation with folded three-dimensional time, we obtain

$$-R = -\frac{8\pi G}{c^4}(\Lambda) \tag{6.60}$$

$$R = \frac{6}{a^2 c^2}\left(a\ddot{a} + \dot{a}^2 + kc^2\right) \tag{6.61}$$

$$R = \frac{6}{a^2 c^2}\left(a\ddot{a} + \dot{a}^2 + kc^2\right) = \frac{8\pi G}{c^4}(\Lambda) \tag{6.62}$$

$$\frac{\ddot{a}}{ac^2} + \frac{\dot{a}^2}{a^2 c^2} + \frac{k}{a^2} = \frac{8\pi G}{6c^4}(\Lambda) = \frac{8\pi G}{6c^4}(-3\rho + 3p) \tag{6.63}$$

The energy density ρ and pressure p above are the energy density and pressure of the cosmological matter and energy. Simplifying the above equation,we obtain

$$\frac{\ddot{a}}{a}+\frac{\dot{a}^2}{a^2}+\frac{kc^2}{a^2}=\frac{4\pi G}{c^2}(p-\rho)=\mathscr{H}(p-\rho) \qquad (6.64)$$

This equation forms the basis for the curvature in a region of space-time of our Universe without sources of mass or non-gravitational fields, and it contains information about the tenses of time. These are weak-field (far-field) gravitational equations while the Einstein Field Equations with the stress-energy-momentum tensor may be weak-field (near-field).

Where is the missing baryonic matter of the observable universe?

The cosmological constant has been added to the standard FLRW metric of cosmology to yield the Lambda-CDM model, also known as the Standard Model of Cosmology, because the model agrees precisely with observations. Current measurements indicate that the present universe has 68.3% of an unfound type of energy, 26.8% of an unfound type of matter, and 4.9% of ordinary matter, and neutrinos and photons in a very small quantity. The Standard Model of Cosmology assumes that the General Theory of Relativity is the correct theory of gravity on cosmological scales.

The combination of matter and energy in the volume of the universe may be expressed as energy density or pressure. Most inflationary models predict that the total matter and energy of the universe should be very close to 100% of the critical density. As measured from the spectrum of the Cosmic Microwave Background, the unfound type of matter and ordinary matter accounts for only about 30% of the critical density, while it is inferred that a type of unfound energy accounts for the remaining 70% of the critical density. The observed amount of baryonic matter does not match theoretical predictions. It was noted in previous research that the observable matter of the universe (ordinary or baryonic matter) is estimated to be about 10^{53} *Kg*, even though, if we compare the universal gravitational constant of the universe to that measured on earth which has been validated numerous times, the ordinary matter of the observable universe would have to be approximately 1.5 x 10^{53} Kg, or approximately fifty per cent more than our previous estimate, ceteris paribus, if the universal gravitational constant were the same at any point of the observable universe.

Let us derive the four and six-dimensional EFEs from the multidimensional EFE,

$$R_{\mu\nu} - \frac{1}{(n-1)} g_{\mu\nu} R = \frac{8\pi G}{c^4}\left(T_{\mu\nu} - \Lambda_{\mu\nu}\right) \tag{6.65}$$

Where "n" is the number of spatiotemporal dimensions.

If $n = 4$ dimensions, we have

$$R_{\mu\nu} - \frac{1}{3} g_{\mu\nu} R = \frac{8\pi G}{c^4}\left(T_{\mu\nu} - \Lambda_{\mu\nu}\right) \tag{6.66}$$

From the four-dimensional EFE we obtain

$$-6\left(\frac{\ddot{a}}{ac^2} + \frac{\dot{a}^2}{a^2c^2} + \frac{k}{a^2}\right) = \frac{8\pi G}{c^4}\left(-3\rho + 3p\right) \tag{6.67}$$

$$-\left(\frac{\ddot{a}}{ac^2} + \frac{\dot{a}^2}{a^2c^2} + \frac{k}{a^2}\right) = \frac{4\pi G}{c^4}\left(-\rho + p\right) \tag{6.68}$$

Since the total ordinary matter of the observable universe has been validated according to the stress-energy-momentum tensor of the four-dimensional EFE, what if the predicted value of the ordinary matter is less than the actual value?

If we substitute Γ (Gamma) for the curvature in the four-dimensional EFE and increment the density of matter and pressure by 50% according to the current value of the universal gravitational constant as measured on earth, we have the following revised four-dimensional EFE. The Cosmic Microwave Background tells us that should be about 50% more ordinary matter.

$$-\Gamma = \frac{4\pi G}{c^4}(-1.5\rho + 1.5p) = \frac{6\pi G}{c^4}(-\rho + p) \tag{6.69}$$

If $n = 6$ dimensions, we have

$$R_{\mu\nu} - \frac{1}{5}g_{\mu\nu}R = \frac{8\pi G}{c^4}(T_{\mu\nu} - \Lambda_{\mu\nu}) \tag{6.70}$$

From the six-dimensional EFE we obtain

$$-6\left(\frac{\ddot{a}}{ac^2} + \frac{\dot{a}^2}{a^2c^2} + \frac{k}{a^2}\right) = \frac{40\pi G}{c^4}(-3\rho + 3p) \tag{6.71}$$

$$-\left(\frac{\ddot{a}}{ac^2} + \frac{\dot{a}^2}{a^2c^2} + \frac{k}{a^2}\right) = \frac{20\pi G}{c^4}(-\rho + p) \tag{6.72}$$

Similarly, if we substitute Γ (Gamma) for the curvature, and increment the density of matter and pressure by 50% according to the current value of the universal gravitational constant as measured on earth, we have

$$-\Gamma = \frac{20\pi G}{c^4}(-1.5\rho + 1.5p) = \frac{30\pi G}{c^4}(-\rho + p) \tag{6.73}$$

Hence, the Einstein constant in the six-dimensional EFE is about 5 times greater than the Einstein constant in the four-dimensional EFE. Furthermore, $30\pi - 4\pi = 26\pi$, so it is possible to account for 4% of ordinary matter, or for 26% of some other unfound type of matter, or to ask for the whereabouts of the remaining 70% of some other unfound type of energy. It is a great credit to research to postulate when there is something missing. Nevertheless, nature always decides the veracity of every leap of ironclad logic that may come from any theorist.

Let us divide the right-side of the revised four-dimensional EFE by the right-side of the six-dimensional EFE,

$$\frac{\dfrac{6\pi G}{c^4}}{\dfrac{20\pi G}{c^4}} = \frac{6}{20} = 0.30 = 30\% \tag{6.74}$$

Hence, the ordinary matter represented by the four-dimensional EFE is only 30% of the matter represented by the six-dimensional EFE, according to the universal gravitational constant estimated for the universe. So, it is understandable to ask, where is the other 70% of the energy in the observable universe? The other 70% of the energy is in the remainder of the 20π which is 14π since $14\pi/20\pi = 70\%$; the proof is in the pudding. Moreover, these results support the hypothesis that all the ordinary matter and energy in the universe, the critical density, is already there to be observed or accounted for. If we assume that time is three-dimensional, the right-side of the current four-dimensional EFE represents three spatial dimensions and one temporal dimension, the curvature terms would be four out of a possible six, 4/6, which may be reduced to 2/3 of the total six-dimensional curvature.

Reconsidering the four-dimensional EFE, we obtain

$$-\frac{4\cdot 6}{6}\left(\frac{\ddot{a}}{ac^2}+\frac{\dot{a}^2}{a^2c^2}+\frac{k}{a^2}\right)=\frac{8\pi G}{c^4}(-3\rho+3p) \tag{6.75}$$

$$-\left(\frac{\ddot{a}}{ac^2}+\frac{\dot{a}^2}{a^2c^2}+\frac{k}{a^2}\right)=\frac{8\pi G}{4c^4}(-3\rho+3p)=\frac{2\pi G}{c^4}(-3\rho+3p)=\frac{6\pi G}{c^4}(-\rho+p) \tag{6.76}$$

It is interesting to note that astronomers measure the mass of a cluster by measuring how background galaxies are distorted by a foreground cluster through gravitational lensing. The mass in a cluster is 5 times greater that the inferred mass in observable stars, gas, and dust.

There are other methods that infer the unobserved mass outweighs the visible by approximately 5 to 1. For example, the scatter in the radial velocities of the galaxies within clusters, and from the x-rays emitted by hot gas in the clusters. The gas temperature and density can be estimated to yield the pressure, since the mass profile of the cluster is determined by the balance of pressure and gravity. These results support the six-dimensionality of space-time and examine the total ordinary matter in the observable universe as it exists as space, time, and energy. The field equations of the General Theory of Relativity describe the spatiotemporal curvature and the distribution of matter throughout the universe.

The effect of the interference of spatiotemporal waves around matter, or mass, results in spatiotemporal curvature that produces a measurable gravitational field. The set of field equations define the gravitational relationship between matter, energy, and space-time. The gravitational relationship between matter, energy, and space-time, demonstrates that the strain on space-time is proportional to the stress of mass and energy affecting that space-time.

6.2. Intrinsic Curvature and Torsion

When two or more dimensions are twisted simultaneously into another dimension, the surface area defined by these two or more dimensions changes, and the new surface area no longer maps into the original surface area of flat space. The new surface area is referred to as intrinsic curvature, and it does not transform as flat space.

External and internal torsion are closely related to external or internal intrinsic curvature. Let us consider the parallel transport of a first order tensor (a vector) around the exterior curved surface of a sphere as shown in figure 4. The convex surface has positive curvature or external intrinsic curvature. Let us start the parallel transport of our first order tensor from the left corner towards the right in a counterclockwise direction around the external perimeter, to finish at the starting point, to measure the curvature bounded by the parallel transport trajectory, and the re-orientation of the tensor, given by an angle of torsion.

Curvature is manifested in the re-orientation of the first order tensor, not in changes to the tensor's magnitude. Torsion is manifested in the angular extent of the re-orientation. The angle of torsion is proportional to the surface area inside the loop of the trajectory. Thus, torsion and curvature are proportional and complementary during parallel transport. Therefore, as a particle approaches an external intrinsic curvature, such as the curvature of a celestial body, the particle would experience both the effects of external intrinsic curvature and torsion.

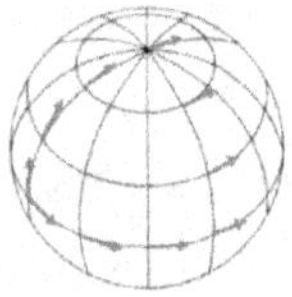

Figure 4.

For example, a vector parallel transported around the perimeter of a curved area, that is one eighth of the surface of a sphere with radius R, has an angle of rotation of 90 degrees as it circumvents the area. The curvature of the area is given by

$$Curvature = \frac{Angle\ of\ Rotation}{Area\ Circumvented} = \frac{\frac{\pi}{2}}{\frac{1}{8}\left(4\pi R^2\right)} = \frac{1}{R^2} \quad (6.77)$$

$$Curvature \equiv \frac{\partial \theta}{\partial S} \equiv \frac{1}{d^2} \quad (6.78)$$

Hence, the curvature of a Riemannian manifold may be obtained from the parallel transport of a vector along a curve. The curvature depends on the path, the curve, and the angle of rotation of the initial vector, where d^2 is the area of the dimensions of curvature, and θ is the radians of rotation of the parallel transported vector. (Ciufolini, 1995)

Let us consider the concave surface with negative curvature or internal intrinsic curvature of the same manifold from the perspective of the center of the sphere facing the interior surface. We start the parallel transport at the same starting point with the same first order tensor, following the same trajectory, and ending at the starting point. We can observe that both tensors end up at the same angle of torsion. However, the external tensor has rotated counterclockwise while the internal tensor has rotated clockwise. Thus, the external and internal first order tensors are equal in terms of torsion but opposite in sign. As a particle approaches an internal intrinsic curvature, the particle would experience both the effects of internal intrinsic curvature and torsion. However, the internal torsion and the external torsion are equal in angular extent and opposite in sign, even though both exist on the same manifold. The external intrinsic curvature and external torsion are supplementary to the external Ricci curvature from an external perspective, such as the perspective of an incident external particle or an external body of mass, while the internal intrinsic curvature and internal torsion are supplementary to the internal Ricci curvature from an internal perspective such as the perspective of the internal body of mass or an incident internal particle.

Both curvature and torsion are crucial to the General Theory of Relativity, and both exert effects on the spacetime about a celestial body. Curvature and torsion are components of the spacetime in the gravitational field of a celestial body of mass.

Hence, let us express the Einstein field equations with external and internal intrinsic curvature and torsion.

$$G_{\mu\nu} + \nabla_{\omega}\theta^{\omega}{}_{\mu\nu} = \kappa T_{\mu\nu} + \nabla_{\omega}\theta^{\omega}{}_{\mu\nu} \tag{6.79}$$

Hence, the external torsion and the internal torsion are equal in angular extent but opposite in sign from the perspective of the Einstein tensor and the stress-energy-momentum tensor of the near-field EFEs, such as the EFEs of a celestial body or a local solar system. For the far-field EFEs of distant galaxies there might be remanent Riemannian torsion.

$$\theta^{\omega}{}_{\mu\nu} = \Gamma^{\omega}{}_{\mu\nu} - \Gamma^{\omega}{}_{\nu\mu} \tag{6.80}$$

where "v" and "μ" are not necessarily equal, and gamma, Γ, is a connection of the Riemannian manifold.

The Levi-Civita connection of the given metric on any pseudo-Riemannian manifold, as used in the General Theory of Relativity, or on any Riemannian manifold, preserves the metric tensor in a unique torsion-free metric connection, as stated in the fundamental theorem of Riemannian geometry. This implies that the metric tensor is preserved by parallel transport, while the resultant torsion tensor is zero.

Furthermore, electromagnetic field torsion may be produced by the resultant electromagnetic field from local and cosmological electromagnetic fields. In such a case, a resultant electromagnetic torsion mass-energy density tensor, $\Psi_{\varepsilon\beta}$, may appear on the right side of the EFEs in addition to the stress-energy-momentum tensor, $T_{\mu\nu}$, that produces a resultant torsion tensor, $\nabla_{\omega}\left(L^{\omega}{}_{\varepsilon\beta} - \Lambda^{\omega}{}_{\varepsilon\beta}\right)$, on the left side of the EFEs in addition to the Einstein curvature tensor.

6.3. On the General Theory of Relativity with Torsion

A region of space-time may have *Torsion,* $L^{\omega}{}_{\varepsilon\beta}$, which would imply that vectors rotate through their path during parallel transport under the General Theory of Relativity as well as in String Theory. Vectors are first order (Rank-1) tensors. Rank-2 tensors may be symmetric (e.g. the Ricci curvature tensor) or antisymmetric (e.g. the electromagnetic tensor).

If a tensor alternates sign (+/−) when any two of its indices are interchanged, then that tensor is antisymmetric. Generally, the subset of indices of such an antisymmetric tensor must be all covariant or all contravariant. If the interchange of any pair of indices of a tensor, or any pair of indices on a square matrix of a tensor on either side of the diagonal, makes the tensor alternate sign, then the tensor is totally antisymmetric. A set of basis of a vector, or tensor, may be conceptualized as a set of reference axes. In curved spacetime, a set of basis vectors changes orientation from point to point, but in flat spacetime, a set of basis vectors would be constant from point to point. If there are changes in the scale of the basis, a vector or tensor that exhibits a behavior of changing scale inversely to the changes in the scale of the basis is contravariant, e.g. velocity, acceleration, and jolt. A vector or tensor that exhibits a behavior of changing scale in the same way as the scale of the basis is covariant, e.g. the gradient of a function. Tensors may exhibit covariant and/or contravariant changes of scale on their components, and in the position of their indices. Upper tensor indices are contravariant and lower indices are covariant.

A tensor is a versatile multidimensional generalization of the concept of a vector. Torsion may be described as a tensor that provides an intrinsic representation of how tangent spaces twist about a curve when they are parallel transported. On the other hand, curvature describes how the tangent spaces roll along the curve. Thus, torsion can be treated as an independent tensor field or as part of the geometry, the geometrical approach can provide greater insight into the theory. (Wald, 1984)

Using the six-dimensional field equation for a charged body of mass,

$$G_{\mu\nu} + \overline{G}_{\varepsilon\beta} = \frac{8\pi G}{c^4}\left(T_{\mu\nu} - \Lambda_{\mu\nu} + \Phi_{\varepsilon\beta}\right) \tag{6.81}$$

In a space with torsion, the Ricci tensor need not be symmetric so that an asymmetric torsion tensor can appear on the right side of the Einstein field equation. A Riemann–Cartan space is a metric-affine space with a connection that is metric, $D_\mu g^{\mu\nu} = 0$. In a Riemann–Cartan space the connection is determined by its torsion and the metric tensor. (Cartan, 1922, 1923)

It is worth mentioning that gravitational theories of relativity rest on conservation laws that proceed from the Bianchi identities; hence, if space has torsion, the divergence of the stress-energy-momentum tensor does not have to vanish. Hence, let us express our metric as $g_{\mu\nu}$, and a torsion as $L^{\omega}{}_{\mu\nu}$, so that there exists a unique operator ∇_μ with torsion, $L^{\omega}{}_{\mu\nu}$, that satisfies $\nabla_\mu g_{\nu\omega} = 0$. (Wald, 1984)

The electromagnetic torsion mass-energy density tensor is defined as the difference between the local and the nonlocal electromagnetic torsion mass-energy density tensors. The local torsion mass-energy density tensor emerges from the local electromagnetic field and the nonlocal or cosmological torsion mass-energy density tensor, if present, that comes from the cosmological electromagnetic field. The local torsion mass-energy density tensor, $\Psi(L)_{\varepsilon\beta}$, is related to the local Riemann curvature tensor and the resultant electrogravitic tensor through the mass-energy density of the local electromagnetic field through space-time and is part of the local gravitational field. The cosmological electromagnetic torsion mass-energy density tensor, $\Psi(\Lambda)_{\varepsilon\beta}$, offsets the local electromagnetic torsion mass-energy density tensor.

$$\Psi_{\varepsilon\beta} = \Psi(L)_{\varepsilon\beta} - \Psi(\Lambda)_{\varepsilon\beta} \tag{6.82}$$

In terms of the torsion exerted by the local and cosmological electromagnetic torsion mass-energy density tensors we obtain

$$\nabla_{\omega}\left(L^{\omega}{}_{\varepsilon\beta} - \Lambda^{\omega}{}_{\varepsilon\beta}\right) = \frac{8\pi G}{c^{4}}\Psi_{\varepsilon\beta} \tag{6.83}$$

The resultant electromagnetic torsion mass-energy density tensor, $\Psi_{\varepsilon\beta}$, may either boost or offset the torsion effect on the local Riemann curvature depending on the strength of either the local or cosmological electromagnetic torsion tensor. The local torsion tensor boosts the local torsion effect.

With an electromagnetic torsion mass-energy density tensor, $\Psi_{\varepsilon\beta}$, that is totally antisymmetric we have,

$$\tilde{G}_{\mu\nu\varepsilon\beta} = G_{\mu\nu} + \bar{G}_{\varepsilon\beta} + \nabla_{\omega}\left(L^{\omega}{}_{\varepsilon\beta} - \Lambda^{\omega}{}_{\varepsilon\beta}\right) = \frac{8\pi G}{c^{4}}\left\{T_{\mu\nu} - \Lambda_{\mu\nu} + \Phi_{\varepsilon\beta} + \Psi_{\varepsilon\beta}\right\} = \frac{8\pi G}{c^{4}}\tilde{T}_{\mu\nu\varepsilon\beta} \tag{6.84}$$

$$R_{\nu\mu} + \bar{R}_{\beta\varepsilon} - \frac{1}{(n-1)}\left(g_{\nu\mu}R + g_{\beta\varepsilon}\bar{R}\right) + \nabla_{\omega}\left(L^{\omega}{}_{\varepsilon\beta} - \Lambda^{\omega}{}_{\varepsilon\beta}\right) = \frac{8\pi G}{c^{4}}\left(T_{\mu\nu} - \Lambda_{\mu\nu} + \Phi_{\varepsilon\beta} + \Psi_{\varepsilon\beta}\right) \tag{6.85}$$

If the resultant electromagnetic torsion mass-energy density tensor, $\Psi_{\varepsilon\beta}$, is partially antisymmetric,

$$R_{\nu\mu} + \bar{R}_{\beta\varepsilon} - \frac{1}{(n-1)}\left(g_{\nu\mu}R + g_{\beta\varepsilon}\bar{R}\right) - 3\nabla_{[\varepsilon} L^{\omega}_{\omega\beta]} + L^{\omega}{}_{\omega\lambda} L^{\lambda}{}_{\varepsilon\beta} - 3\nabla_{[\mu} \Lambda^{\omega}_{\omega\nu]} + \Lambda^{\omega}{}_{\omega\lambda} \Lambda^{\lambda}{}_{\mu\nu} \quad (6.86)$$

$$= \frac{8\pi G}{c^4}\left(T_{\mu\nu} - \Lambda_{\mu\nu} + \Phi_{\varepsilon\beta} + \Psi_{\varepsilon\beta}\right)$$

If the stress-energy-momentum tensor and the electrogravitic tensor were to instantly vanish, $T_{\mu\nu} = 0$, and $\Phi_{\varepsilon\beta} = 0$, we would obtain the following cosmological totally-antisymmetric torsion equation,

$$-\left\{\bar{R}_{\nu\mu} - \frac{1}{(n-1)}\bar{g}_{\nu\mu}\bar{R}\right\} + \nabla_{\omega}\left(L^{\omega}{}_{\varepsilon\beta} - \Lambda^{\omega}{}_{\mu\nu}\right) = \frac{8\pi G}{c^4}\left(\Psi_{\varepsilon\beta} - \Lambda_{\mu\nu}\right) \quad (6.87)$$

Torsion may be present in all three parts of the Riemann curvature tensor since every part has a second order Ricci curvature tensor associated with it, and each part may undergo antisymmetrization.

If we apply g^{ac} to contract the Riemann curvature tensor, R_{abcd}, we can obtain a torsion-free vacuum solution,

$$R_{bd} - \frac{1}{(n-1)} g_{bd} R = 0 \quad (6.88)$$

In the absence of torsion in the Riemann curvature tensor, we have the Interchange Symmetry,

$$R_{abcd} = \frac{1}{2}\left\{R_{cdab} - R_{dacb} - R_{acdb}\right\} = R_{cdab} \quad (6.89)$$

The Riemann curvature tensor, R_{abcd}, is no longer symmetric under the exchange of the first pair of indices with the second pair of indices, R_{cdab}, when torsion is present.

If a totally antisymmetric electromagnetic torsion mass-energy density tensor exists in the curvature of the electrogravitic tensor with negligible cosmological curvature or torsion, we can express the torsion equation as

$$\overline{R}_{\beta\varepsilon} - \frac{1}{(n-1)} g_{\beta\varepsilon} \overline{R} + \nabla_{\omega} L^{\omega}{}_{\varepsilon\beta} = \frac{8\pi G}{c^4}\left(\Phi_{\varepsilon\beta} + \Psi_{\varepsilon\beta}\right) \tag{6.90}$$

We can write the Riemann curvature tensor without torsion as

$$R_{abcd} = S_{abcd} + E_{abcd} + C_{abcd} \tag{6.91}$$

And with a totally antisymmetric Riemann curvature torsion tensor as

$$R_{abcd} = S_{cdab} + E_{cdab} + C_{cdab} - \nabla_{[a}\,\theta_{b]cd} + \nabla_{[c}\,\theta_{d]ab} \tag{6.92}$$

$$R_{abcd} + \nabla_{[a}\,\theta_{b]cd} = R_{cdab} + \nabla_{[c}\,\theta_{d]ab} \tag{6.93}$$

The pressure, or energy density, of charged matter, is excluded from the neutral, or uncharged, local or cosmological, energy density.

The energy density, $\rho_{\chi\chi}$, of the time-time components, $\left(Kg/m^3\right)$, of the electromagnetic torsion mass-energy density tensor defined as the value of the Poynting vector divided by c^3, where "c" is the speed of light.

The time-time components become, $c^2\rho_{\chi\chi}$, which is equal to $\vec{S}_{\chi\chi}/c$ in $\left(J/m^3\right)$. The Poynting vector propagates in every sense of direction of three-dimensional time.

$$\rho_{\chi\chi} = \frac{\vec{S}_{\chi\chi}}{c^3} = \frac{1}{\mu_0 c^3}\left(\vec{E}_{\chi\chi} \times \vec{B}_{\chi\chi}\right) \tag{6.94}$$

The pressure of the space-space components, $\left(N/m^2 \text{ or } J/m^3\right)$, is defined as the exterior product of the electric field and the magnetic flux density divided by the product of the speed of light "c" with the permeability of the medium.

The exterior product is taken in every sense of direction of three-dimensional space.

$$p_{\chi\chi} = \frac{1}{\mu_0 c}\left(\vec{E}_{\chi\chi} \wedge \vec{B}_{\chi\chi}\right) \tag{6.95}$$

Both the Poynting vector and the exterior product of $\vec{E}_{\chi\chi}$ and $\vec{B}_{\chi\chi}$ are antisymmetric tensors. For a region of space-time about a charged celestial body of mass, the six-dimensional electromagnetic torsion mass-energy density tensor may be expressed as

$$\Psi_{\varepsilon\beta} = \left|\begin{matrix} \left\{\frac{\vec{S}_{t_x t_x}}{c}\right\} & \left\{\frac{\vec{S}_{t_x t_y}}{c}\right\} & \left\{\frac{\vec{S}_{t_x t_z}}{c}\right\} & \frac{\vec{S}_{t_x x}}{c^2} & \frac{\vec{S}_{t_x y}}{c^2} & \frac{\vec{S}_{t_x z}}{c^2} \\ \left\{\frac{\vec{S}_{t_y t_x}}{c}\right\} & \left\{\frac{\vec{S}_{t_y t_y}}{c}\right\} & \left\{\frac{\vec{S}_{t_y t_z}}{c}\right\} & \frac{\vec{S}_{t_y x}}{c^2} & \frac{\vec{S}_{t_y y}}{c^2} & \frac{\vec{S}_{t_y z}}{c^2} \\ \left\{\frac{\vec{S}_{t_z t_x}}{c}\right\} & \left\{\frac{\vec{S}_{t_z t_y}}{c}\right\} & \left\{\frac{\vec{S}_{t_z t_z}}{c}\right\} & \frac{\vec{S}_{t_z x}}{c^2} & \frac{\vec{S}_{t_z y}}{c^2} & \frac{\vec{S}_{t_z z}}{c^2} \\ \frac{\vec{S}_{x t_x}}{c^2} & \frac{\vec{S}_{x t_y}}{c^2} & \frac{\vec{S}_{x t_z}}{c^2} & \left(\frac{\vec{E}_{xx} \wedge \vec{B}_{xx}}{\mu_0 c}\right) & \left(\frac{\vec{E}_{xy} \wedge \vec{B}_{xy}}{\mu_0 c}\right) & \left(\frac{\vec{E}_{xz} \wedge \vec{B}_{xz}}{\mu_0 c}\right) \\ \frac{\vec{S}_{y t_x}}{c^2} & \frac{\vec{S}_{y t_y}}{c^2} & \frac{\vec{S}_{y t_z}}{c^2} & \left(\frac{\vec{E}_{yx} \wedge \vec{B}_{yx}}{\mu_0 c}\right) & \left(\frac{\vec{E}_{yy} \wedge \vec{B}_{yy}}{\mu_0 c}\right) & \left(\frac{\vec{E}_{yz} \wedge \vec{B}_{yz}}{\mu_0 c}\right) \\ \frac{\vec{S}_{z t_x}}{c^2} & \frac{\vec{S}_{z t_y}}{c^2} & \frac{\vec{S}_{z t_z}}{c^2} & \left(\frac{\vec{E}_{zx} \wedge \vec{B}_{zx}}{\mu_0 c}\right) & \left(\frac{\vec{E}_{zy} \wedge \vec{B}_{zy}}{\mu_0 c}\right) & \left(\frac{\vec{E}_{zz} \wedge \vec{B}_{zz}}{\mu_0 c}\right) \end{matrix}\right| \tag{6.96}$$

Thus, the time-space components (momentum density), or space-time components (momentum density) are defined as the value of the Poynting vector divided by c^2, given by $\Psi_{ij} = \vec{S}_{\chi\chi}/c^2$ or $\Psi_{ji} = \vec{S}_{\chi\chi}/c^2$ when $i \neq j$, the time-time components are restricted to

$$\Psi_{\varepsilon\beta}(\vec{e}_i)^{\varepsilon}(\vec{e}_j)^{\beta} = \left(\frac{\vec{S}_{\chi\chi}}{c^3}\right)\partial_{ij} \tag{6.97}$$

and the space-space components are restricted to

$$\Psi_{\varepsilon\beta}(\vec{e}_i)^{\varepsilon}(\vec{e}_j)^{\beta} = \left(\frac{\vec{E}_{\chi\chi} \wedge \vec{B}_{\chi\chi}}{\mu_0 c}\right)\partial_{ij} \tag{6.98}$$

The $\vec{S}_{\chi\chi}$ in time-space or space-time components is the value of the Poynting vector.

$$\vec{S}_{\chi\chi} = \varepsilon_0 c^2 \left(\vec{E}_{\chi\chi} \times \vec{B}_{\chi\chi}\right) \tag{6.99}$$

Where $\vec{E}_{\chi\chi}$ is the electric field, $\vec{B}_{\chi\chi}$ is the magnetic flux density, ε_0 is the permittivity of the medium, and "c" is the speed of light. Let us consider a decomposition of foliated space-time hypersurfaces so that tangent spaces may be divided into spatial and temporal components, where spatial tangent spaces orthogonal to a tangent unit vector field, $\vec{n}$, temporal tangent spaces parallel to a tangent unit vector field, $\vec{n}$, so that an outward- pointing temporal unit vector field, n^μ, is normal to the geodesics of an $(n\text{–}1)$–dimensional spatial hypersurface, Σ, in an n–dimensional manifold, M, such that $n^\mu n_\mu = -1$, induces the defoliation.

The metric, $g_{\mu\nu}$, is the first fundamental form with a compatible derivative equal to ∇_μ, that induces a three-dimensional Riemannian metric, $h_{\mu\nu}$, on the spatial hypersurface, Σ, as an object dwelling on space-time. (Baumgarte, 2010)

$$g_{\mu\nu} = h_{\mu\nu} - n_\mu n_\nu \tag{6.100}$$

The spatial metric, $h_{\mu\nu}$, is induced on the spatial hypersurface with a compatible derivative equal to D_μ, that is, $h_{\mu\nu}$ is purely spatial, it has no component along n^μ. Contracting with the normal,

$$n^\mu h_{\mu\nu} = n^\mu g_{\mu\nu} + n_\mu n^\mu n_\nu = n_\nu - n_\nu = 0 \tag{6.101}$$

The spatial metric can be used to project all geometric objects along the direction given by n^μ. The spatial metric effectively decomposes tensors into a purely spatial curvature part which lies on the hypersurface Σ, and a temporal curvature part normal to the hypersurface. For instance, if tensor $R_{\mu\nu}$ is a spatial curvature part, each free index has to be contracted with a projection operator, denoted $\perp$, then it follows that,

$$\perp R_{\mu\nu} \equiv h_{\mu}{}^{\varepsilon} h_{\nu}{}^{\beta} R_{\varepsilon\beta} \tag{6.102}$$

Where $R_{\varepsilon\beta}$ is a temporal curvature part. Thus, given an *(n–1)*–dimensional spatial hypersurface, Σ, in an *n*–dimensional manifold,

M, its extrinsic curvature, $K_{\mu\nu}$, is the rate of change of the temporal unit vector, n^{μ}, normal to the hypersurface, Σ.

The extrinsic curvature is the second fundamental form may be expressed as

$$K_{\mu\nu} = \vec{e}_{\nu} \cdot \nabla_{\mu} \vec{n} \tag{6.103}$$

The Lie derivative, L_n, may also be used to evaluate the change of a tangent unit vector field tensor for the extrinsic curvature, $K_{\mu\nu} = -(1/2) L_n h_{\mu\nu}$, along the flow of another vector field, $\vec{e}_{\nu}$. This change is coordinate invariant and therefore the Lie derivative is defined on any differentiable manifold.

The extrinsic curvature tensor, $K_{\mu\nu}$, has information on the metric that is intrinsic to the surface, as well as on the curvature due to the embedding of the surface. The extrinsic curvature is like the acceleration of a surface. The covariant derivative of the normal unit vector field with respect to a tangent vector to the hypersurface is itself tangent to the hypersurface. Hence, the connection is compatible with the metric.

Let us define the left dual Riemann tensor and the *Chern-Pontryagin scalar* invariant, *P*, for a six-dimensional Lorentzian manifold,

$$^{*}R^{\mu\nu}{}_{\varepsilon\beta} \equiv \frac{1}{2} \epsilon^{\mu\nu\lambda\rho} R_{\lambda\rho\varepsilon\beta} \tag{6.104}$$

$$P \equiv {}^{*}R_{\mu\nu\varepsilon\beta} R^{\mu\nu\varepsilon\beta} \tag{6.105}$$

Let us also define a dual complex n-dimensional electrogravitic Riemann tensor, where the real part is the Riemann tensor and the imaginary part is the dual of the Riemann tensor,

$$\widetilde{R}_{\mu\nu\varepsilon\beta} \equiv R_{\mu\nu\varepsilon\beta} + i\, {}^{*}R_{\mu\nu\varepsilon\beta} \tag{6.106}$$

It follows that the complex n-dimensional electrogravitic Riemann tensor may be decomposed into its electric and magnetic parts, two second rank tensors.
The electric part of the complex electrogravitic Riemann tensor is defined by

$$E_{\varepsilon\beta} \equiv R_{\mu\nu\varepsilon\beta}\, n^{\mu} n^{\nu} \tag{6.107}$$

The magnetic part of the complex electrogravitic Riemann tensor is defined by

$$B_{\varepsilon\beta} \equiv {}^{*}R_{\mu\nu\varepsilon\beta}\, n^{\mu} n^{\nu} \tag{6.108}$$

Hence, from the above two definitions we have

$$\widetilde{R}_{\mu\nu\varepsilon\beta}\, n^{\mu} n^{\nu} = E_{\varepsilon\beta} + iB_{\varepsilon\beta} \tag{6.109}$$

The electromagnetic second rank tensors in complex electrogravitic form may be expressed as

$$\widetilde{R}_{\varepsilon\beta} = \sqrt[2]{E_{\varepsilon\beta}{}^{2} + B_{\varepsilon\beta}{}^{2}}\ \angle Tan^{-1}\left(\frac{B_{\varepsilon\beta}}{E_{\varepsilon\beta}}\right) = \widetilde{R}\angle\theta_{\varepsilon\beta} \tag{6.110}$$

$$\widetilde{R}\angle\theta_{\varepsilon\beta} = \widetilde{R}Cos\,\theta_{\varepsilon\beta} + i\widetilde{R}Sin\,\theta_{\varepsilon\beta} = \widetilde{R}e^{i\theta_{\varepsilon\beta}} \tag{6.111}$$

Therefore, the magnitude of the combined electromagnetic tensors is equivalent to the complex electrogravitic Riemann scalar, $\widetilde{R}$, which represents the curvature of the hypersurface, Σ, on the manifold M.

Thus, let us represent this curvature by the six-dimensional electrogravitic curvature tensor, $\widetilde{R}_{\varepsilon\beta}$, which is equivalent to the n-dimensional Hilbert tensor $\overline{G}_{\varepsilon\beta}$.

$$\widetilde{R}_{\varepsilon\beta} = \overline{R}_{\varepsilon\beta} - \frac{1}{(n-1)} g_{\varepsilon\beta} \overline{R} \tag{6.112}$$

Moreover, the electric and magnetic parts of the complex electrogravitic Riemann tensor in the General Theory of Relativity are analogous to those that occur in the Theory of Electromagnetism. The analysis of the complex electrogravitic Riemann tensor is useful in the understanding of the gravitational field.

Consequently, how do we calculate $E_{\varepsilon\beta}$ and $B_{\varepsilon\beta}$?

First, the Gauss-Codazzi equation relates the orthogonal decomposition of the temporal curvature to the intrinsic and extrinsic curvature of the spatial hypersurface, Σ.

The Gauss-Codazzi equation relates the entirely spatial projection of the space-time curvature tensor to the three-dimensional curvature.

$$h_{\mu}{}^{\alpha} h_{\nu}{}^{\rho} h_{\varepsilon}{}^{\sigma} h_{\beta}{}^{\omega\,(6)}R_{\alpha\rho\sigma\omega} = {}^{(3)}R_{\mu\nu\varepsilon\beta} + K_{\mu\varepsilon}K_{\nu\beta} - K_{\mu\beta}K_{\varepsilon\nu} \tag{6.113}$$

Since by Interchange Symmetry, $R_{\mu\nu\varepsilon\beta} = R_{\varepsilon\beta\mu\nu}$, let us apply the metric $g^{\mu\nu}$ in the above equation to trace over indices μ and ν, substitute ${}^{(6)}R_{\alpha\rho\sigma\omega}$ with its decomposition, and the electrical and magnetic decomposition, to obtain the electrical part. (Garcia-Parrado, 2007)

The electrical part is given by

$$E_{\varepsilon\beta} = K_{\varepsilon\beta}K^{\mu}{}_{\mu} - K_{\varepsilon}{}^{\mu}K_{\beta\mu} + {}^{(3)}R_{\varepsilon\beta} - \frac{1}{2}h_{\varepsilon}{}^{\mu}h_{\beta}{}^{\nu\,(6)}R_{\mu\nu} - \frac{1}{2}h_{\varepsilon\beta}h^{\mu\nu\,(6)}R_{\mu\nu} + \frac{1}{3}h_{\varepsilon\beta}{}^{(6)}R \tag{6.114}$$

The last three terms of the above equation vanish in a

six-dimensional Ricci-flat hypersurface, $R_{\mu\nu} \to 0$. The Codazzi-Mainardi equation involves a decomposition of the six-dimensional Riemann tensor when one index is contracted from the Gauss equation with n^{β} and the remaining three indices are projected onto spatial directions.

$$h_{\mu}{}^{\alpha} h_{\nu}{}^{\rho} h_{\varepsilon}{}^{\sigma} n^{\omega (6)} R_{\alpha\rho\sigma\omega} = \mathrm{D}_{\nu} K_{\mu\varepsilon} - D_{\mu} K_{\nu\varepsilon} \tag{6.115}$$

The magnetic part is given by

$$B_{\varepsilon\beta} = \in_{\mu\nu(\varepsilon} D^{\mu} K_{\beta)}{}^{\nu} \tag{6.116}$$

For a nearly-flat region of space-time about a charged celestial body of mass, let us describe the electromagnetic torsion mass-energy density tensor, $\Psi_{\varepsilon\beta}$, as follows:

$$\Psi_{\varepsilon\beta} = \left| \begin{matrix} \left\{\frac{\vec{S}_{t_x t_x}}{c}\right\} & 0 & 0 & 0 & 0 & 0 \\ 0 & \left\{\frac{\vec{S}_{t_y t_y}}{c}\right\} & 0 & 0 & 0 & 0 \\ 0 & 0 & \left\{\frac{\vec{S}_{t_z t_z}}{c}\right\} & 0 & 0 & 0 \\ 0 & 0 & 0 & \left(\frac{\vec{E}_{xx} \wedge \vec{B}_{xx}}{\mu_0 c}\right) & 0 & 0 \\ 0 & 0 & 0 & 0 & \left(\frac{\vec{E}_{yy} \wedge \vec{B}_{yy}}{\mu_0 c}\right) & 0 \\ 0 & 0 & 0 & 0 & 0 & \left(\frac{\vec{E}_{zz} \wedge \vec{B}_{zz}}{\mu_0 c}\right) \end{matrix} \right| \tag{6.117}$$

Thus, the space-time, or time-space components are $\Psi_{ij} = \Psi_{ji} = 0$, when $i \neq j$.

Hence, with equal terms for the energy densities and pressures, the trace of the electromagnetic torsion mass-energy density tensor is

$$\Psi = g^{\varepsilon\beta} \Psi_{\varepsilon\beta} = -3\left\{\frac{\vec{S}_{xx}}{c}\right\} + 3\left(\frac{\vec{E}_{xx} \wedge \vec{B}_{xx}}{\mu_0 c}\right) \tag{6.118}$$

The torsion mass-energy density component represents the ratio of the alternating area, in a region of space-time, of the parallelogram formed by the cross product of the electric field vector and the magnetic flux density vector to the product of the speed of light with the permeability of the medium (inductance per second).

The torsion pressure component represents a contravariant second-order antisymmetric tensor that alternates at the angular frequency of the electromagnetic field through the region of space-time under consideration.

Moreover, torsion (or torsion-geist) may exist in addition to curvature in the electrogravitic vacuum solution if the resultant electrogravitic tensor, and the resultant stress-energy-momentum tensor, were to instantly vanish, as long as the resultant electromagnetic field endures in the region of space-time under consideration. Conversely, electrogravitic curvature may exist in curvature without electromagnetic torsion.

6.4. Constructing the six-dimensional metric of space-time

The metric expansion of space-time is a feature of solutions to Einstein's Field Equations of General Relativity. The metric of space-time has been defined between points with coordinates that grow with time, rather than remain constant. It has been proposed as an explanation to Hubble's Law that galaxies that are more distant from the earth are receding faster than closer galaxies. The locally visible effect of an accelerating expansion is the disappearance, by runaway red shift, of distant galaxies.

The metric acts as an infinitesimal space-time interval, or line element, squared. For this reason, one often sees the notation ds^2 for the metric. The space-time interval or line element ds^2 conveys information about the causal structure of the space-time. When $ds^2 < 0$, the interval is time-like, $i^2 ds^2$, and the square root of the absolute value of ds^2 is an incremental proper time, ids. Only time-like intervals can be physically traversed by an object of mass. When $ds^2 = 0$, the interval is light-like, and can only be traversed by light. When $ds^2 > 0$, the interval is space-like and the square root of ds^2 acts as an incremental proper length. Space-like intervals cannot be traversed, since they connect events that are out of each other's light cones. Events can be causally related only if they are within each other's light cones. Furthermore, the metric interval between events in Lorentz geometry of space-time is an invariant for any observer, so let us denote for homogeneous and isotropic six-dimensional space-time, or Einsteinian space-time, the six-dimensional space-time intervals for an observer traveling near the speed of space-time in uniform relative motion as follows:

a. Time-like Interval

$$v_s^{2}(\Delta t_x^{\ 2} + \Delta t_y^{\ 2} + \Delta t_z^{\ 2}) > (\Delta x^2 + \Delta y^2 + \Delta z^2) \qquad (6.119)$$

Therefore, $s^2 < 0$

$$\Delta\tau = \sqrt[2]{\left(\Delta t_x^{\ 2} + \Delta t_y^{\ 2} + \Delta t_z^{\ 2}\right) - \left(\frac{\Delta x^2 + \Delta y^2 + \Delta z^2}{v_s^{\ 2}}\right)} \qquad (6.120)$$

where $\Delta\tau$ equals the proper time

The proper time interval would be measured by an observer with a clock traveling between two events, when the observer's path intersects each event as that event occurs. The proper time of a time-like interval is a real number value.

b. Light-like Interval

$$v_s^{2}(\Delta t_x^{\ 2} + \Delta t_y^{\ 2} + \Delta t_z^{\ 2}) = (\Delta x^2 + \Delta y^2 + \Delta z^2) \qquad (6.121)$$

Therefore, $s^2 = 0$

In a light-like interval, the spatial distance between two events is exactly balanced by the time between the two events. The events define a squared space-time interval of zero.

c. Space-like Interval

$$(\Delta x^2 + \Delta y^2 + \Delta z^2) > v_s^{2}(\Delta t_x^{\ 2} + \Delta t_y^{\ 2} + \Delta t_z^{\ 2}) \qquad (6.122)$$

Therefore, $s^2 > 0$

$$\Delta\sigma = \sqrt[2]{\left(\Delta x^2 + \Delta y^2 + \Delta z^2\right) - v_s^{\,2}\left(\Delta t_x^{\,2} + \Delta t_y^{\,2} + \Delta t_z^{\,2}\right)} \qquad (6.123)$$

where $\Delta\sigma$ equals the proper distance

For these space-like event pairs with a positive squared space-time interval, $s^2 > 0$, the measurement of space-like separation is the proper distance. Like the proper time of a time-like interval, the proper distance, $\Delta\sigma$, of a space-like interval, is a real number value. When a space-like interval separates two events, not enough time passes between their occurrences for there to exist a causal relationship crossing the spatial distance between the two events at the speed of light or slower.

Generally, the events are considered not to occur in each other's future or past. There exists a reference frame such that the two events are observed to occur at the same time, but there is no reference frame in which the two events can occur in the same spatial location.

Let us denote the simplest example for the metric in homogeneous and isotropic Minkowski space-time, or flat space-time, with six-dimensional coordinates (t_x, t_y, t_z, x, y, z) to be

$$ds^2 = -c^2 dt_x^{\,2} - c^2 dt_y^{\,2} - c^2 dt_z^{\,2} + dx^2 + dy^2 + dz^2 \qquad (6.124)$$

$$ds^2 = \eta_{\mu\nu} dt^{\mu} dr^{\nu} \qquad (6.125)$$

Let us denote for six-dimensional homogeneous and isotropic space-time the Einsteinian metric tensor for special relativity that can be represented by the symbol $\eta_{\mu\nu}$, in six-dimensional flat space-time. (Naber, 1992)

$$\eta_{\mu\nu} = \begin{vmatrix} -c^2 & 0 & 0 & 0 & 0 & 0 \\ 0 & -c^2 & 0 & 0 & 0 & 0 \\ 0 & 0 & -c^2 & 0 & 0 & 0 \\ 0 & 0 & 0 & 1 & 0 & 0 \\ 0 & 0 & 0 & 0 & 1 & 0 \\ 0 & 0 & 0 & 0 & 0 & 1 \end{vmatrix} \qquad (6.126)$$

Let us reconsider the Friedman-Lemaitre-Robertson-Walker metric for homogeneous and isotropic space-time in terms of the proper time, with the time dependent scalar factor, $c = 1$, for near-flat expanding space-time, such that

$$-c^2 d\tau^2 = -c^2 dt^2 + a(t)^2 d\Sigma^2 \tag{6.127}$$

$$-d\tau^2 = -dt^2 + a(t)^2 d\Sigma^2 \tag{6.128}$$

$$-d\tau^2 = -dt_x{}^2 - dt_y{}^2 - dt_z{}^2 + a(t)^2(dx^2 + dy^2 + dz^2) \tag{6.129}$$

$$-d\tau^2 = g_{\mu\nu} dt^\mu dr^\nu \tag{6.130}$$

where the spatial curvature radius *a(t)* is often chosen to equal 1 in the present cosmological era.

Thus, we find the Cartesian coordinate metric tensor and its inverse metric tensor for near-flat expanding space-time, with $c = 1$, to be

$$g_{\mu\nu} = \begin{vmatrix} -c^2 & 0 & 0 & 0 & 0 & 0 \\ 0 & -c^2 & 0 & 0 & 0 & 0 \\ 0 & 0 & -c^2 & 0 & 0 & 0 \\ 0 & 0 & 0 & 1 & 0 & 0 \\ 0 & 0 & 0 & 0 & 1 & 0 \\ 0 & 0 & 0 & 0 & 0 & 1 \end{vmatrix} \tag{6.131}$$

$$g_{\mu\nu} = \begin{vmatrix} -1 & 0 & 0 & 0 & 0 & 0 \\ 0 & -1 & 0 & 0 & 0 & 0 \\ 0 & 0 & -1 & 0 & 0 & 0 \\ 0 & 0 & 0 & 1 & 0 & 0 \\ 0 & 0 & 0 & 0 & 1 & 0 \\ 0 & 0 & 0 & 0 & 0 & 1 \end{vmatrix} \tag{6.132}$$

The metric tensor, $g_{\mu\nu}$, is a transform for some curved space-time to flat space-time. The metric tensor transform preserves the distance between points and the angles between lines.

If the metric tensor were to vary, it would fail to preserve the distance between points and the angles between lines. The metric tensor does not vary from point to point when space-time is flat.

$$g^{\mu\nu} = \begin{vmatrix} -\frac{1}{c^2} & 0 & 0 & 0 & 0 & 0 \\ 0 & -\frac{1}{c^2} & 0 & 0 & 0 & 0 \\ 0 & 0 & -\frac{1}{c^2} & 0 & 0 & 0 \\ 0 & 0 & 0 & 1 & 0 & 0 \\ 0 & 0 & 0 & 0 & 1 & 0 \\ 0 & 0 & 0 & 0 & 0 & 1 \end{vmatrix} \quad (6.133)$$

$$g^{\mu\nu} = \begin{vmatrix} -1 & 0 & 0 & 0 & 0 & 0 \\ 0 & -1 & 0 & 0 & 0 & 0 \\ 0 & 0 & -1 & 0 & 0 & 0 \\ 0 & 0 & 0 & 1 & 0 & 0 \\ 0 & 0 & 0 & 0 & 1 & 0 \\ 0 & 0 & 0 & 0 & 0 & 1 \end{vmatrix} \quad (6.134)$$

6.5. The Einstein Field Equations in six-dimensional curved space-time

Let us now consider the Einstein Field Equations for six-dimensional space-time in general relativity, i.e., three spatial dimensions and three temporal dimensions, about a spherical object with mass "*m*", volume "*V*", and a radius of "*r*", such that we find the EFE about the mass to be

$$R_{\mu\nu} - g_{\mu\nu}\left(\frac{1}{(n-1)}R - \Lambda\right) = \left(\frac{\frac{\partial^2 V}{\partial t^2}}{mc^4}\right) T_{\mu\nu} \quad (6.135)$$

$$\rho_{vac} = \frac{\Lambda c^2}{8\pi G} \quad (6.136)$$

$$\Lambda = \frac{8\pi G(\rho_{vac})}{c^2} \tag{6.137}$$

$$\Lambda = \frac{\ddot{a}\rho_{vac}}{mc^2} \tag{6.138}$$

$$\Lambda = \frac{\ddot{a}}{ac^2} = \frac{\nabla \cdot \vec{g}}{c^2} \tag{6.139}$$

where $\ddot{a}$ is the acceleration of curved space-time, ρ_{vac} is the spatiotemporal (vacuum) energy density (J/m^3), $\nabla \cdot \vec{g}/c^2$ is the cosmological curvature $(1/m^2)$, and Λ is the cosmological constant $(1/m^2)$. The gravitational field vector of the spatiotemporal curvature is $-\nabla^2 \ddot{a} = -\vec{g}$ in (m/s^2) which is the divergence of the gradient of $\ddot{a}$. The divergence of the gravitational field is $\nabla \cdot \vec{g}$.

A positive spatiotemporal energy density ρ_{vac} resulting from $\Lambda c^2/8\pi G$ implies a negative spatiotemporal pressure, $-p_{vac}$, resulting from $-\Lambda c^4/8\pi G$, $\left(in\ N/m^2\ or\ J/m^3\right)$, and vice versa. The negative spatiotemporal pressure will drive an accelerated expansion of the universe.

6.6. Obtaining the six-dimensional metric, Ricci and Einstein tensors for curved space-time

First, let us consider the Friedman-Lemaitre-Robertson-Walker metric in spherical polar coordinates for curved space-time, with a spatial curvature of k_s, a temporal curvature of k_t, and a time dependent scalar factor $a(t)^2 = 1$, for expanding space-time, such that

$$-c^2 d\tau^2 = -c^2\left(1 - k_t t^2\right)dt^2 - c^2 t^2 d\theta_t^2 - c^2 t^2 \,\mathrm{Sin}^2\,\theta_t d\phi_t^2 + \tag{6.140}$$

$$a(t)^2\left(\frac{dr^2}{1-k_s r^2}+r^2 d\theta_s{}^2+r^2\,\mathrm{Sin}^2\,\theta_s d\phi_s{}^2\right)$$

We find the spherical-coordinate spacetime curvature tensor to be

$$\Omega_{\mu\nu}=\begin{vmatrix}-(1-k_t t^2) & 0 & 0 & 0 & 0 & 0\\ 0 & -t^2 & 0 & 0 & 0 & 0\\ 0 & 0 & -t^2 Sin^2\theta_t & 0 & 0 & 0\\ 0 & 0 & 0 & 1/(1-k_s r^2) & 0 & 0\\ 0 & 0 & 0 & 0 & r^2 & 0\\ 0 & 0 & 0 & 0 & 0 & r^2 Sin^2\theta_s\end{vmatrix} \qquad (6.141)$$

Secondly, let us describe the six-dimensional Ricci Tensor to be

$$R_{\mu\nu}=\begin{vmatrix}R_{t_x t_x} & R_{t_x t_y} & R_{t_x t_z} & R_{t_x x} & R_{t_x y} & R_{t_x z}\\ R_{t_y t_x} & R_{t_y t_y} & R_{t_y t_z} & R_{t_y x} & R_{t_y y} & R_{t_y z}\\ R_{t_z t_x} & R_{t_z t_y} & R_{t_z t_z} & R_{t_z x} & R_{t_z y} & R_{t_z z}\\ R_{x t_x} & R_{x t_y} & R_{x t_z} & R_{xx} & R_{xy} & R_{xz}\\ R_{y t_x} & R_{y t_y} & R_{y t_z} & R_{yx} & R_{yy} & R_{yz}\\ R_{z t_x} & R_{z t_y} & R_{z t_z} & R_{zx} & R_{zy} & R_{zz}\end{vmatrix} \qquad (6.142)$$

The Ricci curvature tensor of the EFEs, $R_{\mu\nu}$, is a rank-2 tensor. Rank-2 tensors are symmetric in four-dimensional or six-dimensional space-time.

For instance, if you fly around the curvature of the earth between surface points A and B, does the curvature of the earth change if you were to fly in the opposite direction between surface points B and A? Clearly, not.

Hence, the indices of rank-2 tensors used to represent curvature must be symmetric.

A tensor is symmetric in a pair of indices when the element indicated by those indices on the matrix of that tensor is the same as the element indicated by the transpose of those indices, $R_{\mu\nu} = R_{\nu\mu}$. All metric tensors in four-dimensional (collapsed time, 3 + 1) or six-dimensional spacetime (3 + 3) are symmetric in their indices. After transposing indices on a symmetric tensor in a term, there would be no change in the sign of the term that includes the symmetric tensor with the transposed indices.

If the tensor is antisymmetric and the indices of the tensor are transposed, there would be a reversal in the direction of parallel-transport, and a change of sign in the term that includes the antisymmetric tensor.

The Ricci tensor may be expressed using Christoffel symbols. The Christoffel symbol, of the first kind or second kind, describes curvature in some space-time defined by the metric tensor, $g_{\mu\nu}$.

To obtain the time-time and space-space components of the Ricci tensor

$$R_{\tau\tau} = \Gamma^{\lambda}{}_{\tau\tau,\lambda} - \Gamma^{\lambda}{}_{\tau\lambda,\tau} + \Gamma^{\lambda}{}_{\tau\tau}\Gamma^{\sigma}{}_{\lambda\sigma} - \Gamma^{\sigma}{}_{\tau\lambda}\Gamma^{\lambda}{}_{\tau\sigma} \tag{6.143}$$

By isotropy, the Ricci tensor's time-time, space-time, or time-space components are $R_{ij} = R_{ji} = 0$ when $i \neq j$, and the space-space components are restricted to $R_{\mu\nu}(\vec{e}_i)^{\mu}(\vec{e}_j)^{\nu} = (R_{ss})\delta_{ij}$, such that for the time-time components when $i = j$, using Einstein summation notation, we find

$$R_{\tau\tau} = -\Gamma^{\lambda}{}_{\tau\lambda,\tau} - \Gamma^{\sigma}{}_{\tau\lambda}\Gamma^{\lambda}{}_{\tau\sigma} \tag{6.144}$$

$$\Gamma^{\lambda}{}_{\tau\lambda} = \frac{1}{2} g^{\lambda\sigma}(g_{\sigma\tau,\lambda} + g_{\sigma\lambda,\tau} - g_{\tau\lambda,\sigma}) \tag{6.145}$$

$$\Gamma^{\lambda}{}_{\tau\lambda} = \frac{1}{2} g^{\lambda\sigma}(g_{\sigma\lambda,\tau}) = \frac{1}{2} g^{\lambda\sigma}\partial_{\tau}(g_{\sigma\lambda}) \tag{6.146}$$

If $g_{\sigma\lambda} = a^2\widetilde{g}_{\sigma\lambda}$ and $g^{\lambda\sigma} = \dfrac{\widetilde{g}^{\lambda\sigma}}{a^2}$ then

$$\Gamma^{\lambda}{}_{\tau\lambda} = \frac{1}{2}\left(\frac{\widetilde{g}^{\lambda\sigma}}{a^2}\right)\frac{1}{c}\partial_t(a^2\widetilde{g}_{\sigma\lambda}) \tag{6.147}$$

$$\Gamma^{\lambda}{}_{\tau\lambda} = \frac{1}{2c}\left(\frac{\widetilde{g}^{\lambda\sigma}}{a^2}\right)(\widetilde{g}_{\sigma\lambda}2a\dot{a}) = \frac{1}{c}\left(\frac{\dot{a}}{a}\right)\widetilde{g}^{\lambda\sigma}\widetilde{g}_{\sigma\lambda} \tag{6.148}$$

$$\Gamma^{\lambda}{}_{\tau\lambda} = \frac{1}{c}\left(\frac{\dot{a}}{a}\right) \tag{6.149}$$

$$\Gamma^{\lambda}{}_{\tau\lambda,\tau} = \partial_\tau\Gamma^{\lambda}{}_{\tau\lambda} = \frac{1}{c^2}\partial_t\left(\frac{\dot{a}}{a}\right) = \frac{1}{c^2}\left(\frac{a\ddot{a}-\dot{a}^2}{a^2}\right) \tag{6.150}$$

$$\Gamma^{\lambda}{}_{\tau\lambda,\tau} = \frac{1}{c^2}\left(\frac{\ddot{a}}{a} - \frac{\dot{a}^2}{a^2}\right) \tag{6.151}$$

As $\sigma = \lambda$, we have

$$\Gamma^{\sigma}{}_{\tau\lambda} = \Gamma^{\lambda}{}_{\tau\sigma} = \frac{1}{c}\left(\frac{\dot{a}}{a}\right) \tag{6.152}$$

$$R_{\tau\tau} = -\Gamma^{\lambda}{}_{\tau\lambda,\tau} - \Gamma^{\sigma}{}_{\tau\lambda}\Gamma^{\lambda}{}_{\tau\sigma} \tag{6.153}$$

$$R_{\tau\tau} = \frac{1}{c^2}\left(-\frac{\ddot{a}}{a} + \frac{\dot{a}^2}{a^2} - \frac{\dot{a}^2}{a^2}\right) = -\frac{1}{c^2}\left(\frac{\ddot{a}}{a}\right) \tag{6.154}$$

Thus, we find that

$$R_{t_x t_x} = R_{t_y t_y} = R_{t_z t_z} = -\frac{1}{c^2}\left(\frac{\ddot{a}}{a}\right) = -\frac{\ddot{a}}{ac^2} \tag{6.155}$$

To obtain the space-space components R_{ss} of the Ricci tensor, we have

$$R_{ij} = \Gamma^{\lambda}{}_{ij,\lambda} - \Gamma^{\lambda}{}_{i\lambda,j} + \Gamma^{\lambda}{}_{ij}\Gamma^{\sigma}{}_{\lambda\sigma} - \Gamma^{\sigma}{}_{i\lambda}\Gamma^{\lambda}{}_{j\sigma} \quad (6.156)$$

$$R_{ij} = \widetilde{R}_{ij} + \partial_{\lambda}\Gamma^{\lambda}{}_{ij} + \Gamma^{\lambda}{}_{ij}\Gamma^{\sigma}{}_{\lambda\sigma} \quad (6.157)$$

$$\widetilde{R}_{ij} = 2k\widetilde{g}_{ij} = \frac{2kg_{ij}}{a^2} \quad (6.158)$$

Where *"k"* is the intrinsic curvature of the universe. (Ludvigsen, 1999)

As $\lambda = \sigma$ and $\sigma = i$ or j, and space-space metric components $g^{\lambda\lambda} = g^{\sigma\sigma} = 1,\ g_{ij} = a^2\widetilde{g}_{ij}$, and $g^{ji} = \dfrac{\widetilde{g}^{ji}}{a^2}$

$$\Gamma^{\lambda}{}_{ij} = \frac{1}{2}g^{\lambda\sigma}(g_{\sigma i,j} + g_{\sigma j,i} - g_{ij,\sigma}) \quad (6.159)$$

$$\Gamma^{\lambda}{}_{ij} = \frac{1}{2}g^{\lambda\sigma}(g_{\sigma j,i}) = \frac{1}{2}(1)(g_{\sigma j,i}) = \frac{1}{2}(1)\partial_i(g_{ij}) = \frac{1}{2}\left(\frac{1}{c}\right)\partial_t(a^2\widetilde{g}_{ij}) \quad (6.160)$$

$$\Gamma^{\lambda}{}_{ij} = \frac{1}{2c}(\widetilde{g}_{ij}2a\dot{a}) \quad (6.161)$$

$$\Gamma^{\lambda}{}_{ij} = \frac{1}{c}\widetilde{g}_{ij}(a\dot{a}) = \frac{1}{c}\frac{g_{ij}}{a^2}(a\dot{a}) = \frac{g_{ij}}{c}\left(\frac{\dot{a}}{a}\right) \quad (6.162)$$

$$\partial_{\lambda}\Gamma^{\lambda}{}_{ij} = \frac{1}{c}\widetilde{g}_{ij}\partial_{\lambda}(a\dot{a}) = \frac{1}{c^2}\widetilde{g}_{ij}\partial_t(a\dot{a}) = \frac{1}{c^2}\left(\frac{g_{ij}}{a^2}\right)(a\ddot{a} + \dot{a}^2) \quad (6.163)$$

$$\partial_\lambda \Gamma^\lambda{}_{ij} = \frac{g_{ij}}{c^2}\left(\frac{\ddot{a}}{a} + \frac{\dot{a}^2}{a^2}\right) \tag{6.164}$$

As $\sigma = j$ or i, we have

$$\Gamma^\sigma{}_{\lambda\sigma} = \Gamma^i{}_{\lambda j} = \frac{1}{2} g^{i\sigma}(g_{\sigma\lambda,j} + g_{\sigma j,\lambda} - g_{\lambda j,\sigma}) \tag{6.165}$$

$$\Gamma^\sigma{}_{\lambda\sigma} = \Gamma^i{}_{\lambda j} = \frac{1}{2} g^{ii}(g_{i\lambda,j} + g_{ij,\lambda} - g_{\lambda j,i}) \tag{6.166}$$

$$\Gamma^\sigma{}_{\lambda\sigma} = \Gamma^i{}_{\lambda j} = \frac{1}{2} g^{ij}(g_{ij,\lambda}) = \frac{1}{2}\left(\frac{\tilde{g}^{ij}}{a^2}\right)\partial_\lambda(a^2 \tilde{g}_{ij}) \tag{6.167}$$

$$\Gamma^\sigma{}_{\lambda\sigma} = \Gamma^i{}_{\lambda j} = \frac{1}{2c}\left(\frac{\tilde{g}^{ij}}{a^2}\right)\partial_t(a^2 \tilde{g}_{ij}) \tag{6.168}$$

$$\Gamma^\sigma{}_{\lambda\sigma} = \Gamma^i{}_{\lambda j} = \frac{1}{2c}\left(\frac{\tilde{g}^{ij}}{a^2}\right)(\tilde{g}_{ij} 2a\dot{a}) = \frac{1}{c}\left(\frac{\dot{a}}{a}\right)\tilde{g}^{ij}\tilde{g}_{ij} \tag{6.169}$$

We substitute the above expression with the Kronecker tensor $\delta^i{}_j = \tilde{g}^{ij}\tilde{g}_{ij} = 1$ to obtain

$$\Gamma^\sigma{}_{\lambda\sigma} = \Gamma^i{}_{\lambda j} = \frac{1}{c}\left(\frac{\dot{a}}{a}\right)\delta^i{}_j = \frac{1}{c}\left(\frac{\dot{a}}{a}\right) \tag{6.170}$$

Multiplying terms, we have

$$\Gamma^\lambda{}_{ij}\Gamma^\sigma{}_{\lambda\sigma} = \Gamma^\lambda{}_{ij}\Gamma^i{}_{\lambda j} = \left\{\frac{g_{ij}}{c}\left(\frac{\dot{a}}{a}\right)\right\}\frac{1}{c}\left(\frac{\dot{a}}{a}\right) = \frac{g_{ij}}{c^2}\left(\frac{\dot{a}}{a}\right)^2 \tag{6.171}$$

Substituting terms to obtain the Ricci tensor

$$R_{ij} = \widetilde{R}_{ij} + \partial_{\lambda}\Gamma^{\lambda}{}_{ij} + \Gamma^{\lambda}{}_{ij}\Gamma^{\sigma}{}_{\lambda\sigma} \qquad (6.172)$$

$$R_{ij} = \frac{2kg_{ij}}{a^2} + \frac{g_{ij}}{c^2}\left(\frac{\ddot{a}}{a} + \frac{\dot{a}^2}{a^2}\right) + \frac{g_{ij}}{c^2}\left(\frac{\dot{a}}{a}\right)^2 = g_{ij}\left(\frac{\ddot{a}}{ac^2} + \frac{2\dot{a}^2}{a^2c^2} + \frac{2k}{a^2}\right) \qquad (6.173)$$

$$g^{ij}R_{ij} = g^{ij}g_{ij}\left(\frac{\ddot{a}}{ac^2} + \frac{2\dot{a}^2}{a^2c^2} + \frac{2k}{a^2}\right) = \frac{\ddot{a}}{ac^2} + \frac{2\dot{a}^2}{a^2c^2} + \frac{2k}{a^2} \qquad (6.174)$$

Hence, we find the space-space components R_{ss} for curved space-time to be

$$R_{xx} = R_{yy} = R_{zz} = R_{ss} = \frac{\ddot{a}}{ac^2} + \frac{2\dot{a}^2}{a^2c^2} + \frac{2k}{a^2} \qquad (6.175)$$

We can describe the six-dimensional Ricci tensor as follows:

$$R_{\mu\nu} = \begin{vmatrix} -\frac{\ddot{a}}{ac^2} & 0 & 0 & 0 & 0 & 0 \\ 0 & -\frac{\ddot{a}}{ac^2} & 0 & 0 & 0 & 0 \\ 0 & 0 & -\frac{\ddot{a}}{ac^2} & 0 & 0 & 0 \\ 0 & 0 & 0 & R_{ss} & 0 & 0 \\ 0 & 0 & 0 & 0 & R_{ss} & 0 \\ 0 & 0 & 0 & 0 & 0 & R_{ss} \end{vmatrix} \qquad (6.176)$$

Let us describe the spherical-coordinate inverse spacetime curvature tensor as

$$\Omega^{\mu\nu} = \begin{vmatrix} -\frac{1}{(1-k_t t^2)} & 0 & 0 & 0 & 0 & 0 \\ 0 & -\frac{1}{t^2} & 0 & 0 & 0 & 0 \\ 0 & 0 & -\frac{1}{t^2 Sin^2\theta_t} & 0 & 0 & 0 \\ 0 & 0 & 0 & (1-k_s r^2) & 0 & 0 \\ 0 & 0 & 0 & 0 & \frac{1}{r^2} & 0 \\ 0 & 0 & 0 & 0 & 0 & \frac{1}{r^2 Sin^2\theta_s} \end{vmatrix} \quad (6.177)$$

$$R = \Omega^{\mu\nu} R_{\mu\nu} = \frac{\ddot{a}}{ac^2(1-k_t t^2)} + \frac{\ddot{a}}{ac^2 t^2} + \frac{\ddot{a}}{ac^2 t^2 Sin^2\theta_t} + (1-k_s r^2)(R_{ss}) + \frac{(R_{ss})}{r^2} + \frac{(R_{ss})}{r^2 Sin^2\theta_s} \quad (6.178)$$

Hence, during an infinitesimal spatiotemporal interval in nearly-flat homogeneous and isotropic space-time, we find the trace of the Ricci tensor to be

$$R = g^{\mu\nu} R_{\mu\nu} = (3)\frac{\ddot{a}}{ac^2} + (3)\left(\frac{\ddot{a}}{ac^2} + \frac{2\dot{a}^2}{a^2c^2} + \frac{2k}{a^2}\right) = (3)\left(\frac{2\ddot{a}}{ac^2} + \frac{2\dot{a}^2}{a^2c^2} + \frac{2k}{a^2}\right) = \frac{6}{a^2c^2}\left(a\ddot{a} + \dot{a}^2 + kc^2\right) \quad (6.179)$$

Letting the space curvature component Ω_S equal the sum of the spatial coefficients and the time curvature component Ω_t equal the sum of the temporal coefficients of the six-dimensional metric we have

$$\Omega_S = \left(1 - k_s r^2\right) + \frac{1}{r^2} + \frac{1}{r^2 Sin^2\theta_s} \quad (6.180)$$

$$\Omega_t = \frac{1}{\left(1 - k_t t^2\right)} + \frac{1}{t^2} + \frac{1}{t^2 Sin^2\theta_t} \quad (6.181)$$

$$R = \frac{3\ddot{a}}{ac^2}\Omega_t + \frac{3\ddot{a}}{ac^2}\Omega_S + \frac{6\dot{a}^2}{a^2c^2}\Omega_S + \frac{6k}{a^2}\Omega_S \quad (6.182)$$

Let us now describe the six-dimensional Einstein Tensor to be

$$G_{\mu\nu} = R_{\mu\nu} - \frac{1}{(n-1)} R g_{\mu\nu} \qquad (6.183)$$

we obtain the following Trace for the Ricci tensor

$$R = g^{\mu\nu} R_{\mu\nu} \qquad (6.184)$$

$$R g_{\mu\nu} = g^{\mu\nu} g_{\mu\nu} R_{\mu\nu} = 6 R_{\mu\nu} \qquad (6.185)$$

For four-dimensional (folded) space-time, with $n = 4$, a six-dimensional Ricci tensor, and a six-dimensional stress-energy-momentum tensor,

$$G_{\mu\nu} = R_{\mu\nu} - \frac{1}{3}(6) R_{\mu\nu} = R_{\mu\nu} - 2R_{\mu\nu} = -R_{\mu\nu} \qquad (6.186)$$

$$G_{\mu\nu} + \Lambda g_{\mu\nu} = \frac{8\pi G}{c^4} T_{\mu\nu} \qquad (6.187)$$

Contracting the Einstein tensor and the Ricci tensor with $g^{\mu\nu}$, we find

$$G = -R = -\frac{3\ddot{a}}{ac^2}\Omega_t - \frac{3\ddot{a}}{ac^2}\Omega_S - \frac{6\dot{a}^2}{a^2c^2}\Omega_S - \frac{6k}{a^2}\Omega_S \qquad (6.188)$$

$$G_{\mu\nu} = \begin{vmatrix} G_{11} & 0 & 0 & 0 & 0 & 0 \\ 0 & G_{22} & 0 & 0 & 0 & 0 \\ 0 & 0 & G_{33} & 0 & 0 & 0 \\ 0 & 0 & 0 & G_{44} & 0 & 0 \\ 0 & 0 & 0 & 0 & G_{55} & 0 \\ 0 & 0 & 0 & 0 & 0 & G_{66} \end{vmatrix} \qquad (6.189)$$

where the non-zero six-dimensional Einstein Tensor components in spherical polar coordinates are

$$-\frac{\ddot{a}}{ac^2}\Omega_t = -\frac{\ddot{a}}{ac^2}\left(\frac{1}{\left(1-k_t t^2\right)}+\frac{1}{t^2}+\frac{1}{t^2 Sin^2\theta_t}\right) \quad (6.190)$$

$$\left(-\frac{\ddot{a}}{ac^2}-\frac{2\dot{a}^2}{a^2c^2}-\frac{2k}{a^2}\right)\Omega_S = \left(-\frac{\ddot{a}}{ac^2}-\frac{2\dot{a}^2}{a^2c^2}-\frac{2k}{a^2}\right)\left\{\left(1-k_s r^2\right)+\frac{1}{r^2}+\frac{1}{r^2 Sin^2\theta_s}\right\} \quad (6.191)$$

$$G_{11} = G_{22} = G_{33} = -\frac{\ddot{a}}{ac^2}\Omega_t \quad (6.192)$$

$$G_{44} = G_{55} = G_{66} = -\frac{\ddot{a}}{ac^2}\Omega_S - \frac{2\dot{a}^2}{a^2c^2}\Omega_S - \frac{2k}{a^2}\Omega_S \quad (6.193)$$

If the stress-energy-momentum tensor is isotropic and homogeneous, and the spatiotemporal curvature from components Ω_t and Ω_s is negligible, we have

$$G_{11} = G_{22} = G_{33} = -\frac{\ddot{a}}{ac^2} \quad (6.194)$$

$$G_{44} = G_{55} = G_{66} = -\frac{\ddot{a}}{ac^2} - \frac{2\dot{a}^2}{a^2c^2} - \frac{2k}{a^2} \quad (6.195)$$

Reconsidering the EFE, in a very large region of universal space-time, with no significant local mass or energy present as an object, or at a very long distance from a point mass: $T_{\mu\nu} = 0$, assuming $n > 3$ dimensions, and the cosmological constant is nonzero, one can rewrite the above EFE in the form:

$$R_{\mu\nu} = \Lambda g_{\mu\nu} \quad (6.196)$$

6.7. The pressure and density continuity equations and the trace of the stress-energy-momentum tensor in six-dimensional curved space-time

Let us omit the cosmological constant term for now by making the following replacement

$$\rho \to \rho + \frac{\Lambda c^2}{8\pi G} \tag{6.197}$$

$$p \to p - \frac{\Lambda c^4}{8\pi G} \tag{6.198}$$

Thus, for diagonal components of energy density ρ and pressure "p" in the stress-energy-momentum tensor matrix, we use the Ricci tensor, for an infinitesimal time interval, to find

The energy density equation

$$\frac{\dot{a}^2}{a^2 c^2} + \frac{k}{a^2} = -\frac{4\pi G}{c^2}(-\rho) + \frac{\Lambda}{4} \tag{6.199}$$

$$\frac{\dot{a}^2}{a^2 c^2} + \frac{k}{a^2} = \frac{4\pi G\rho}{c^2} + \frac{\Lambda}{4} \tag{6.200}$$

$$\frac{\dot{a}^2}{a^2} + \frac{kc^2}{a^2} - \frac{\Lambda c^2}{4} = 4\pi G\rho \tag{6.201}$$

The pressure equation

$$\frac{2\ddot{a}}{ac^2} + \frac{\dot{a}^2}{a^2 c^2} + \frac{k}{a^2} = -\frac{4\pi G p}{c^4} + \frac{3\Lambda}{4} \tag{6.202}$$

$$\frac{2\ddot{a}}{ac^2} + \frac{4\pi G\rho}{c^2} + \frac{\Lambda}{4} = -\frac{4\pi G p}{c^4} + \frac{3\Lambda}{4} \tag{6.203}$$

$$\frac{2\ddot{a}}{ac^2} = -\frac{4\pi G}{c^2}\left(\rho + \frac{p}{c^2}\right) + \frac{\Lambda}{2} \tag{6.204}$$

$$\frac{\ddot{a}}{a} = -2\pi G\left(\rho + \frac{p}{c^2}\right) + \frac{\Lambda c^2}{4} \tag{6.205}$$

The EFE reveals, that combining the density and the pressure equations and adding the cosmological constant term back in, we find

$$G_{\mu\nu} + \Lambda g_{\mu\nu} = \frac{8\pi G}{c^4} T_{\mu\nu} \tag{6.206}$$

$$R - \frac{1}{3}(6)R + 6\Lambda = \frac{8\pi G}{c^4} T \tag{6.207}$$

$$-R = \frac{8\pi G}{c^4} T - 6\Lambda \tag{6.208}$$

$$R = -\frac{8\pi G}{c^4}(-3\rho + 3p) + 6\Lambda \tag{6.209}$$

$$\frac{3\ddot{a}}{ac^2}\Omega_t + \frac{3\ddot{a}}{ac^2}\Omega_s + \frac{6\dot{a}^2}{a^2c^2}\Omega_s + \frac{6k}{a^2}\Omega_s = \frac{8\pi G(3\rho)}{c^4} - \frac{8\pi G(3p)}{c^4} + 6\Lambda \tag{6.210}$$

$$-\frac{\ddot{a}}{ac^2}\Omega_t - \frac{\ddot{a}}{ac^2}\Omega_s - \frac{2\dot{a}^2}{a^2c^2}\Omega_s - \frac{2k}{a^2}\Omega_s = -\frac{8\pi G(\rho - p)}{c^4} - 2\Lambda \tag{6.211}$$

If we consider the equation above when the stress-energy-momentum tensor is assumed to be isotropic and homogeneous, and spatiotemporal curvatures Ω_t and Ω_s are negligible, we obtain

$$-\frac{\ddot{a}}{ac^2} - \frac{\ddot{a}}{ac^2} - \frac{2\dot{a}^2}{a^2c^2} - \frac{2k}{a^2} = -\frac{8\pi G(\rho - p)}{c^4} - 2\Lambda \tag{6.212}$$

$$-\frac{2\ddot{a}}{ac^2} - \frac{2\dot{a}^2}{a^2c^2} - \frac{2k}{a^2} = -\frac{8\pi G(\rho - p)}{c^4} - 2\Lambda \tag{6.213}$$

$$\frac{\ddot{a}}{a}+\frac{\dot{a}^2}{a^2}+\frac{kc^2}{a^2}=\frac{4\pi G(\rho-p)}{c^2}+\Lambda c^2 \qquad (6.214)$$

Where the constant $4\pi G/c^4$ is defined here as Hilbert's constant $\mathscr{H}$. These equations form the basis for extracting knowledge of the Universe. (Wald, 1977)

They contain information about the past, present, and future of the Universe.

Substituting for G in terms of the spatial acceleration about the mass in the EFE, with nonzero spatial or temporal curvature of the volume, we find

$$|G| = \frac{\frac{\partial^2 V}{\partial t^2}}{8\pi m} = \frac{\ddot{a}}{8\pi m} \qquad (6.215)$$

$$-\frac{6\ddot{a}}{ac^2}-\frac{6\dot{a}^2}{a^2c^2}-\frac{6k}{a^2}+6\Lambda=-\frac{8\pi\ddot{a}(-3\rho+3p)}{8\pi mc^4} \qquad (6.216)$$

$$-\frac{6\ddot{a}}{ac^2}-\frac{6\dot{a}^2}{a^2c^2}-\frac{6k}{a^2}+6\Lambda=\frac{\ddot{a}(-3m+3mc^2)}{amc^4}=-\frac{3\ddot{a}}{ac^2}\left(\frac{c^2-1}{c^2}\right)\approx-\frac{3\ddot{a}}{ac^2} \qquad (6.217)$$

$$-\frac{2\ddot{a}}{ac^2}-\frac{2\dot{a}^2}{a^2c^2}-\frac{2k}{a^2}+2\Lambda\approx-\frac{\ddot{a}}{ac^2} \qquad (6.218)$$

$$-\frac{\ddot{a}}{ac^2}-\frac{2\dot{a}^2}{a^2c^2}-\frac{2k}{a^2}\approx-2\Lambda \qquad (6.219)$$

$$\frac{\ddot{a}}{2ac^2}+\frac{\dot{a}^2}{a^2c^2}+\frac{k}{a^2}\approx\Lambda \qquad (6.220)$$

The above equation demonstrates that the cosmological curvature acceleration ratio $\ddot{a}/2ac^2$ is equal to the stress-energy-momentum curvature acceleration ratio $\ddot{a}/2ac^2$ in the manifold about or within the system of mass.

The stress-energy-momentum curvature acceleration ratio does not contribute to the spatial velocity square ratio or to the spatial curvature *"k"* ratio.

Moreover, the spatiotemporal negative curvature about the system of mass plus the stress-energy-momentum curvature of the system of mass is counteracted by the cosmological-constant positive curvature.

The stress-energy-momentum curvature acceleration ratio is equal to one-half of the spatiotemporal curvature acceleration ratio of the Ricci tensor.

$$\frac{3\ddot{a}}{ac^2} \approx \frac{8\pi G}{c^4} T \tag{6.221}$$

$$\frac{\ddot{a}}{a} \approx \frac{8\pi G}{c^2}\left(-\rho + p\right) \tag{6.222}$$

Let us describe the cosmological stress-energy-momentum tensor, or Zwicky tensor, to be

$$\Lambda_{\mu\nu} = \begin{vmatrix} \Lambda_{t_x t_x} & \Lambda_{t_x t_y} & \Lambda_{t_x t_z} & \Lambda_{t_x x} & \Lambda_{t_x y} & \Lambda_{t_x z} \\ \Lambda_{t_y t_x} & \Lambda_{t_y t_y} & \Lambda_{t_y t_z} & \Lambda_{t_y x} & \Lambda_{t_y y} & \Lambda_{t_y z} \\ \Lambda_{t_z t_x} & \Lambda_{t_z t_y} & \Lambda_{t_z t_z} & \Lambda_{t_z x} & \Lambda_{t_z y} & \Lambda_{t_z z} \\ \Lambda_{x t_x} & \Lambda_{x t_y} & \Lambda_{x t_z} & \Lambda_{xx} & \Lambda_{xy} & \Lambda_{xz} \\ \Lambda_{y t_x} & \Lambda_{y t_y} & \Lambda_{y t_z} & \Lambda_{yx} & \Lambda_{yy} & \Lambda_{yz} \\ \Lambda_{z t_x} & \Lambda_{z t_y} & \Lambda_{z t_z} & \Lambda_{zx} & \Lambda_{zy} & \Lambda_{zz} \end{vmatrix} \tag{6.223}$$

And the cosmological stress-energy-momentum tensor, or Zwicky-flat tensor, for nearly-flat expanding spacetime may be expressed as

$$\Lambda_{\mu\nu} = \begin{vmatrix} -c^2\overline{\rho}_{t_x t_x} & 0 & 0 & 0 & 0 & 0 \\ 0 & -c^2\overline{\rho}_{t_y t_y} & 0 & 0 & 0 & 0 \\ 0 & 0 & -c^2\overline{\rho}_{t_z t_z} & 0 & 0 & 0 \\ 0 & 0 & 0 & \overline{p}_{xx} & 0 & 0 \\ 0 & 0 & 0 & 0 & \overline{p}_{yy} & 0 \\ 0 & 0 & 0 & 0 & 0 & \overline{p}_{zz} \end{vmatrix} \qquad (6.224)$$

The temporal components of the Zwicky-flat tensor represent cosmological mass density and the spatial components represent cosmological energy pressure. The cosmological mass components and the cosmological energy components contribute a spatiotemporal positive curvature from the perspective of the curvature tensor side (left side) of the six-dimensional EFE.

And the reverse cosmological stress-energy-momentum tensor, or reverse Zwicky-flat tensor, for nearly-flat expanding spacetime may be expressed as

$$\Lambda^{\mu\nu} = \begin{vmatrix} -\frac{1}{c^2\overline{\rho}_{t_x t_x}} & 0 & 0 & 0 & 0 & 0 \\ 0 & -\frac{1}{c^2\overline{\rho}_{t_y t_y}} & 0 & 0 & 0 & 0 \\ 0 & 0 & -\frac{1}{c^2\overline{\rho}_{t_z t_z}} & 0 & 0 & 0 \\ 0 & 0 & 0 & \frac{1}{\overline{p}_{xx}} & 0 & 0 \\ 0 & 0 & 0 & 0 & \frac{1}{\overline{p}_{yy}} & 0 \\ 0 & 0 & 0 & 0 & 0 & \frac{1}{\overline{p}_{zz}} \end{vmatrix} \qquad (6.225)$$

Hence, we can describe the six-dimensional EFE which incorporates cosmological mass, $-c^2\overline{\rho}dv$, and cosmological energy, $\overline{p}dv$, as follows:

$$G_{\mu\nu} = \frac{8\pi G}{c^4}\left(T_{\mu\nu} - \Lambda_{\mu\nu}\right) \tag{6.226}$$

Reconsidering the above EFE, in a very large region of universal space-time, with no significant local mass or energy present as an object, $T_{\mu\nu} = 0$, assuming $n > 3$ dimensions, one can write the six-dimensional EFE for cosmological mass and cosmological energy in the form:

$$G_{\mu\nu} = -\frac{8\pi G}{c^4}\Lambda_{\mu\nu} \tag{6.227}$$

Contracting indices,

$$R = \frac{8\pi G}{c^4}\Lambda \tag{6.228}$$

Let us reconsider the cosmological stress-energy-momentum metric for a homogeneous and isotropic system of cosmological mass, in terms of the cosmological mass density and cosmological pressure,

embedded in near-flat space-time, such that

$$d\overline{\lambda}^2 = -c^2 d\overline{\rho}_{t_x t_x}{}^2 - c^2 d\overline{\rho}_{t_y t_y}{}^2 - c^2 d\overline{\rho}_{t_z t_z}{}^2 + d\overline{p}_{xx}{}^2 + d\overline{p}_{yy}{}^2 + d\overline{p}_{zz}{}^2 \tag{6.229}$$

$$d\overline{\lambda}^2 = g_{\mu\nu} d\overline{\rho}^{\mu} d\overline{p}^{\nu} \tag{6.230}$$

The concept of cosmological pressure is based on spacetime being a field in itself. As time emerges, the waves of the temporal field of spacetime interfere constructively or destructively to expand space or to contract space.

The actions of the temporal field in spacetime provide measurable cosmological pressure at every point. The cosmological pressure is the energy density of the spacetime field. Cosmological mass and cosmological energy are properties of spacetime.

The eminent physicist Albert Einstein expounded that it is possible for more space to emerge, and his cosmological constant predicted that empty space possesses its own energy.

Thus, cosmological pressure is a property of spacetime itself that is not diffused as spacetime expands in all directions. As time emerges, more space may be created, and more cosmological energy may be generated. The greater the cosmological energy content of the spacetime in our universe, the faster the cosmic acceleration. Hence, the cosmic acceleration is a consequence of the cosmological pressure produced by the spacetime field.

The cosmological pressure differential between areas of low cosmological pressure and high cosmological pressure would produce spatiotemporal curvature that may be observed as gravitational lensing. Therefore, curvature may be a result of the presence of a large system of mass, or a pressure differential between regions of spacetime.

A spatial manifold that is near and about the mass of a celestial body has lower time passing rate than a nearly-flat spatial manifold that is far away from the mass. Nearly-flat space has a faster cosmological time passing rate than curved space. Faster cosmological time passing rate may produce higher cosmological pressure. The General Theory of Relativity predicts that curvature changes the cosmological time passing rate of spacetime.

Let us describe the six-dimensional stress-energy-momentum tensor to be

$$T_{\mu\nu} = \begin{vmatrix} T_{t_x t_x} & T_{t_x t_y} & T_{t_x t_z} & T_{t_x x} & T_{t_x y} & T_{t_x z} \\ T_{t_y t_x} & T_{t_y t_y} & T_{t_y t_z} & T_{t_y x} & T_{t_y y} & T_{t_y z} \\ T_{t_z t_x} & T_{t_z t_y} & T_{t_z t_z} & T_{t_z x} & T_{t_z y} & T_{t_z z} \\ T_{x t_x} & T_{x t_y} & T_{x t_z} & T_{xx} & T_{xy} & T_{xz} \\ T_{y t_x} & T_{y t_y} & T_{y t_z} & T_{yx} & T_{yy} & T_{yz} \\ T_{z t_x} & T_{z t_y} & T_{z t_z} & T_{zx} & T_{zy} & T_{zz} \end{vmatrix} \tag{6.231}$$

For the FLRW universe, the stress-energy-momentum tensor is assumed to be isotropic and homogeneous with the energy density components and the pressure components being non-trivial because of symmetry. Thus, the time-time, space-time, or time-space components are $T_{ij} = T_{ji} = 0$ when $i \neq j$ and the space-space components are restricted to $T_{\mu\nu}(\vec{e}_i)^{\mu}(\vec{e}_j)^{\nu} = p\partial_{ij}$ such that

$$T_{\mu\nu} = \begin{vmatrix} c^2\rho_{t_x t_x} & 0 & 0 & 0 & 0 & 0 \\ 0 & c^2\rho_{t_y t_y} & 0 & 0 & 0 & 0 \\ 0 & 0 & c^2\rho_{t_z t_z} & 0 & 0 & 0 \\ 0 & 0 & 0 & p_{xx} & 0 & 0 \\ 0 & 0 & 0 & 0 & p_{yy} & 0 \\ 0 & 0 & 0 & 0 & 0 & p_{zz} \end{vmatrix} \tag{6.232}$$

Thus, if the stress-energy-momentum tensor has equal densities and pressures, the trace of the stress-energy-momentum tensor is $T = g^{\mu\nu}T_{\mu\nu} = -3\rho + 3p$.

The continuity equation relates the pressure and density functions and it is usually represented by the four-dimensional covariant derivative of the stress-energy-momentum tensor $T_{\mu\nu}$, in the direction of μ, $\nabla_\mu T^{\mu\nu} = 0$, but let us introduce the six-dimensional covariant derivative acting on $T_{\mu\nu}$ such as

$$\vec{\Re}_\mu T^{\mu\nu} = 0 \tag{6.233}$$

Hence, this means that the continuity equation no longer implies that the momentum energy and non-gravitational field energy expressed by the stress-energy-momentum tensor are absolutely conserved, i.e. the gravitational field can do work on mass and vice versa.

Thus, let us now derive the continuity equation for the speed of the energy density as follows

$$\rho = \frac{\Lambda c^2}{8\pi G} = \frac{m}{a} \tag{6.234}$$

$$\frac{\partial\rho}{\partial a} = -\frac{m}{a^2} \tag{6.235}$$

From the trace of $T_{\mu\nu}$, we can obtain the sum of the absolute values of the mass density components and the pressure components for near-flat spacetime so that

$$\frac{m}{a} = |-3\rho| + \left|3\frac{p}{c^2}\right| = 3\rho + 3\frac{p}{c^2} \tag{6.236}$$

$$\frac{\partial\rho}{\partial t} = \left(\frac{\partial a}{\partial t}\right)\left(\frac{\partial\rho}{\partial a}\right) = \dot{a}\left(-\frac{m}{a^2}\right) = \left(-\frac{\dot{a}}{a}\right)\left(\frac{m}{a}\right) \tag{6.237}$$

Thus, we find the continuity equation to be

$$\frac{\partial\rho}{\partial t} = -3\left(\frac{\dot{a}}{a}\right)\left(\rho + \frac{p}{c^2}\right) \tag{6.238}$$

In the cosmological case where the ratio of pressure over density is nearly constant, we find

$$\frac{p}{\rho} = \varepsilon \tag{6.239}$$

$$\frac{\partial\rho}{\partial t} = -3\left(\frac{\dot{a}}{ac^2}\right)\left(c^2\rho + p\right) \tag{6.240}$$

$$\frac{\partial\rho}{\partial t} = -3\left(\frac{\dot{a}}{ac^2}\right)\left(c^2 + \varepsilon\right)\rho \tag{6.241}$$

$$\frac{\frac{\partial\rho}{\partial t}}{\rho} = -3\left(\frac{\dot{a}}{ac^2}\right)\left(c^2 + \varepsilon\right) \tag{6.242}$$

Thus, we obtain the solution to this continuity equation to be

$$\rho = \rho_1 \frac{a}{c^2}^{-\left(3c^2 + 3\varepsilon\right)} \tag{6.243}$$

here the density is ρ_1 at a = 1, and ε is the pressure-over-density constant.

6.8. Reformulating the six-dimensional EFE equation for dynamic curved space-time and mass

Let us express now the six-dimensional EFE in trace-reversed form as follows

$$R_{\mu\nu} - \frac{\Lambda}{2} g_{\mu\nu} = \frac{8\pi G}{c^4}\left[T_{\mu\nu} - \frac{1}{4} g_{\mu\nu} T\right] \tag{6.244}$$

Hence, the right side of the EFE is no longer a static expression leading to non-linearity and back reaction problems, but it is now a dynamic expression for the energy and mass. Contracting terms on all tensors with $g^{\mu\nu}$, we find

$$(R - 3\Lambda) = \frac{8\pi\ddot{a}}{8\pi mc^4}\left[T - \frac{6}{4}T\right] \tag{6.245}$$

$$(R - 3\Lambda) = \frac{\ddot{a}}{mc^4}\left[-\frac{T}{2}\right] \tag{6.246}$$

Thus, we find the scalar curvature to be

$$R = \frac{\ddot{a}}{mc^4}\left[-\frac{T}{2}\right] + 3\Lambda \tag{6.247}$$

For a distance where *"r"* is very small compared to the radius of the universe we have

$$R \approx \frac{-\ddot{a}T}{2mc^4} \approx -\frac{\ddot{a}}{2c^4}\left(\frac{T}{m}\right) \approx -\frac{\ddot{a}}{2c^4}\left(\frac{1}{m}\left[\frac{3m}{a} + \frac{3mc^2}{a}\right]\right) \tag{6.248}$$

$$R \approx -\frac{\ddot{a}}{2c^4}\left(\frac{3}{a}\left[c^2 + 1\right]\right) \approx -\frac{3\ddot{a}}{2ac^2} \tag{6.249}$$

The above trace-reversed Ricci curvature scalar is equal to one-half of the stress-energy-momentum curvature acceleration ratio which produces spatiotemporal curvature within the system of mass.

6.9. On the anatomy of the stress-energy-momentum tensor

Let us illustrate the stress-energy-momentum tensor matrix for general relativity and its components by areas of energy, momentum and stress to ease the analysis, differentiation and discussion of actions and attributes.

Two areas, the energy momentum-per-area flux and the energy density shear stress, are proposed based on isotropy and the characteristics and actions of the time-time components.

According to current interpretation, the stress-energy-momentum tensor has attributes of mass, energy and non-gravitational force fields. In the following illustration of the tensor, areas are indicated for energy, momentum, stress, and pressure.

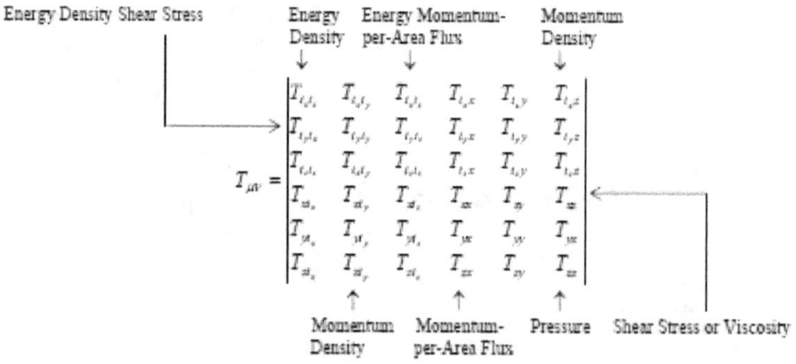

Figure 5. The stress-energy-momentum tensor

However, let us now analyze the symmetrical attributes of the stress-energy-momentum tensor according to the properties of the different actions of pressure.

Energy refers to *Joules*, or *Kg*, if *c*, the speed of light, equals one. Pressure (*p*) is force divided by area.

Thus, if we examine each of the tensor areas, the component units of measurement, and the actions of the components in those areas in relation to pressure we find:

Type of stress-energy-momentum	*Pressure, c (speed of light)*
Space-Time Momentum Density	*p/c*
Momentum-per-Area Flux	*p*
Pressure	*p*
Shear Stress	*p*
Time-Space Momentum Density	*p/c*
Energy Momentum-per-Area Flux	*p*
Energy Density	*p*
Energy Density Shear Stress	*p*

Figure 6. Stress Energy Tensor Pressures

If $c = 1$, as in general relativity terms, all areas are pressure areas, as if the tensor were a *dynamic energy-space-time pressure tensor*. Let us illustrate this concept further.

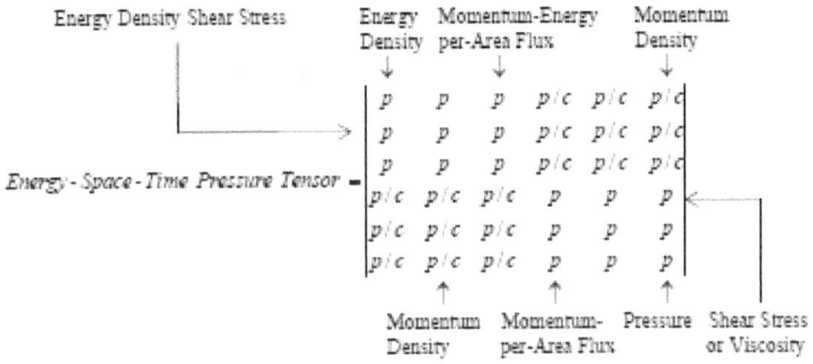

Figure 7. The dynamic energy-space-time pressure tensor

Hence, from this illustration the components still have the same attributes and actions, but the symmetric properties of the stress-energy-momentum tensor are now more apparent. *The actions of the stress-energy-momentum tensor are pressure actions by any physical manifestation of energy*. Thus, these actions result in momentum-density, kinematic energy, and non-gravitational force fields, as the result of the active pressure of the energy or mass on space-time and the reactive pressure of space-time on the energy or mass. By General Relativity, the reactive pressure of space-time on energy or mass is directly related to the gravitational field.

It is worth noting that for a smooth object of mass that is moving very slow with respect to the speed of light in isotropic and homogeneous space-time, the components of momentum density are going to be very small when compared to the time-time energy density components and the space-space pressure components of the stress-energy-momentum tensor of the object.

§ 7. On the direct square law of space-time

Space-time expands, or contracts, directly proportional to the square of the radius of the distance.

$$\frac{1}{r^2}=\frac{1}{c^2t^2}=\frac{1}{\left(\frac{1}{c^2}\frac{\partial^2 a}{\partial t^2}\right)^2}=\frac{1}{\left(\frac{\ddot{a}}{c^2}\right)^2} \tag{7.1}$$

If a volume of space-time, *a*, is treated as a point source of the medium, with spherical waveforms of propagation in homogeneous and isotropic space-time, then its surface area being that of an infinitesimal Planck sphere in continuous space-time would be $4\pi c^2 t_p{}^2$. The intensity of space-time would be the new surface area of the sphere, *P*, after propagation, divided by the surface area, $4\pi c^2 t_p{}^2$, after the spatial propagation interval was measured, such that

$$I_{st} = \frac{P}{4\pi c^2 t_p{}^2} \tag{7.2}$$

The gravitational force, electromagnetic radiation intensity, acoustic radiation intensity and Coulomb's Law are all inverse square laws that derive this characteristic or quality from the property of the direct square law of space-time for all these fields are embedded in the field and medium of space-time. Thus, *it is the spatiotemporal field, as a medium for the effect of gravitation, electromagnetic radiation or acoustic emanation that in its expansion or contraction exerts the inverse square law property on other physical fields traversing and existing in its dimensions.*

Every tensor transformation in the General Theory of Relativity that is a function of spacetime may be promoted as a field. All components of tensors and vectors are functions of spacetime. Spacetime is the underlying frame or coframe field to every tensor field, vector field, vector space, or dual vector space, in the manifold.

Since the components or concepts of the inverse law property were put forward in the seventeenth century by Ismael Bullialdus, Giovanni Alfonso Borelli, Robert Hooke, Halley, Isaac Newton, Wren et alia, the intensity of energy propagation, $I = (E/A)/r^2$, the intensity of light with no loss due to absorption or scattering, the intensity of sound or acoustic radiation, or the radiation from an isotropic antenna, $I = w/4\pi r^2$, have all been based on this property to expound how the intensity or the force would vary at a distance from a point source. Nonetheless, it was unknown and later overlooked that the true effect or pattern of the inverse square law property has been related to the propagating properties of space-time and the fact that space-time is a field in itself as well as a medium that exerts the inverse square law property to other physical fields or physical phenomena.

An example of the inverse square law property in Thermodynamics is the Bekenstein-Hawking equation for Entropy, $S_{BH} = k_B A/4l_p^2$, for the event horizon of black holes, where *"A"* is the area of the event horizon. This equation equalizes the entropy ratio, S_{BH}/k_B , to the ratio of spatiotemporal areas, $A/4l_p^2$, whose spatiotemporal dimensions represent extended space-time. If we express the equation as $(k_B/4S_{BH}) = l_p^2/4\pi R_{BH}^2$, we can recognize how the spatiotemporal dimensions on the right side of the equation vary by the inverse square law.

Furthermore, let us express the static Bekenstein-Hawking equation for Entropy as a function of the dynamic expansion of space-time.

$$S_{BH} = \frac{k_B \pi [R_{BH}]^2}{l_p^2} = \frac{k_B \pi [R_{BH}]^2 c^3}{G\hbar} = \frac{k_B \pi c^4 [R_{BH}]^2}{\left(\frac{\ddot{a}}{\sqrt[2]{8\pi}}\right)^2} \qquad (7.3)$$

$$S_{BH} = 8\pi^2 k_B \left(\frac{R_{BH}}{\frac{\ddot{a}}{c^2}}\right)^2 = 8\pi^2 k_B \left(\frac{\ddot{a}}{8\pi g_{BH}}\right)\left(\frac{c^4}{\ddot{a}^2}\right) = 8\pi^2 k_B \left(\frac{\ddot{a}}{8\pi \nabla^2 \ddot{a}}\right)\left(\frac{c^4}{\ddot{a}^2}\right) \qquad (7.4)$$

$$S_{BH} = \frac{\pi k_B}{\left(\frac{\ddot{a}}{c^2}\right)\left(\frac{\nabla^2\ddot{a}}{c^2}\right)} = \frac{\pi m_{BH}\left(\frac{\ddot{a}}{c^2}\right)^2\left(\frac{\dot{a}}{a}\right)^2}{T_{BH}\left(\frac{\ddot{a}}{c^2}\right)\left(\frac{\nabla^2\ddot{a}}{c^2}\right)} = \frac{\pi m_{BH}(\ddot{a})\left(\frac{\dot{a}}{a}\right)^2}{T_{BH}\nabla^2\ddot{a}} \qquad (7.5)$$

The acceleration of space-time is $\ddot{a} = 8\pi G m_p = 8\pi g r^2 = \hbar c / 8\pi m_p$. Furthermore, for the strong-field regime of a black hole about the outer event horizon we have $\ddot{a}/8\pi \leq R_{BH}c^2$ and $g_{BH} = \nabla^2\ddot{a}$. From the above dynamic equation of the Entropy for the event horizon of a black hole, it is theorized that Entropy is a function of space-time as space-time extends or contracts about the volume of a hypothetical spherical black hole as the radius of the black hole may also change.

Thus, the Entropy of the black hole is directly proportional to the mass and to the time rate of change of the space-time about the black hole, and inversely proportional to the temperature.

§ 8. On the introduction and applicability of six-dimensional space-time vector differential operators

We begin by considering vector field and scalar field vector differential operators that are continuously defined in homogeneous and isotropic Einsteinian space-time for multi-dimensional applications in a Cartesian or a Spherical Polar coordinate system. These operators are the offspring of the Del vector differential operator, ∇, and of the d'Alembertian four-dimensional vector differential operator, $\Box$, developed by the eminent physicists William Rowan Hamilton and Jean le Rond d'Alembert.

8.1. Defining and formulating Einsteinian operators

The Einsteinian vector operators are in partial differential form and are reflective of the properties of space and time in multi-dimensional space-time.

For Einsteinian space-time in a Cartesian or a Spherical Polar coordinate system, the multi-dimensional differential operators are:

The Robertonian operator $\vec{\mathfrak{R}}$ or covariant derivative for six-dimensional spacetime

$$\vec{\mathfrak{R}} = -\frac{1}{c}\frac{\partial}{\partial t_x}\vec{a}_{t_x} - \frac{1}{c}\frac{\partial}{\partial t_y}\vec{a}_{t_y} - \frac{1}{c}\frac{\partial}{\partial t_z}\vec{a}_{t_z} + \frac{\partial}{\partial x}\vec{a}_x + \frac{\partial}{\partial y}\vec{a}_y + \frac{\partial}{\partial z}\vec{a}_z \quad (8.1)$$

$$\vec{\mathfrak{R}} = -\frac{1}{c}\frac{\partial}{\partial t}\vec{t} - \left(\frac{1}{ct}\right)\frac{\partial}{\partial \theta}\vec{\theta}_t - \left(\frac{1}{ctSin\theta}\right)\frac{\partial}{\partial \varphi}\vec{\varphi}_t + \frac{\partial}{\partial r}\vec{r} + \left(\frac{1}{r}\right)\frac{\partial}{\partial \theta}\vec{\theta}_r + \left(\frac{1}{rSin\theta}\right)\frac{\partial}{\partial \varphi}\vec{\varphi}_r \quad (8.2)$$

The d'Robertonian Operator $\Diamond$ for six-dimensional space-time

$$\Diamond = -\frac{1}{c}\frac{\partial}{\partial t_x} - \frac{1}{c}\frac{\partial}{\partial t_y} - \frac{1}{c}\frac{\partial}{\partial t_z} + \frac{\partial}{\partial x} + \frac{\partial}{\partial y} + \frac{\partial}{\partial z} \quad (8.3)$$

$$\Diamond = -\frac{1}{c}\frac{\partial}{\partial t} - \left(\frac{1}{ct}\right)\frac{\partial}{\partial \theta} - \left(\frac{1}{ctSin\theta}\right)\frac{\partial}{\partial \varphi} + \frac{\partial}{\partial r} + \left(\frac{1}{r}\right)\frac{\partial}{\partial \theta} + \left(\frac{1}{rSin\theta}\right)\frac{\partial}{\partial \varphi} \quad (8.4)$$

The Double Robertonian Operator $\Diamond^2$ for six-dimensional space-time

$$\Diamond^2 = \vec{\mathfrak{R}}^2 = \vec{\mathfrak{R}} \cdot \vec{\mathfrak{R}} = \frac{1}{c^2}\frac{\partial^2}{\partial t_x^{\,2}} + \frac{1}{c^2}\frac{\partial^2}{\partial t_y^{\,2}} + \frac{1}{c^2}\frac{\partial^2}{\partial t_z^{\,2}} + \frac{\partial^2}{\partial x^2} + \frac{\partial^2}{\partial y^2} + \frac{\partial^2}{\partial z^2} \quad (8.5)$$

$$\Diamond^2 = \vec{\mathfrak{R}}^2 = \frac{1}{c^2}\frac{\partial^2}{\partial t^2} + \frac{1}{c^2 t^2}\frac{\partial^2}{\partial \theta^2} + \frac{1}{c^2 t^2 Sin^2\theta}\frac{\partial^2}{\partial \varphi^2} + \frac{\partial^2}{\partial r^2} + \frac{1}{r^2}\frac{\partial^2}{\partial \theta^2} + \frac{1}{r^2 Sin^2\theta}\frac{\partial^2}{\partial \varphi^2} \quad (8.6)$$

The Tempus Operator, or Tem Operator, for three-dimensional time

$$\odot = \vec{\mathfrak{R}}_\tau = -\frac{1}{c}\frac{\partial}{\partial t_x}\vec{a}_{t_x} - \frac{1}{c}\frac{\partial}{\partial t_y}\vec{a}_{t_y} - \frac{1}{c}\frac{\partial}{\partial t_z}\vec{a}_{t_z} \quad (8.7)$$

$$\odot = \vec{\mathfrak{R}}_\tau = -\frac{1}{c}\frac{\partial}{\partial t}\vec{t} - \left(\frac{1}{ct}\right)\frac{\partial}{\partial \theta}\vec{\theta} - \left(\frac{1}{ctSin\theta}\right)\frac{\partial}{\partial \varphi}\vec{\varphi} \quad (8.8)$$

The Double Tem for three-dimensional time

$$\odot^2 = \vec{\mathfrak{R}}_\tau^{\,2} = \vec{\mathfrak{R}}_\tau \cdot \vec{\mathfrak{R}}_\tau = \frac{1}{c^2}\frac{\partial^2}{\partial t_x^{\,2}} + \frac{1}{c^2}\frac{\partial^2}{\partial t_y^{\,2}} + \frac{1}{c^2}\frac{\partial^2}{\partial t_z^{\,2}} \qquad (8.9)$$

$$\odot^2 = \frac{1}{c^2}\frac{\partial^2}{\partial t^2} + \frac{1}{c^2 t^2}\frac{\partial^2}{\partial \theta^2} + \frac{1}{c^2 t^2 Sin^2\theta}\frac{\partial^2}{\partial \varphi^2} \qquad (8.10)$$

The Furl of a three-dimensional temporal vector or spatial vector $\vec{V}$

$$\odot \times \vec{V} = furl\vec{V} = \begin{vmatrix} -\vec{a}_{t_x} & -\vec{a}_{t_y} & -\vec{a}_{t_z} \\ \frac{1}{c}\frac{\partial}{\partial t_x} & \frac{1}{c}\frac{\partial}{\partial t_y} & \frac{1}{c}\frac{\partial}{\partial t_z} \\ V_{t_x} & V_{t_y} & V_{t_z} \end{vmatrix} \qquad (8.11)$$

where $\vec{V} = V_{t_x}\vec{a}_{t_x} + V_{t_y}\vec{a}_{t_y} + V_{t_z}\vec{a}_{t_z}$ *or* $\vec{V} = V_x\vec{a}_x + V_y\vec{a}_y + V_z\vec{a}_z$

The Furl of a three-dimensional vector $\vec{V}$

$$\odot \times \vec{V} = furl\,\vec{V} = \frac{1}{c}\left(\frac{\partial V_y}{\partial t_z} - \frac{\partial V_z}{\partial t_y}\right)a_{t_x} + \frac{1}{c}\left(\frac{\partial V_z}{\partial t_x} - \frac{\partial V_x}{\partial t_z}\right)a_{t_y} + \frac{1}{c}\left(\frac{\partial V_x}{\partial t_y} - \frac{\partial V_y}{\partial t_x}\right)a_{t_z} \qquad (8.12)$$

The Swirl of a three-dimensional temporal vector or spatial vector $\vec{V}$

$$\eth \times \vec{V} = swirl\ \vec{V} = \begin{vmatrix} -\vec{a}_{t_x} & -\vec{a}_{t_y} & -\vec{a}_{t_z} \\ \frac{1}{c^2}\frac{\partial^2}{\partial t_x^{\,2}} & \frac{1}{c^2}\frac{\partial^2}{\partial t_y^{\,2}} & \frac{1}{c^2}\frac{\partial^2}{\partial t_z^{\,2}} \\ V_{t_x} & V_{t_y} & V_{t_z} \end{vmatrix} \qquad (8.13)$$

The Swirl of a three-dimensional vector $\vec{V}$

$$\eth \times \vec{V} = swirl\ \vec{V} = \frac{1}{c^2}\left(\frac{\partial^2 V_y}{\partial t_z^{\ 2}} - \frac{\partial^2 V_z}{\partial t_y^{\ 2}}\right)a_{t_x} + \frac{1}{c^2}\left(\frac{\partial^2 V_z}{\partial t_x^{\ 2}} - \frac{\partial^2 V_x}{\partial t_z^{\ 2}}\right)a_{t_y} + \frac{1}{c^2}\left(\frac{\partial^2 V_x}{\partial t_y^{\ 2}} - \frac{\partial^2 V_y}{\partial t_x^{\ 2}}\right)a_{t_z} \quad (8.14)$$

The direction of either the furl (or time rotor) or swirl is the axis of rotation, as determined by the left-hand rule, and the magnitude of the furl is the magnitude of rotation. The magnitude of the swirl is the acceleration in the rotation of the furl.

Let us now define the tem operator as a multidimensional tempus operator

$$\odot = -\sum_{i=1}^{n} \frac{1}{c}\frac{\partial}{\partial t_i}\hat{e}_{t_i} \quad where \ \left(\hat{e}_{t_i} : 1 \le i \le n\right) \quad (8.15)$$

More compactly using Einstein summation notation, the tem operator is written as

$$\odot = -\vec{e}_{t_i}\frac{\partial t_i}{c} \quad (8.16)$$

8.2. The n-divergence of six-dimensional space-time

A characteristic of the space-time field is that its *n*-divergence outside its physical source for an irrotational vector field in six-dimensional space-time is not zero, $div^n\ \vec{\Psi}_{st} \neq 0$,

where $\vec{\Psi}_{st}$ is a six-dimensional space-time vector field $-\Psi_{t_x}\vec{a}_{t_x} - \Psi_{t_y}\vec{a}_{t_y} - \Psi_{t_z}\vec{a}_{t_z} + \Psi_x\vec{a}_x + \Psi_y\vec{a}_y + \Psi_z\vec{a}_z$ and div^n is the n-divergence. This characteristic is due to the fact that space-time is n-diverging, or varying in all directions of space and time. When and where space-time is homogeneous or isotropic or not, $div^n = \vec{\Psi}_{st} \neq 0$, and its n-divergence would be positive if space-time is expanding or negative if it is contracting.

Let us apply the Robertonian operator $(\vec{\Re})$ such that

$$div^n\ \vec{\Psi}_{st} = \vec{\Re} \cdot \vec{\Psi}_{st} \quad (8.17)$$

where $\vec{\Re}$ is the Robertonian operator specified as

$$\vec{\Re} = -\frac{1}{c}\frac{\partial}{\partial t_x}\vec{a}_{t_x} - \frac{1}{c}\frac{\partial}{\partial t_y}\vec{a}_{t_y} - \frac{1}{c}\frac{\partial}{\partial t_z}\vec{a}_{t_z} + \frac{\partial}{\partial x}\vec{a}_x + \frac{\partial}{\partial y}\vec{a}_y + \frac{\partial}{\partial z}\vec{a}_z \qquad (8.18)$$

Thus, we find the n-divergence applying the Robertonian operator to be

$$\vec{\Re}\cdot\vec{\Psi}_{st} = \frac{1}{c}\frac{\partial \Psi_{t_x}}{\partial t_x} + \frac{1}{c}\frac{\partial \Psi_{t_y}}{\partial t_y} + \frac{1}{c}\frac{\partial \Psi_{t_z}}{\partial t_z} + \frac{\partial \Psi_x}{\partial x} + \frac{\partial \Psi_y}{\partial y} + \frac{\partial \Psi_z}{\partial z} \qquad (8.19)$$

Let us imagine that during a specified space-time interval of an inertial coordinate system in Einsteinian space-time for a closed surface or for a volume of a body, an evenly and uniformly distributed physical property such as charge, or mass, exists in the closed boundary of the surface or body. The divergence of the field of the physical property would not be zero if the net time rate of change of space-time in the direction of the axes of the specified coordinate system is either a positive or a negative real number. This result or condition implies that space-time acts as the propagating medium in all directions and dimensions of the Einsteinian space-time in the neighborhood of the surface or body in the specified coordinate system.

Thus, in that imaginary physical condition of space-time we find

$$div^n\,\vec{\Psi}_{st} = \vec{\Re}\cdot\vec{\Psi}_{st} \neq 0 \qquad (8.20)$$

8.3. Applying Einsteinian operators on scalar and vector fields

The Einsteinian space-time operators are useful in dealing with both space-time vector fields and scalar fields. Given any space-time scalar field Ψ we may form a vector field, called the n-gradient of Ψ and written as n-grad Ψ, simply by applying the Robertonian operator $\vec{\Re}$ to Ψ. Given a space-time vector field, $\vec{U} = -U_{t_x}\vec{a}_{t_x} - U_{t_y}\vec{a}_{t_y} - U_{t_z}\vec{a}_{t_z} + U_x\vec{a}_x + U_y\vec{a}_y + U_z\vec{a}_z$, we may apply the Robertonian operator $\vec{\Re}$ in two different ways. One way is to take the dot product of $\vec{\Re}$ and $\vec{U}$, yielding the scalar field called the n-divergence of $\vec{U}$ and written as n-div $\vec{U}$ or $\mathfrak{R}\cdot\vec{U}$. The other way is to take the twirl of $\vec{U}$ and written twirl $\vec{U}$, or ◎$\vec{U}$.

The above operations may be summarized as

$$n-grad\,\Psi = \vec{\mathfrak{R}}\Psi = -\frac{1}{c}\frac{\partial\Psi}{\partial t_x}\vec{a}_{t_x} - \frac{1}{c}\frac{\partial\Psi}{\partial t_y}\vec{a}_{t_y} - \frac{1}{c}\frac{\partial\Psi}{\partial t_z}\vec{a}_{t_z} + \frac{\partial\Psi}{\partial x}\vec{a}_x + \frac{\partial\Psi}{\partial y}\vec{a}_y + \frac{\partial\Psi}{\partial z}\vec{a}_z \quad (8.21)$$

$$\text{n-div}\,\vec{U} = \vec{\mathfrak{R}}\cdot\vec{U} = \frac{1}{c}\frac{\partial Ut_x}{\partial t_x} + \frac{1}{c}\frac{\partial Ut_y}{\partial t_y} + \frac{1}{c}\frac{\partial Ut_z}{\partial t_z} + \frac{\partial U_x}{\partial x} + \frac{\partial U_y}{\partial y} + \frac{\partial U_z}{\partial z} \quad (8.22)$$

Given that $\vec{U} = -U_{t_x}\vec{a}_{t_x} - U_{t_y}\vec{a}_{t_y} - U_{t_z}\vec{a}_{t_z} + U_x\vec{a}_x + U_y\vec{a}_y + U_z\vec{a}_z$ *is a space-time vector field.*

The Twirl of a six-dimensional space-time vector $\vec{U}$

$$◎\vec{U} = \begin{vmatrix} -\vec{a}_{t_x} & -\vec{a}_{t_y} & -\vec{a}_{t_z} \\ \frac{1}{c}\frac{\partial}{\partial t_x} & \frac{1}{c}\frac{\partial}{\partial t_y} & \frac{1}{c}\frac{\partial}{\partial t_z} \\ U_{t_x} & U_{t_y} & U_{t_z} \end{vmatrix} + \begin{vmatrix} \vec{a}_x & \vec{a}_y & \vec{a}_z \\ \frac{\partial}{\partial x} & \frac{\partial}{\partial y} & \frac{\partial}{\partial z} \\ U_x & U_y & U_z \end{vmatrix} \quad (8.23)$$

$$twirl\,\vec{U} = -\frac{1}{c}\left(\frac{\partial U_{t_z}}{\partial t_y} - \frac{\partial U_{t_y}}{\partial t_z}\right)\vec{a}_{t_x} - \frac{1}{c}\left(\frac{\partial U_{t_x}}{\partial t_z} - \frac{\partial U_{t_z}}{\partial t_x}\right)\vec{a}_{t_y} - \frac{1}{c}\left(\frac{\partial U_{t_y}}{\partial t_x} - \frac{\partial U_{t_x}}{\partial t_y}\right)\vec{a}_{t_z} \quad (8.24)$$
$$+\left(\frac{\partial U_z}{\partial y} - \frac{\partial U_y}{\partial z}\right)\vec{a}_x + \left(\frac{\partial U_x}{\partial z} - \frac{\partial U_z}{\partial x}\right)\vec{a}_y + \left(\frac{\partial U_y}{\partial x} - \frac{\partial U_x}{\partial y}\right)\vec{a}_z$$

The Whirl of a six-dimensional space-time vector $\vec{U}$

$$⊛\vec{U} = \begin{vmatrix} -\vec{a}_{t_x} & -\vec{a}_{t_y} & -\vec{a}_{t_z} \\ \frac{1}{c^2}\frac{\partial^2}{\partial t_x^{\,2}} & \frac{1}{c^2}\frac{\partial^2}{\partial t_y^{\,2}} & \frac{1}{c^2}\frac{\partial^2}{\partial t_z^{\,2}} \\ U_{t_x} & U_{t_y} & U_{t_z} \end{vmatrix} + \begin{vmatrix} \vec{a}_x & \vec{a}_y & \vec{a}_z \\ \frac{\partial^2}{\partial x^2} & \frac{\partial^2}{\partial y^2} & \frac{\partial^2}{\partial z^2} \\ U_x & U_y & U_z \end{vmatrix} \quad (8.25)$$

$$whirl\,\vec{U} = -\frac{1}{c^2}\left(\frac{\partial^2 U_{t_z}}{\partial t_y^{\,2}} - \frac{\partial^2 U_{t_y}}{\partial t_z^{\,2}}\right)\vec{a}_{t_x} - \frac{1}{c^2}\left(\frac{\partial^2 U_{t_x}}{\partial t_z^{\,2}} - \frac{\partial^2 U_{t_z}}{\partial t_x^{\,2}}\right)\vec{a}_{t_y} - \frac{1}{c^2}\left(\frac{\partial^2 U_{t_y}}{\partial t_x^{\,2}} - \frac{\partial^2 U_{t_x}}{\partial t_y^{\,2}}\right)\vec{a}_{t_z}$$
$$+\left(\frac{\partial^2 U_z}{\partial y^2} - \frac{\partial^2 U_y}{\partial z^2}\right)\vec{a}_x + \left(\frac{\partial^2 U_x}{\partial z^2} - \frac{\partial^2 U_z}{\partial x^2}\right)\vec{a}_y + \left(\frac{\partial^2 U_y}{\partial x^2} - \frac{\partial^2 U_x}{\partial y^2}\right)\vec{a}_z \quad (8.26)$$

Note that n-grad Ψ, the twirl $\vec{U}$, and the whirl $\vec{U}$ are vectors, whereas n-div $\vec{U}$ is a scalar.

The twirl and the whirl represent the velocity and acceleration of a six-dimensional vector field in space-time.

Hence, let us apply the Einsteinian vector differential operators to a scalar field Ψ or to vector fields $\vec{U}$ or $\vec{V}$ to obtain the Double Robertonian operator $\vec{\Re}^2$ of a scalar field Ψ:

$$\vec{\Re}^2\,\Psi = \frac{1}{c^2}\frac{\partial^2\Psi}{\partial t_x^{\,2}} + \frac{1}{c^2}\frac{\partial^2\Psi}{\partial t_y^{\,2}} + \frac{1}{c^2}\frac{\partial^2\Psi}{\partial t_z^{\,2}} + \frac{\partial^2\Psi}{\partial x^2} + \frac{\partial^2\Psi}{\partial y^2} + \frac{\partial^2\Psi}{\partial z^2} \quad (8.27)$$

The Double Robertonian operator $\vec{\Re}^2$ applied to a vector field $\vec{U}$:

$$\vec{\Re}^2\cdot\vec{U} = -\frac{1}{c^2}\frac{\partial^2 U_{t_x}}{\partial t_x^{\,2}}\vec{a}_{t_x} - \frac{1}{c^2}\frac{\partial^2 U_{t_y}}{\partial t_y^{\,2}}\vec{a}_{t_y} - \frac{1}{c^2}\frac{\partial^2 U_{t_z}}{\partial t_z^{\,2}}\vec{a}_{t_z} + \frac{\partial^2 U_x}{\partial x^2}\vec{a}_x + \frac{\partial^2 U_y}{\partial y^2}\vec{a}_y + \frac{\partial^2 U_z}{\partial z^2}\vec{a}_z \quad (8.28)$$

$$where \quad \Diamond^2 = \vec{\Re}^2 = \frac{1}{c^2}\frac{\partial^2}{\partial t_x^{\,2}} + \frac{1}{c^2}\frac{\partial^2}{\partial t_y^{\,2}} + \frac{1}{c^2}\frac{\partial^2}{\partial t_z^{\,2}} + \frac{\partial^2}{\partial x^2} + \frac{\partial^2}{\partial y^2} + \frac{\partial^2}{\partial z^2}$$

In general-curvilinear coordinates $\left(\zeta^1,\zeta^2,\zeta^3,\zeta^4,\zeta^5,\zeta^6\right)$ of six-dimensional spacetime we have

$$\Diamond^2\vec{U} = \sum_{m=1}^{6}\Diamond\zeta^m\cdot\sum_{m=1}^{6}\sum_{n=1}^{6}\Diamond\zeta^n\frac{\partial^2\vec{U}}{\partial\zeta^m\partial\zeta^n} + \sum_{m=1}^{6}\Diamond^2\zeta^m\frac{\partial\vec{U}}{\partial\zeta^m} \quad (8.29)$$

Where summation over the repeated indices is implied by the Einstein summation convention.

$$\Diamond^2 \vec{U} = \Diamond\zeta^m \cdot \Diamond\zeta^n \frac{\partial^2 \vec{U}}{\partial\zeta^m \partial\zeta^n} + \Diamond^2\zeta^m \frac{\partial \vec{U}}{\partial\zeta^m} \tag{8.30}$$

The Tem Operator of a scalar field Ψ :

$$\vec{\mathfrak{R}}_\tau \Psi = -\frac{1}{c}\frac{\partial\Psi}{\partial t_x}\vec{a}_{t_x} - \frac{1}{c}\frac{\partial\Psi}{\partial t_y}\vec{a}_{t_y} - \frac{1}{c}\frac{\partial\Psi}{\partial t_z}\vec{a}_{t_z} \tag{8.31}$$

The Double Tem Operator of a three-dimensional vector field $\vec{V}$:

$$\vec{\mathfrak{R}}_\tau^{\,2} \cdot \vec{V} = -\frac{1}{c^2}\frac{\partial^2 V_x}{\partial t_x^{\,2}}\vec{a}_{t_x} - \frac{1}{c^2}\frac{\partial^2 V_y}{\partial t_y^{\,2}}\vec{a}_{t_y} - \frac{1}{c^2}\frac{\partial^2 V_z}{\partial t_z^{\,2}}\vec{a}_{t_z} \tag{8.32}$$

Lastly, we may reformulate some Heaviside-Maxwell equations in point form using the Tem Operator:

$$\begin{aligned} \nabla x\vec{E} &= \vec{\mathfrak{R}}_\tau x\vec{B} \\ \nabla x\vec{H} &= \vec{J} - \left(\vec{\mathfrak{R}}_\tau x\vec{D}\right) \\ \nabla \cdot \vec{D} &= \rho \\ \nabla \cdot \vec{B} &= 0 \end{aligned} \tag{8.33}$$

§ 9. Conclusion

In conclusion, we have seen that time is a property of space-time that exhibits a similar three-dimensional structure as space, two sides of the same coin, interacting together to form the past, present, and future. Time is measured as a speed and it is a path-dependent quality of space-time. It is hoped that some experimenter, or enquirer, with the proper resources may succeed in conducting the physical experiments that validate the principles and theories of space-time expressed in this document.

Chapter 8

On the natures of gravity, light, and space-time

§ 1. Introduction: the gravitational field

Gravity has been described as the natural force of attraction exerted by a celestial body, such as Earth, upon objects at or near its surface, tending to draw them toward the center of the body or the force of attraction that moves or tends to move bodies towards the center of a celestial body, such as the earth or moon.

Gravity is considered the fundamental force of attraction that all objects with mass have for each other. Like the electromagnetic force, gravity has effectively infinite range and obeys the inverse-square law. At the atomic level, where masses are very small, the force of gravity is negligible, but for objects that have very large masses such as planets, stars, and galaxies, gravity is a predominant force, and it plays an important role in theories of the structure of the universe. Gravity is believed to be mediated by the graviton, although the graviton has yet to be isolated by experiment. Gravity is weaker than the strong force, the electromagnetic force, and the weak force.

Sir Isaac Newton described gravity, with his law of universal gravitation, as the mutual attraction between any two bodies in the universe. He developed an equation describing an instantaneous gravitational effect that any two objects, no matter how far apart or how small, exert on each other. These effects diminish as the distance between the objects gets larger and as the masses of the objects get smaller. As formulated by Newton, this natural force of attraction between any two objects is directly proportional to the product of their masses and inversely proportional to the square of the distance between them. Newton's theory explained both the motion of celestial bodies and the trajectory of the legendary falling apple, up to that time completely unconnected natural phenomena, using the same equations.

Albert Einstein developed the first revision of these ideas. Einstein needed to extend his theory of Special Relativity to be able to understand cases in which bodies were subject to forces and acceleration, as in the case of gravity. According to Special Relativity, however, the instantaneous gravitational effects in

Newton's theory would not be possible, because for gravity to act instantaneously, it would have to travel at infinite speeds, faster than the speed of light, the upper speed limit in the theory of Special Relativity. Thus, Einstein developed the General Theory of Relativity to overcome these inconsistencies, which connected gravity, mass, and acceleration in a new way. The General Theory of Relativity demonstrates that space, time, and mass are connected, and related by the dimensions of space-time-mass that constitute an existential coherent continuum. (Einstein, 1952)

Einstein developed the principle of equivalence explaining that as objects with mass, or massless objects such as photons, travel in their trajectories through space-time, relativistic mass curves space-time, creating the effects of gravity. For Einstein, the effects of gravity felt by an astronaut standing in a stationary rocket on the Earth and the same effects of gravity felt by the same astronaut and rocket accelerating in outer space, far from any significant gravitational field, were indistinguishable gravitational effects.

Thus far, the gravitational forces of Newton's universal law of gravitation or the effects of Einstein's curvature of space-time are the gravitational effects on space-time, mass or between masses of objects. Hence, what is gravity itself? How does gravity produce such effects between masses or on space-time? Is there a particle, such as the graviton, mediating gravity? Where is the graviton? Is gravity discrete or continuous?

These are some of the questions that physicists have asked in the pursuit of the elusive nature of gravity. However elusive the nature of gravity has been, every discovery or development has brought us closer to understanding gravity and the gravitational field.

§ 2. On the nature of gravity

The nature of gravity is the nature of space-time for mass itself is not the source of the gravitational field, but mass is one of its causes. The gravitational force does not emanate from the mass, but it is exerted upon the surface of the mass by space-time. Thus, the space-time between two masses is curved and that curvature of space-time causes a mutual gravitational effect. This is the nature of the force pushing on our bodies and on the objects on or near the earth towards the center of our planet. That force or pressure upon us is the force or pressure of space-time. *Gravity is the contraction of space and the slowdown of time, conjugates of space-time, acting perpendicular to the surface of an object with rest mass or relativistic mass.*

Throughout this document, the property of mass may be referred to as rest mass or relativistic mass. Thus, before going further let us clarify what is meant by these labels for the property of mass. Rest mass, m_0 , is a system of mass or the mass of an object that is not moving relative to an observer's frame of reference. Relativistic mass, m' , is the dilating mass of an object traveling at relativistic speed relative to an observer's inertial frame of reference. Thus, it is possible to define the complex mass as $\tilde{m} = m_0 \pm im'$, depending on the retarded, advanced, or null, direction of movement of the object of mass.

However, if we consider these labels for the properties of mass, we see that even rest mass translates at some speed through space-time as a system or as part of a much larger moving system in the universe. Moreover, rest mass has some of its particles traveling or spinning relativistically within the system of mass itself. Then, isn't rest mass relativistic in nature?

Hence, *the principle of relativistic mass asserts that the masses of particles or objects in a dynamic motional universe are relativistic in nature, e.g.* $\tilde{m} = m_0' \pm im'$ *, or relativistic as a result of relativistic translation as measured from an inertial frame of reference.*

The nature of gravity infers a very important quality of space-time, the reciprocal relationship of space and time as conjugates in space-time. This reciprocal relationship is reflected in the simultaneous and compensating way that space contracts and time dilates. This effective relationship keeps the speed of space-time on even keel. Thus, the speed of space-time remains uniform.

In the absence of all physical manifestations of mass and energy in isotropic and homogeneous space-time, every point in space-time expands in all directions. In a region of expanding space-time, the relative magnitude of space-time may increase as space-time expands, preserving the conjugate qualities of space and time.

2.1. On the space wave function

The mathematician Walter Craig and the physicist Steven Weinstein in 2008 proved that under a nonlocal constraint, the initial value problem is well-posed for periodic initial data given on a codimension-one hypersurface. The following illustrations and descriptions are proposed to adapt their significant findings and conceptualizations to a three-dimensional temporal hypersurface model in 3T-Physics that folds or collapses into our familiar one-dimensional time model in 1T-Physics. It is proposed that periodic initial data on a mixed hypersurface (space-like and time-like) obeying a particular nonlocal constraint evolves deterministically in the remaining resultant time dimension *t*.

Let us illustrate space-time as a generalization of a massless scalar field, $\Psi = \Psi(x, y, z, t_X, t_Y, t_Z)$, described by a hyperbolic space-time wave function with three spatial dimensions and three temporal dimensions as follows

$$\left(\frac{\partial^2}{\partial x^2}+\frac{\partial^2}{\partial y^2}+\frac{\partial^2}{\partial z^2}\right)\Psi=\left(\frac{\partial^2}{\partial t^2}\right)\Psi \qquad (2.1)$$

$$\left(\frac{\partial^2}{\partial t^2}\right)\Psi=\left(\frac{\partial^2}{\partial t_X{}^2}+\frac{\partial^2}{\partial t_Y{}^2}+\frac{\partial^2}{\partial t_Z{}^2}\right)\Psi \qquad (2.2)$$

$$\left(\frac{\partial^2}{\partial x^2}+\frac{\partial^2}{\partial y^2}+\frac{\partial^2}{\partial z^2}\right)\Psi=\left(\frac{\partial^2}{\partial t_X{}^2}+\frac{\partial^2}{\partial t_Y{}^2}+\frac{\partial^2}{\partial t_Z{}^2}\right)\Psi \qquad (2.3)$$

The above equations may be describing either a spatiotemporal medium in four dimensions or six dimensions since the four-dimensional model is conceptualized as the collapsed version of the six-dimensional model. In the four-dimensional model, the resultant temporal metric element dt^2 or the temporal magnitude $|t|$ used in equations of 1T-Physics represent the intersection of three temporal planes, or hypersurfaces in 3T-Physics, that are conjugates to the familiar three spatial planes of a Cartesian coordinate system. These equations may describe the propagation of the components of the electromagnetic field. If we are given sufficient information about the electromagnetic field at a given point $t(t_X, t_Y, t_Z)$ in time, a stable and unique solution of the equation exists. It is well known that the initial value of this equation is well-posed. Hence, the periodic initial data completely determine the data at all other times, in such a way that those small errors in the specification of the initial data do not lead to uncontrollable errors in the solution of the equation.

Let us consider the case in which the periodic initial data lies on the codimension-one hypersurface described by the above equations as in a Cauchy problem. A Cauchy surface is a complete, connected spatial surface, that intersects every temporal geodesic or null geodesic, once and only once. Any surface of constant "t" in Minkowski space-time is a Cauchy surface. A Cauchy two-manifold is a temporal instant, that given initial conditions can determine the future and the past uniquely. The periodic initial data of the above equations consist of the electromagnetic field and its first normal temporal derivative, perpendicular to the hypersurface at each point, since the above equations consists of only second derivatives.

The periodic initial data is as follows

$$u(p) = \Psi(p,t) = \Psi(p,0) = \Psi(x,y,z,0,0,0) \qquad (2.4)$$

$$v(p) = \frac{\partial \Psi(p,t)}{\partial t} = \frac{\partial \Psi(p,0)}{\partial t} = \frac{\partial (x,y,z,0,0,0)}{\partial t} \qquad (2.5)$$

Where p is equivalent to a spatial point at (x, y, z) and t is equivalent to a temporal point at (t_X, t_Y, t_Z).

A spatial point "p" may be defined as the intersection of three spatial planes. Similarly, let us define a temporal point "t" as the intersection of three temporal planes in 3T-Physics. (Born et al, 1999)

In this case, if appropriately differentiable functions $u(p)$ and $v(p)$, that represent the relevant properties of the electromagnetic field at some time "t", are given, then, a unique, stable solution exists for all times. Thus, with that condition, the Cauchy problem for the above wave equation is well-posed. It is noted that there have been concerns about the initial value problems of a wave function with multiple temporal dimensions.

Therefore, let us recognize that in the case of temporal multidimensionality, where it has been shown by using the mean value theorem of Asgeirsson that for the arbitrary choices of initial data, solutions for the wave equation do not exist, it has also been shown by the Holmgren-John Uniqueness Theorem that there are unique solutions everywhere on the hypersurface that do exist as long as that periodic initial data is consistent with some solution. (Courant, 1962)

The Holmgren-John Uniqueness Theorem tells us that domains of dependence and influence are compact, so that we only need to know the solution on a compact region "*R*" of the hypersurface in order to determine the solution of the Cauchy problem at a given point "*P*" above the hypersurface. Thus, data in the compact region "*R*" determines data at point "*P*" above the hypersurface. Hence, when a constraint is imposed on the periodic initial data, the data yields a well-posed Cauchy problem after all. (Craig, 2009)

Let us now consider a constraint on the periodic initial data of *u(p, t)* and *v(p, t)* so that only the Fourier transforms of the periodic initial data satisfying the constraint will lead to a stable solution.

$$|\theta|^2 \leq |d|^2 \tag{2.6}$$

$$\hat{u}(d_1, d_2, d_3, \theta_1, \theta_2, \theta_3) = F\{u(x, y, z, t_X, t_Y, t_Z)\} \tag{2.7}$$

$$\hat{v}(d_1, d_2, d_3, \theta_1, \theta_2, \theta_3) = F\{v(x, y, z, t_X, t_Y, t_Z)\} \tag{2.8}$$

Where "*d*" is equivalent to a spatial point at (d_1, d_2, d_3) and "θ" is equivalent to a temporal point at $(\theta_1, \theta_2, \theta_3)$.

The allowable sets of periodic initial data by the constraint are given by the inverse Fourier transforms of the functions $\hat{u}(d_1, d_2, d_3, \theta_1, \theta_2, \theta_3)$ and $\hat{v}(d_1, d_2, d_3, \theta_1, \theta_2, \theta_3)$ as shown below.

$$u(x, y, z, t_X, t_Y, t_Z) = F^{-1}\hat{u}(d_1, d_2, d_3, \theta_1, \theta_2, \theta_3) \tag{2.9}$$

$$v(x, y, z, t_X, t_Y, t_Z) = F^{-1}\hat{v}(d_1, d_2, d_3, \theta_1, \theta_2, \theta_3) \tag{2.10}$$

The initial data constraint has a non-local property; it establishes nontrivial correlations between the values of the wave function of the field at different points on the hypersurface. The non-locality is considered causally benign since there is no sense in which changes in the compact region "*R*" may bring instantaneous changes in a larger region of the hypersurface.

Therefore, the features of 1T-Physics, in three dimensions of space and one of time, transition well to the realm of 3T-Physics, when we posit the initial value problem as a Cauchy problem. Moreover, there is a well-defined Hamiltonian, or energy functional, which is conserved with respect to the chosen temporal variable.

The effect of the constraint may be visualized as the infinitesimal increase in the extent of the temporal dimensions as compared to the spatial dimensions of the codimension hypersurfaces as time emerges and expands three-dimensionally over existing three-dimensional space. Each of the three Cartesian coordinate planes visualized as a layer of space conjugated to a thinner layer of time. Hence, the above constraint originates a well-posed Cauchy problem, where solutions that exist are unique and close to 1T-Physics.

An observer in the above Cartesian coordinate system may attain temporal orientation from each temporal coordinate (t_X, t_Y, t_Z) as an extension of the spatial coordinates (x, y, z) in any direction and sense of motion in space-time. Space-time may be represented with the signature $(-,-,-,+,+,+)$ with the negative sign for temporal dimensions. If the observer considers a finite Cartesian coordinate spatial plane, the temporal layer may be considered to exist on either side of that spatial plane as well as on all its edges or sides.

Consequently, as the wave equation is visualized in 1T-Physics, the collapse of the three-dimensionality of time into one-dimensional time effects the anisotropy of time and results in the linear temporal perspective of space-time. From our present perspective, the 3T-Physics model theorized above may help explain non-locality phenomena in cosmology such as but not limited to: the improvised constraints for the states of the universe, the fine-tuned parameters to address the low entropy and near-homogeneity of the early universe, or the entanglement of field properties at different spatial locations predicted by quantum mechanics. The 3T-Physics model is not being furthered to replace 1T-Physics where and when the latter is effective in its methods and predictions but as a conceptual possibility to expound what is physically possible in the realm of space-time.

2.2. On the relativistic effects of fast-moving clocks

In the absence of translational or rotational acceleration effects, let us imagine a very long rod with two synchronous and simultaneous atomic clocks and two observers, one observer and a clock at each end of the rod, entering the curved space-time very near an object of mass, like an arrow, head first in a radial direction to the center of the object, for a time interval $\Delta\tau_{ab}$, we would observe the deceleration of time to be

$$a_{t_{ab}} = -\frac{\left(\omega_{t_a} - \omega_{t_b}\right)}{\Delta\tau_{ab}} \qquad (2.11)$$

where $a_{t_{ab}}$ is the deceleration of time during the time interval $\Delta\tau_{ab}$ and ω_{t_a} is the instantaneous angular speed of the front clock (arrow's head) and ω_{t_b} is the instantaneous angular speed of the rear clock (arrow's tail) as the arrow of time traverses the time interval. As our very long imaginary rod enters the gravitational field of the object the effects of the time deceleration between the atomic clocks at the ends would reveal, after precise measurement by the observers, a time deceleration $\left(\omega_{t_a} - \omega_{t_b}\right)$ during the relativistic temporal interval $\Delta\tau_{ab}$.

For relativistic motion in the space-time in a gravitational field, we find the deceleration of time to be

$$a_{t_{gk}} = \frac{\omega_{t_{gk1}} - \omega_{t_{gk2}}}{\Delta\tau_g - \Delta\tau_k} \qquad (2.12)$$

where $\left(\omega_{t_{gk1}} - \omega_{t_{gk2}}\right)$ is the temporal angular velocity difference as measured by two local synchronous clocks between two events during the time interval $\left(\Delta_{\tau_g} - \Delta_{\tau_k}\right)$ measured by a moving clock such that

$$\Delta\tau_g = -\left(\frac{a_g}{c^2}\right)\sum_{i=1}^{k} r\cdot\Delta t_i \qquad \textit{gravitational time effects} \quad (2.13)$$

$$\Delta\tau_k = \sum_{i=1}^{k}\left[\left(\sqrt[2]{1-\frac{v_i^{\,2}}{c^2}}\right)\cdot\Delta t_i\right] \qquad \textit{kinematical time effects} \quad (2.14)$$

where "*c*" is the speed of light, and v_i is the speed of the moving clock. These temporal effects result from the deceleration and the slowdown of time which in turn effects the gravitational acceleration between two objects.

The acceleration of time is significant for a synchronous clock traveling relativistically through homogeneous and isotropic space-time in the absence of a gravitational field. As the clock travels through space-time, it will experience acceleration or deceleration of time depending on its speed and path, regardless of the type of motion along its trajectory. In the presence of a gravitational field at slow speed, the acceleration or deceleration of the synchronous clock will depend on how it travels perpendicular to the surface of the object of mass, through the gravitational field of the object. The velocity of time changes with respect to the distance "*r*" to the center of the mass, and depends on the gravitational acceleration a_g and the speed of light "*c*". (Taylor et al, 1966)

2.3. The acceleration and speed of space-time

Let us now consider the Friedman-Lemaitre-Robertson-Walker metric for homogeneous and isotropic space-time, for a uniform curvature in an elliptical space-time, such that

$$-c^2 d\tau^2 = -c^2 dt^2 + a(t)^2\, d\Sigma^2 \qquad (2.15)$$

where the current value of the scalar factor equals one, $a(t)^2 = (l_p/l_{t_0})^2 = 1$, such that we have

$$d\Sigma^2 = c^2dt^2 - c^2d\tau^2 \tag{2.16}$$

$$\frac{d\Sigma^2}{dt^2} = c^2 - c^2\frac{d\tau^2}{dt^2} \tag{2.17}$$

Thus, the determinable speed of light is equal to the speed of time in homogeneous and isotropic space-time. The speed of time is both consistent and determinable. Thus, let us acknowledge the squared speed of space-time to be

$$v_\Sigma^{\ 2} = c^2 - c^2 v_\tau^{\ 2} \tag{2.18}$$

$$v_\Sigma^{\ 2} = c^2\left(1 - v_\tau^{\ 2}\right) \tag{2.19}$$

$$v_\Sigma = c\left(\sqrt[2]{1 - v_\tau^{\ 2}}\right) \tag{2.20}$$

Then, if the velocity of proper time is zero with respect to coordinate time, then the velocity of space-time is equal to the speed of light. Space-time expands at the speed of light.

$$v_\Sigma = c \tag{2.21}$$

Hence, we also can express the velocity of space-time as the acceleration times the spatial distance "*r*" such that

$$a_\Sigma r = c^2 - c^2 v_\tau^{\ 2} \tag{2.22}$$

$$a_\Sigma r = c^2(1 - v_\tau^{\ 2}) \tag{2.23}$$

$$\frac{a_\Sigma r}{c^2} = 1 - v_\tau^{\ 2} \tag{2.24}$$

$$v_\tau = \sqrt[2]{1 - \frac{a_\Sigma r}{c^2}} = \sqrt[2]{1 - \frac{Gm}{rc^2}} \tag{2.25}$$

2.4. The acceleration and speed of proper time

Let us consider the time-like interval of an object moving in space-time such that

$$c^2 d\tau^2 = c^2 dt^2 - dr^2 \tag{2.26}$$

$$c^2 \frac{d\tau^2}{dt^2} = c^2 - \frac{dr^2}{dt^2} \tag{2.27}$$

$$c^2 v_\tau^{\ 2} = c^2 - v_r^{\ 2} \tag{2.28}$$

$$v_\tau^{\ 2} = \frac{c^2 - v_r^{\ 2}}{c^2} \tag{2.29}$$

$$v_\tau = \frac{\sqrt[2]{c^2 - v_r^{\ 2}}}{c} = \sqrt[2]{1 - \frac{v_r^{\ 2}}{c^2}} \tag{2.30}$$

$$c^2 \frac{d\left[v_\tau^{\ 2}\right]}{dt} = -\frac{d\left[v_r^{\ 2}\right]}{dt} \tag{2.31}$$

$$\frac{d\left[v_\tau^{\ 2}\right]}{dt} = -\frac{1}{c^2}\left(\frac{d\left[v_r^{\ 2}\right]}{dt}\right) = -\frac{1}{c^2}\left(\frac{d(a_r \cdot r)}{dt}\right) \tag{2.32}$$

$$\frac{d\left[v_\tau^{\ 2}\right]}{dt} = -\frac{1}{c^2}\left[a_r \frac{dr}{dt} + r\frac{da_r}{dt}\right] \tag{2.33}$$

Thus, the speed of proper time equals the reciprocal of the Lorentz factor, and the acceleration of proper time is directly proportional to the derivative of the acceleration, or the jolt, of the moving object of mass times the distance "*r*" plus the product of the velocity of the moving object times the acceleration, and indirectly proportional to the speed of light squared. If we consider the acceleration of a moving object in space-time and express that acceleration in terms of the accelerations of proper time and coordinate time we would have

$$v_r^{\,2} = c^2 - c^2 v_\tau^{\,2} \tag{2.34}$$

$$\frac{v_r^{\,2}}{r} = \frac{c^2}{r}\frac{dt^2}{dt^2} - \frac{c^2}{r}\frac{d\tau^2}{dt^2} \tag{2.35}$$

$$\frac{c^2}{r}\left[\frac{v_r^{\,2}}{c^2}\right] = \frac{c^2}{r}\left[\frac{dt^2}{dt^2}\right] - \frac{c^2}{r}\left[\frac{d\tau^2}{dt^2}\right] \tag{2.36}$$

$$\frac{c^2}{r}\left[\frac{1}{c^2}\frac{dr^2}{dt^2}\right] = \frac{c^2}{r}\left[\frac{dt^2}{dt^2}\right] - \frac{c^2}{r}\left[\frac{d\tau^2}{dt^2}\right] \tag{2.37}$$

Therefore, both the acceleration of proper time and the acceleration of a moving object in space-time may be expressed as fractions of, and may be based upon, the acceleration of coordinate space-time, c/r^2 , at the locality in space-time where the object is moving.

2.5. The acceleration and luminal speed of an object in space-time

Let us consider the light-like interval for a massless object moving at the speed of time

$$c^2 d\tau^2 \;=\; c^2 dt^2 - dr^2 \;=\; 0 \tag{2.38}$$

$$c^2 dt^2 \;=\; dr^2 \tag{2.39}$$

$$c^2 \;=\; \frac{dr^2}{dt^2} \tag{2.40}$$

$$v_t^{\,2} \;=\; v_r^{\,2} \tag{2.41}$$

$$\frac{dv_t}{dt} \;=\; \frac{dv_r}{dt} \tag{2.42}$$

$$a_t \;=\; a_r \tag{2.43}$$

Thus, the acceleration or speed of a massless object in space-time during a light-like interval equals the acceleration or speed of time, if time is traveling at light speed when unobstructed in homogenous and isotropic space-time. Moreover, for an object that is not traveling through space when the speed of the object is zero, $v_r = 0$, and $v_\tau = \sqrt[2]{1-\left(v_r^{\,2}/c^2\right)} = 1$, as expected in General Relativity because proper time equals coordinate time, we find the speed of proper time is equal to the speed of light.

$$v_\tau \;=\; v_t \;=\; c \tag{2.44}$$

2.6. The gravitational acceleration of space-time-mass

As space-time curves in the presence of relativistic mass, space decelerates as it contracts and time slows down as it dilates proportionally, as the magnitude of space-time compresses about the geometry of the mass and perpendicular to its surface, to maintain an equal and opposite uniform acceleration. The acceleration of time of the space-time about the mass has the opposite directional orientation as the gravitational acceleration of the gravitational field about the mass. Thus, *gravity is proportional and directly related to the deceleration of space-time and the slowdown of the speed of space-time in a gravitational field about an object with rest mass or relativistic mass.*

Per Einstein's equivalence principle over small times for a very small space-time locality and without the obstruction of tidal forces, let us now consider the indistinguishable gravitational effects on a non-rotating object of mass accelerating relativistically in outer space toward a non-rotating massive celestial body, to the gravitational effects felt by the same object of mass as it enters at relativistic speed the gravitational field of the non-rotating massive celestial body, far from any other significant gravitational field. Thus, when we consider the velocity of proper time for a relativistic event in a gravitational field using a Maclaurin series, we find

$$\frac{d\tau}{dt} = \sqrt[2]{1-\frac{v_r^{\,2}}{c^2}} \approx \frac{\left(a_g \cdot r\right)}{c^2} - \frac{v_r^{\,2}}{2c^2} - \frac{v_r^{\,4}}{8c^4} - \frac{v_r^{\,6}}{16c^6} - \frac{v_r^{\,8}}{128c^8} + \ldots \tag{2.45}$$

$$\sqrt[2]{1-\frac{v_r^{\,2}}{c^2}} \approx \frac{(a_g \cdot r)}{c^2} - \frac{v_r^{\,2}}{2c^2} \tag{2.46}$$

$$a_g \approx \frac{v_r^{\,2} + 2c\left(\sqrt[2]{c^2 - v_r^{\,2}}\right)}{2r} \tag{2.47}$$

Hence, the gravitational acceleration between two non-rotating bodies of mass is a function of the speed of proper time, v_τ, the spatial distance "*r*", and the speeds of the bodies, within the local gravitational field.

Therefore, *gravity is the effect that space-time has on the property of mass through the deceleration of space-time and the slowdown of the speed of space-time. Gravity is strongest where space is most contracted and time is most dilated.* As space contracts and time dilates, the deceleration of space-time results in a deceleration, or negative acceleration of gravity, towards the center of mass perpendicular to the surface of the geometry of the object. The changes in space-time result in the gravitational effect and in the curvature of space-time as described by the General Theory of Relativity of Albert Einstein.

§ 3. On the effects of mass dilation

As a moving particle or an object approaches the speed of light in homogeneous and isotropic space-time, space contracts in the direction of motion, creating a kinematical gravitational field that adds to the mass' gravitational field, contracting space-time and increasing gravity, while dilating both time and the mass of the particle or object.

By Newton's third law, "the internal forces acting between particles inside a mass cancel in pairs," length contracts in the direction of relativistic travel, as the mass presses against the space-time in front of it in the direction of motion, causing the space-time to contract, and making a spherical object appear to flatten or pancake in the direction of propagation. As space-time contracts, rest mass dilates to relativistic mass because of the energy increase of the mass due to its velocity or momentum, with no change in the internal structure of the mass of the object. In fact, the energy increase of the mass with velocity or momentum originates from the geometrical properties of space-time, not from the mass of the object. (Feynman, 1988) Mass dilation originates from the momentum speed of space and time acting directly on the mass of an object. For the temporal momentum we find

$$p_t = m\left(\frac{dt}{d\tau}\right) = \frac{m}{\sqrt[2]{1-\frac{v_r^{\ 2}}{c^2}}} = \frac{mc}{\sqrt[2]{c^2 - v_r^{\ 2}}} \tag{3.1}$$

where the speed is given as the increment of the coordinate time over the proper time.

Similarly, for the three-dimensional spatial momentum when $v_r < c$ we have

$$p_r = = m\left(\frac{dr}{d\tau}\right) = m\left(\frac{dr}{d\tau}\right)\left(\frac{dt}{dt}\right) = m\left(\frac{dr}{dt}\right)\left(\frac{dt}{d\tau}\right) = \frac{mv_r}{\sqrt[2]{1-\frac{v_r^{\ 2}}{c^2}}} = \frac{mc}{\sqrt[2]{\frac{c^2}{v_r^{\ 2}}-1}} \tag{3.2}$$

where the speed is given as the increment of the coordinate space over the proper time.

Hence, the energy of the temporal momentum of the mass is less than the energy of the spatial momentum by a factor equal to the spatial speed such that

$$\frac{E_r}{E_\tau} = v_r \tag{3.3}$$

Similarly, in terms of the Lagrangian action *S*, and the Hamiltonian, $\hat{H}$, for the relativistic motion of the mass traveling in homogeneous and isotropic space-time during a time interval we find

$$S = \int_{t_1}^{t_2} L dt = \int_{t_1}^{t_2} -mc^2\left(\frac{d\tau}{dt}\right)dt = \int_{t_1}^{t_2} -mc^2\left(\sqrt[2]{1-\frac{v_r^{\ 2}}{c^2}}\right)dt \tag{3.4}$$

$$p = \frac{\partial L}{\partial v_r} = -\frac{1}{2}\left[\frac{mc^2(-2v_r)}{c^2 \sqrt[2]{1-\frac{v_r^{\,2}}{c^2}}}\right] = \frac{mv_r}{\sqrt[2]{1-\frac{v_r^{\,2}}{c^2}}} = \frac{mv_r}{v_\tau} \tag{3.5}$$

$$\hat{H} = v_r p - L = \frac{mv_r^{\,2}}{\sqrt[2]{1-\frac{v_r^{\,2}}{c^2}}} - \left[-mc^2\left(\sqrt[2]{1-\frac{v_r^{\,2}}{c^2}}\right)\right] = \frac{mc^2}{\sqrt[2]{1-\frac{v_r^{\,2}}{c^2}}} = \frac{mc^2}{v_\tau} \tag{3.6}$$

Hence, we find that mass dilates as a result of the velocity of time decreasing as the object speeds up through space, and this effect increases the momentum of the object and its kinetic energy, as the object travels against the increasing gravitational field about the mass opposite to the direction of motion.

As the resultant force of the relativistic mass increases in the direction of motion, the internal forces acting between atomic particles cancel in pairs as the internal space between particles contracts. The contraction of internal space in the direction of motion allows the positional equilibrium of particles at closer spatial distances, maintaining the balance and relationship of all the quantum forces at atomic and subatomic levels while preserving the laws of particle physics. As the internal space inside the relativistic mass contracts to its new boundary, the system of forces of the mass is again at equilibrium with space-time. Thus, mass has dilated and length has contracted proportionally. Hence, *for relativistic motion in homogeneous and isotropic space-time, mass and time dilate and space contracts to obey the laws of conservation of spatial momenta.*

§ 4. On the universal gravitational constant

The universal gravitational constant, denoted *G*, is an empirical physical constant involved in the calculation of the gravitational attraction between objects with the property of mass. It first appeared in Isaac Newton's law of universal gravitation and subsequently in Albert Einstein's general theory of relativity.

From the law of universal gravitation,

$$G = \frac{F \cdot r^2}{m_1 \cdot m_2} \quad (4.1)$$

where G has been given a value of most nearly $(6.67428 \pm 0.00067)E - 11 \quad m^3/(k_g \cdot s^2)$ at the present time. F is the attractive force between two objects of rest mass m_1 and m_2, and "r" is the distance between them.

The universal gravitational constant was measured in 1798, seventy one years after Isaac Newton's death, by Henry Cavendish, with a torsion balance invented and built by John Mitchell, as published in *Philosophical Transactions* in the year 1798. Cavendish's aim was to measure the Earth's density relative to water, through the precise knowledge of the gravitational interaction. Amazingly, the accuracy of the measured value of *G* has increased only modestly since the original Cavendish experiment.

Most recently the published values of *G* obtained from high precision experiments have varied rather broadly. The upper and lower boundary limits of $6.645E - 11 \;\; to \;\; 6.715E - 11$ are recent results for all free-fall-type and Cavendish-type experiments. Therefore, *G* is a variable quantity instead of being a constant value as originally assumed by Newton and later by Einstein in their theories. Thus, if *G* is not the universal gravitational constant, then what is it? What makes it vary? Why does it vary?

4.1. The nature of G

Let us start our investigation by looking at the nature of *G*. Gravitation about a mass is the result of the interaction of space-time and mass. Neither space-time nor mass is a constant property, so the gravitational field resulting from the curvature of space-time about the mass is a dynamic field effected by the dynamic properties of space-time-mass. Both mass and *G* are a function of motion, any change in mass causes a change in *G*. *Thus, G is the ratio of the dynamic interaction of space-time and the property of mass of the object that is subjected to a dynamic gravitational field.*

Let us consider a curved space-time volume displaced by a massive celestial body as the space-time volume contracts and accelerates over time creating a gravitational field. In the special case of spherical object with rest mass, m_0, let us consider a variation of Gauss' flux theorem for gravity as follows

$$\int_{V_1}^{V_2} \vec{\nabla} \cdot \vec{g} \, dV = -4\pi \int_{r_1}^{r_2} \overrightarrow{v_g} \cdot \overrightarrow{v_g} \, dr \tag{4.2}$$

$$\int_{V_1}^{V_2} \vec{\nabla} \cdot \vec{g} \, dV = -4\pi G \rho V = -4\pi G m_0 \tag{4.3}$$

$$V = \frac{4}{3} \pi r^3 \tag{4.4}$$

$$\frac{dV}{dr} = 4\pi r^2 \tag{4.5}$$

$$\frac{d^2V}{dr^2} = 8\pi r \tag{4.6}$$

$$\overrightarrow{v_g} \cdot \overrightarrow{v_g} = v_g^{\ 2} = \frac{dr^2}{dt^2} = gr \tag{4.7}$$

$$-4\pi \int_{r_1}^{r_2} \left(\overrightarrow{v_g} \cdot \overrightarrow{v_g} \right) dr = -4\pi \left(\frac{\frac{d^2V}{dt^2}}{\frac{d^2V}{dr^2}} \right) r = -4\pi \left(\frac{r \frac{d^2V}{dt^2}}{8\pi r} \right) \tag{4.8}$$

$$-4\pi G m_0 = -4\pi \left(\frac{\frac{d^2V}{dt^2}}{8\pi} \right) \tag{4.9}$$

Then, we find the acceleration of space-time to be

$$-\frac{d^2V}{dt^2} = -\ddot{a} = 8\pi G m_0 = 8\pi g r^2 \qquad (4.10)$$

Thus, we find G as the dynamic universal space-time-mass ratio to be

$$G = -\frac{\frac{d^2V}{dt^2}}{8\pi m_0} \qquad (4.11)$$

$$= \frac{g \cdot r^2}{m_0} \qquad (4.12)$$

and the gravitational acceleration as a function of space-time becomes

$$g = -\left(\frac{1}{8\pi r^2}\right)\frac{d^2V}{dt^2} \qquad (4.13)$$

We can express the convergence modulus of expanding space-time about a massive spherical celestial body as

$$\frac{-\frac{d^2V}{dt^2}}{4\pi r^2} = \frac{8\pi G m_0}{4\pi r^2} = \frac{2Gm_0}{r^2} = 2g \qquad (4.14)$$

4.2. What are the big G and small g of the earth?

Let us derive G for an oblate spheroid approximating a celestial body like the earth, in the weak-field regime $\ddot{a}/8\pi << rc^2$ where we find

$$G = -\frac{g \cdot r_1 \cdot r_3}{m_0} = -\frac{\frac{d^2V}{dt^2}}{8\pi m_0} \qquad (4.15)$$

where r_1 is the equator radius and r_3 is the pole radius and the gravitational acceleration is expressed as

$$g = -\frac{Gm_0}{r_1 \cdot r_3} = -\left(\frac{1}{8\pi r_1 r_3}\right)\frac{d^2V}{dt^2} \qquad (4.16)$$

4.3. On the relativistic big G and small g of fast-moving objects

For a spherical object traveling at relativistic speed in homogeneous and isotropic space-time, length contracts and mass dilates along the path of the trajectory, where we find *G and g* to be

$$r = r_0 \sqrt[2]{1 - {v_r}^2 / c^2} \qquad (4.17)$$

$$-\frac{d^2V}{dt^2} = 8\pi g r^2 = 8\pi r^2 \frac{d^2r}{dt^2} \qquad (4.18)$$

$$\frac{d^2r}{dt^2} = g \qquad (4.19)$$

$$m = \frac{4}{3}\pi r^3 \rho = \frac{m_0}{\sqrt[2]{1 - {v_r}^2 / c^2}} \qquad (4.20)$$

$$\frac{dm}{dr} = 4\pi r^2 \rho \qquad (4.21)$$

$$\frac{d^2m}{dr^2} = 8\pi r\rho = \frac{(8\pi r)(3m)}{4\pi r^3} = \frac{6m}{r^2} \qquad (4.22)$$

$$G = -\frac{8\pi r^2\left(\frac{d^2r}{dt^2}\right)}{8\pi m} = -\frac{gr^2}{m} \qquad (4.23)$$

$$\left(\frac{d^2r}{dt^2}\right) = g = -\frac{Gm}{r^2} \qquad (4.24)$$

$$\left|\frac{g}{G}\right| = \frac{m}{r^2} = \frac{1}{6}\frac{d^2m}{dr^2} \qquad (4.25)$$

As the relativistic mass dilates, space contracts and time dilates, increasing the gravitational field of the object. *The dynamic universal space-time-mass ratio, G, better known as the gravitational constant, is susceptible to the oscillatory changes or variations of both the acceleration of space with respect to time or the acceleration of mass with respect to space, as a dynamic ratio of these variables which are functions of the speed of the object or mass.*

§ 5. On the specific enthalpy of a gravitational system

Enthalpy is a property of a substance, like pressure, temperature, and volume, but it cannot be measured directly. Normally, the enthalpy of a substance is given with respect to some reference value. It is the change in specific enthalpy, Δh, and not the absolute value, that is important in specific enthalpy applications.

Specific enthalpy "*h*", is given as $h = u + Pv$, where "*u*" is the specific internal energy (*kJ/kg*) of the system being studied, "*P*" is the pressure of the system $\left(N/m^2\right)$, and "*v*" is the specific volume $\left(m^3/k_g\right)$ of the system. Enthalpy is usually applied in connection with the analysis of an open system in the field of thermodynamics.

The amount of energy used to form a large system of mass or object and its gravitational field is directly proportional to the amount of energy required to displace and compress the space-time medium to be occupied by the system of mass to obey the law of conservation of energy during large mass formation.

For the expanding mass of a spherical object to be formed, we find

$$\frac{\partial^2 V}{\partial t^2} = 8\pi G m_0 = 8\pi h r_0 \qquad (5.1)$$

where "h" is the intrinsic energy per kilogram $\left(J/k_g\right)$ constituting space-time displacement energy and the energy of formation for a large system of rest mass m_0 with a radius of r_0.

Thus, in terms of potential and internal energy, we find

$$h_g = \frac{Gm_0}{r_0} \tag{5.2}$$

$$h = \frac{U_i}{m_0} + \frac{U_g}{m_0} = \frac{U_i}{m_0} + \frac{Gm_0}{r_0} \tag{5.3}$$

$$U_g = kU_i \tag{5.4}$$

$$h_i = \frac{U_g}{km_0} = \frac{Gm_0}{kr_0} \tag{5.5}$$

$$h = \frac{Gm_0}{kr_0} + \frac{Gm_0}{r_0} \tag{5.6}$$

Therefore, we obtain

$$h = \left(\frac{Gm_0}{r_0}\right)\left[1 + \frac{1}{k}\right] = h_g\left[1 + \frac{1}{k}\right] \tag{5.7}$$

$$\sqrt[k]{e} = \lim_{k \to \infty} \sqrt[k]{\left(1 + \frac{1}{k}\right)^k} \text{ and } k > 0 \tag{5.8}$$

$$h = h_g \sqrt[k]{e} \tag{5.9}$$

The intrinsic energy "h" of a system of mass, m_0, is proportional to the intrinsic potential energy of the system "h_g" by a factor of $\sqrt[k]{e}$, or $1+(1/k)$, where "k" is the potential energy to internal energy ratio for a system of mass, m_0 , and $k>0$.

§ 6. On the dichotomy of gravitational theory

A theory is needed that will reconcile the curvature of space-time with a quantum field for gravity. General relativity is a classical theory that does not address gravity on a smaller than Planck scale. In general terms, the discreteness of quantum theory is not compatible with the smoothness of Einstein's general relativity. A theory more comprehensive than general relativity must explain the behavior of gravity near the Planck scale. Furthermore, the Standard Model of Quantum Field Theory is not space-time background independent, but General Relativity is regarded to be space-time background independent. A theory of quantum gravity is needed in order to reconcile these differences.

6.1. On quantum gravity theory and infinitesimal space-time

Certainly, per our previous investigation, the properties of space-time, time and space, effect the gravitational field at any scale. The contraction of space-time results in the gravitational field about an object whether the object is a point particle or a celestial body. Even though, space-time is believed to be frothy near or at some level below the Planck scale, space-time may be churning, curving in a non-Euclidean and non-commutative way.

If atomic dimensions of fundamental particles are of magnitudes much greater than those dimensions at or below the Plank level, a quantum theory of gravity need not exclude the essence of general relativity at the particle level. (Bohr et al, 1958)

Furthermore, it appears that there is a limit to the amount of detail contained in a volume of space-time. Structure becomes simpler at smaller distances. Surely there must be some minimum length at which the simplest elements of natural structure are found and surely this must mean space-time is discrete. Recent theoretical results from String Theories and the Loop-representation of Gravity do suggest that space-time has some discrete aspects at the Planck scale. There seems to be good reasons to suppose that space-time is discrete in some sense at the Planck Scale. Theories of quantum gravity suggest that there are a finite number of significant degrees of freedom and there is also a minimum length beyond which measurement cannot be performed. Planck length represents a minimum size beyond which the Heisenberg Uncertainty Principle prevents measurement if applied to the metric field of an Einstein gravitational field. Reportedly, the nearest thing physicists have to an experimental result in quantum gravity is the black hole information loss paradox which originates from semi-classical treatments of quantum gravity. Therefore, a future technological breakthrough may offer hope for experimental input into quantum gravity research because the Planck energy is currently so far beyond reach.

Let us imagine a technology that allows us to peer into the frothy environment of space-time, where space-time suds with waves and time may become delocalized. At that scale of space-time, time may be dynamically fractured, but possibly reincorporated or continuously blended into space-time. The gravitational field of the space-time waves would equally blend with a dynamic oscillatory nature. Then, our technology would help to investigate further if for the most infinitesimal elementary particles or manifestations of mass, the space-time waves delimits the boundary between space-time and mass or space-time and its dimensions objectify the manifestations of mass at the elementary scale.

The nature of space-time, being the same at any locality of the space-time waves, separate space and time regions, would possibly emerge, co-exist, and recombine. In essence, space-time may be infinitesimally coarse, but to its greatest extent, space-time is smooth and continuous, for it is in the smooth space-time continuum that most interactions occur between elementary particles, objects, or systems of mass. The effect of these dynamic gravitational oscillations of the space-time waves on an elementary particle would greatly depend on the comparative scale of the particle to the wave, for if there are very large magnitudes of scale, the nature of space-time would smooth and dampen these infinitesimal waves, or effervescence, and the smoothness of space-time would prevail.

6.2. The Chronon and the Chronino: quanta of time

A comprehensive theory of gravity avails to theorize and encompass a quantum of time, such as the Chronon, and its gravitational manifestation, such as a Graviton, as objects that give existence to the quantum gravitational field. The Graviton would mediate the force of gravity in the framework of quantum field theory. The existence of the Graviton would be connected to the curvature of space-time through the interactions of elementary Chronons. *The gravitational field couples the space-time structure with matter, energy and quantum fields. Since quantum fields consist of mass, energy and space-time, quantum fields affect the curvature of space-time, and in turn, space-time affects the characteristics of quantum fields.* Neither the Chronon nor the Graviton needs to be a direct manifestation of mass, but infinitesimal incorporeal constructs for quantum space-time and quantum gravity as manifestations of quantum field effects of space-time-mass. Well-defined interactions between Chronons and Gravitons must model the behavior of the field quanta and lead to renormalizable gravitation. The description of the Quantum Field Theory in terms of Chronons and Gravitons must serve as a low-energy effective theory close to or above the Planck scale so that infinities do not arise due to quantum effects.

Unlike the Graviton or Chronon, the recently found Higgs Boson is an object of mass, electrically neutral, and spinless. The Higgs Field is thought to be the field by which particles acquire mass. Thus, through interaction with the Higgs field, particles acquire their specific masses. If a Graviton or a Chronon interacted with the Higgs field, then gravitational waves would be observed to propagate slower than the speed of light, which would imply that the Graviton or Chronon has acquired mass.

A Chronon is a proposed quantum of time, that is, a discrete and indivisible unit of time as part of a theory that proposes that time is not continuous. In Piero Caldirola's model from 1980, one Chronon corresponds to about 6.97×10^{-24} seconds for an electron. This is much larger than the Planck time, which is only about 5.39×10^{-44} seconds. The Planck time is a universal quantization of time itself, whereas the value of the Chronon is a function of the system and its boundary conditions, for the Chronon is a system's quantum evolution along its world line. (Caldirola, 1980)

Allegedly, the Chronon allows for a lucid answer to the quantum mechanical question of whether a free-falling charged particle does or does not emit radiation. The Chronon model allegedly prevents the difficulties encountered by the approaches of Dirac and Abraham-Lorentz to the question, and provides a natural explanation of quantum decoherence. In quantum mechanics, quantum decoherence is the mechanism by which quantum systems interact with their environments to exhibit probabilistically additive behavior. Quantum decoherence gives the appearance of wave function collapse and justifies the intuition and framework of classical physics as an acceptable approximation. Decoherence is the mechanism by which the classical limit emerges out of a quantum starting point and it determines the location of the quantum-classical boundary. For example, decoherence occurs when a system interacts with its environment in a thermodynamically irreversible way.

In a theory of quantum gravity, time should not be considered a fixed background parameter for the convenience of the theory or its mathematical expressions, for we know that time is both fluid and dynamic even for the theorist.

In Caldirola's model, the value for a Chronon, θ_0, associated with a particle is given by

$$\theta_0 = \frac{q^2}{6\pi\varepsilon_0 m_0 c^3} \tag{6.1}$$

Since the value of the Chronon depends on the associated particle's charge q and mass m_0, the nature of the particle being considered must be specified.

Hence, let us specify the Planck temporal volume density of a reduced Chronino, ϕ_θ, to be

$$\phi_\theta = \frac{\frac{4}{3}\pi c^2 t_p^{\,4}}{\hbar} \quad \left({Sec^3}/{Kg} \right) \tag{6.2}$$

$$\frac{\partial \phi_\theta}{\partial t_p} = \frac{16\pi c^2 t_p^{\,3}}{3\hbar} \quad \left({Sec^2}/{Kg} \right) \tag{6.3}$$

$$\frac{\partial^2 \phi_\theta}{\partial t_p^{\,2}} = \frac{16\pi c^2 t_p^{\,2}}{\hbar} \quad \left({Sec}/{Kg} \right) \tag{6.4}$$

where "*c*" is the speed of light, and consequently we find the reduced relativistic mass of the Chronino to be

$$m_{\phi_\theta} = \frac{\hbar \nu_{\phi_\theta}}{c^2} \tag{6.5}$$

where $\hbar$ is the reduced Planck's constant, ν_{ϕ_θ} is the frequency of the Chronino, and "c" is the speed of light.

The value of the temporal volume of the Chronino is of the same order as the temporal volume of a sphere, $(4/3)\pi t_p{}^3$, with a radius equal to the Planck time which is the spatiotemporal distance ct_p that a photon would travel along the radius of the sphere in homogenous and isotropic space-time; as space-time expands or contracts, the temporal volume of the Chronino, as defined here, would remain a constant temporal quantity in an inertial frame of reference. Thus, if we assume that the Chronino acquires mass, $m_{\varphi\theta}$, then its two-dimensional momentum of expansion, or contraction, about its equator is the reduced Planck's constant $\hbar$.

6.3. The Gravitino: a quantum of Planck space-time-mass acceleration

The universal gravitational constant has been recently defined to a highly accurate value using the quantum nature of matter. The measurement of the gravitational pull between rubidium atoms and very heavy tungsten cylinders has an uncertainty of a hundred and fifty parts per million, 0.015%. The accuracy of the quantum measurement method, that uses atom interferometers that exploit the wave-like nature of matter, is slightly larger than the most accurate conventional measurement methods.

Despite an increase in the precision of recent conventional measurements, the discrepancy of measured values of the universal gravitational constant has widened, and the source of the discrepancy of conventional measurement methods is still unidentified. The quantum measurement method is unlikely to have the same errors of conventional measurement methods. Hence, these results prove the validity of the universal gravitational constant even at the quantum level. Consequently, the quantum application of the universal gravitational constant in EFEs of the General Theory of Relativity is validated since the universal gravitational constant is applicable to space-time-mass regardless of scale.

To find the value for a Gravitino, g_0, let us specify G_0 as the Planck space-time-mass ratio to be

$$G_0 = \frac{\hbar c}{8\pi m_p^{\,2}} = \frac{\hbar c}{8\pi \hbar c / 8\pi G} = G = 6.67428 \times 10^{-11} \left(\text{meter}^3 / \text{Kg} \cdot \text{sec}^2 \right) \quad (6.6)$$

The reduced Planck mass m_p may be expressed using the alternative normalization as

$$m_p = \sqrt[2]{\hbar c / 8\pi G} \quad (6.7)$$

Thus, for a spherical elementary particle of mass,

$$\ddot{a} = \frac{\partial^2 V}{\partial t^2} = \frac{\hbar c}{8\pi m_p} \approx 2.897791811 \text{ x } 10^{-19} \text{ meter}^3 / \text{sec}^2 \quad (6.8)$$

$$\dot{a} = \frac{\hbar l_p}{8\pi m_p} \approx 1.562217765 \text{ x } 10^{-62} \text{ meter}^3 / \text{sec} \quad (6.9)$$

Where $\ddot{a}$ is the acceleration of the volume of space-time about a reduced Planck mass, and $\dot{a}$ is its velocity.

Similarly, the value of the reduced Planck mass of the Gravitino may be specified to be

$$m_{g_0} = \frac{\hbar \nu_{g_0}}{c^2} \quad (6.10)$$

Where "c" is the speed of light and ν_{g_0} is the frequency of the Gravitino. Thus, let us specify the Gravitino g_0 and its Jolt and Snap to be

$$g_0 = \frac{G_0 m_{g_0}}{l_p^{\,2}} \quad \left(\text{meter} / \text{sec}^2 \right) \quad (6.11)$$

$$\frac{\partial g_0}{\partial l_p} = -\frac{2G_0 m_{g_0}}{l_p^{\ 3}} \qquad \left(1/\sec^2\right) \tag{6.12}$$

$$\frac{\partial^2 g_0}{\partial l_p^{\ 2}} = \frac{6G_0 m_{g_0}}{l_p^{\ 4}} \qquad \left(1/meter \cdot \sec^2\right) \tag{6.13}$$

Hence, we specify the chrono-gravitonic acceleration ratio to be

$$\Delta_{g_0}^{\phi_\theta} = \frac{\dfrac{\partial^2 \phi_\theta}{\partial t_p^{\ 2}}}{\dfrac{\partial^2 g_0}{\partial l_p^{\ 2}}} = \frac{8\pi c^2 t_p^{\ 2} l_p^{\ 4}}{3G_0 \hbar m_{g_0}} = \frac{2\phi_\theta l_p^{\ 2}}{g_0 t_p^{\ 2}} = \frac{2\phi_\theta c^2}{g_0} \qquad \left(meter \cdot \sec^3/Kg\right) \tag{6.14}$$

Where the proposed gravitino is a hypothetical massless elementary particle or Fermion, in the Fermion Gauge group, predicted to have spin 3/2 and zero charge by some quantum gravity theories, and $\hbar$ is the reduced Plank constant. General Relativity and Spatiotemporal Supersymmetry are combined in supergravity theories, the gravitino is the supersymmetric partner and gauge fermion of the hypothesized graviton.

Thus, the gravitino is specified here as the hypothetical fundamental particle with a quantum gravitational field acting at the center of energy and mass of a particle. The gravitino acts as if all the energy and mass of the associated particle were concentrated in a spherical particle-like space-time structure in the particle's center of energy and mass. For the small g_λ of a specific particle in terms of G_0 or $\hbar$, we find

$$g_\lambda = \frac{G_0 m_\lambda}{l_p^{\ 2}} = \frac{\hbar c m_\lambda}{8\pi m_p^{\ 2} l_p^{\ 2}} = \frac{\ddot{a}}{\left(\dfrac{m_p}{m_\lambda}\right) l_p^{\ 2}} = \frac{\ddot{a}\alpha_\lambda}{l_p^{\ 2}} \qquad \left(meter/\sec^2\right) \tag{6.15}$$

Where l_p is the Planck length, m_λ is the mass of a specific particle, α_λ is the coupling constant, and $\hbar$ is the reduced Planck constant. The gravitational force is considered to have unlimited range which effects the gravitino to be massless, supersymmetric, and to have a spin of 3/2, where the stress-energy tensor, which is a second-rank tensor, is the source of the gravitational field in the Einstein Field Equations of General Relativity, just as mass is the source of such a field in Newtonian gravity.

6.3.(a) Luxons or massless Tardyons: Geons and Gravitinos

It has been previously suggested that the concept of a physical body of a subparticle in agreement with the General Theory of Relativity may, in essence, be manifested out of electromagnetic radiation or gravitational radiation, or as a mixture of electrogravitic radiation that coalesces through its gravitational field as a physical state of an object.

A cluster of radiant electrogravitic energy is a purely classical object, not an elementary particle, with a structure that may be treated as a massless virtual object that may interact with other objects or systems of mass. As originally proposed the radiant electrogravitic object is a gravitational electromagnetic entity, or Geon, whose size and structure are sufficiently great that quantum effects may be involved. The gravitational field would be indistinguishable from a comparable object of mass. It has been suggested that the rotating loop of radiant electromagnetic energy, or rotating loop of radiant gravitational energy, would have a tendency, as a Geon gets larger, to weaken and collapse into a singularity, or micro blackhole. (Wheeler, 1955)

There are solutions to the EFE for a Brill-Hartle type of Geon. It is possible to hypothesize that if a virtual Geon is structurally complex at a stable scale, it may not decay, as the gravitational wave and the electromagnetic energy are conserved through time dilation, while the Geon is acted upon by quantum effects, whereas if it is of sufficient scale, less complex, with less time dilation, the Geon may decay, collapse, and dissipate. A virtual subparticle that appears and disappears. Would a stable Geon, that may emerge to scale as a quark to form a hadron with other quarks, eventually produce an elementary particle? The Geon would be, in principle, a viable classical model for the subparticles, or building blocks, of elementary particles. It may not be coincidental that the Geon resembles the virtual structure of the gluonic confinement of a quark. The cluster of radiant electrogravitic energy may translate, spin, and pulsate, through space-time, and be attracted or deflected by varying fields of force like any other particle or system of mass. However, the cluster is, in essence, a body of mass without mass, or a massless manifestation of energy, according to the conventional definition of mass. Moreover, if the cluster of radiant electrogravitic energy were to consist exclusively of gravitational radiation, as in a gravitational cluster, without any local manifestation of energy from mass, there would still be an overall manifestation of energy. The cluster of radiant gravitational energy would exist only through the localized, contracted, and structured spatiotemporal curvature through adequate time dilation.

It is interesting to propose that the radiant electromagnetic energy within the confinement of a gravitino may manifest quantum effects like an emerging virtual massless particle. The electromagnetic characteristics of the emerging gravitino originate from the electromagnetic qualities of the curvature and stress of the spatiotemporal medium. Consequently, it is possible to hypothesize that a gravitino may emerge from a Riemannian manifold, e.g. $R_{\varepsilon\beta}$, in space-time, where the Riemann tensor, $R_{\mu\nu\varepsilon\beta}$, is the electromagnetic source of the emerging electrogravitic virtual particle. As the spatiotemporal stress of a Riemannian manifold exceeds a threshold, it may spring gluonic energy and fields, that lead to radiant electromagnetic energy, charges, and carriers. From this perspective, quarks may evolve from Geons, whereas Geons may evolve from gravitinos, although, conceptualization of virtual subparticles requests empirical evidence.

From the perspective of a rotating frame of reference, the outward acceleration for the electromagnetic radiation is c^2/r, where r is the radius of the circular orbit that holds the radiation, while the inward gravitational acceleration for the circular radiant energy of mass M, as if it were located at the center of the radiation loop, is GM/r^2. Whenever the two accelerations are equal and opposite, it is possible for the electrogravitic physical state of the theoretical object to be in a state of equilibrium, or stable, through time dilation and length contraction. In the event that the stable spatiotemporal state of the object decouples from the surrounding space-time, the object may exist as an electrogravitic entity. If the angular phase of spin of the object were positive, negative, or null, then expansion, contraction, or stasis, of the volume of the object may occur. The possible accelerations of the Euler, Coriolis, and centrifugal forces, are described for a rotating frame of reference of a translating body of circular radiant energy.

$$-\frac{v^2}{r}\vec{a}_r = -\frac{d\vec{\omega}}{dt}\times\vec{r} - 2\vec{\omega}\times\left[\frac{d\vec{r}}{dt}\right] - \vec{\omega}\times\left(\vec{\omega}\times\vec{r}\right) \quad (6.16)$$

In the case of the circular radiant energy, when there is no translation of the center of the rotating frame of reference, or there is no variation in the angular velocity of the reference frame's axes, the Euler and Coriolis accelerations are equal to zero, although, the centrifugal acceleration would be c^2/r. At a state of equilibrium, the magnitudes of the accelerations are equal and opposite,

$$\frac{c^2}{r} = \frac{GM}{r^2} \quad (6.17)$$

$$r = \frac{GM}{c^2} \rightarrow \frac{r_s}{2} \quad (6.18)$$

The Schwarzschild radius, r_s, is the radius of the event horizon surrounding a non-rotating black hole. Any object of mass with a physical radius smaller than its Schwarzschild radius will be a black hole. The Schwarzschild radius for a mass like our sun is approximately 3 *km*. However, since our object of interest is not of mass, it is possible for the physical property of the radius to be half the Schwarzschild radius since there is no mass at the center of mass to collapse the object into an infinitesimal black hole. The event horizon would be a porous boundary.

Infinitesimal black holes are virtual, even though some miniature acoustic black holes simulations have been empirically demonstrated. During the early universe, as some matter expanded very rapidly, it was possible that some slower-expanding matter may have been contracted into infinitesimal black holes. A black hole may have charge and rotation. The Kerr-Newman metric is a general solution to the EFE that describes a black hole with electromagnetic charge and angular momentum.

The circular radiant energy may be visualized as a discrete loop of infinitesimal radiant segments or contiguous packets. Each segment, or packet, has an equal and opposite segment, or packet, on the opposite side of the loop.

Thus, it is possible to consider an electrically charged pair of opposite segments, or packets, and the Coulombic force interacting between them, with each segment, or packet, having an electrical charge Q. As proposed in previous research, the Coulombic force has a temporal nature.

Let us consider an example of an electrogravitic object with some of the properties of a Kerr-Newman black hole. As a neutral and static body of mass is electrically charged and/or spun, energy has to be applied to the body of mass. The energy applied has mass-equivalence, so M is always greater than the irreducible mass, M_{irr}.

The length-scale of the electric charge Q of the mass of a Kerr-Newman black hole is given by

$$r_Q^2 = \frac{Q^2 G}{4\pi\varepsilon_0 c^4} \tag{6.19}$$

The total mass-equivalence, M, containing the energy of the electric field, the rotational energy, and the irreducible mass, M_{irr}, of a Kerr-Newman black hole, are related by

$$M_{irr} = \frac{\sqrt{2M^2 - r_Q^2 c^4 / G^4 + 2M\sqrt{M^2 - \left(r_Q^2 + a^2\right)c^4 / G^2}}}{2} \tag{6.20}$$

Where $a^2 = J^2 / M^2 c^2$, and the equivalence-mass, M, may be expressed as

$$M = \sqrt{\frac{16M_{irr}^4 + 8M_{irr}^2 r_Q^2 c^4 / G^2 + r_Q^4 c^8 / G^4}{16M_{irr}^2 - 4a^2 c^4 / G^2}} \tag{6.21}$$

If the irreducible mass is zero for an electrogravitic object with some of the properties of a Kerr-Newman black hole, we have

$$M^2 = \frac{\dfrac{r_Q^4 c^8}{G^4}}{\dfrac{-4a^2 c^4}{G^2}} = i^2 \frac{r_Q^4 c^4}{4a^2 G^2} = i^2 \frac{r_Q^4 M^2 c^6}{4J^2 G^2} \tag{6.22}$$

$$4J^2G^2 = i^2 r_Q^4 c^6 \tag{6.23}$$

$$J = \sqrt{\frac{i^2 r_Q^4 c^6}{4G^2}} = \frac{i r_Q^2 c^3}{2G} = \frac{iQ^2}{8\pi\varepsilon_0 c} \tag{6.24}$$

The complex angular momentum of the product of two electric charges, Q^2, is not a function of the mass-equivalence of the object. The complex angular momentum is directly proportional to the square of the electric charge, Q, or directly proportional to the square of the length-scale.

The complex angular momentum per electromagnetic Planck charge is given by

$$\frac{J}{Q_p} = \frac{iQ_p^2}{8\pi\varepsilon_0 c Q_p} = \frac{iQ_p^2 \cdot t_p}{8\pi\varepsilon_0 \cdot l_p \cdot l_p \cdot t_p} = \frac{iQ_p^2}{8\pi\varepsilon_0 l_p{}^2} = \frac{iQ_p^2}{8\pi\varepsilon_0 c^2 t_p{}^2} \tag{6.25}$$

Thus, the complex angular momentum of the object per electromagnetic Planck charge is equivalent to Coulomb's law for the force interacting between two static Planck charges within an oblate spheroidal spatiotemporal volume with permittivity ε_0, separated by a distance ct_p. In the above example, from a quantum perspective, the laterality of the angular motion of the charged object per unit of electromagnetic charge is equivalent to an attractive or repelling force.

6.4. On the general theory of relativity and its underlying principles

In 1915, the renowned physicist Albert Einstein published the General Theory of Relativity as a geometric theory of gravitation. The General Theory of Relativity is the current description of gravitation in modern physics. It has unified Isaac Newton's law of universal gravitation and the theory of special relativity, and describes gravity as a geometric property of space-time. The curvature of space-time is directly related to the properties of mass, energy, gravity and spatial momentum of whatever matter and radiation are present. The Einstein Field Equations specify the relation with a system of partial differential equations.

The General Theory of Relativity is the simplest theory that is consistent with experimental data. The predictions of general relativity have been confirmed in all observations and experiments up to the present time. Many predictions of general relativity differ significantly from the predictions of classical physics, especially predictions involving the propagation of light, the geometry of space, the passage of time, and the motion of bodies in free fall. The General Theory of Relativity has also pointed toward the existence of an expanding universe, gravitational waves, gravitational lensing and black holes. (Greene, 1999)

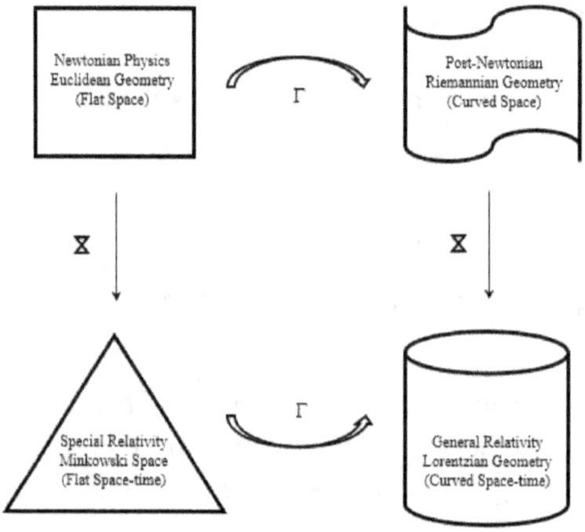

Figure 1.

Since general relativity is not the only relativistic theory of gravity, a fundamental question is how general relativity can be reconciled with the laws of quantum physics to yield a full and self-consistent quantum gravity theory. General relativity has evolved as a highly successful gravitational and osmological model that has passed rigorous and precise experimental examinations. However, there are strong indications the theory is incomplete.

The question of the reality of space-time singularities and the problem of quantum gravity remain open. Even when space-time is suggested to have a discrete structure, should we ask the rhetorical question: should the continuum theories that have been so successful be abandoned? A comprehensive theory of gravity that encompasses a dual theory of space-time with both discrete and continuous aspects may resolve the present perplexity in gravitational physics without abandoning successful theories.

Gravitation plays a special role in general relativity in defining the space-time in which events take place. General relativity operates in continuous time. Time is continuous to human perception, but in the practice of human affairs it is continuous at a rate. Human beings have split time into values such as seconds, hours, minutes, days, months and years to facilitate time keeping and scheduling tasks, for most human tasks are discrete in time. Mass has been divided by volume to get density, so naturally time has also been divided into some manageable and useful quantitative values.

The relativistic effects on mass, length, and time, plus the nature of light are at the heart of the general theory of relativity. The structure of space-time provides the underlying framework for the interactions of mass, energy, and light with space and time. A better understanding of the nature of light and its duality, the coupling constant for electromagnetic interaction between particles, and the gravitational coupling constant, will lead us into a more effective application of general relativity and a better understanding of electromagnetic and gravitational phenomena.

Let us consider the Sommerfeld's constant, fine structure constant, α , or coupling constant, characterizing the strength of the electromagnetic interaction between elementary or non-elementary charged particles.

$$\alpha = k_e \frac{e^2}{\hbar c} \approx \frac{1}{137.036} \tag{6.26}$$

$$\alpha = k_e \frac{e^2}{8\pi m_p \ddot{a}} = \left(\frac{1}{4\pi\varepsilon_0}\right)\left(\frac{e^2}{8\pi m_p \ddot{a}}\right) = \left(\frac{\hbar c}{q_p^{\,2}}\right)\left(\frac{e^2}{8\pi m_p \ddot{a}}\right) = \left(\frac{8\pi m_p \ddot{a}}{q_p^{\,2}}\right)\left(\frac{e^2}{8\pi m_p \ddot{a}}\right) = \left(\frac{e}{q_p}\right)^2 \tag{6.27}$$

$$\frac{e^2}{q_p^{\,2}} = \frac{4\pi\varepsilon\hbar c}{4\pi\varepsilon_0 \hbar c} = \frac{\varepsilon}{\varepsilon_0} = \frac{\varepsilon_0 \varepsilon_r}{\varepsilon_0} = \varepsilon_r \tag{6.28}$$

Therefore, the structure constant α represents the relative permittivity of an elementary particle as given by

$$\varepsilon_r \approx \frac{1}{137.036} \tag{6.29}$$

Where ε_r is the relative permittivity of the electron, ε is the actual permittivity, "*e*" is the charge of an electron, k_e is the Coulomb's constant $1/4\pi\varepsilon_0$, $\ddot{a}$ is the acceleration of space-time, q_p is the Planck charge $\left(\sqrt{4\pi\varepsilon_0\hbar c}\right)$, m_p is the Planck mass. Hence, the Coulomb's constant k_e may be denoted as the amount of Coulomb force per unit of surface charge density. The absolute dielectric permittivity ε_0 of free space may be described as the attribute of space-time that enables space-time to act as a capacitor, to hold charges per unit of length, or as the surface charge density per unit of Coulomb force for a spherical charged mass. Hence, the relative permittivity of an elementary particle represents how the particle holds charge per unit of length as compared to how space-time holds charge per unit of length.

The coupling constant may be alternatively defined by the charge of a non-elementary particle, such as a proton; however, the square of the ratio of charge, "*e*" or "*p*", to q_p, illustrates that the polarity of the charge, "*e*" or "*p*", may be inconsequential to the definition of the coupling constant artifact.

The fine structure constant α may be described as the ratio of the square of the charge of a spherical particle (electron) divided by the product of the spherical Planck mass to the acceleration of space-time, times the ratio of the product of the spherical Planck mass to the acceleration of space-time divided by the square of the spherical Planck charge. Thus, the Coulomb force between charges, the volumes of particles of mass, the uniform distribution of electrical charges on the surface of the volumes of particles of mass, and the acceleration of space-time, provide the underlying framework for the coupling constant.

Perhaps now we are able to ask some rhetorical questions about the fine structure constant: How was the acceleration of space-time in the past history of the universe? Was the coupling constant less in the early universe? Would a changing acceleration of space-time from the early stages of the universe to the present era vary the coupling constant over time through space? Is there a direction through the space-time of the universe toward the past value of the coupling constant or toward its future value based on how and from whence space-time expands or how the volume of the mass of particles changes through the passage of time? The answers to these questions lie in the vastness of the Cosmos. Researchers are able to examine the electromagnetic waves from distant galaxies to uncover the value of the coupling constant from the frequency shift of electrons in atoms as atoms absorb or emit electromagnetic waves as the universe expands over time. From the above equations, it is possible that a change in acceleration of space-time with respect to time, as well as a change in the acceleration of the mass of particles with respect to the radius, may account for the variation in the coupling constant from the early stages of the universe to the present era, in the direction of the expansion and acceleration of space-time.

Similarly, let us consider the gravitational coupling constant α_G characterizing the gravitational attraction between spherical elementary particles of mass.

$$\alpha_G^{*} = \frac{Gm_e^2}{\hbar c} = \frac{Gm_e^2}{8\pi G m_p^2} = \frac{m_e^2}{8\pi m_p^2} = \frac{1}{8\pi}\left(\frac{m_e}{m_p}\right)^2 \approx \frac{1.751751119 \text{ x } 10^{-45}}{8\pi} \quad (6.30)$$

Then, using the alternative normalization for the gravitational coupling constant α_G^{*} we find

$$\alpha_G^{*} \approx 6.969996241 \text{ x } 10^{-47} \quad (6.31)$$

Let us compare the ratio of the strengths of the electromagnetic interaction to the gravitational attraction between charged and elementary particles of mass, such as electrons, to obtain

$$\frac{\alpha}{\alpha_G^{*}} = \frac{\left(\frac{e}{q_p}\right)^2}{\frac{1}{8\pi}\left(\frac{m_e}{m_p}\right)^2} \approx 8\pi \cdot \left(4.165747315 \text{ x } 10^{42}\right) \approx 1.0469665 \text{ x } 10^{44} \quad (6.32)$$

The above ratio clearly illustrates that the square of the magnitude of the charge of an electron to the Planck charge is orders of magnitude greater than the square of the ratio of the mass of an electron to the Planck mass. Moreover, the acceleration of space-time plays an equivalent role upon the mass of an electron and the Planck mass, which factors out of the ratio.

Electromagnetism dominates the realm of quantum objects that are highly charged and polarized, whereas gravitation dominates the macroscopic realm of objects that are electrostatically neutral to a very high degree and are very massive compared to the Planck mass. Furthermore, the gravitational constant, *G*, is a function of, and directly proportional to, the acceleration of space-time, where the acceleration of space-time has orders of magnitude of lesser surface area on the volume of a quantum particle to act upon, than it would have on a macroscopic object, or celestial body, with respect to the Planck volume.

Relativists in Physics seek to understand the fundamental physical constants, the properties of Einstein's field equations, and the nature of singularities. The direct detection of gravitational waves validates the General Theory of Relativity for very strong gravitational fields. The General Theory of Relativity is over a hundred years old, but it refuses to relent from being a very active research area in current physics.

§ 7. On the nature of light

The arrow of space-time directs the causality of events, pointing from cause to effect. Light radiates from its source instead of converging because of the expansion of space around the light source and its geometry. As space expands the electromagnetic radiation of light, in waves and quanta, expand to fill the available space in all directions possible at the speed of light. As space-time expands, the information transmitted by a ray of light may enter, or exit, a closed system to be retained if future information is allowed to be

internalized by the system. Thus, space-time is also a key driver behind the law of knowledge, so the level of information within a closed system may increase or decrease over time.

7.1. On the duality of light

A photon is said to be a massless particle traveling at the speed of light "*c*" in vacuum. For the photon, the time interval between events is always light-like. However, a photon has momentum and as a result it has relativistic mass m_p. From the work of Max Planck and Albert Einstein we learned, among other things, that *the energy of the photon* in free space-time was equal to Plank's constant "*h*" times the frequency of the photon υ, such that

$$E = h\upsilon \tag{7.1}$$

A photon is specified here as a bundle or quantum of electromagnetic energy condensed in an infinitesimal volume of space-time. The photon acts like a point-like particle or condensation of electromagnetic energy that invokes an electric field and a magnetic field simultaneously as it translates in wave-like motion on a trajectory plane passing through the axis of the induced electric field and the magnetic field waveforms, or through the axis of the Poynting vector. The photonic waveform on its trajectory plane is equidistant to the plane of the induced electric field waveform and to the plane of the magnetic field waveform. Moreover, a photon may be unpolarized, with all planes of propagation being equally probable as in natural light, or it may be linearly, circularly or elliptically polarized depending on the orientation of its orbital plane in space-time.

If the photon is linearly polarized on its photonic plane of propagation, its linear polarization determines its amplitude which is orthogonal to the direction of propagation of the Poynting vector. Similarly, if the photon is linearly polarized along the axis of propagation, its linear polarization determines its wavelength which is parallel to the direction of the Poynting vector. The photonic cycles of oscillation over time determine the photonic frequency. The linearity of polarization of the photonic waveform results as the photon propagates on its photonic plane in the direction of its Poynting vector. At relativistic speed, as length contracts and amplitude dilates in the direction of propagation of the source, the orbit of the photon becomes more elliptical, with its major axis in the direction of the amplitude.

It is speculated here that the nature of a photon may be illustrated as an energy minimum quantum loop of electromagnetic energy spinning with an angular momentum and translating in an orbit, on the trajectory plane described above, as it oscillates in space-time along a major axis equal to the peak-to-peak length of its projected waveform. The magnitude of its spin is $\left(\sqrt[2]{2}\right)\hbar$ and the component measured along its direction of motion, its helicity is $\pm\hbar$. These two possible helicities called right-handed and left-handed, correspond to the two possible circular polarization states of the photon. The width of the photonic orbit or its minor axis, or its projection along the axis of the Poynting vector, determines the wavelength of the photonic waveform. Regardless of the illustrated form of the light quantum oscillator, if the speed of oscillation is relativistic, then the photon would manifest an oscillatory relativistic mass in addition to the translational relativistic mass effected by translation through space-time in wave-like motion. The mass relativistic effects would be additive.

As space-time expands, it is proposed here that the photon's path becomes a wave trajectory along the axis of the electromagnetic field, or the resulting Poynting vector, invoked by the photon's motion and its electromagnetic field at every point along its trajectory. These effects are explainable by current electromagnetic theory in Maxwell's equations. The photonic wave is the result of a periodic disturbance or deformation of the space-time medium the wave is in. *The very existence of photonic or electromagnetic waves is a consequence of the space-time medium expansion, or contraction, as it propagates, in all its dimensions.* This is observable when light follows the curvature of space-time as it travels through a gravitational field, or why we have been able to observe gravitational lensing of distant celestial objects in the field of cosmology. Thus, *it is proposed here that the duality of light is a direct result of the oscillatory nature of the photon in its electromagnetic state and its motion through space-time due to the propagation of space-time.*

As the photon travels its path, it has relativistic mass, no charge, and it observes the conservation of energy principle at every point in space-time. The conservation of energy requires that the space-time medium move the energy somewhere else as it propagates or comes to rest, for the kinetic energy of the photon and its related electromagnetic field are conserved in the direction of movement. (Jackson, 1999)

As the photon assumes relativistic mass, it will be able, unlike any other object with mass, to travel as fast as space-time propagates. Thus, *photons and their electromagnetic waves, such as light, travel, in isotropic and homogeneous space-time, at the speed that space-time propagates to obey the principle of conservation of spatiotemporal momentum of the photon.* The fastest thing in the universe is space-time itself. Light is the fastest thing that our human senses can perceive, even though its speed is instantaneous to our senses.

7.2. The wavelengths of space-time and light

Hence, for a photon oscillating at relativistic speed, the wavelength of space λ_s decreases proportionally, and the wavelength of time λ_t dilates, as length contracts in the direction of motion of the photon. At a large scale, the wavelength of space-time remains invariant in homogenous and isotropic space-time as given by

$$\lambda_{st} \rightarrow \mp\lambda_t \pm \lambda_s \quad (7.2)$$

As the photon gains relativistic mass, length contracts, in the direction of motion, the wavelength of the photon, λ_C, decreases, its period, T_C, dilates, and the speed of light remains constant for any observer regardless of the motion of the source of light or the inertial frame of reference of the observer. However, the temporal wavelength of the photon is the wavelength of time, λ_t, the spatial wavelength of the photon is the wavelength of space, λ_s, for the spatiotemporal wavelength of a photon is the sum of the wavelengths of space and time. Thus, we find the wavelength of the photon to be

$$\lambda_p \rightarrow \lambda_{st} \quad (7.3)$$

As time propagates at light speed, its temporal wavelength, λ_t, equals the temporal wavelength of light, λ_c, in homogeneous and isotropic space-time. An electromagnetic wave, e.g. Light, will travel at "c", due to the principle of conservation of energy. The velocity of light will always be at its maximum value c as measured by any observer in an inertial frame of reference. Let us call this light phenomenon *the light barrier principle.*

7.2.(a) The Cosmological Wave Shift

There are several types of redshifts such as: gravitational, doppler, and cosmological. As an electromagnetic signal, or a ray of visible light, escapes the gravitational well of a celestial body of mass in space-time, there is a reddening of the light, or gravitational redshift in the electromagnetic spectrum. If a spacecraft has a headlight and a taillight, the wavelength of the light coming from the headlight, or taillight, would change, or have a Doppler shift, due to the relative motion and direction of the spacecraft, from the perspective of an observer in an inertial frame of reference. The change would be a blueshift, if the spacecraft is moving toward the observer, or a redshift, if the spacecraft is moving away from the observer. The cosmological redshift is a consequence of the spatiotemporal distance. The cosmological redshift from other galaxies is observable. A star has a small but detectable cosmological redshift. If the star is moving, it also has a doppler shift.

A gravitational field stretches out electromagnetic waves of any frequency away from the center of mass of a celestial body, producing a gravitational redshift similar to the Doppler effect on incident sound waves traveling through air.

There was an experiment performed by Robert Pound and Glen Rebka in 1959 at Harvard University, where 14 keV gamma-rays of radioactive iron-57 were sent between an emitter and a receiver in the left tower of the Jefferson Lab, that found a minute difference in the natural frequency of the gamma-rays due to the local gravitational field distortion. The experiment was performed with the emitter at the bottom and the receiver at the top of the tower, then with the emitter at the top and the receiver at the bottom of the tower, and the results demonstrated that the frequency shift had the same magnitude but opposite sign. (Pound, 1959)

Spatiotemporal waves travel in both directions between emitter and receiver as every point in space-time may expand, or contract. The more expanded temporal waves near the surface of the mass, or matter, of a celestial body, contract away from the surface, whereas the more contracted spatial waves expand away from the surface. Conversely, the more contracted temporal waves expand toward the surface, or center of mass, of a celestial body, whereas the more expanded spatial waves contract toward the surface, or center of mass. Hence, the difference in the wavelength and frequency of the spatiotemporal wave between two distant points, in either direction of the gravitational field along the same path, would have the same magnitude, ceteris paribus, but opposite sign.

The results of the Pound-Rebka experiment did more than confirm the General Theory of Relativity, the experiment also confirmed that the wavelength of an electromagnetic signal expands as a spatiotemporal wave expands out of a gravitational well. As the gamma-rays are emitted upward toward a top receiver in the left tower, they are redshifted. As the gamma-rays are emitted downward toward a bottom receiver in the tower, they are blueshifted. Thus, the frequency of a gamma-ray decreases as the spatiotemporal wave expands, as the gamma-rays are emitted upward.

The spatiotemporal wave is the carrier of an electromagnetic signal, or a gravitational signal. The spatiotemporal wave consists of a temporal wave and a spatial wave. The non-linear proportional relationship between the spatial and temporal waves holds the speed of light, "c", constant. Therefore, it is possible to propose that the gravitational, doppler, and cosmological redshifts are all spatiotemporal redshifts. Redshifts or blueshifts are spatiotemporal in nature. As the spatiotemporal wave expands out of a gravitational well, its wavelength extends and it redshifts. As a spacecraft with a taillight travels away from an observer, the receding movement of the spacecraft through space expands its spatiotemporal wavelength from its location to the observer, and it redshifts. As a spatiotemporal extent between an observer and a distant arbitrary point in the observer's universe expands, it redshifts. This is the principle of spatiotemporal redshift or blueshift, or the principle of cosmological wave shift. The relationship between the temporal wavelength, period, or frequency, and the spatial wavelength, period, or frequency, of the spatiotemporal wave is mathematically reciprocal.

$$\lambda_s = \frac{1}{\lambda_t} \tag{7.4}$$

$$T_s = \frac{1}{T_t} \tag{7.5}$$

$$f_s = \frac{1}{f_t} \tag{7.6}$$

Where the magnitude of the wavelength, period, and frequency, of light is normalized to "1".

The observable redshift-effect of the emitter and receiver frequencies between emitter and receiver is given by

$$f_r = \sqrt{\frac{1 - v/c}{1 + v/c}} = f_e \tag{7.7}$$

$$v \approx \frac{gh}{c} \tag{7.8}$$

where "v" is the relative wave speed between the emitter and receiver, "h" is the vertical distance between emitter and receiver, "g" is the gravitational acceleration, and "c" is the speed of light.

7.3. On the Electrophononic Effect

A phonon is an elementary energy packet of quantum-mechanical vibrational motion in which a lattice of molecules or atoms oscillate uniformly at a specific frequency. The electrophononic effect relates and underscores the frequency and the intensity of a mechanical quantum oscillator to its quantum energy when an exact packet of energy $\hbar\omega$ is supplied to the harmonic oscillator lattice to raise it to the next energy level.

The quantum of vibrational motion includes the phonon and its related electromagnetic field waves of a certain frequency, amplitude, and intensity; its quantum energy depends on the frequency of the phonon as a wave packet. This concept unites wave theory and particle theory as one and the same, for it explains the dual wave-particle nature of the phonon. In terms of the emission of vibrational motion energy, or phononic quanta, it is the conjugate concept to the photoelectric effect expounded by Albert Einstein.

The photoelectric effect as expounded by Albert Einstein in 1905 exemplifies the emission of an electron by a metallic element. The frequency of the photonic radiation is crucial, not the intensity, when incident photons cause electrons to be ejected from a metal surface. Conversely, the intensity of a direct current and its related DC voltage potential, not the frequency, is crucial when an electronic current and voltage potential cause the oscillation of phonons in an incandescent filament. The electrophononic effect was exemplified by Thomas Edison's successful incandescent light bulbs using zero frequency DC current and DC voltage as a power source. (Einstein, 1952)

Thus, assuming no heat conduction or convection, the energy transferred in packets or quanta of phonons from a DC power source to the harmonic oscillator lattice of a superconducting filament to raise it to the next energy level, for the electrophononic effect, or Tesla-Edison effect, may be expressed as

$$E = \sqrt[2]{\hbar VI} \tag{7.9}$$

Where $\hbar$ is the reduced Planck constant, "V" is the voltage potential from the DC source and "I" the electronic fluid intensity of the Cooper pairs without heat transfer $\partial Q_h / \partial t = IR^2 = 0$ from the emitting source due to no resistance in the superconductor. Thus, the angular frequency of the energy transfer in "n" packets of phonons by the DC power source, may be expressed to be

$$\omega = \sqrt[2]{\frac{VI}{\hbar n^2}} \tag{7.10}$$

Otherwise, heat transfer through electromagnetic radiation or other means would have to be subtracted from the energy, E.

$$E = \sqrt[2]{\hbar VI - \left(\partial_t Q_h\right)^2} \tag{7.11}$$

The energy transfer from the DC power source for a single phonon is given by

$$\hbar\omega = \sqrt[2]{\hbar \partial V \partial I} = \sqrt[2]{\hbar \frac{\partial(\hbar\omega)}{\partial t}} = \hbar\sqrt[2]{\dot{\omega}} \tag{7.12}$$

and the angular frequency of the transfer of energy for a single phonon may be expressed as

$$\omega = \sqrt[2]{\dot{\omega}} \tag{7.13}$$

Thus, the angular frequency, or the speed of the angular frequency, of the energy transfer for a single phonon, is multimode.

7.4. On the energy, mass and characteristics of a photonic wave

Furthermore, as the quantum energy of light depends on its intensity and frequency, the energy of the Poynting vector is directly related to the amplitude of the electromagnetic or photonic wave for the energy of the intensity of light, and represents the rate of energy transfer per unit area of the electromagnetic field of the quantum of light in free space-time. The amplitude of the photonic wave is the distance between the farthest point in the path of the photon and the directional axis of propagation. The energy and momentum of a photon depend only on its frequency, $h^2\upsilon^2 = E_p^{\ 2} + p^2c^2$. The energy of a photonic wave is proportional to its amplitude squared, A^2, and its intensity is the power it delivers per unit area of its orbital wave path.

The orbital wave path is the closed surface the photon would create in one revolution about its center of motion on the directional axis of propagation. Thus, when the amplitude of a photonic wave is doubled, $(2A)^2 = 4A^2$, its energy is quadrupled. Similarly, the area of a circular wave path is πr^2, if the radius is doubled, $\pi(2r)^2 = 4\pi r^2$, the area is quadrupled. Thus, *the energy of a photonic wave is directly related to the surface area of the photon's orbital wave path.* The energy levels of electrons in atoms are discrete; each element emits or absorbs its own characteristic frequencies. The frequency of light results from the angular velocity of the photon, $\upsilon = \omega/2\pi$, and its period is $T = 1/\upsilon$. Thus, the frequency and wavelength of the light quantum is that of the photon. The peak of the wave varies proportional to the electromagnetic field strength or to the kinetic energy of the orbit of the photon. Therefore, we find the energy E_q of the light quantum as it propagates to be

$$E_q^{\ 2} = h^2\upsilon^2 - \frac{l_p^{\ 4} t_p^{\ 2} E_m^{\ 2} B_m^{\ 2}}{\mu_0^{\ 2}} = h^2\upsilon^2 - \frac{\left(l_p^{\ 2}\right)^2 E_m^{\ 2} B_m^{\ 2}}{\dot{\mu}_0^{\ 2}} \quad (7.14)$$

and we find the momentum of the light quantum to be

$$p_q = \frac{E_q}{c} = m_q c \tag{7.15}$$

where m_q is the relativistic mass of the light quantum, "c" is the speed of light; E_m and B_m are maximum amplitudes. $E(t,x) = E_m Cos(\omega t - kx)$ represents the electric field and $B(t,x) = B_m Cos(\omega t - kx)$ represents the magnetic field.

The centripetal acceleration of the photon in its orbit keeps it on path about its center of motion on the directional axis of propagation while the translational speed of propagation of space-time keeps the photon moving onwards on trajectory.

Let us imagine a photonic standing wave that may be illustrated as a photonic wave that does not translate in space-time. Let a photon be oscillating in place about its center of motion with its related induced electromagnetic fields in a medium of space-time that is oscillating, but not propagating. In the standing wave case, the electromagnetic fields will increase and decrease on one half of the orbit, then increase and decrease on the opposite side of the axis passing through the center of motion on the plane of the closed orbital surface during the other half of the orbit.

If a photon is oscillating relativistically, there is an imbalance between the energy and momentum of the photon, where that imbalance is the relativistic mass of the photon.

$$\Delta E_i^2 = h^2\upsilon^2 - \left\{E_p^2 + \left(p^2c^2 - \Delta p^2c^2\right)\right\} = \Delta p^2c^2 = m_q^2 v_q^2 v_t^2 \tag{7.16}$$

$$m_q = \sqrt[2]{\frac{\Delta E_i^2}{v_q^2 v_t^2}} = \frac{\Delta E_i}{v_q v_t} \tag{7.17}$$

Thus, the speed of the photon is equal to

$$v_q = v_t = c \tag{7.18}$$

Then, we find the relativistic mass of the photon to be

$$m_q = \frac{\Delta E_i}{c^2} \tag{7.19}$$

Thus, we express the speed of light as

$$c^2 = \frac{\Delta E_i}{m_q} \tag{7.20}$$

Where we can see how *the speed of light "c" depends on the imbalance between the energy and momentum of the photon, on the relativistic mass of the photon, and on the wavelength and frequency of space-time.*

The imbalance between the energy and momentum of a photon stems from the difference in energy states of the photon as it travels relativistically in homogeneous and isotropic space-time. As a photon oscillates in the directions and relativistic dimensions perpendicular to the direction of propagation (amplitude dilates) of the photonic waveform, the angular frequency energy is conserved. The width of the orbit of the photon (period) contracts in the direction of propagation, but the oscillatory energy is conserved. In the direction of motion, the kinetic energy is relativistic, wavelength and period contract to maintain light speed, which in turn decreases the momentum of the photon. The frequency energy, or oscillatory energy, perpendicular to the direction of motion, is conserved. As a result, there is an imbalance between the energy and momentum of the photon, and the photon gains relativistic mass.

7.5. On the double-slit experiment for light

The double slit experiment dates back to 1801 when Thomas Young conducted the first experiment to prove his theory that light consisted of waves instead of particles. Light spreads out in spherical waves from the source according to Huygens' Principle. The patterns of interference of light were accepted as evidence for the wave theory of light. It was not until the twentieth century that modified double slit experiments re-validated the particle theory of light. Young's experiment of 1803 used the edge of a thin card to split a sunlight beam which resulted in a double slit effect.

The double slit experiment uses a coherent light source to illuminate a thin plate, with two parallel slits cut in it, and the light passing through the slits strikes the screen behind the slits. The wave nature of light causes the light waves passing through both slits to interfere with each other. An interference pattern of bright and dark bands of light is observed on the screen. However, the light is absorbed at the screen, as if light were made of discrete particles or photons. (Jönsson, 1961)

According to classical particle physics, if light travels from the source to the screen as particles, the brightness at any point on the screen should be the sum of the brightness distribution when the right slit is blocked and the brightness distribution when the left slit is blocked. However, when both slits are unblocked, some points or bands on the screen are brighter and other points or bands are darker. This is explained by the wave nature of light and the additive and subtractive interference of waves, not by the exclusively additive nature of particles such as photons. The bright regions show constructive interference where a crest of a wave meets a crest from another wave. The dark regions show destructive interference where a crest of a wave meets a trough from another wave. Therefore, light must have a particle-wave duality. The double slit experiment has been performed with photons, electrons, and atoms, and each time the same result is found.

In the process of measuring the which-path information for the

photon in the double slit experiment, any change that is made to the apparatus or any device that intrudes upon either path of the photon, or both paths simultaneously, such as a photon detector used to collect information about which slit a photon passes through, results in the destruction of the interference pattern of the waves of light. However, an experiment performed in 1987 produced results that demonstrated that which-path information could be obtained without destroying the possibility of interference. These results showed that there may be technological ways to measure which-path information that are not so intrusive to the photon, apparatus, or the experimental medium, that allow both the detection of which-path information and the existence of the interference pattern of light. (Mittelstaedt et al, 1987)

Thus, the complementarity principle of light, which states that during an experiment light can demonstrate both particle and wave characteristics, but not both characteristics at the same time, needs to be reviewed. By Phillippe Eberhard's proven theorem, if the accepted equations of quantum theory are correct, it should never be possible to experimentally violate causality using quantum effects. (Eberhard et al, 1989)

According to quantum eraser experiments, the experimenter arranges to detect which-path information, a beam splitter is used to provide two paths for the photon, and then the which-path information is erased after the fact. However, the erasure of information after the fact means that another beam splitter is used to direct the light toward the screen, removing the path separation or which-path information. Only then, the interference of light is restored. Total phase differences are introduced along the paths in these experiments because of reflective surface effects and the effects of passing through the mediums of the beam splitters. When each path has its separate exit, there is a constructive or a destructive wave interference effect for the wave output of each separate path. Moreover, a quantum eraser experiment done in the year 2000, with an erase-delay until after the photon has struck the detector screen, showed that an interference pattern can be recovered. (Scully et al, 2000)

7.6. A space-time wave double slit thought experiment

Let us imagine a double slit experiment for space-time waves, photons and their related electromagnetic waves, also called light quanta, which are emitted from a source of coherent monochromatic light in a region of space-time that is both homogeneous and isotropic. Let us set up the apparatus of our experiment such that our light source emits photons one at a time toward a thin plate with two parallel slits cut in it at a very small specified distance, and a screen a specified distance behind the slits for the photon to strike. Two photon detectors will be available for us to install by any one of the two slits, one at a time, or both slits simultaneously if needed. These detectors will be able to detect a photon passing through the slits we are monitoring so that we can record the which-path information. The screen will be a detector screen that will detect each photon's strike and will relay that information to a computer and monitor that will show a visual representation of the bright dots of light on the screen as they form bright regions along the width and height of the detector screen.

Once the equipment is set up with a detector by the right slit only, space-time waves are already propagating along the direction of travel of the light quantum even before the light source is turned on.

Let us consider what is happening throughout the space-time medium of the apparatus between the emitter and the end screen. As a space-time wave propagates through the space-time medium in the direction of the slits and the screen, the wave is coherent and unobstructed in any way. As soon as the space-time wave nears the slits, on the side of the right slit, let us call it slit *B*, the physical presence and interaction of the detector with the space-time wave creates a disturbance or obstruction and the waveform decoheres to the right as it tries to pass through slit *B* of the thin plate. All the possible consistent states of the measured system and the measuring detector, including the observer, are present in a real physical, not just formally mathematical, quantum superposition. Such a superposition of consistent state combinations of different systems becomes an entangled state.

Meanwhile, on the left slit, slit *A*, there is no detector installed and the space wave propagates through slit *A* unobstructed and undisturbed by anything. On the left side of the screen, the wave *A* that propagates out of slit *A* reaches the screen first, then the wave *B* propagating out slit *B* reconstitutes as a decoherent wave, with potential changes to its phase, frequency, or polarization, as it strikes the screen. As a result of the decoherence, wave *A* and wave *B* do not interfere with each other. The wave through slit *A* has followed a critical path from the emitter to the end screen. *The critical path of a spatiotemporal wave would be the path of least action or the path of the greatest conservation of energy with the least possible disturbance to the wave along its path.*

The six-dimensional Einstein-Hilbert action is a functional of the six-dimensional metric tensor of General Relativity that yields the six-dimensional Einstein Field Equations through the principle of least action through one dimension of coordinate time in smooth space-time given by

$$S = \int \left(\frac{c^4}{16\pi G t_p^{\,2}} R + \frac{L_M}{t_p^{\,2}} \right) \sqrt{-g}\, d^6 x \qquad (7.21)$$

$$S=\int\left(\frac{1}{2}\frac{mc^4R}{\ddot{a}t_p^{\,2}}+\frac{L_M}{t_p^{\,2}}\right)\sqrt{-g}\;d^6x \tag{7.22}$$

The determinant of the metric $\det\left(g_{\mu\nu}\right)$ is useful as a factor of scaling, rotation, expansion, contraction, shearing, or reflection, for length, area, or volume, in the six-dimensional manifold. For a nearly-flat region of six-dimensional space-time, $\det\left(g_{\mu\nu}\right)$ is "–1" and "$-g$" is the square root of a positive scalar that corresponds to the characteristics of the curvature of the interval.

Where L_M is a Lagrangian density term of kinetic energy $\left(J/m^3\right)$ describing any matter fields appearing in the six-dimensional theory, "R" is the trace of the six-dimensional Ricci tensor, "g" is the determinant of the six-dimensional metric tensor, and "$\ddot{a}$" is the acceleration of space-time as it extends or contracts.

In four-dimensional space-time (with folded time), we have

$$d^\tau x=\sqrt[2]{\left(dx^4\right)^2+\left(dx^5\right)^2+\left(dx^6\right)^2} \tag{7.23}$$

$$d^4x=d^3x\cdot d^\tau x \tag{7.24}$$

$$S=\int\left(\frac{c^4}{16\pi G}R+L_M\right)\sqrt{-g}\;d^4x \tag{7.25}$$

The Einstein-Hilbert action $\left(J\cdot s\right)$ is a mathematical functional for the General Theory of Relativity that takes the trajectory of the system as its argument to render a result that is a real number. The equations of motion of the system, or the EFEs, can be derived from the action as a physical attribute of the dynamics of the system. The development and proposal of the four-dimensional action in 1915 was a significant contribution to the General Theory of Relativity by the eminent mathematician David Hilbert.

Next, let us consider what happens when the experimenter turns on the switch for the light source and a single light quantum is emitted toward the slits and screen. Let us follow the first light quantum along its path. As the light quantum is emitted indeterminably it has its own characteristic position and momentum in space-time, the light quantum travels in the direction of the propagating space-time wave toward the slits and screen. As the light quantum nears the slits, the space-time is disturbed towards the right slit, or slit *B*, and the space-time towards slit *A* is undisturbed, more homogeneous and isotropic, leading the light quantum, or photon, to the critical path through the slit *A* and onwards to the detector screen.

The photon follows the critical path of the greatest conservation of energy through the slit *A* and onwards to strike the detector screen. Depending on the physical dimensions of the slits, the distance of separation of the slits, the distance from the slits to the center of the screen and the wavelength of the light quantum, the photonic wave of the light quantum may travel through both slits, but when it does, the photonic wave *B* through the right slit would decohere as it follows the space-time wave through the slit *B* and onwards to the detector screen. However, photonic wave *B* would not interfere with photonic wave *A* due to its decoherence and its potential difference in frequency and polarization.

Thus, over time the experimenter would only see on the computer screen the growing visual representation of a light wave without interference. If the experimenter tries the same experiment by only changing the position of the photon detector from slit *B* to slit *A*, the same result would be shown but shifted toward the center of slit *A*. (Tonomura et al, 1989)

If the experimenter tries the experiment with a detector installed by the slit *A* and another by the slit *B*, all other variables being equal, the space-time waves propagating to each slit will decohere and not interfere with one another as they propagate out of the slits toward the end screen. As the light quantum nears the slits, the photon will follow the critical path through a slit according to its position and momentum and onwards to strike the detector screen. By the uncertainty principle it would not be possible to measure the location of a particle in space-time and its momentum at the same time. In other words, position and momentum are complementary. Thus, the path of the photon through a slit is indeterminate. Therefore, photons do not arrive at the screen in any predictable order. Thus, knowing where all the previous photons appeared on the experimenter's computer monitor, or where the photons struck the detector screen, and in what order they appeared will not tell us anything about where the next photon will strike, even though it might be possible to estimate an approximate probability of where the next photon might strike on the detector screen.

As the experimenter tries the double slit experiment without detectors, all other variables being equal, the interference pattern appears on the experimenter's computer monitor. As the space-time waves pass through the slits, both waves are coherent and interfere as they propagate onwards toward the end screen. As the light quantum travels near the slits, the photon follows the critical path in an indeterminate way through the slit and onward to strike the detector screen. Depending on the physical dimensions of the slits, the distance of separation of the slits, the distance from the slits to the center of the screen and the wavelength of the light quantum, the photonic wave of the light quantum may travel through both slits.

Thus, the photonic wave passes though both slits, wave *A* and wave *B* interfere with each other onward to the detector screen, and over time, the experimenter observes the interference pattern of light waves for the experiment on the computer's monitor.

The principle of complementarity of space-time and particles in quantum mechanics states that the physical manifestations of energy and mass as waves and particles, or wave-particle duality, are complementary because of the behavior and properties of space-time waves as they propagate and serve as the medium for the waves and particles of energy with or without mass. Space-time wave functions are the underlying structure and principle of propagation for all other energy and mass wave functions or physical fields.

7.7. The probability space-time wave function

Let us imagine two emerging space-time wave functions, in homogeneous and isotropic space-time, of an energy wave packet passing through slits *A* and *B* as

$$\Psi_a(y)+\Psi_b(y) = \frac{e^{i\left(\frac{p}{\hbar}\right)y}}{\sqrt[2]{2\pi r}}+\frac{e^{i\left(\frac{m}{\hbar}\right)y}}{\sqrt[2]{2\pi r}} \tag{7.26}$$

where p and m are the momenta of the alternative paths of the energy wave packet traveling with the space-time wave through slit *A* or through slit *B*, "y" is the variable of a vertical point on the screen, the length of the path is $2\pi r$, and "$e^{i(p/\hbar)y}$" and "$e^{i(m/\hbar)y}$" are oscillating waves.

Taking the complex conjugates of the sum $\Psi_a(y)+\Psi_b(y)$ and integrating from 0 to $2\pi r$ we find

$$P_{ab}(y) = \frac{2}{\sqrt[2]{2\pi r}} \int_0^{2\pi r} \left\{1 + Cos\left[\frac{y(p-m)}{\hbar}\right]\right\} dy \qquad (7.27)$$

Hence, we observe from the above probability wave function that there are places on the horizontal axis, or x-axis, where the space-time wave has probabilistic values of zero.

Thus, the probability is uniform and constant for any experiment with only one slit, but when the experiment is done with two slits without disturbance there are places where the probability is zero. This is what happens in a double slit experiment without detectors when the interference pattern results on the screen.

In a single slit experiment, there would only be one momentum to consider. Therefore, let us specify the wave function as

$$\Psi_s(y) = \frac{e^{i(p/\hbar)y}}{\sqrt[2]{2\pi r}} \qquad (7.28)$$

Taking the complex conjugate of $\Psi_s(y)$ and integrating from 0 to $2\pi r$ we find

$$P_s(y) = \frac{1}{2\pi r} \int_0^{2\pi r} e^{i(p/\hbar)y} e^{-i(p/\hbar)y} dy = 1 \qquad (7.29)$$

Hence, the space-time wave function of the single slit experiment has a uniform and constant probability of 100% in finding the energy wave packet at a value of "y".

7.8. Space-time wave interference characteristics

For the interference pattern of space-time waves in the double slit experiment, the phase angle difference ϕ_s between waves is found to be

$$\phi_s = w\sigma \, Sin\,\theta_s \qquad (7.30)$$

where $\phi_s = \phi_A - \phi_B$, *w* is the wave number, σ is the distance between the slits, and θ_s is the angle that a line intersecting both waves *A* and *B* through the origin at the start of each wave cycle makes with a line perpendicular to the axis of propagation of the waves.

During destructive interference we find

$$\theta_s = ArcSin\left(\frac{2k\pi + \pi}{w\sigma}\right) \qquad for\; k \geq 0 \qquad (7.31)$$

During constructive interference we find

$$\theta_s = ArcSin\left(\frac{2k\pi}{w\sigma}\right) \qquad for\; k \geq 0 \qquad (7.32)$$

and the *slit ratio* is specified to be

$$Sin\,\theta_s = \frac{\delta}{\rho} \qquad (7.33)$$

where δ is the distance between the center lines of the slits and ρ is the distance from either slit center line to the center point of the screen.

Thus, during the space-time wave interference pattern of the double slit experiment, high fringes occur at positions denoted by β_{hf} to be

$$\beta_{hf} = \frac{2\pi k\rho}{w\sigma} \qquad (7.34)$$

and the distance between two successive high fringes on the space-time wave interference pattern is found to be

$$\Lambda_{hf} = \frac{2\pi\rho}{w\sigma} \tag{7.35}$$

where Λ_{hf} and β_{hf} are inversely proportional to δ .

On the space-time wave interference pattern of the double slit experiment where the superposition principle applies, the high bands of space-time waves, or high fringes, show constructive interference where a crest from wave *A* meets a crest from wave *B*. The low bands show destructive interference where a crest from wave *A* meets a trough from wave *B* or vice versa.

7.9. The critical path of a photon or particle

On the photonic double slit experiment, the detection of a photon involves a physical interaction between the photon and the detector of the sort that physically changes the photon's space-time medium and the outcome of the experiment. The time of interaction between the photon and the detector may not be exactly the same for each trial through the same slit or between slits during the same trial. In a double slit experiment, if a path of a photon is disturbed by the presence of a detector, the time required for interaction between photon and detector would cause the photon to decohere from its undisturbed space-time wave critical path. *The critical path of a photon or particle would be the path of least action with the minimum expenditure of energy, or the path of the greatest conservation of energy with the least possible disturbance to the photon or particle along its path.*

The probability of arrival of a single photon at various points on the detection screen is a function of the wavelength of the photon λ_c and of the distance δ between the center lines of the slits. For a space-time wave, or an electromagnetic wave, the wavelength and frequency are inversely proportional to each other, and the wavelength and the momentum are inversely proportional to each other, as the wave follows the critical path according to the principle of conservation of energy. (de Broglie, 1953)

The de Broglie wavelength and frequency were given as

$$\lambda = \frac{h}{p} \tag{7.36}$$

$$\upsilon = \frac{E}{h} \tag{7.37}$$

Which leads us to *the conservation of space-time momentum energy equation* for any quantum wave function

$$E_Q = \lambda p \upsilon \tag{7.38}$$

Conclusively, the probabilistic values of a probability wave pass through space-time and are a function of space-time. It is proposed here that space-time waves are manifested as propagating probability waves that lead the way or path for particles, as particles travel through space-time. *Space-time waves are preemptive waves to electromagnetic waves and other types of waves*. They travel at light speed and exhibit all the characteristics of waves, such as constructive and destructive wave interference.

Space-time may expand or contract, by Huygens-Fresnel Principle, as space-time wavelets emerge and interfere. Expansion and acceleration may happen as the space-time wave effect is stronger at long range without obstruction, and weaker at short range between gravitational nodes of obstruction (mass or energy), as spatiotemporal wavelets emerge and evolve over the passage of three-dimensional time. The expansion and acceleration effect of space-time due to spatiotemporal wave interference may be regarded as repulsive gravity.

§ 8. Conclusion

The nature of gravity is the nature of space-time itself. As space-time presses on an object of mass, space-time decelerates and slows down exerting a gravitational field perpendicular to the surface of the object. Historically, objects were thought of as being attracted to the earth's gravity, but it is now very clear the objects have always been pushed down instead onto the earth by the curvature of space-time.

Gravity is strongest where space is most curved and when time is slowest. Gravity tells Time how to go, Time tells Gravity how to pull. During relativistic motion in homogeneous and isotropic space-time, mass and time dilate and length contracts to obey the law of conservation of spatiotemporal momentum. The dynamic relationship between mass and space-time is intrinsically contained within the concept of the universal gravitational constant. Space-time needs to be examined as a preemptive propagating wave field to physical fields, dynamic and eventful, not a static background to physical phenomena.

PART IV

WAVES OF THE FIELDS OF FORCE

Chapter 9

On the electromagnetic and electrogravitic fields of charges and masses in space-time

§ 1. Introduction: the electromagnetic field

A moving electrically charged object produces a physical field that is called an electromagnetic field. This physical field affects other charged objects that are near the field. The electromagnetic interaction between two objects is possible because an electromagnetic field may extend boundlessly in free space.

The electromagnetic field is one of the four known fundamental field forces of nature such as the gravitational force, the strong interaction and the weak interaction.

The electromagnetic field of a current of charges comprises the electric field and the magnetic field. Presently, it is thought that the magnetic field is produced by moving charges while the electric field is produced by stationary charges and their combined effect is viewed as the source of the electromagnetic field. The Lorentz force law and Maxwell's equations describe the interactions of electromagnetic fields. (Feynman, 1988)

The electromagnetic field propagates smoothly as a wave, but it is also quantized of individual particles called photons. All electromagnetic fields are force fields, carrying energy and capable of producing an action at a distance. These fields have characteristics of both waves and particles.

Two main characteristics which define an electromagnetic field are its frequency and its corresponding wavelength. The frequency simply describes the number of oscillations or cycles per second, while the term wavelength describes the distance between one wave and the next. Thus, frequency and wavelength are interrelated, the longer the wavelength the shorter the frequency of the wave. Electromagnetic waves travel at the speed of light. (Born et al, 1999)

1.1. Electric fields

Electric fields exist in the presence of electrically charged particles, exerting forces on other particles within the influence of the electric field. The electric field strength is measured in volts per meter (V/m). Any conductor of electricity that carries a charge will produce an electric field even if there is no current flowing through it. The electric field produced by a conductor is proportional to the voltage potential of the conductor with respect to a reference point, the higher the voltage potential the stronger the electric field produced at a distance from the conductor.

1.2. Magnetic fields

Magnetic fields are conceptualized to exist from the motion of electrical charges through a conducting medium. The magnetic field strength is measured in amperes per meter (A/m). Any current flowing through a conductor will effect a magnetic field about the conductor. The magnetic field effected by a flow of charges is directly proportional to the flow of current through the conductor, the higher the current the greater the strength of the magnetic field. However, the strength of the magnetic field will lessen as the distance from the source increases. The flux density of a magnetic field is measured in Teslas. Magnetic fields are not as easily attenuated as electric fields by different types of matter. Magnetic domains are microscopic magnetic regions composed of atoms whose magnetic fields are aligned in a common direction. Most substances or conducting mediums are not magnets. A spinning electron is a moving charge that produces a magnetic field. Electrons pair up with their spins opposite each other, so their fields cancel each other out. In ferromagnetic materials, such as Iron, Cobalt, and Nickel, magnetic fields produced by electron spins do not cancel completely. Strong coupling between nearby atoms form large magnetic domains of atoms whose net spins are aligned.

Magnetic fields about a current-carrying conductor are the direct product of the electromagnetic field of the source and magnetic domains (dipoles) in the conducting material. The magnetic domains are aligned with the electromagnetic field of the source of voltage potential, and the electric field produces the current of charges through the conducting medium. The greater the source of the magnetic field, the greater the number of magnetic domains aligned with the polarity of the source. The electric field of the magnetic domains would act as a counter electric field, or back emf, to the source electric field, extending from positive to negative. If the source electric field, from positive to negative, is assumed to be in a clockwise rotation, then the magnetic domains are rotating counterclockwise as seen from positive to negative of the back emf.

1.3. Electromagnetic fields: discrete or continuous?

The present conjecture on the structure of the electromagnetic field appears contrarian since the electromagnetic field may be viewed as having both a continuous and a discrete structure. These distinct ways of viewing the electromagnetic field are a legacy of the evolution of the theory of particles and waves. The continuous structure of smoothly propagating electromagnetic waves develops from the oscillations of electrical charges. Energy is transmitted smoothly and continuously through the electromagnetic field from one location to another. This structural view is very successful for low frequency radiation sources.

The discrete structure of the electromagnetic field provides a more rugged view of the composition of the force field. The discrete structure of the propagating electromagnetic field develops from the packets or quanta of electromagnetic energy in its composition. Energy is transferred in packets or quanta of photons with a characteristic frequency such that

$$E = h\nu \qquad (1.1)$$

Where h is Planck's constant, and "ν" is the frequency of the photon. (Einstein, 1952)

The discrete view of the electromagnetic field has been successful in producing the current quantum field theory of electrodynamics which provides a description of the interaction of moving charges and electromagnetic fields. (Jackson, 1999)

§ 2. On the dynamic characteristics of the electromagnetic field

As the electromagnetic theory evolved, electrical charges were thought to produce an electric field when stationary and a magnetic field when moving through a conductor. That view changed over time into a unified view where the field was considered as composed of both the electric field and the magnetic field, the electromagnetic field. A charge distribution produces the magnetic field as other charge distributions within the field experience a force. The net electromagnetic field produced is the cumulative effect of all fields present. Thus, the electromagnetic field is dynamic and interactive between charge distributions.

The electromagnetic field resolves into four main characteristics. First, the electric field and the magnetic field, of a net electromagnetic field in a region of space-time, only interact with one another. Second, the electric field and magnetic field of a charge distribution exert forces on other charge distributions that are within the influence of those fields. Third, charge distributions can move through space-time. Fourth, all charge distributions can produce electromagnetic fields.

A charge distribution produces the quanta of the electromagnetic field as a separate packet of energy than the charges themselves. The electromagnetic field quantum is a packet of energy translating through space-time at relativistic speed with respect to an inertial observer while the field-producing charges may not be, as in the case of man-made electromagnetic field from an electric current through a conductor. However, a field-producing charged particle can translate, approaching the speed of light, through space-time with respect to an inertial observer with the assistance of an enormous amount of energy. (Taylor et al, 1966)

2.1. The resultant magnetic field

Charges traveling along a conductor are oriented in the direction of the electric field according to the polarity of the charge. Negative charges will travel along the conductor in the direction of the positive charge distribution, or the source, producing the electric field, through the conducting medium. A positive charge will travel away from the positive field-producing charge distribution. In the presence of a strong electric field some atomic dipoles align with the magnetic south pole toward the positive field-producing charge distribution. As charges and atomic dipoles line up with the source electric field flowing through a conductor, charges move through the conductor and current flows. As the electric field of the source propagates, the electric fields of charge distributions and atomic dipoles manifest themselves as a magnetic flux density $\vec{B}$ about the conductor. Thus, *a magnetic field springs from the curl of a resultant magnetic field that is perpendicular to the direction of propagation of the source electric field within the conducting medium of the current.*

The magnetic loops about a conductor, when current flows through a conductor, are regions of equal magnetic field potential for the external resultant magnetic field. The external resultant magnetic field is perpendicular at a point in the *B*-loop, to the internal source electric field lines in the direction of propagation. Within the conductor, the resultant magnetic field components may add or subtract with components from the source electromagnetic field. However, in the absence of interference from other electromagnetic fields, on the outside space about the conductor, the resultant magnetic field components are unopposed and manifest themselves in what has been called magnetic flux density, $\vec{B}$. These external magnetic field lines are nothing more than resultant magnetic field lines, or resultant magnetic field remanence, that were perpendicular to the internal source time-variable electric field lines in the direction of propagation, and were able to travel outside the conductor without interference, in the absence of any other field.

Hence, electric fields and magnetic fields, related to a conducting medium, are a manifestation of the same physical phenomenon of an electromagnetic field. For a conducting medium, the direction of the magnetic flux density field about the conductor is perpendicular to the direction of propagation of its complementary electric field within the conductor and vice versa. The source electric field lines follow the return path back to the source within the physical boundaries of the conducting medium. In the absence of interference from other fields, the action of the curl of the resultant magnetic field outside the conductor is boundless, and as the curl of the resultant magnetic field travels, it can penetrate matter easily due to the fact that it is unopposed unless matter is properly shielded. This principle was instrumental in the design and application of antennae.

The causality of electric fields and magnetic fields, under the present paradigm, has raised the issue of either field causing the other to change in time, propagating the electromagnetic wave through its medium. Neither Maxwell equations nor their solutions provide a causal link between the electric field and its conjugate the magnetic field, but a simultaneous manifestation by time-variable electromagnetic fields of charges and time-variable electric currents. However, fields move at the speed of light and currents may move slower than the speed of light. Hence, as fields propagate at the speed of light, the simultaneity assumption becomes counter-intuitive, because fields are faster than charges.

As an electron comes into existence, the electron tries to move at the speed of light since it is fundamentally massless. The omnipresent Higgs field of virtual Higgs particles, that briefly and continuously appear and disappear, obstructs and detours the electronic trajectory. Thus, the electron zigzags rather than follow a linear trajectory, moving slower than the speed of light through the Higgs field, with an imbalance of energy and momentum, gaining mass.

Therefore, there is a causal link, a time delay, under the present paradigm, between the manifestation of the magnetic loop field about a conductor, and the flow of current and the action of the electric fields of negative charges. Nevertheless, under the concept of a resultant magnetic field, it is intuitive to consider the incident electric field, $\overrightarrow{E}$, and the resultant magnetic field, $\overrightarrow{B_{RE}}$, to be simultaneous and causally-linked through the actions of electromagnetic fields at the speed of light. Under this conceptualization, Maxwell equations in a resultant magnetic field form would be causally-linked and simultaneous.

§ 3. On the electromagnetic field of the photon

In the case of a photon's electromagnetic field, the electric field and the magnetic field are viewed as two separate fields at quadrature, that is, at ninety degrees from each other. A photon is specified here as a bundle or quantum of electromagnetic energy condensed in an infinitesimal volume of space-time.

The photon acts like a point-like particle, or condensation of electromagnetic energy, that invokes an electric field and a magnetic field simultaneously as it translates in wave-like motion on a trajectory plane passing through the axis of the induced electric field and the magnetic field waveforms, or through the axis of the Poynting vector. The electric field and the magnetic field of a photon are proposed here to be two orthogonal electromagnetic field components of a singular electromagnetic-photonic field.

3.1. The electromagnetic-photonic field and force

The photonic waveform on its trajectory plane is unequal to the plane of the electric field waveform and to the plane of the magnetic field waveform. The photon propagates on its trajectory plane in the direction of its Poynting vector. Thus, the electric field and the magnetic field of a photon are the orthogonal components of the electromagnetic-photonic field.

The electromagnetic-photonic force results from the kinetic energy of the photon as it propagates through space-time. The rate of energy acting on a unit area equals the power density (*Watts/area*) of the work performed by the waveform of the photon per unit area times the distance, as it propagates over space.

As the photon propagates relativistically, it gains mass, and this photonic mass performs work, as it dilates, and as it compresses space-time along the direction of motion. Currently, we refer to the power density as the Poynting vector.

$$\frac{\partial \overrightarrow{F_{EP}}}{\partial t} = r\left(\vec{E} \times \vec{H}\right) \tag{3.1}$$

Thus, we find the electromagnetic-photonic force to be

$$\overrightarrow{F_{EP}} = \int_{t_1}^{t_2} r\left(\vec{E} \times \vec{H}\right)\partial t \tag{3.2}$$

Hence, the electromagnetic-photonic field $\vec{E}_{EP}$ resolves into an expression consisting of two orthogonal field components, $\vec{E}$ and $\vec{H}$, or $\vec{E}$ and $\vec{B}/\mu_0$, "*r*" is the distance where work is performed by the force over time along the direction of motion, and μ_0 is the permeability of free space. Thus, we find the electromagnetic-photonic field in the presence of a charge *Q* to be

$$\overrightarrow{E_{EP}} = \int_{t_1}^{t_2} \left(\frac{r}{Q}\right)\left[\vec{E} \times \vec{H}\right]\partial t \tag{3.3}$$

§ 4. On the electromagnetic fields of moving charges

For a unit charge Q, or a photon, freely moving at a speed v, when a magnetic flux density field $\vec{B}$ is being generated by an electric field $\vec{E}$ due to the angle of propagation of the photon as it orbits, space-time expands propagating the photon forward in time, manifesting an electromagnetic field in the dimensions of isotropic and homogeneous space-time in the absence of other electromagnetic fields.

From both sides of Faraday's Law, and dividing through by a unit of area dA, we find the magnitude of the electric field in terms of the magnitudes of the magnetic flux density and the velocity of propagation.

$$\oint \vec{E} \cdot d\vec{l} = -\frac{d}{dt} \iint \vec{B} \cdot d\vec{A} \tag{4.1}$$

$$\frac{\partial E}{\partial x} = -\frac{\partial B}{\partial t} \tag{4.2}$$

$$\partial E = -\frac{\partial x}{\partial t} \partial B \tag{4.3}$$

$$E = -vB \tag{4.4}$$

$$E = -v\mu_0 H \tag{4.5}$$

Hence, we find the speed of a unit charge, or photon, at a point in the trajectory to be

$$-v = \frac{E}{B} = \frac{E}{\mu_0 H} \tag{4.6}$$

and the magnitude of the magnetic flux density field $\vec{B}$ is found to be

$$B = -\frac{E}{v} \tag{4.7}$$

Thus, the magnetic flux density of an electromagnetic field is equivalent to the electric field divided by the velocity of the photon, or unit charge, through isotropic and homogeneous space-time in the direction of propagation. The velocity of the photon as it orbits and propagates through space-time, in the absence of other fields, is the slope of a tangent line at any point of its trajectory, given by the ratio of its electric field to its magnetic flux density.

So, the magnitude of the magnetic field strength $\vec{H}$ is constructed to be

$$H = -\frac{E}{\mu_0 v} \tag{4.8}$$

Therefore, a distance element $I\partial\vec{l}$ of a current through a conducting medium moving freely at speed of "v" across a magnetic flux density field $\vec{B}$ in the above space-time experiences a force of

$$\partial\vec{F} = I\partial\vec{\ell} \times \overrightarrow{\left(-\frac{E}{v}\right)} \tag{4.9}$$

4.1. Deriving the resultant electric field and force from present constructs

Hence, we are able to similarly derive the resultant electric field to be

$$\overrightarrow{E_{RE}} = \frac{\partial\left(\vec{L} \times \vec{B}\right)}{\partial t} \tag{4.10}$$

where $\vec{B}$ is the resultant magnetic flux density field, the distance vector $\vec{L}$ is the radial distance vector from the centerline of a round conductor, to a point on the $\vec{B}\ loop$ about the conductor, and the curl of the components of $\vec{E}_{RE}$ is directionally parallel, on any point of the $\vec{B}\ loop$, to the direction of propagation of the source electric field in the conductor.

Moreover, $\vec{E}$ and $\vec{B}$ are the fundamental fields in present electromagnetic theory, and $\vec{H}$ and $\vec{D}$, the magnetic field strength and the electric displacement field, are very useful constructs in the solution of electromagnetic field problems.

Thus, we find the *resultant electric field force* for a moving charge Q, in the direction of unit vector $\vec{a}_{RE}$, where L is the radial distance from the centerline of a round conductor to a point on the loop of $\vec{B}$, in the presence of a resultant magnetic flux density field $\vec{B}$, and a source electric field $\vec{E}$, to be

$$\overrightarrow{F_{RE}} = \left(-\frac{QE}{B}\right)\overrightarrow{a_{RE}} \times \vec{B} \qquad (4.11)$$

and the resultant magnetic flux density field is

$$\vec{B} = \int_{t_1}^{t_2}\left(\frac{\overrightarrow{E_{RE}}}{L}\right)\partial t \qquad (4.12)$$

4.2. Expressing the source and resultant electric fields in terms of each other

Thus, we find the source electric field invoking the resultant electric field to be

$$\vec{E} = -v\int_{t_1}^{t_2}\left(\frac{\overrightarrow{E_{RE}}}{L}\right)\partial t \qquad (4.13)$$

Similarly, we express the source electric field and the resultant electric field, in terms of each other, in vector field notation as

$$\frac{\partial \vec{E}}{\partial t} = \left(-\frac{v}{L}\right)\overrightarrow{E_{RE}} \tag{4.14}$$

$$\overrightarrow{E_{RE}} = \left(-\frac{L}{v}\right)\frac{\partial \vec{E}}{\partial t} \tag{4.15}$$

where "v" is the speed of the resultant electromagnetic field, L is the distance from the centerline of the conductor to a point on the loop of $\vec{B}$ about the conductor, and $\vec{E}$ is the source electric field within the conductor.

4.3. Electromagnetic field speed and other constructs in terms of the resultant electric field

Hence, the speed of the simultaneous electromagnetic field in terms of the resultant electric field $\vec{E}_{RE}$ and the source electric field equals

$$|v| = \frac{E}{\int_{t_1}^{t_2}\left(\frac{E_{RE}}{L}\right)\partial t} \tag{4.16}$$

By integrating one of Maxwell's equation and substituting for $\vec{B}$ in terms of $\vec{E}_{RE}$ we find

$$\vec{\nabla}\times\vec{E} = -\frac{\partial\left\langle\int_{t_1}^{t_2}\left(\frac{\overrightarrow{E_{RE}}}{L}\right)\partial t\right\rangle}{\partial t} \tag{4.17}$$

$$-\left(\vec{\nabla}\times\vec{E}\right) = \frac{\overrightarrow{E_{RE}}}{L} \tag{4.18}$$

$$\overrightarrow{E_{RE}} = -L\left(\vec{\nabla}\times\vec{E}\right) \tag{4.19}$$

Similarly, other magnitudes of electromagnetic constructs are expressed as

$$\overrightarrow{H} = \left(\frac{1}{\mu_0}\right)\int_{t_1}^{t_2}\left(\frac{\overrightarrow{E_{RE}}}{L}\right)\partial t \tag{4.20}$$

$$\overrightarrow{E_{RE}} = \mu_0 L \frac{\partial \overrightarrow{H}}{\partial t} \tag{4.21}$$

$$\overrightarrow{E_{RE}} = \frac{\overrightarrow{D_{RE}}}{\varepsilon_0} \tag{4.22}$$

where $\overrightarrow{D_{RE}}$ is the resultant electric displacement field and ε_0 is the permittivity of free space-time.

4.4. Deriving the Lorentz force on a moving charge

The Lorentz force $\overrightarrow{F_L}$ on a moving charge Q equals the sum of the source electric field force and the resultant electric field force present at a point, at a distance L, such that

$$\overrightarrow{F_L} = \overrightarrow{F_{SE}} + \overrightarrow{F_{RE}} = Q\vec{E} + \left[\left(-\frac{QE}{B}\right)\overrightarrow{a_{RE}} \times \vec{B}\right] \tag{4.23}$$

$$\overrightarrow{F_L} = Q\left\{\vec{E} + \left[\left(-\frac{E}{B}\right)\overrightarrow{a_{RE}} \times \vec{B}\right]\right\} \tag{4.24}$$

$$\overrightarrow{F_L} = Q\left\langle \vec{E} + \left\{\left[\left[\left(-\frac{E}{\int_{t_1}^{t_2}\left(\frac{E_{RE}}{L}\right)\partial t}\right)\overrightarrow{a_{RE}}\right] \times \int_{t_1}^{t_2}\left(\frac{\overrightarrow{E_{RE}}}{L}\right)\partial t\right]\right\}\right\rangle \tag{4.25}$$

where $\overrightarrow{a_{RE}}$ is a unit vector for the electric field $\vec{E}$ in the direction of the resultant electric field, and L is the distance from the centerline of the conductor to a point where Q intersects the loop of $\vec{B}$.

The Lorentz force law governs the interaction of the electromagnetic field with charged matter. Thus, the Lorentz force is shown above to be a result of the actions of the source electric field and the resultant electric field of an electromagnetic field on a unit charge Q.

The six-dimensional covariant form of the Lorentz force may be expressed as

$$\frac{dp^{\varepsilon\beta}}{d\tau^{\varepsilon\beta}} = q_{\chi\chi} F^{\varepsilon\beta} u_{\varepsilon\beta} \tag{4.26}$$

where $p^{\varepsilon\beta}$ is a six-momentum, $\tau^{\varepsilon\beta}$ is proper time, $q_{\chi\chi}$ is the charge of a particle, $F^{\varepsilon\beta}$ is the six-dimensional contravariant electromagnetic field strength, and $u_{\varepsilon\beta}$ is a covariant six-velocity of a particle. For metric signature of (−, −, −, +, +, +), the six-dimensional contravariant electromagnetic field strength is

$$F^{\varepsilon\beta} = \begin{vmatrix} 0 & E_{t_x t_y} & -E_{t_x t_z} & \frac{E_{t_x x}}{c} & \frac{E_{t_x y}}{c} & \frac{E_{t_x z}}{c} \\ -E_{t_y t_x} & 0 & E_{t_y t_z} & \frac{E_{t_y x}}{c} & \frac{E_{t_y y}}{c} & \frac{E_{t_y z}}{c} \\ E_{t_z t_x} & -E_{t_z t_y} & 0 & \frac{E_{t_z x}}{c} & \frac{E_{t_z y}}{c} & \frac{E_{t_z z}}{c} \\ -\frac{E_{xt_x}}{c} & -\frac{E_{xt_y}}{c} & -\frac{E_{xt_z}}{c} & 0 & B_{xy} & -B_{xz} \\ -\frac{E_{yt_x}}{c} & -\frac{E_{yt_y}}{c} & -\frac{E_{yt_z}}{c} & -B_{yx} & 0 & B_{yz} \\ -\frac{E_{zt_x}}{c} & -\frac{E_{zt_y}}{c} & -\frac{E_{zt_z}}{c} & B_{zx} & -B_{zy} & 0 \end{vmatrix} \tag{4.27}$$

Let us express the Lorentz force as the combination of the electric force (Coulomb force) and the resultant electric force (magnetic force or Laplace force) of an electromagnetic field exerting a force on a particle of charge, $q_{\chi\chi}$, as the particle travels at a velocity, $u_{\varepsilon\beta}$, in the six-dimensional field potential of space-time-charge, $\Phi_{\varepsilon\beta}$.

The six-dimensional electrogravitic field equation may be expressed as

$$F^{\varepsilon\beta} u_{\varepsilon\beta} \overline{G}_{\varepsilon\beta} = \frac{8\pi}{q_{\chi\chi}} \Phi_{\varepsilon\beta} \tag{4.28}$$

$$F^{\varepsilon\beta} u_{\varepsilon\beta} \overline{G}_{\varepsilon\beta} = \frac{\ddot{a}}{r q_{\chi\chi}} \Phi_{\varepsilon\beta} \tag{4.29}$$

$$F^{\varepsilon\beta} u_{\varepsilon\beta} \left(\overline{R}_{\varepsilon\beta} - \frac{1}{(n-1)} g_{\varepsilon\beta} \overline{R} \right) = \frac{\ddot{a}}{r q_{\chi\chi}} \Phi_{\varepsilon\beta} \tag{4.30}$$

where n is the number of dimensions and $n > 1$, $\overline{G}_{\varepsilon\beta}$ is the electrogravitic curvature tensor, $\ddot{a}$ is the acceleration of space-time, r is the magnitude of a spatial distance, $\Phi_{\varepsilon\beta}$ is the electrogravitic tensor.

The six-dimensional electrogravitic field equation may represent the electrogravitic curvature of a manifold about a rotating charged supermassive black hole; e.g., a Kerr-Newman supermassive black hole.

If we consider mathematically a rotating charged supermassive blackhole with a physically significant electromagnetic field, the six-dimensional electrogravitic field equation is an electrovacuum field equation, for a six-dimensional manifold outside the ergosphere of the supermassive blackhole, in the absence of all other external non-electrogravitic fields.

4.5. Summary of electro-resultant field equations

Summarizing the expressions for the resultant electric field, in terms of other electromagnetic constructs, in a vector field notation we have

$$\overrightarrow{E_{RE}} = \frac{\partial\left(\vec{L} \times \vec{B}\right)}{\partial t} \tag{4.31}$$

$$\overrightarrow{E_{RE}} = -L\left(\vec{\nabla} \times \vec{E}\right) \tag{4.32}$$

$$\overrightarrow{E_{RE}} = \mu_0 L \frac{\partial \vec{H}}{\partial t} \tag{4.33}$$

$$\overrightarrow{E_{RE}} = \frac{\overrightarrow{D_{RE}}}{\varepsilon_0} \tag{4.34}$$

4.6. Maxwell's equations in terms of electric field and resultant electric field notation

The behavior of electric and electro-resultant fields in free space-time, whether in cases of electrostatics, magnetostatics, or electrodynamics, is described by Maxwell's equations.

Let us now represent Maxwell's equations in a combined vector-integral notation in terms of the electric field $\vec{E}$ and the resultant electric field $\overrightarrow{E_{RE}}$ as follows:

$$\vec{\nabla} \cdot \overrightarrow{E_{RE}} = \frac{\rho_{RE}}{\varepsilon} \tag{4.35}$$

$$\vec{\nabla} \cdot \left[\int_{t_1}^{t_2} \left(\frac{\overrightarrow{E_{RE}}}{L}\right) \partial t\right] = 0 \tag{4.36}$$

$$\vec{\nabla} \times \vec{E} = -\frac{\partial \left[\int_{t_1}^{t_2} \left(\frac{\overrightarrow{E_{RE}}}{L} \right) \partial t \right]}{\partial t} \tag{4.37}$$

$$\vec{\nabla} \times \left[\int_{t_1}^{t_2} \left(\frac{\overrightarrow{E_{RE}}}{L} \right) \partial t \right] = \mu \left(\vec{J} + \varepsilon \frac{\partial \vec{E}}{\partial t} \right) \tag{4.38}$$

where ρ_{RE} is the resultant charge density, which may depend on time and position, $\mathcal{E}$ is the permittivity of the medium, μ is the permeability of the medium, and $\vec{J}$ is the current density vector, also a function of time and position. Inside materials which possess complex responses to electromagnetic fields, the permittivity and permeability terms may be represented by complex numbers, or tensors.

§ 5. On the electrogravitic force and the refractive force of space-time-mass

5.1. The electrogravitic force equivalence

The force exerted on a charged particle *Q*, such as a proton, by a strong external electric field $\vec{E}$, in this case by another proton, is significantly larger than the force of gravity acting on the minute mass of the charged atomic particle *Q*, indicated by the following approximate force ratio.

$$\frac{F_{EXT}}{F_g} \approx 10^{36} \tag{5.1}$$

Otherwise, the electric force between two uniform charge distributions on two masses, viewed as two point charges, is given by the Coulomb's force law as

$$F_E = \frac{Q_1 Q_2}{4\pi\varepsilon_0 r^2} \tag{5.2}$$

The gravitational force between the same two masses is given by Isaac Newton's equation as

$$F_g = \frac{Gm_1m_2}{r^2} \tag{5.3}$$

Thus, we find the magnitude of the electrogravitic force ratio between two charged masses to be

$$\frac{F_E}{F_g} = \frac{Q_1Q_2}{4\pi Gm_1m_2\varepsilon_0} \tag{5.4}$$

Now, in the absence of any other electromagnetic fields or gravitational fields, in terms of the magnitudes of the electric force and the gravitational force for two equally charged identical masses, $v << c$, we have the following Coulomb-Newton equivalence

$$F_E = \frac{Q^2F_g}{4\pi Gm^2\varepsilon_0} \tag{5.5}$$

$$F_g = \frac{4\pi Gm^2\varepsilon_0F_E}{Q^2} \tag{5.6}$$

5.2. The electrogravitic acceleration of a charged mass

For a uniformly charged spherical mass moving much slower than the speed of light ($v << c$) in a gravitational field, the gravitational acceleration felt by the charged mass is found to be

$$\overrightarrow{a_g} = -\frac{4\pi Gm_Q\varepsilon_0\overrightarrow{F_E}}{Q^2} \tag{5.7}$$

Substituting for G, from previous investigation, in the above equation $\left(G = \ddot{a}/8\pi m_Q\right)$ for v << c, where $\ddot{a}$ is the acceleration of space felt by the charged spherical mass, we have

$$\vec{a}_g = -\frac{\ddot{a}\varepsilon_0\vec{F}_E}{2Q^2} = -\frac{\ddot{a}\varepsilon_0\vec{E}}{2Q} \tag{5.8}$$

Similarly, for a uniformly charged spherical mass moving relativistically in space-time (v $\rightarrow$ c), the gravitational acceleration felt by the mass of the charge is

$$\overrightarrow{a_g} = -\frac{4\pi G m_Q \varepsilon_0 \overrightarrow{F_E}}{Q^2 \sqrt[2]{1-\frac{v^2}{c^2}}} \quad (5.9)$$

$$\overrightarrow{a_g} = -\frac{4\pi G m'_Q \varepsilon_0 \overrightarrow{E}}{Q} \quad (5.10)$$

$$\vec{a}_g = -\frac{\ddot{a}\varepsilon_0 \vec{E}}{2Q \sqrt[2]{1-\frac{v^2}{c^2}}} \quad (5.11)$$

5.3. The electrogravitic force in terms of the electric field

Hence, the electrogravitic relationship for a uniformly charged spherical mass in a gravitational field, in terms of the gravitational force and the electric force, or the electric field, of the charged mass, is

$$\vec{F}_g = -\left(\frac{\ddot{a}\varepsilon_0 m_Q}{2Q^2}\right)\vec{F}_E = -\left(\frac{\ddot{a}\varepsilon_0 m_Q}{2Q}\right)\vec{E} \quad (5.12)$$

In the absence of other electromagnetic fields or gravitational fields, the force of gravity on a uniformly charged spherical mass has a direction opposite to the direction of the electric field of a positively charged mass.

5.4. The refractive electromagnetic force and acceleration of free space-time on a point charge

Let us express the magnitude of the electric field $\overrightarrow{E}$ of a positive charge to be

$$E = \frac{Q}{4\pi\varepsilon_0 r^2} \quad (5.13)$$

Hence, the magnitude of the refractive electric force of free space-time, acting on a positive point charge, conceptualized as a virtual mass, is given by

$$F_{RE} = \frac{Q^2}{\varepsilon_0 r^2} \tag{5.14}$$

$$F_E = -\frac{F_{RE}}{4\pi} \tag{5.15}$$

Solving for the refractive electric force of free space-time, $\overrightarrow{F_{RE}}$, in the presence of an electric field, we find

$$\overrightarrow{F_{RE}} = -4\pi \overrightarrow{F_E} \tag{5.16}$$

Thus, the refractive electric force of free space-time, $\overrightarrow{F_{RE}}$, on a positively charged mass is about twelve orders of magnitude larger than the force of the electric field, $\overrightarrow{F_E}$, of the charged mass at a point in isotropic and homogenous space-time, and the refractive electric force of free space-time, $\overrightarrow{F_{RE}}$, acts in the opposite direction to the electric field, $\vec{E}$, of the positively charged mass.

Therefore, in terms of the refractive electric acceleration of free space, $\overrightarrow{a_{RE}}$, upon a positively charged mass and the electric force, $\overrightarrow{F_E}$, exerted by the charged mass upon other charged masses we have

$$\overrightarrow{F_{RE}} = m_Q \overrightarrow{a_{RE}} = -4\pi \overrightarrow{F_E} \tag{5.17}$$

$$m_Q \overrightarrow{a_{RE}} = -4\pi Q \vec{E} \tag{5.18}$$

$$\overrightarrow{a_{RE}} = -\frac{4\pi Q \vec{E}}{m_Q} \tag{5.19}$$

$$|a_{RE}| = \frac{Q^2}{m_Q \varepsilon_0 r^2} \tag{5.20}$$

Thus, we can see that the refractive electric acceleration of free space-time acts in the opposite direction to the electric field $\vec{E}$ of the positively charged mass m_Q. Hence, in the absence of other fields, it is proposed here that the refractive electric force field is a consequence of a virtual electromagnetic symmetrical partner, or an equivalent virtual charged mass, to every particle existing in homogeneous and isotropic space-time. The boundary between the particle and space-time acts as a mirror due to space-time distortion, or a reflector, manifesting the image of an equal and opposing virtual symmetrical partner to the existing particle in space-time.

The only electric field lines from the particle that cross the boundary between the particle and space-time are at ninety degrees to the surface of the boundary. Those orthogonal electric field lines extend radially outward in space-time, while all other non-perpendicular electric field lines are reflected back toward the particle, manifesting a virtual charged mass or symmetrical partner, on the space-time boundary mirror. This exerts a refractive electric field from free space-time on a point charge Q. Thus, the permittivity of free space-time is the effect of the refractive electric force of the point charge, or charged mass, at a distance from its center.

Similarly, let us express the magnitude of the magnetic field $\vec{B}$ of a point charge to be

$$\vec{B} = \frac{\mu_0}{4\pi} \frac{Q\vec{v} \times \vec{r}}{|\vec{r}|^2} \tag{5.21}$$

where "v" is the velocity of the point charge Q, and "r" is the distance from the center of the point charge to the spatiotemporal position where the magnetic field $\vec{B}$ is being measured. (Heaviside, 1888)

A similar concept is proposed here for the refractive magnetic field that is exerted from free space-time on a moving magnetic dipole treated as two joined hemispherical symmetrical point charges.

$$F_{RM} = \mu_0 I^2 \tag{5.22}$$

where "I" is the current passing through the loop of the curve C, and μ_0 is the permeability of free space-time.

In the case of a point charge moving at a constant speed, where $v << c$, the refractive force is given by

$$F_{RM} = 4\pi QvB = 4\pi F_m \tag{5.23}$$

$$\vec{F}_{RM} = -4\pi \vec{F}_m \tag{5.24}$$

where v is the velocity of the point charge Q, B is the magnetic flux density, and Fm is the magnitude of the orthogonal magnetic field force exerted by the moving magnetic dipole, or hemispherical symmetrical charges, radially outward, or inward, from each pole through the space-time boundary mirror. *Hence, the permeability of free space-time is the effect of the refractive magnetic force of a point charge, or charged mass, at a distance from its center.* As a result of the *Principle of Symmetrical Refraction*, free space-time exhibits qualities of permittivity and permeability to any point charge, charged body, magnetic dipole, or electromagnetic radiant energy in space-time.

5.5. The refractive magnetic field of a uniform spherical magnetic dipole in space-time

The magnitude of the refractive magnetic force $\overrightarrow{F_{RM}}$ of a spherical and symmetrical magnetic dipole in homogeneous and isotropic space-time, in the absence of other fields, is given by

$$F_{RM} = \mu_0 I^2 \tag{5.25}$$

The permeability of free space-time, μ_0, is defined in this document as the force $\overrightarrow{F_{RM}}$ of the refractive magnetic field, $\overrightarrow{B_{RM}}$, of a spherical and symmetrical magnetic dipole divided by the square of the current "I" at the space-time-mass boundary as shown below:

$$\mu_0 = \frac{F_{RM}}{I^2} \tag{5.26}$$

where we find the force of the refractive magnetic field to be

$$\vec{F}_{RM} = \int_c \vec{B} \cdot I \, \vec{\partial l} = Q\vec{v} \times \vec{B} \tag{5.27}$$

and v is the velocity of the charge Q, the inner distance vector $\vec{\partial l}$ is an infinitesimal line element in the loop of the curve C. That is, $\vec{\partial l}$ is a vector with magnitude equal to the length of an infinitesimal line element, and with a direction tangential to the curve C, on each hemisphere of the spherical and symmetrical magnetic dipole.

Therefore, we can express the external magnetic field $\overrightarrow{B_{EXT}}$ in terms of the internal, or source magnetic field $\overrightarrow{B_{INT}}$ of a spherical and symmetrical magnetic dipole in free space-time, and the refractive magnetic field $\overrightarrow{B_{RE}}$ for each hemisphere at the space-time boundary as

$$\overrightarrow{B_{EXT}} = \overrightarrow{B_{INT}} - \overrightarrow{B_{RE}} \tag{5.28}$$

Multiplying each magnetic field by the product of the magnitudes of charge and velocity, Qv_Q , to obtain the external force per unit area able to act on an external moving point charge, Q_{EXT}, we find

$$\overrightarrow{F_{EXT}} = \overrightarrow{F_{INT}} - \overrightarrow{F_{RE}} \tag{5.29}$$

Thus, the magnitude of the total refractive electromagnetic force of homogeneous and isotropic free space-time, in the absence of other fields, can be expressed as

$$F_{TREM} = \frac{Q^2}{\varepsilon_0 r^2} + \mu_0 I^2 \tag{5.30}$$

Hence, we find the magnitude of the total refractive electromagnetic field force per unit charge of free space-time able to act on a charged point particle in homogeneous and isotropic space-time to be

$$\frac{F_{TREM}}{Q} = \frac{Q}{\varepsilon_0 r^2} + \mu_0 a_Q \tag{5.31}$$

Where a_Q is the acceleration of the charged point particle. Thus, we can see that as the spatial distance r becomes larger, the total refractive electromagnetic field force becomes less electric and more magnetic on a spherical and symmetrical point charge in free space-time.

5.6. On the unification of gravitation and electromagnetism

Let us consider a source of gravitational field that is also the source of an electromagnetic field in a region of space-time in such a way that the primary role of the electromagnetic gravitation is the formation and conservation of the structure of matter.

The density of the product of charge and voltage potential is energy density, $\overline{\rho}_{\chi\chi}$. The density of electromagnetic flux, $\left(Webers/m^2\right)$, may be defined as the product of flux, $\left(Kg \cdot m^2/s^2 \cdot A\right)$, and curvature, $\left(1/m^2\right)$. From previous research, charge was defined as the product of spatial length and temporal extent, and consequently, electromagnetic flux $\left(Kg \cdot m \cdot m \cdot s/s^2 \cdot C\right)$ is equivalent to force.

Therefore, flux density is equivalent to pressure $\left(\overline{\phi}_{\chi\chi} = \overline{p}_{\chi\chi}\right)$ for a charged system of mass. The pressure, or energy density, of charged matter is excluded from the neutral, or uncharged, local or cosmological, pressure or energy density.

For a region of space-time about a charged celestial body of mass, the six-dimensional electrogravitic tensor may be expressed as

$$\Phi_{\varepsilon\beta} = \begin{vmatrix} c^2\overline{\rho}_{q_xq_x} & c^2\overline{\rho}_{q_xq_y} & c^2\overline{\rho}_{q_xq_z} & \frac{\vec{S}_{q_xx}}{c^2} & \frac{\vec{S}_{q_xy}}{c^2} & \frac{\vec{S}_{q_xz}}{c^2} \\ c^2\overline{\rho}_{q_yq_x} & c^2\overline{\rho}_{q_yq_y} & c^2\overline{\rho}_{q_yq_z} & \frac{\vec{S}_{q_yx}}{c^2} & \frac{\vec{S}_{q_yy}}{c^2} & \frac{\vec{S}_{q_yz}}{c^2} \\ c^2\overline{\rho}_{q_zq_x} & c^2\overline{\rho}_{q_zq_y} & c^2\overline{\rho}_{q_zq_z} & \frac{\vec{S}_{q_zx}}{c^2} & \frac{\vec{S}_{q_zy}}{c^2} & \frac{\vec{S}_{q_zz}}{c^2} \\ \frac{\vec{S}_{xq_x}}{c^2} & \frac{\vec{S}_{xq_y}}{c^2} & \frac{\vec{S}_{xq_z}}{c^2} & \overline{\phi}_{xx} & \overline{\phi}_{xy} & \overline{\phi}_{xz} \\ \frac{\vec{S}_{yq_x}}{c^2} & \frac{\vec{S}_{yq_y}}{c^2} & \frac{\vec{S}_{yq_z}}{c^2} & \overline{\phi}_{yx} & \overline{\phi}_{yy} & \overline{\phi}_{yz} \\ \frac{\vec{S}_{zq_x}}{c^2} & \frac{\vec{S}_{zq_y}}{c^2} & \frac{\vec{S}_{zq_z}}{c^2} & \overline{\phi}_{zx} & \overline{\phi}_{zy} & \overline{\phi}_{zz} \end{vmatrix} \tag{5.32}$$

Thus, the charge-space components (charge energy momentum density), or space-charge components (charge energy momentum density) are $\Phi_{ij} = \vec{S}_{\chi\chi}/c^2$ or $\Phi_{ji} = \vec{S}_{\chi\chi}/c^2$ when $i \neq j$, the charge-charge components are restricted to

$$\Phi_{\varepsilon\beta}(\vec{e}_i)^{\varepsilon}(\vec{e}_j)^{\beta} = \overline{\rho}\partial_{q_iq_j} \tag{5.33}$$

and the space-space components are restricted to

$$\Phi_{\varepsilon\beta}(\vec{e}_i)^{\varepsilon}(\vec{e}_j)^{\beta} = \overline{\phi}\partial_{ij} \tag{5.34}$$

and the $\vec{S}_{\chi\chi}$ in charge-space or space-charge components is the Poynting vector.

$$\vec{S}_{\chi\chi} = \varepsilon_0 c^2\left(\vec{E}_{\chi\chi} \times \vec{B}_{\chi\chi}\right) \tag{5.35}$$

Where $\vec{E}_{\chi\chi}$ is the electric field, $\vec{B}_{\chi\chi}$ is the magnetic flux density, ε_0 is the permittivity of the medium, and "c" is the speed of light. For a nearly-flat region of space-time about a charged celestial body of mass, the six-dimensional electrogravitic tensor may be expressed as

$$
\Phi_{\varepsilon\beta} = \begin{vmatrix} c^2\bar{\rho}_{q_x q_x} & 0 & 0 & 0 & 0 & 0 \\ 0 & c^2\bar{\rho}_{q_y q_y} & 0 & 0 & 0 & 0 \\ 0 & 0 & c^2\bar{\rho}_{q_z q_z} & 0 & 0 & 0 \\ 0 & 0 & 0 & \bar{\phi}_{xx} & 0 & 0 \\ 0 & 0 & 0 & 0 & \bar{\phi}_{yy} & 0 \\ 0 & 0 & 0 & 0 & 0 & \bar{\phi}_{zz} \end{vmatrix} \tag{5.36}
$$

Thus, the charge-charge, space-space, space-charge, or charge-space components, $\Phi_{ij} = \Phi_{ji} = 0$, are zero when $i \neq j$.

If the electrogravitic tensor has equal charge energy densities and pressures, the trace of the electrogravitic tensor is

$$
\Phi = g^{\varepsilon\beta}\Phi_{\varepsilon\beta} = -3c^2\bar{\rho}_{q_\chi q_\chi} + 3\bar{\varphi}_{\chi\chi} \tag{5.37}
$$

The electrogravitic tensor is the difference between the local and the nonlocal electrogravitic tensors. The local electrogravitic tensor emerges from local charged matter and the nonlocal or cosmological electrogravitic tensor, if present, comes from cosmological charged matter. The local electrogravitic tensor $\Phi(L)_{\varepsilon\beta}$ is related to the stress-energy-momentum tensor through the energy density of the local system of mass, and boosts the local curvature of space-time and the local gravitational field. The cosmological electrogravitic tensor $\Phi(\Lambda)_{\varepsilon\beta}$ offsets the local electrogravitic tensor.

$$
\Phi_{\varepsilon\beta} = \Phi(L)_{\varepsilon\beta} - \Phi(\Lambda)_{\varepsilon\beta} \tag{5.38}
$$

In terms of the electrogravitic curvature exerted by the electrogravitic tensor we obtain

$$
\bar{R}_{\varepsilon\beta} - \frac{1}{(n-1)} g_{\varepsilon\beta}\bar{R} = \frac{8\pi G}{c^4}\Phi_{\varepsilon\beta} \tag{5.39}
$$

The electrogravitic tensor $\Phi_{\varepsilon\beta}$ may either boost or offset the curvature effect of the stress-energy-momentum tensor depending on the strength of either the local or cosmological electrogravitic tensor.

The presence of the resultant electromagnetic field within the gravitational field near the charged celestial body of mass changes curvature in the nearly-flat region of space-time under consideration. The local electrogravitic tensor boosts local curvature.

The six-dimensional electrogravitic curvature tensor for a nearly-flat region of space-time may be expressed as

$$\overline{R}_{\varepsilon\beta} = \begin{vmatrix} \overline{R}_{q_x q_x} & 0 & 0 & 0 & 0 & 0 \\ 0 & \overline{R}_{q_y q_y} & 0 & 0 & 0 & 0 \\ 0 & 0 & \overline{R}_{q_z q_z} & 0 & 0 & 0 \\ 0 & 0 & 0 & \overline{R}_{xx} & 0 & 0 \\ 0 & 0 & 0 & 0 & \overline{R}_{yy} & 0 \\ 0 & 0 & 0 & 0 & 0 & \overline{R}_{zz} \end{vmatrix} \qquad (5.40)$$

The six-dimensional electrogravitic curvature metric tensor is

$$g_{\varepsilon\beta} = \begin{vmatrix} -c^2 & 0 & 0 & 0 & 0 & 0 \\ 0 & -c^2 & 0 & 0 & 0 & 0 \\ 0 & 0 & -c^2 & 0 & 0 & 0 \\ 0 & 0 & 0 & 1 & 0 & 0 \\ 0 & 0 & 0 & 0 & 1 & 0 \\ 0 & 0 & 0 & 0 & 0 & 1 \end{vmatrix} \qquad (5.41)$$

and the six-dimensional reverse electrogravitic curvature metric tensor is

$$g^{\varepsilon\beta} = \begin{vmatrix} -\frac{1}{c^2} & 0 & 0 & 0 & 0 & 0 \\ 0 & -\frac{1}{c^2} & 0 & 0 & 0 & 0 \\ 0 & 0 & -\frac{1}{c^2} & 0 & 0 & 0 \\ 0 & 0 & 0 & 1 & 0 & 0 \\ 0 & 0 & 0 & 0 & 1 & 0 \\ 0 & 0 & 0 & 0 & 0 & 1 \end{vmatrix} \qquad (5.42)$$

The n-dimensional field equations in the presence of an

electromagnetic field are given by

$$R_{\mu\nu} - \frac{1}{(n-1)} g_{\mu\nu} R + \overline{R}_{\varepsilon\beta} - \frac{1}{(n-1)} g_{\varepsilon\beta} \overline{R} = \frac{8\pi G}{c^4} \left(T_{\mu\nu} - \Lambda_{\mu\nu} + \Phi_{\varepsilon\beta} \right) \quad (5.43)$$

$$R_{\mu\nu} - \frac{1}{(n-1)} \left(g_{\mu\nu} R + g_{\varepsilon\beta} \overline{R} \right) + \overline{R}_{\varepsilon\beta} = \frac{8\pi G}{c^4} \left(T_{\mu\nu} - \Lambda_{\mu\nu} + \Phi_{\varepsilon\beta} \right) \quad (5.44)$$

Where $\overline{R}$ is the scalar of the electrogravitic curvature tensor.

Substituting with the n-dimensional Einstein tensor $G_{\mu\nu}$ and introducing the n-dimensional Hilbert tensor $\overline{G}_{\varepsilon\beta}$ we have

$$G_{\mu\nu} + \overline{G}_{\varepsilon\beta} = \frac{8\pi G}{c^4} \left(T_{\mu\nu} - \Lambda_{\mu\nu} + \Phi_{\varepsilon\beta} \right) \quad (5.45)$$

$$G_{\mu\nu} + \overline{G}_{\varepsilon\beta} = \frac{8\pi G}{c^4} \left(\Phi_{\mu\nu} + \Phi_{\varepsilon\beta} \right) \quad (5.46)$$

Where $\Phi_{\mu\nu}$ is the stress-energy-momentum difference tensor between the local stress-energy-momentum tensor and the cosmological stress-energy-momentum tensor.

Let us denote the n-dimensional electrogravitic Riemann tensor as

$$\widetilde{R}_{\mu\nu\varepsilon\beta} = \frac{1}{n-2} \left(g_{\mu\varepsilon} S_{\nu\beta} - g_{\mu\beta} S_{\nu\varepsilon} + g_{\nu\beta} S_{\mu\varepsilon} - g_{\nu\varepsilon} S_{\mu\beta} \right) \quad (5.47)$$

Where for any $S_{\chi\chi}$ term, we find

$$S_{ab} = \overline{R}_{ab} - \frac{1}{n} g_{ab} \overline{R} \quad (5.48)$$

Contracting with $g^{\mu\nu}$ we have

$$\widetilde{R}_{\varepsilon\beta} = \overline{R}_{\varepsilon\beta} - \frac{1}{(n-1)} g_{\varepsilon\beta} \overline{R} \tag{5.49}$$

The n-dimensional electrogravitic Riemann tensor can be defined as a fourth order tensor, $\widetilde{R}_{\mu\nu\varepsilon\beta}$, for the curvature of space-time-charge and mass.

$$\widetilde{R}_{\mu\nu\varepsilon\beta} = \begin{vmatrix} \overline{R}_{q_xq_x} & \overline{R}_{q_xq_y} & \overline{R}_{q_xq_z} & \overline{R}_{q_xt_x} & \overline{R}_{q_xt_y} & \overline{R}_{q_xt_z} & \overline{R}_{q_xx} & \overline{R}_{q_xy} & \overline{R}_{q_xz} \\ \overline{R}_{q_yq_x} & \overline{R}_{q_yq_y} & \overline{R}_{q_yq_z} & \overline{R}_{q_yt_x} & \overline{R}_{q_yt_y} & \overline{R}_{q_yt_z} & \overline{R}_{q_yx} & \overline{R}_{q_yy} & \overline{R}_{q_zz} \\ \overline{R}_{q_zq_x} & \overline{R}_{q_zq_y} & \overline{R}_{q_zq_z} & \overline{R}_{q_zt_x} & \overline{R}_{q_zt_y} & \overline{R}_{q_zt_z} & \overline{R}_{q_zx} & \overline{R}_{q_zy} & \overline{R}_{q_zz} \\ \overline{R}_{t_xq_x} & \overline{R}_{t_yq_x} & \overline{R}_{t_zq_x} & R_{t_xt_x} & R_{t_xt_y} & R_{t_xt_z} & R_{t_xx} & R_{t_xy} & R_{t_xz} \\ \overline{R}_{t_xq_y} & \overline{R}_{t_yq_y} & \overline{R}_{t_zq_y} & R_{t_yt_x} & R_{t_yt_y} & R_{t_yt_z} & R_{t_yx} & R_{t_yy} & R_{t_yz} \\ \overline{R}_{t_xq_z} & \overline{R}_{t_yq_z} & \overline{R}_{t_zq_z} & R_{t_zt_x} & R_{t_zt_y} & R_{t_zt_z} & R_{t_zx} & R_{t_zy} & R_{t_zz} \\ \overline{R}_{xq_x} & \overline{R}_{yq_x} & \overline{R}_{zq_x} & R_{xt_x} & R_{xt_y} & R_{xt_z} & \overline{R}_{xx} & \overline{R}_{xy} & \overline{R}_{xz} \\ \overline{R}_{xq_y} & \overline{R}_{yq_y} & \overline{R}_{zq_y} & R_{yt_x} & R_{yt_y} & R_{yt_z} & \overline{R}_{yx} & \overline{R}_{yy} & \overline{R}_{yz} \\ \overline{R}_{xq_z} & \overline{R}_{yq_z} & \overline{R}_{zq_z} & R_{zt_x} & R_{zt_y} & R_{zt_z} & \overline{R}_{zx} & \overline{R}_{zy} & \overline{R}_{zz} \end{vmatrix} \tag{5.50}$$

Similarly, the n-dimensional electrogravitic tensor can be defined as a fourth order tensor, $\widetilde{\Phi}_{\mu\nu\varepsilon\beta}$, for space-time-charge and mass.

$$\widetilde{\Phi}_{\mu\nu\varepsilon\beta} = \begin{vmatrix} c^2\overline{\rho}_{q_xq_x} & c^2\overline{\rho}_{q_xq_y} & c^2\overline{\rho}_{q_xq_z} & \overline{T}_{q_xt_x} & \overline{T}_{q_xt_y} & \overline{T}_{q_xt_z} & \overline{T}_{q_xx} & \overline{T}_{q_xy} & \overline{T}_{q_xz} \\ c^2\overline{\rho}_{q_yq_x} & c^2\overline{\rho}_{q_yq_y} & c^2\overline{\rho}_{q_yq_z} & \overline{T}_{q_yt_x} & \overline{T}_{q_yt_y} & \overline{T}_{q_yt_z} & \overline{T}_{q_yx} & \overline{T}_{q_yy} & \overline{T}_{q_yz} \\ c^2\overline{\rho}_{q_zq_x} & c^2\overline{\rho}_{q_zq_y} & c^2\overline{\rho}_{q_zq_z} & \overline{T}_{q_zt_x} & \overline{T}_{q_zt_y} & \overline{T}_{q_zt_z} & \overline{T}_{q_zx} & \overline{T}_{q_zy} & \overline{T}_{q_zz} \\ \overline{T}_{t_xq_x} & \overline{T}_{t_yq_x} & \overline{T}_{t_zq_x} & c^2\rho_{t_xt_x} & c^2\rho_{t_xt_y} & c^2\rho_{t_xt_z} & T_{t_xx} & T_{t_xy} & T_{t_xz} \\ \overline{T}_{t_xq_y} & \overline{T}_{t_yq_y} & \overline{T}_{t_zq_y} & c^2\rho_{t_yt_x} & c^2\rho_{t_yt_y} & c^2\rho_{t_yt_z} & T_{t_yx} & T_{t_yy} & T_{t_yz} \\ \overline{T}_{t_xq_z} & \overline{T}_{t_yq_z} & \overline{T}_{t_zq_z} & c^2\rho_{t_zt_x} & c^2\rho_{t_zt_y} & c^2\rho_{t_zt_z} & T_{t_zx} & T_{t_zy} & T_{t_zz} \\ \overline{T}_{xq_x} & \overline{T}_{yq_x} & \overline{T}_{zq_x} & T_{xt_x} & T_{xt_y} & T_{xt_z} & \overline{\phi}_{xx} & \overline{\phi}_{xy} & \overline{\phi}_{xz} \\ \overline{T}_{xq_y} & \overline{T}_{yq_y} & \overline{T}_{zq_y} & T_{yt_x} & T_{yt_y} & T_{yt_z} & \overline{\phi}_{yx} & \overline{\phi}_{yy} & \overline{\phi}_{yz} \\ \overline{T}_{xq_z} & \overline{T}_{yq_z} & \overline{T}_{zq_z} & T_{zt_x} & T_{zt_y} & T_{zt_z} & \overline{\phi}_{zx} & \overline{\phi}_{zy} & \overline{\phi}_{zz} \end{vmatrix} \tag{5.51}$$

Thus, the charge-time, or time-charge components (charge-energy momentum density) are $\Phi_{ij} = k\left(\vec{S}_{\chi\chi}/c^2\right)$ or $\Phi_{ji} = (1-k)\left(\vec{S}_{\chi\chi}/c^2\right)$ when $i \neq j$, where k is a constant, $0 \leq k \leq 1$, the charge-charge components are restricted to

$$\Phi_{\varepsilon\beta}(\vec{e}_i)^{\varepsilon}(\vec{e}_j)^{\beta} = \overline{\rho}\partial_{q_i q_j} \tag{5.52}$$

During the reduction of $\widetilde{\Phi}_{\mu\nu\varepsilon\beta}$ with $g^{\mu\nu}$, charge-time, charge-space, time-charge, and space-charge components combine; e.g. $\overline{T}_{q_x t_x} + \overline{T}_{q_x x} \rightarrow \vec{S}_{t_x x}/c^2$, or $\overline{T}_{x q_x} + \overline{T}_{t_x q_x} \rightarrow \vec{S}_{x t_x}/c^2$. These components contract to form the components of the second order electrogravitic tensor, $\Phi_{\varepsilon\beta}$. Thus, charge-space, or space-charge components (charge-energy momentum density) are $\Phi_{ij} = (1-k)(\vec{S}_{\chi\chi}/c^2)$ or $\Phi_{ji} = k(\vec{S}_{\chi\chi}/c^2)$ when $i \neq j$, where k is a constant, $0 \leq k \leq 1$, and the space-space components are restricted to

$$\Phi_{\varepsilon\beta}(\vec{e}_i)^{\varepsilon}(\vec{e}_j)^{\beta} = \overline{\phi}\partial_{ij} \tag{5.53}$$

Thus, the time-space components (momentum density), or space-time components (momentum density) are $\Phi_{ij} = \vec{S}_{\chi\chi}/c^2$ or $\Phi_{ji} = \vec{S}_{\chi\chi}/c^2$ when $i \neq j$, the time-time components are restricted to

$$\Phi_{\varepsilon\beta}(\vec{e}_i)^{\varepsilon}(\vec{e}_j)^{\beta} = \rho\partial_{ij} \tag{5.54}$$

For a nearly-flat region of space-time about a charged celestial body of mass, the n-dimensional electrogravitic tensor may be expressed as

$$\widetilde{\Phi}_{\mu\nu\varepsilon\beta} = \begin{vmatrix} c^2\overline{\rho}_{q_x q_x} & 0 & 0 & 0 & 0 & 0 & 0 & 0 & 0 \\ 0 & c^2\overline{\rho}_{q_y q_y} & 0 & 0 & 0 & 0 & 0 & 0 & 0 \\ 0 & 0 & c^2\overline{\rho}_{q_z q_z} & 0 & 0 & 0 & 0 & 0 & 0 \\ 0 & 0 & 0 & c^2\rho_{t_x t_x} & 0 & 0 & 0 & 0 & 0 \\ 0 & 0 & 0 & 0 & c^2\rho_{t_y t_y} & 0 & 0 & 0 & 0 \\ 0 & 0 & 0 & 0 & 0 & c^2\rho_{t_z t_z} & 0 & 0 & 0 \\ 0 & 0 & 0 & 0 & 0 & 0 & \overline{\phi}_{xx} & 0 & 0 \\ 0 & 0 & 0 & 0 & 0 & 0 & 0 & \overline{\phi}_{yy} & 0 \\ 0 & 0 & 0 & 0 & 0 & 0 & 0 & 0 & \overline{\phi}_{zz} \end{vmatrix} \tag{5.55}$$

Thus, the time-charge, charge-time, space-charge, charge-space, space-time, time-space, charge-charge, time-time, or space-space components are zero, $\Phi_{ij} = \Phi_{ji} = 0$, when $i \neq j$.

If the n-dimensional electrogravitic tensor has equal densities and pressures, the trace of the n-dimensional electrogravitic tensor is

$$\widetilde{\Phi} = g^{\mu\nu\varepsilon\beta}\widetilde{\Phi}_{\mu\nu\varepsilon\beta} = -3c^2\bar{\rho}_{q_\chi q_\chi} - 3c^2\rho_{t_\chi t_\chi} + 3\bar{\varphi}_{\chi\chi} \tag{5.56}$$

Contracting with $g^{\mu\nu}$ for a nearly-flat region of space-time, we get

$$\Phi_{\varepsilon\beta} = \begin{vmatrix} c^2\bar{\rho}_{q_x q_x} & 0 & 0 & 0 & 0 & 0 \\ 0 & c^2\bar{\rho}_{q_y q_y} & 0 & 0 & 0 & 0 \\ 0 & 0 & c^2\bar{\rho}_{q_z q_z} & 0 & 0 & 0 \\ 0 & 0 & 0 & \bar{\phi}_{xx} & 0 & 0 \\ 0 & 0 & 0 & 0 & \bar{\phi}_{yy} & 0 \\ 0 & 0 & 0 & 0 & 0 & \bar{\phi}_{zz} \end{vmatrix} \tag{5.57}$$

Substituting for the *n*-dimensional electrogravitic Riemann tensor and the *n*-dimensional electrogravitic tensor we obtain

$$\widetilde{R}_{\mu\nu\varepsilon\beta} = \frac{8\pi G}{c^4}\widetilde{\Phi}_{\mu\nu\varepsilon\beta} \tag{5.58}$$

Where K is Einstein's constant for the *n*-dimensional electrogravitic field equations.

$$\widetilde{R}_{\mu\nu\varepsilon\beta} = \kappa \cdot \widetilde{\Phi}_{\mu\nu\varepsilon\beta} \tag{5.59}$$

The above *n*-dimensional electrogravitic field equations form an electromagnetic theory of matter based on fundamental equations for Physics and Cosmology.

§ 6. On the nature of complex space-time.

The concept of complex space-time broadens the impression of real-valued spatial and temporal coordinates to a dynamic complex wave medium of complex-valued spatial and temporal coordinates.
In quantum mechanics, wave functions are complex-valued functions that describe particles in real-valued spatial and temporal coordinates, and the set of a system's wave functions is an infinite-

dimensional complex Hilbert space. Wave functions and fields extend to complex space-time, and this complex space-time may be interpreted as an extended complex spatial and temporal coordinate system. Consequently, complex space-time is the wave medium for the extended fields of complex spatiotemporal wavelets.

The notion of complex space geometry has been previously considered by prominent researchers such as Albert Einstein, with a complex metric tensor, but without complex space-time. Hence, it seems that the notion of complex space-time has not received an equal amount of attention. Imaginary time has real physical meaning and it allows mathematical analysis to endure the analytic extension of the real temporal variable onto the complex plane, which makes some solutions practicable.

Time is dimensional in nature; every spatial dimension has an orthogonal conjugate temporal dimension. A temporal dimension has a dimension of one because only one coordinate is needed to specify any point within the temporal dimension. The temporal coordinate on the measurable extent of a temporal dimension may be specified by an imaginary number. In this conception of complex space-time, every event may be specified by complex coordinates.

Each complex coordinate has a spatial part (real part) and a temporal part (an imaginary part), where both parts are associated with an apparent or complex value. A single complex coordinate system may be applied to a spatiotemporal surface with two existing dimensions of space and time. Complex coordinates may specify spatiotemporal distance, surface, or volume. A tesseract may be represented using a single complex coordinate system where the corners of the inner cube represents a volume of "now" with complex coordinates that have imaginary parts with zero values, and the outer cube represents a volume of a "future now" with complex coordinates that have non-zero spatial and temporal parts. In a complex coordinate system, the spatiotemporal complex dimension of the tesseract is six, with three spatial dimensions (real parts) and three temporal dimensions (imaginary parts).

In the above example of a tesseract, if all six sides of the tesseract are considered adjacent to other identical tesseracts, then the spatiotemporal wave on each side and adjacent side of the tesseract would interfere with each other. If space-time is homogeneous and isotropic on each side, then the waves would interfere on each side and subtract. The imaginary parts of each adjacent pair of complex spatiotemporal coordinates, on each side of the tesseract, would cancel each other, and the resultant expansion of space would be null. If the imaginary parts of each adjacent pair of complex spatiotemporal coordinates on each side of the tesseract do not cancel each other, then space expands into a larger volume of a "future now".

Time is a dimensional process. Time is the outgrowth, or ingrowth, of every dimensional extent of space-time. Thus, time, as it relates to every spatial dimension, is both a process and a dimension. The properties of time include the passage of time at different rates where slower rates in gravitational fields are continuously connected to faster rates as a function of distance from the object of mass or gravitational source. Time in all its dimensions is an emergent phenomenon. Time emerges from the expansion of space. The expansion of space is the direct cause of time. The passage of time is a function of the rate of expansion of space. In turn, the rate of expansion of space depends on the spatiotemporal pressure at any point in space-time which is a function of mass and energy. The rate of expansion of the spatiotemporal wave medium, or the passage of time, manifests a gravitational field near an object of mass. Electromagnetic waves propagate at the speed of light through complex space-time because that is the speed of expanding space, or the speed of the passage of time, at any point or event in complex space-time. Time, the emergent property of the expansion of space, endows motion to particles and fields at any point in complex space-time. *All motion is a function of the dimensions of space-time, and motion exists due to the expansion, or contraction, of space.*

As an object of mass travels through space, it would travel less through time because the object would have to move through less distance within the temporal wave, as the temporal wave expands in the local reference frame of the moving object. Hence, it is space-time that endows the passage of time, the motion of objects of mass and fields, and perpetuates motion through its property of inertia.

§ 7. The impedance of free space-time

Space-time has impedance characteristics at every expanding point, the impedance of space-time is the physical characteristic that attenuates, amplifies, or phase-shifts the propagation of an electromagnetic plane wave, or the electromagnetic field of a charged particle or a charged object, traveling through space-time. An electromagnetic wave travels at 299,792,458.0 meters per second. The spatiotemporal impedance Z_0 is usually described as the ratio of the magnitudes of the electric and magnetic fields of electromagnetic radiation traveling through free space-time, the square root of the ratio of the permeability to the permittivity of free space-time, the reciprocal of the product of the permittivity and the speed of light though vacuum, or as the product of the permeability of free space-time and the speed of light through vacuum.

$$Z_0 = \frac{|E|}{|H|} = \sqrt{\frac{\mu_0}{\varepsilon_0}} = \frac{1}{\varepsilon_0 c_0} = \mu_0 c_0 \approx 376.73031\ldots \ (\text{Ohms}) \qquad (7.1)$$

Therefore, the impedance of free space-time involves both the permittivity and the permeability of the medium. These physical properties are the properties of inductors and capacitors. The permittivity describes how space-time behaves as a capacitor and the permeability describes how space-time behaves as an inductor. Space-time is able to hold charge per unit of voltage potential to establish an electric field and is also able to conduct charges to establish a magnetic field.

Let us illustrate the impedance of free space-time as the parallel circuit of resistance, inductance, capacitance and conductance, as a single *RLGC* network at an angular frequency ω. The parallel impedance of the *RLGC* circuit resonates at the critical angular frequency. *G* is the conductance in the circuit representing the leakage current across the spatiotemporal dielectric of the capacitance, and *R* is the parallel resistance of the circuit. As the angular frequency of the *RLGC* circuit increases by orders of magnitude to equal the critical angular frequency, the resistance *R* and the conductance *G* approach infinity, as the *LC* circuit begins to resonate. The circuit would behave as parallel inductance and capacitance in its impedance as an antiresonator. Thus, we may represent each expanding point, or node, of space-time as consisting of a spatiotemporal antiresonator such as in Figure 1.

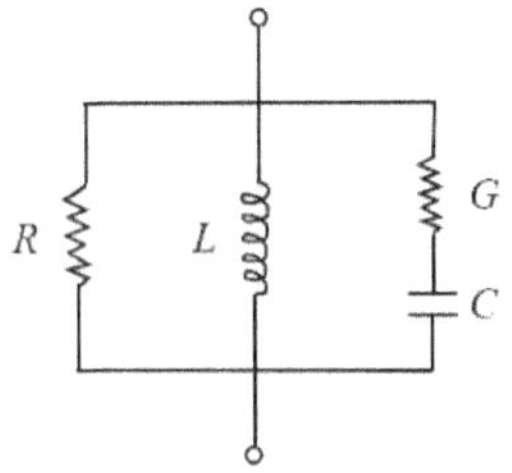

Figure 1.

$$Z(j\omega) = R \,||\, j\omega L \,||\, \frac{1}{j\omega C} \,||\, G = \frac{1}{\frac{1}{R} + \frac{1}{j\omega L} + j\omega C + \frac{1}{G}} \quad (7.2)$$

$$Z(j\omega_0) \approx \lim_{\omega \to \omega_0} \frac{1}{\frac{1}{\left|e^{j\omega_0}\right| R} + \frac{1}{j\omega_0 L} + j\omega_0 C + \frac{1}{\left|e^{j\omega_0}\right| G}} \quad (7.3)$$

where $\omega_0 >> \omega$

$$Z(j\omega_0) = j\omega_0 L \,||\, \frac{1}{j\omega_0 C} = \frac{1}{\frac{1}{j\omega_0 L} + j\omega_0 C} \quad (7.4)$$

Thus, we can depict space-time as having the properties of a resistor-inductor-conductor-capacitor network, an *RLGC* network, at every expanding point, for an electromagnetic wave at the critical angular frequency.

Let us imagine a single spatiotemporal cube with eight nodes, such as in Figure 2. Each node represents a corner of the single spatiotemporal cube, and each line (or circuit) between two nodes of the single spatiotemporal cube is shared by four identical spatiotemporal cubes. However, each node is common to, or shared by, eight identical spatiotemporal cubes. Thus, there is a *RLGC* circuit (a resistor, an inductor, a conductor, and a capacitor) in each sense of direction of space. For each of the six senses of spatial direction, let us assign one sixth of the total resistance, one sixth of the total inductance, one sixth of the total conductance, and one sixth of the total capacitance of a single node. Then, at any of the nodes of a single spatiotemporal cube, we have three antiresonator circuits located at the single spatiotemporal cube and three antiresonator circuits located on adjacent spatiotemporal cubes. Each antiresonator circuit provides one sixth of the total *R, L, G,* and *C* at every common node as shown in Figure 3.

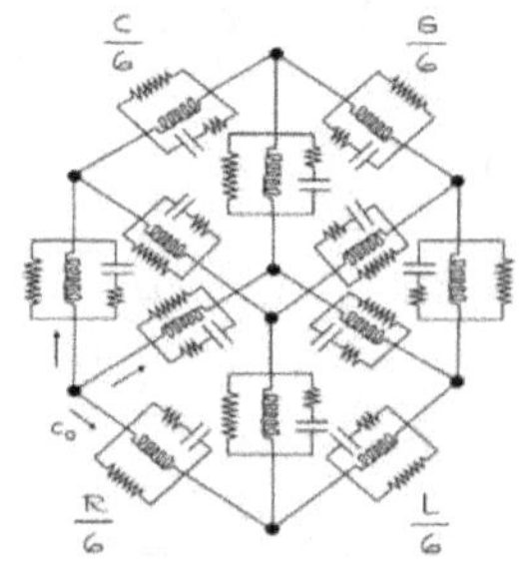

Figure 2.

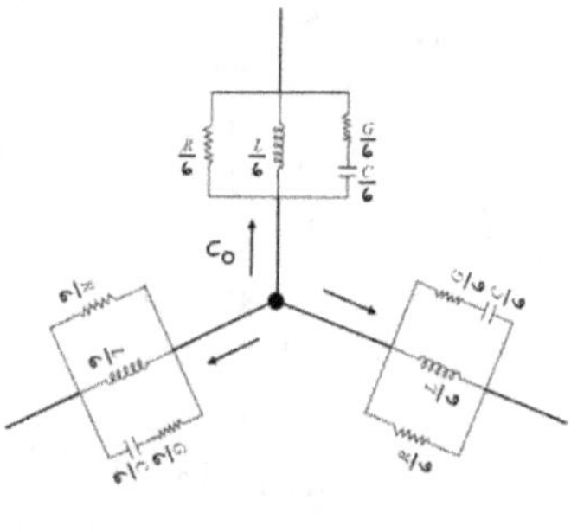

Figure 3.

Light as an electromagnetic wave travels through every

spatiotemporal node in every sense of direction attenuated, amplified, phase-shifted, or obstructed by the impedance of free space, energy, or matter. Thus, the speed of light depends on the critical angular frequency that amplifies, attenuates, or phase-shifts electromagnetic waves passing through the spatiotemporal dimensions of the *RLGC* network when the inductive reactance and the capacitive reactance are at electrical resonance.

The spatiotemporal electromagnetic resonant angular frequency is the reciprocal of the square root of the product of the capacitance and the inductance of free space-time, or the reciprocal of the square root of the product of the permittivity, the permeability, and a spatial distance ℓ_0 through the *LC* circuit between nodes in free space-time.

$$\omega_0 = \frac{1}{\sqrt{LC}} = \frac{1}{\ell_0\sqrt{\varepsilon_0\mu_0}} = \frac{1}{\sqrt{(4\pi \text{ x } 10^{-7}\, H)(8.541878176 \text{ x } 10^{-12}\, F)}} \tag{7.5}$$

$$\omega_0 \approx 299792458.0 \; rads/\sec \tag{7.6}$$

$$f_0 = \frac{\omega_0}{2\pi} \approx 47713451.59236942258889\, Hz \approx 47.71 MHz \tag{7.7}$$

The magnitude of the critical angular frequency $|\omega_0|$ is equal to the magnitude of the speed of light in vacuum $|c_0|$ as light, as an electromagnetic wave, travels through the spatiotemporal medium.

The critical linear frequency is a low frequency and low energy wave in the VHF radio range of the electromagnetic spectrum, with a long wavelength λ_0 of approximately 6.28 meters.

$$\omega_0 = \frac{c_0}{\ell_0} \tag{7.8}$$

$$c_0 = \omega_0 \ell_0 \tag{7.9}$$

Therefore, we may interpret the magnitude of the speed of light in vacuum as the magnitude of the angular frequency of the *RLGC* network of free space times the distance ℓ_0 associated with the spatial angular frequency, in meters per radian, from node to node, that light would travel to traverse the *RLGC* circuit in the vacuum. The temporal angular frequency ω_0 is the critical angular frequency of the *LC* network for the reactive values associated with the permittivity and permeability of free space. Electromagnetic waves traverse the *RLGC* network spatial distances at temporal angular frequencies above, equal to, or below the critical value.

Moreover, the speed of light in vacuum may be expressed in terms of the frequencies of space and time through the *RLGC* network. As an electromagnetic wave traverses the spatiotemporal *RLGC* network its speed depends on the ratio of its temporal angular frequency to its spatial angular frequency.

$$c_0 = \frac{\partial r}{\partial t} = \frac{\frac{\partial \theta}{\partial t}}{\frac{\partial \theta}{\partial r}} = \frac{\omega_t}{\omega_s} \tag{7.10}$$

$$c_0 = \frac{\omega_t}{\omega_s} = \frac{\omega_0}{\omega_{\ell_0}} = \frac{1}{\sqrt{\varepsilon_0 \mu_0}} \tag{7.11}$$

$$\omega_{s_0} = \omega_0 \sqrt{\varepsilon_0 \mu_0} = \sqrt{\frac{\varepsilon_0 \mu_0}{LC}} = \sqrt{\frac{LC}{LC}} = 1.00 \; rads/meter \tag{7.12}$$

Hence, it is apparent that the larger temporal angular frequency of the spatiotemporal medium, when compared to the spatial angular frequency, is directly related by orders of magnitude to the very high speed of light through the vacuum. It is reasonable to acknowledge that electromagnetic waves, or photons, owe their great speed through space-time to the expansion and dimensions of time. A source of light at an expanding spatiotemporal point radiates photons and electromagnetic waves in all senses of directions, as time expands simultaneously in all its senses of directions from that same point.

Consequently, electromagnetic waves in vacuum traverse the spatial distance of the *RLGC* network from node to node to some extent at the spatial angular frequency, through the impedance of the *LGC* circuit in parallel with the infinite impedance of the resistance *R* that is equivalent to approximately the impedance of the resonant *LC* circuit. The critical spatial frequency of light equals the critical temporal frequency times the reciprocal of the speed of light which is the square root of the product of the permittivity and the permeability of free space.

Let us express the capacitance, inductance, impedance, admittance, and the speed of light in vacuum as the product of permeability and permittivity within a distance of one meter. The admittance of free space-time becomes Y_0 for any electromagnetic wave with a resonant linear frequency equal to f_0 .

$$C = \varepsilon_0 \ell_0 \quad \left(Faradays \ \ or \ \ \frac{s}{\Omega} \right) \tag{7.13}$$

$$L = \mu_0 \ell_0 \quad \left(Henries \ \ or \ \ \frac{N}{m} \right) \tag{7.14}$$

$$|Z_0| = \sqrt{\frac{\mu_0 \ell_0}{\varepsilon_0 \ell_0}} = \sqrt{\frac{\mu_0}{\varepsilon_0}} = \sqrt{\frac{L}{C}} = 376.7303133968621 \approx 120\pi \approx 377 \ \ (\text{Ohms}) \tag{7.15}$$

$$Y_0 = \sqrt{\frac{\varepsilon_0}{\mu_0}} = \sqrt{\frac{C}{L}} \tag{7.16}$$

$$c_0 = \frac{1}{\sqrt{LC}} \tag{7.17}$$

So far, we have assumed that the temporal and spatial angular frequencies of free space-time are independent of the expansion or contraction of space-time, and are not relative to any observer's frame of reference which is contrary to the General Theory of Relativity.

Thus, if we assume the principle of absolute spatiotemporal angular frequencies, light may be attenuated, amplified, or phase-shifted by Z_0 , through the homogeneous and isotropic free spatiotemporal wave medium on any frame of reference at its resonant angular frequency to hold the velocity of light at c_0. This resonant property of space-time maintains the constancy of the speed of light at the critical angular frequency through the vastness of the universe. Hence, at the critical angular frequency, the parallel impedance of the inductive reactance and the capacitive reactance of homogeneous and isotropic free space is the parallel impedance of a *LC* network at every expanding node in space-time as shown in Figure 4.

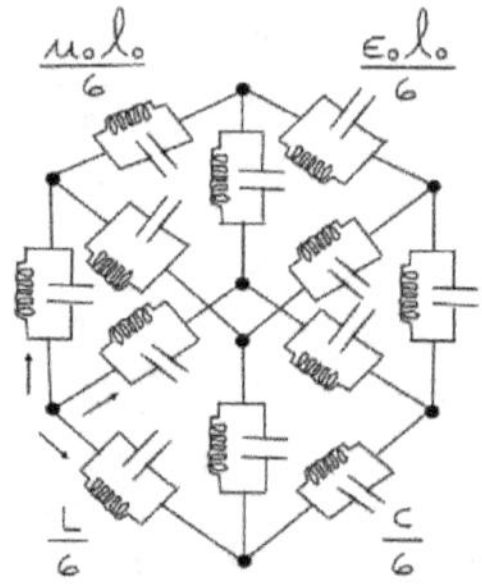

Figure 4.

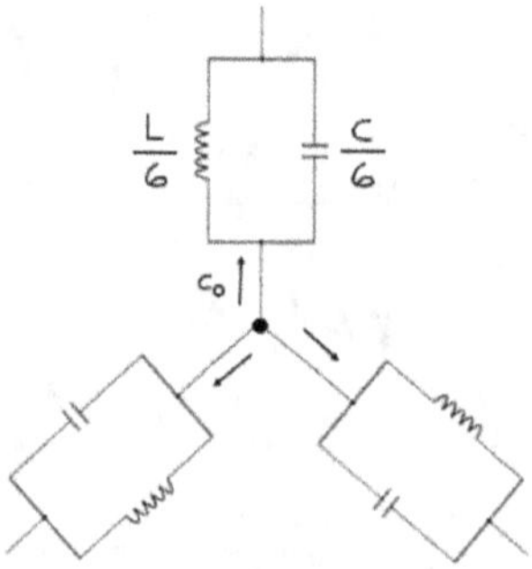

Figure 5.

$$Z_0 = \frac{1}{\frac{1}{j\omega L} + j\omega C} = \frac{1}{j\left(\omega C - \frac{1}{\omega L}\right)} = \frac{\omega L}{j\left(\omega^2 LC - 1\right)} \qquad (7.18)$$

$$jZ_0\left(\omega^2 LC - 1\right) = \omega L \tag{7.19}$$

$$jZ_0 = \frac{\omega L}{\dfrac{\omega^2}{c_0^{\,2}} - 1} \tag{7.20}$$

$$jZ_0 = \frac{\omega_{z_0} L}{\left(\omega_{p_0}\sqrt{LC} + 1\right)\left(\omega_{p_1}\sqrt{LC} - 1\right)} \tag{7.21}$$

$$jZ_0 = \frac{\sqrt{\dfrac{L}{C}}\, s_{z_0}}{\left(s_{p_0} + \dfrac{1}{\sqrt{LC}}\right)\left(s_{p_1} - \dfrac{1}{\sqrt{LC}}\right)} = \frac{377 s_{z_0}}{\left(s_{p_0} + c_0\right)\left(s_{p_1} - c_0\right)} \tag{7.22}$$

Let us consider the impedance transfer function of the *LC* circuit as the parallel output impedance across the *LC* circuit to an input reference base impedance value, Z_{base}, of one Ohm.

$$H(s) = \frac{jZ_0}{Z_{base}} = \frac{jZ_0}{1\,\Omega} = jZ_0 = \frac{377 s_{z_0}}{\left(s_{p_0} + c_0\right)\left(s_{p_1} - c_0\right)} = \frac{\left(\dfrac{377}{c_0^{\,2}}\right) s_{z_0}}{\left(\dfrac{s_{p_0}}{c_0} + 1\right)\left(\dfrac{s_{p_1}}{c_0} - 1\right)} \tag{7.23}$$

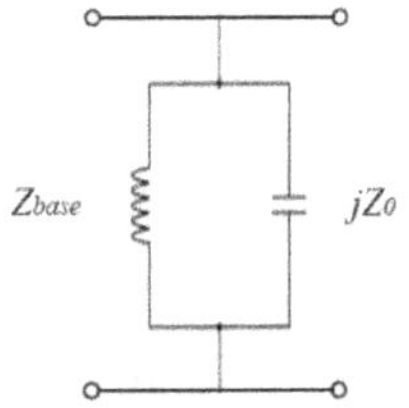

Figure 6.

The spatiotemporal *LC* network transfer function has a gain of $377/c_0^{\,2}$, one zero in the numerator, and two poles in the denominator equal to $-c_0$ and c_0. Thus, the *LC* network is a second order band-pass response containing two conjugate poles and a zero at the origin, where the response always peaks precisely at the temporal critical frequency. The following Bode plot is a graphical representation of a linear, time-invariant spatiotemporal *LC* network transfer function with zero initial conditions and zero-point equilibrium. Any sinusoidal electromagnetic wave that inputs the linear system may be changed in magnitude, when it is amplified or attenuated, and may be changed in phase, when it is advanced or retarded. Hence, the *LC* network response can be described for every frequency, just by its gain and phase shift. The Bode plot traces the gain and phase shift of the *LC* network to a range of frequencies. Let us begin the Bode plot of the transfer function by calculating the DC magnitude of the gain of the *LC* network system, $Z_0/c_0^{\,2}$.

$$20\log\frac{Z_0}{c_0^{\,2}} = 20\log\frac{\mu_0}{c_0} = 20\log\frac{377}{(299792458)^2} \approx -287.55\ \text{dB} \quad (7.24)$$

The zero and the two poles of the transfer function are:

$$s_{z_0} = 0 \quad (7.25)$$

$$s_{p_0} = -299792458 \quad (7.26)$$

$$s_{p_1} = 299792458 \quad (7.27)$$

The impedance magnitudes of the individual elements of the spatiotemporal antiresonator in Figure 1 are illustrated in Figure 7. The asymptotes for the total parallel impedance Z_0 are approximated by simply selecting the smallest individual element impedance. So, the total parallel impedance is dominated by the inductor at low frequency, by the resistor at mid frequency, and by the capacitor at high frequency as shown in Figure 7. Conductance is the reciprocal of resistance, so when resistance approaches infinity conductance approaches zero. If the angular corner frequencies are well-separated in value, the corner frequencies can be found by equating asymptotes as follows:

$$R = \omega_L L \rightarrow \omega_L = \frac{R}{L} \tag{7.28}$$

$$R = \frac{1}{\omega_C C} \rightarrow \omega_C = \frac{1}{RC} \tag{7.29}$$

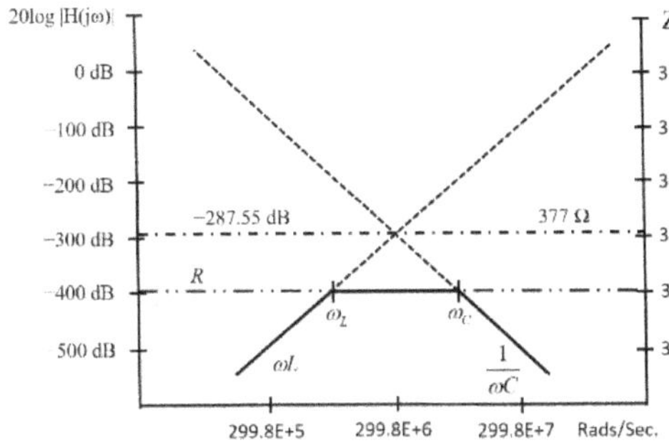

Figure 7.

As the resistance value R approaches infinity at the critical frequency, the asymptotes for the total parallel impedance of the LC network become independent of R, and transition directly from the inductive reactance line to the capacitive reactance line at the critical angular frequency as shown in Figure 8. The critical angular frequency is the frequency where the inductor and capacitor asymptotes have equal value.

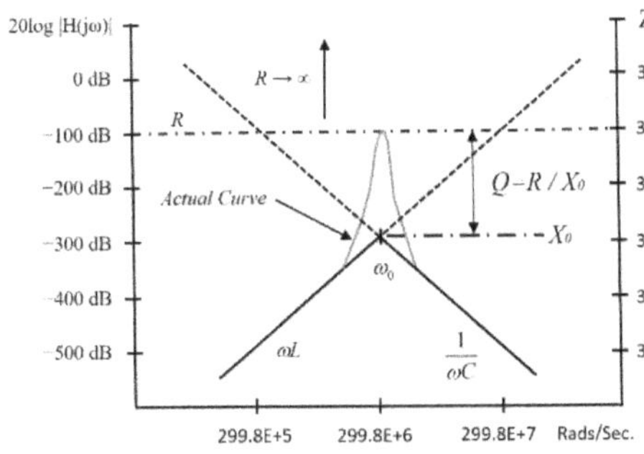

Figure 8.

$$\omega_0 = \frac{1}{\sqrt{LC}} \tag{7.30}$$

At the critical angular frequency, the slope of the asymptotes of Z_0 changes from +20dB/decade to –20dB/decade, and hence, there are two poles and a zero. The actual curve peaks as the value of the resistance approaches infinity, and the value of the total parallel impedance becomes the value of the resistance *R* at the critical angular frequency as shown in Figure 8.

Thus, at the critical angular frequency of the spatiotemporal antiresonator, the inductive reactance and the capacitive reactance cancel out and the total parallel impedance Z_0 equals the value of the resistance *R*. Consequently, the value of the resistance *R* determines the value of the curve at the critical angular frequency as the resistance goes to infinity, but the values of the inductive reactance and capacitive reactance determine the values of the asymptotes, at the critical angular frequency.

The deviation of the actual curve from the asymptotes at the critical angular frequency is given by the quality factor, *Q*, between the value of reactance, X_0 , when the inductive reactance and the capacitive reactance are equal, and the value of resistance *R*. The quality factor, *Q*, of the curve is the ratio of the resistance value *R* to the value of reactance at X_0. The sharpness of the actual curve depends on the quality factor. The quality factor is equal to the ratio of 2π times the maximum energy stored in the capacitor and the inductor to 2π times the total energy lost in the resistance in one period. The voltage *V* is the voltage potential across the resistance *R*.

$$X_0 = \omega_0 L = \frac{1}{\omega_0 C} \tag{7.31}$$

$$Q = \frac{2\pi\left(V^2/\omega_0 R\right)}{2\pi\left(LV^2/R^2\right)} = \frac{R}{\omega_0 L} = \omega_0 RC = R\sqrt{\frac{C}{L}} = \frac{R}{\sqrt{\frac{L}{C}}} = \frac{R}{|Z_0|} \tag{7.32}$$

$$\left|Q\right|_{dB} = \left|R\right|_{dB} - \left|X_0\right|_{dB} \tag{7.33}$$

$$Q = \frac{R}{X_0} \tag{7.34}$$

Illustrating the phase plot of the *LC* network transfer function, we have

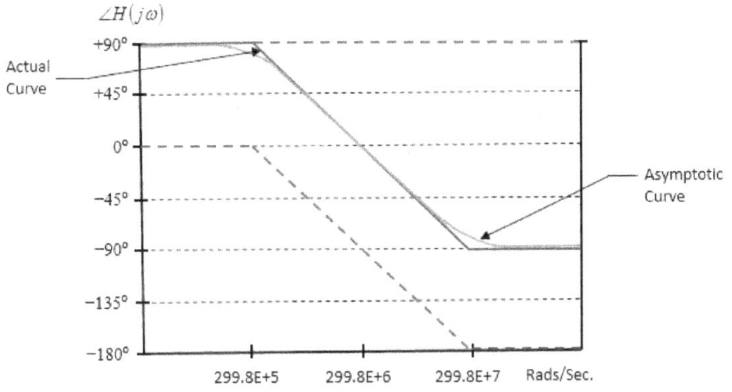

Figure 9.

The phase shift at low frequencies approaches +90 degrees, a phase lead. At the critical angular frequency, the phase shift is zero. The phase shift at high frequencies approaches −90 degrees, a phase lag.

The phase shift is +45 degrees at ω_L and −45 degrees at ω_C as shown in Figures 7 and 9. At the critical temporal frequency of light in vacuum, light, an electromagnetic wave, may traverse the dimensions of time, in forward or reverse motion, in a direction that is perpendicular to all spatial dimensions, as time expands or contracts. This spatiotemporal resonant mechanism forms the underlying concept for the constancy and propagation of light through the vacuum.

Thus, as the temporal angular frequency of an electromagnetic wave, or light, decreases below the critical angular frequency value ω_0 , an electromagnetic wave traverses more and more the spatial distance between nodes through the spatiotemporal *RLGC* network, and less of the temporal distance. The phase of the electromagnetic wave leads the reference angle of the *RLGC* network more. Conversely, as the temporal angular frequency of an electromagnetic wave increases above the critical angular frequency value, the electromagnetic wave traverses less and less the spatial distance through the spatiotemporal *RLGC* network, and more of the temporal distance.

The electromagnetic wave lags the reference angle of the *RLGC* network more. In either scenario, the magnitude of the amplitude of an electromagnetic wave is attenuated away from the temporal angular frequency, but the phase angle of an electromagnetic wave shifts when it lags or leads, respectively, the reference angle of the *RLGC* network. At the temporal critical frequency, the phase angle between the electromagnetic wave and the reference angle of the *RLGC* network is zero. The electromagnetic wave and the reference angle of the *RLGC* network are in phase in the direction of propagation.

The DC magnitude has a maximum gain of μ_0/c_0 for the *LC* network, which is the permeability of free space divided by the magnitude of the speed of light in vacuum. There is attenuation away from the critical angular frequency.

Therefore, when the magnitude of the angular frequency $|\omega|$ of an electromagnetic wave in free space-time is equal to the magnitude of the speed of light $|c_0|$ in free space-time, the impedance of free space-time, Z_0, has a gain of approximately −287.55 dB. Electromagnetic waves with an angular frequency below, or above, the resonant angular frequency of the speed of light, ω_0, are attenuated further when traveling through spatiotemporal dimensions. The spatiotemporal antiresonant mechanism attenuates the amplitude and shifts the phase of electromagnetic waves between +90° and −90° through the spatiotemporal distances.

The inductive or capacitive reactance, or impedance in general, are usually given in units of Ohms. An Ohm may be expressed in units of kilogram, meter, second, and Coulomb. Thus, we may also express an Ohm as a charged mass times its acceleration per unit of charge. Then, it is reasonable to assume than an Ohm is a force per Coulomb felt by a charged particle or object as it moves within an electric field or magnetic field potential. This is the case of electronic current flow within a conductor which may have electric and magnetic fields about its atomic domains, or the case of a charged object of mass or energy, moving through the spatiotemporal structure.

$$\Omega \equiv \frac{Kg \cdot m^2}{s \cdot C^2} \tag{7.35}$$

From previous research, a unit of charge, a Coulomb, was defined as a unit of distance times a unit of time. If we substitute an equivalent quantity of a meter times a second for a quantity of Coulomb in the above units for an Ohm, we simplify the units to a force per charge.

$$\Omega \equiv \frac{Kg \cdot m^2}{s \cdot C^2} \equiv \frac{Kg \cdot m^2}{s \cdot m \cdot s \cdot C} \equiv \frac{Kg \cdot m}{s^2 \cdot C} \equiv \frac{ma}{C} \equiv \frac{N}{C} \equiv \frac{F}{Q} \tag{7.36}$$

$$Z_0 = \frac{|E|}{|H|} \equiv \frac{N/C}{A/m} \equiv \frac{N \cdot m \cdot s}{C \cdot C} \equiv \frac{N \cdot m \cdot s}{C \cdot m \cdot s} \equiv \frac{N}{C} \equiv \Omega \tag{7.37}$$

Hence, an Ohm is a unit of force on a charge, or Newton per Coulomb, an impedance to the free flow or propagation of a charged particle, object, or physical field. It is reasonable to assume that the inductive and capacitive reactance of free space is also a force per Coulomb of charge that is exerted on charged objects and particles by electromagnetic potentials in the substructure of space-time that exhibits permittivity and permeability. The capacitor and inductor elements in the spatiotemporal medium exert a force on physical fields, charged particles and objects, at every expanding or contracting point or node.

Let us also consider light in the spatiotemporal medium in terms of the impedance and the permeability of free space-time. The unit of impedance in Newton per Coulomb is useful to derive the unit of light, or the unit of an electromagnetic wave, which is a meter per second.

$$c_0 = \frac{Z_0}{\mu_0} \equiv \frac{N/C}{N/m^2} \equiv \frac{m^2}{C} \equiv \frac{m^2}{m \cdot s} \equiv \frac{m}{s} \qquad (7.38)$$

7.2. The Evanescent Electromagnetic Wave

The electromagnetic waves near the critical temporal frequency are predominantly evanescent waves. In the electromagnetic near-field of any antenna in the spatiotemporal wave medium, the evanescent waves exhibit exponential decay or attenuation, as a function of distance from the surface of formation, very intensely within one third of a wavelength. Some of the field energy is reabsorbed by the antenna of the source, while the rest is radiated as electromagnetic waves.

The capacitive effect of the spatiotemporal antiresonator exhibits coupling between its spatiotemporal plates as charges and currents are induced on partially reflective surfaces in the far-field where the components of the wave reach the ratio of the impedance of free space-time and the wave propagates radiatively.

In the near-field of any antenna, evanescent waves may be either predominantly magnetic, or inductive, or predominantly electric, or capacitive, depending on the impedance of the radiative source. If the impedance of the radiative source matches either the inductive impedance or the capacitive impedance of the spatiotemporal antiresonator of free space-time, then the emergent evanescent wave would be predominantly magnetic or electric, respectively. Above the inductive corner frequency of the spatiotemporal antiresonator, the impedance is real and resistive and the wave carries energy.

Below the inductive corner frequency, the impedance is inductive reactive and the wave is evanescent. Similarly, the impedance is real and resistive below the capacitive corner frequency. Above the capacitive corner frequency, the impedance is capacitive reactive and the wave is evanescent. Evanescent waves are solutions to the wave equation of an electromagnetic field.

Evanescent waves occur between two spatiotemporal wave media with different properties of motion. Evanescent waves are very useful to power devices wirelessly in the near-field, or in the coupling of Tesla coils with wireless electrical devices.

7.3. The relationship between the impedance of free space-time and the Lorentz Factor

Let us consider the relationship between the speed of an object traveling through space-time and the spatiotemporal frequencies of homogeneous and isotropic space-time through the *RLGC* network.

The product of the speed of light and the permeability of free space-time is equal to the impedance of free space-time.

$$\frac{\omega_0}{\omega_S} = c_0\left(1 - \frac{v^2}{c^2}\right) \tag{7.39}$$

$$\frac{\omega_0}{\omega_S} = \frac{Z_0}{\mu_0}\left(1 - \frac{v^2}{c^2}\right) = \frac{Z_0\sqrt{1 - \frac{v^2}{c^2}}}{\frac{\mu_0}{\sqrt{1 - \frac{v^2}{c^2}}}} \tag{7.40}$$

Consequently, let express the spatiotemporal frequencies in terms of the impedance of free space and the permeability of free space.

$$\omega_0 = Z_0\sqrt{1 - \frac{v^2}{c^2}} \tag{7.41}$$

$$Z_0 = \frac{\omega_0}{\sqrt{1-\frac{v^2}{c^2}}} \tag{7.42}$$

$$\omega_S = \frac{\mu_0}{\sqrt{1-\frac{v^2}{c^2}}} \tag{7.43}$$

$$\mu_0 = \omega_S \sqrt{1-\frac{v^2}{c^2}} \tag{7.44}$$

From the permeability and the impedance of free space-time we may derive the permittivity of free space-time.

$$\varepsilon_0 = \frac{\mu_0}{Z_0^{\;2}} = \frac{\omega_S \sqrt{1-\frac{v^2}{c^2}}}{\left(\frac{\omega_0}{\sqrt{1-v^2/c^2}}\right)^2} \tag{7.45}$$

The Lorentz factor is the factor by which relativistic mass, length, and time change for an object that is moving through the spatiotemporal wave medium.

$$\gamma = \frac{1}{\sqrt{1-\frac{v^2}{c^2}}} = \frac{Z_0}{\omega_0} = \frac{\omega_S}{\mu_0} \tag{7.46}$$

Therefore, the ratio of the impedance of free space-time Z_0 to the temporal angular frequency, or the ratio of the spatial angular frequency to the permeability of free space-time μ_0, is equivalent to the Lorentz factor.

Then, all the relativistic effects exerted on length, time, or mass of an object moving through a homogeneous and isotropic spatiotemporal wave medium are attributed to the impedance characteristics of space-time and the temporal angular frequency, or to the interaction of the object's electromagnetic signature and the spatiotemporal antiresonator impedance elements in the path of propagation. Thus, the spatiotemporal antiresonator impedance is relative.

The impedance of free space-time undergoes relativistic effects similar to the dilation or contraction of time, or mass, while the permeability of free space-time undergoes relativistic effects similar to the dilation or contraction of length. As an object travels through space more and less through time, the permeability and permittivity of free space-time decrease, but the impedance of free space-time increases.

Furthermore, contrary to the General Theory of Relativity, the temporal and spatial angular frequencies of homogeneous and isotropic free space-time are independent of the expansion or contraction of space-time, or mass, and are not relative to any observer's inertial frame of reference. According to the principle of absolute spatiotemporal angular frequencies, an electromagnetic wave maintains the constancy of the speed of light because the wave's radians in its temporal or spatial angular frequencies are not relative.

During the dilation or contraction of the wavelength, or the period, of an electromagnetic wave in homogeneous and isotropic free space-time, contrary to the General Theory of Relativity, the proportional magnitudes of the wave's radians per unit of length, or time, remain constant.

§ 8. On the electric field, charge and angular momentum of a rotating Kerr-Newman supermassive black hole

During the formation of a rotating Kerr-Newman supermassive black hole as the mass of a celestial body collapses to a ring, the space within the forming black hole contracts inward, the electromagnetic forces of the unified electroweak interaction of the mass and charges become stronger than the gravitational field, as the atomic structures, charge distribution, and energy collapse into a denser mass. As the mass coheres, charges condense creating an increasingly cohesive charge distribution about the geometry of the collapsing mass, springing strong electromagnetic fields out of the condensed mass as charges focally direct their electromagnetic field lines radially outward. The electromagnetic field forces of the unified electroweak interaction act on the area of the outer event horizon of the Kerr-Newman black hole preserving its geometry and boundary, creating a space-time differential for the type of space-time within and outside of the black hole. The concentrated electromagnetic fields sustain the process of the collapsing mass and establish a stable gravitational field for the black hole.

As the rotating Kerr-Newman black hole forms, it radiates, so black holes are not really entirely black. The radiation may come from virtual particle pairs or the mass and charges repelled by external electromagnetic fields as an external mass is accreted by the black hole. To an observer watching and monitoring radiation emission at a safe distance from the black hole, it would look like the black hole has just emitted particles or radiated energy. This process may happen repeatedly, in which case the observer would see a continuous stream of radiation from the rotating Kerr-Newman black hole. (Melia, 2007 and 2009)

8.1. The electric field of a rotating supermassive black hole

Let us now envision the electromagnetic field of the rotating supermassive Kerr-Newman black hole as produced by an enclosed charge magnitude, Q_S , depending on the location or point in space-time where we theoretically determine the magnitude of the electromagnetic field that exists regardless of its origin.

Within the outer event horizon of the black hole at a distance ct_{INT} from the origin, the magnitudes of the internal electric field $\overrightarrow{E_{INT}}$ and the magnetic field $\overrightarrow{B_{INT}}$ are proposed to be

$$E_{INT} = \frac{iQ_S}{4\pi\varepsilon_0 \left(ct_{INT}\right)^2} \tag{8.1}$$

$$B_{INT} = \frac{i\mu_0 Q_S v}{4\pi\varepsilon_0 \left(ct_{INT}\right)^2} \tag{8.2}$$

The force per unit of charge of the internal magnetic field and the internal electric field would strengthen due to length dilation and time contraction, as the distance ct_{INT} increases from the origin of the collapsing mass of the rotating and charged Kerr-Newman black hole toward the outer event horizon, as radial distance extends outward.

The spatiotemporal contraction indicates that space-time flows at a velocity greater than c. Hence, the internal electromagnetic equations represent physical fields in the direction of the advanced spatiotemporal wave.

Outside of the outer event horizon, the ergosphere curves space-time with an angular speed about the black hole and an external electric field follows the curved space of the ergosphere around the outer event horizon. As the Kerr-Newman black hole accretes neutral matter, the external electric field prevails. Hence, if the ring singularity is positively charged, there is an outward electromagnetic field between the ring singularity and the outer event horizon. However, the torsion and contraction of the space-time inside the outer event horizon toward the ring singularity may skew and refract internal electromagnetic field lines from the ring due to distortion or lensing of the inner space-time of the blackhole.

Moreover, the electromagnetic field around the ergosphere may be expressed in terms of the external electromagnetic field's density and strength about the outer event horizon, in which case we have the magnitudes of the electric field density $\overrightarrow{D_{EXT}}$ and the magnetic field strength $\overrightarrow{H_{EXT}}$ expressed as

$$D_{EXT} = \frac{Q_s}{4\pi(ct_{EXT})^2} = \varepsilon_0 E_{EXT} \tag{8.3}$$

$$H_{EXT} = \frac{Q_S v}{4\pi(ct_{EXT})^2} = vD_{EXT} = \frac{B_{EXT}}{\mu_0} \tag{8.4}$$

Hence, the magnitude of the external magnetic flux density may be expressed as

$$B_{EXT} = \mu_0 v D_{EXT} = vE_{EXT} \tag{8.5}$$

Thus, at a radial distance ct_{IEH} within the outer event horizon of a rotating Kerr-Newman supermassive black hole, where the electromagnetic field force counterbalances the gravitational force to preserve the geometry and boundary of the outer event horizon, we find

$$\frac{Q_S}{4\pi\varepsilon_0\left(ct_{IEH}\right)^2}+\frac{\mu_0 Q_S v^2}{4\pi\left(ct_{IEH}\right)^2}=\frac{F_g}{Q_S} \tag{8.6}$$

$$F_g=\frac{Q_S^{\ 2}}{4\pi\left(ct_{IEH}\right)^2}\left(\frac{1}{\varepsilon_0}+\mu_0 v^2\right)=Mg \tag{8.7}$$

If the velocity is equal to the speed of light inside the supermassive black hole, we have

$$Q_S^{\ 2}=2\pi\varepsilon_0 Mg\left(ct_{IEH}\right)^2 \tag{8.8}$$

$$Q_S=ct_{IEH}\sqrt[2]{2\pi\varepsilon_0 Mg}=t_{IEH}\sqrt[2]{\frac{2\pi Mg}{\mu_0}} \tag{8.9}$$

Therefore, substituting the mass, the radius of the inner electrogravitic horizon (IEH), $ct_{IEH}=2GM/c^2$, and the gravitational acceleration at the inner electrogravitic horizon of a rotating Kerr-Newman black hole $g_{IEH}=GM/\left(2GM/c^2\right)^2=c^4/4GM=c/2t_{IEH}$, into the above equation for Q_S , we can formulate and determine its charge to be

$$Q_S=\sqrt[2]{\frac{4\pi\varepsilon_0 G^2M^3}{c^3 t_{IEH}}}=M\sqrt[2]{2\pi\varepsilon_0 G} \tag{8.10}$$

Similarly, we express the field-density conjecture at the inner electrogravitic horizon to be

$$Mg = Q_S E \tag{8.11}$$

$$\frac{g}{E} = \frac{\rho_S}{\rho_M} = \sqrt[2]{\frac{4\pi\varepsilon_0 G^2 M}{c^3 t_{IEH}}} = \sqrt[2]{2\pi\varepsilon_0 G} \tag{8.12}$$

Hence, *the electrogravitic ratio of the black hole equals the ratio of the densities of the charge and the mass at the inner electrogravitic horizon to preserve the physical geometry of the black hole.*

Substituting for *G* in the above equation $\left(G = \ddot{a}/8\pi M\right)$ we have

$$G\sqrt[2]{\frac{4\pi\varepsilon_0 M}{c^3 t_{IEH}}} = \frac{\ddot{a}}{4c}\sqrt[2]{\frac{\varepsilon_0}{\pi M c t_{IEH}}} \tag{8.13}$$

Thus, we can see that the physical geometry of the supermassive black hole depends purposely on the acceleration of space $\ddot{a}$, the passage of time, the mass *M* of the black hole, the speed of light, the constant π, and the permittivity ε_0 of free space-time.

8.3. Determining the charge, mass and angular momentum of a supermassive black hole

But how we may ask, do we determine the mass, or the charge, of a faraway celestial object such as a rotating Kerr-Newman supermassive black hole?

For a rotating Kerr-Newman black hole with an angular momentum *J*, mass *M*, and charge *Q*, we have the following relation

$$\frac{J^2}{M^2} + Q^2 \leq M^2 \tag{8.14}$$

where the mass of the black hole can take any positive value.

The total charge *Q* and the total angular momentum *J* are expected to satisfy the above equation for a rotating Kerr-Newman black hole of mass *M*. Black holes saturating this inequality are called extremal. An extremal black hole is a black hole with the minimal possible mass that can be compatible with the given charge and angular momentum. (Newman et al, 1965)

Thus, for a rotating Kerr-Newman black hole that is at least extremal in nature we find

$$\frac{J^2}{M^2} + \frac{4\pi\varepsilon_0 G^2 M^3}{c^3 t_{IEH}} \leq M^2 \qquad (8.15)$$

$$J^2 + \frac{4\pi\varepsilon_0 G^2 M^5}{c^3 t_{IEH}} \leq M^4 \qquad (8.16)$$

$$J^2 \leq M^4 - \frac{4\pi\varepsilon_0 G^2 M^5}{c^3 t_{IEH}} \qquad (8.17)$$

$$J^2 \leq M^4 \left(1 - \frac{4\pi\varepsilon_0 G^2 M}{c^3 t_{IEH}}\right) \qquad (8.18)$$

$$J \leq M^2 \sqrt[2]{\left(1 - \frac{4\pi\varepsilon_0 G^2 M}{c^3 t_{IEH}}\right)} \qquad (8.19)$$

$$J \leq M^2 \sqrt[2]{(1 - 2\pi\varepsilon_0 G)} \qquad (8.20)$$

Hence, as the mass-momentum constant is nearly one, $(1 - 2\pi\varepsilon_0 G) \approx 1$, we can express the magnitude of the angular momentum *J* of a rotating Kerr-Newman black hole, a black hole that is at least extremal, as approximately less than or equal to the squared magnitude of its mass *M*.

$$J \leq M^2 \qquad (8.21)$$

Conversely, the magnitude of the mass *M* is approximately greater than or equal to the square root of the magnitude of the angular momentum *J* of the rotating Kerr-Newman black hole.

$$M \geq \sqrt[2]{J} \qquad (8.22)$$

Hence, the mass of a rotating Kerr-Newman black hole may be approximated from the observation and measurement of its angular momentum. Therefore, we can determine the approximate magnitude of the charge Q_s of a rotating Kerr-Newman supermassive black hole to be

$$Q_S = M \sqrt[2]{2\pi\varepsilon_0 G} \tag{8.23}$$

Substituting for M in the penultimate equation, we have

$$Q_S \geq \sqrt[2]{2\pi\varepsilon_0 GJ} \tag{8.24}$$

where J is the angular momentum, G is the gravitational constant, and ε_0 is the permittivity of free space-time.

§ 9. Conclusion

The electromagnetic field and the resultant electromagnetic field are the true fundamental fields of electromagnetic phenomena, with the present constructs $\vec{D}$ and $\vec{H}$ having their useful applications in electrical engineering theory. Therefore, a resultant magnetic field springs from the curl of the resultant electric field of magnetic dipoles within a conductor perpendicular to the direction of propagation of the source electric field within the conductor of the current flow. The gravitational force and the electric force are distinct physical phenomena. However, we can express the electrogravitic force as a relationship between the gravitational and electrical forces. This electrogravitic relationship becomes useful to describe physical phenomena such as the counterbalance of electrogravitic forces at the inner electrogravitic horizon of a charged and rotating Kerr-Newman supermassive black hole and may play a significant role in other physical phenomena where the gravitational field and the lectromagnetic field may counteract, counterbalance, or coincide with each other.

Chapter 10

A Novel Treatise on Electromagnetism

§ 1. The units of spatiotemporal charge

The area around an electrically charged object or particle has a property which is referred to as an electric field as described more than a hundred years ago by the eminent researcher Michael Faraday. The electric field of a particle or object has the ability to exert a force on other particles or objects at a distance. The strength of such electric field is related to the electrical potential or pressure referred to as voltage potential, and the force of the electric field is projected through space-time from one quantum charge to another.

Conveniently, an electron (*e*) or a gluon (*e/3*) may be considered a basic unit of elementary charge derived from the quantum charge. As a charge moves in space-time, the charge manifests an electric field and a magnetic field. The electric and magnetic fields are conjugate fields like space and time are conjugates of each other. They are two different manifestations of a physical field, but not two separate natural phenomena. They are often referred to as a combined electromagnetic field for a moving charged particle just as space and time are referred to as space-time for a moving object of mass, not two separate natural phenomena. However, a magnetic field and an electric field, or a spatial dimension and a temporal dimension, may often be represented as a dichotomy in the analyses of physics.

The Planck Coulomb force between two Planck charges Q_p is given by

$$F_Q = \frac{\hbar c}{l_p^{\ 2}} = \frac{Q_p^{\ 2}}{4\pi\varepsilon_0 l_p^{\ 2}} \tag{1.1}$$

$$Q_p^{\ 2} = 4\pi\varepsilon_0 \hbar c = 4\pi\varepsilon_0 F_Q l_p^{\ 2} \tag{1.2}$$

A unit of Planck quantum charge q_p may be expressed as a unit of space-time area such as the area of a spatiotemporal plane with a length equal to the Planck length and a width equal to the Planck time. The acceleration of half the unit of area of the unit of charge is equal to π.

$$q_p^2 = t_p^2 l_p^2 \tag{1.3}$$

$$q_p = t_p l_p \approx 8.713036182 \text{ x } 10^{-79} \; m \cdot s \tag{1.4}$$

The following illustrations would help visualize a Planck quantum of charge and quantum light.

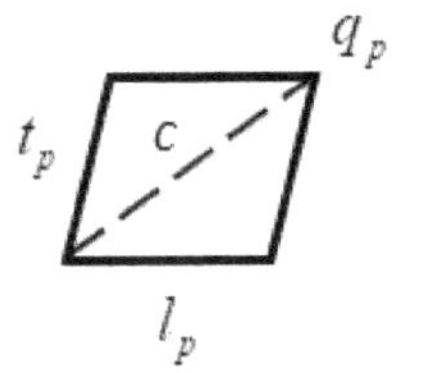

Figure 1.

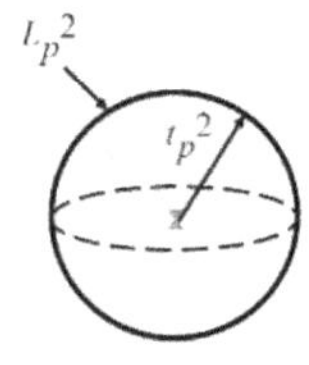

Figure 2.

The Planck Coulomb unit force near an object of mass would be normal to the center point of the unit of area of the temporal plane and would point toward the center of gravitational acceleration of the mass of the object. This is the direction in which the temporal wave expands toward the mass of the object. This force is related to the electric field of a tube of force.

$$F_q = \frac{q_p^2}{4\pi\varepsilon_0 l_p^2} = \frac{(t_p l_p)^2}{4\pi\varepsilon_0 l_p^2} = \frac{t_p^2}{4\pi\varepsilon_0} \tag{1.5}$$

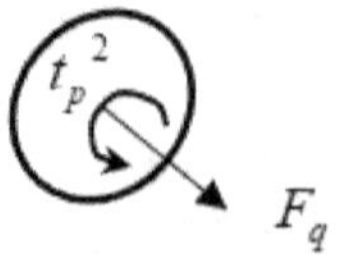

Figure 3.

Thus, the free space-time permittivity may be expressed as

$$\varepsilon_0 = \frac{1}{4\pi}\frac{t_p^2}{F_q} \tag{1.6}$$

To express the Biot-Savart Law for a Planck unit charge in magnitude form we have

$$\vec{B} = \frac{\mu_0 q_p}{4\pi}\vec{v} \times \frac{\vec{r}'}{|r'|^2} \tag{1.7}$$

$$B = \frac{F_q}{l_p^2} = \frac{\mu_0 (t_p l_p)}{4\pi}\left(\frac{l_p}{t_p}\right)\left(\frac{1}{l_p^2}\right) = \frac{\mu_0}{4\pi} \tag{1.8}$$

$$F_q = \frac{\mu_0 l_p^2}{4\pi} \tag{1.9}$$

The free space-time permeability is given by

$$\mu_0 = 4\pi\frac{F_q}{l_p^2} \tag{1.10}$$

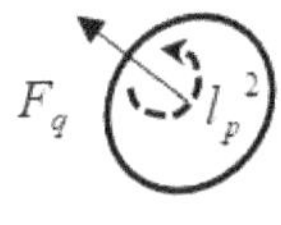

Figure 4.

The Planck Coulomb unit force near an object of mass would be normal to the center point of the unit of area of the spatial plane and would point away from the center of gravitational acceleration of the mass of the object. This is the direction in which the spatial wave expands away from the mass of the object. This force is related to the magnetic field of a tube of force. The speed of light may be expressed in terms of the permittivity and permeability of free space-time.

$$\varepsilon_0\mu_0 = \left(\frac{1}{4\pi}\frac{t_p^2}{F_q}\right)\left(4\pi\frac{F_q}{l_p^2}\right) = \frac{t_p^2}{l_p^2} = \frac{1}{c^2} \qquad (1.11)$$

The photon, or light quantum, may be represented as the linearly polarized complex spatiotemporal unit of area oscillating at an angle, for instance at forty-five degrees, from the vertical plane in the direction of propagation of the electromagnetic field, with an angular frequency about the axis of force, and with a sense of propagation and velocity in the direction of the linear frequency, out of the page.

Figure 5.

The laterality of either the temporal or spatial unit area of charge is the same about the Coulomb force which gives the photon a neutral charge. We may consider the temporal unit area as the future projection of its spatial unit area. The linearly polarized spatiotemporal wave of the photon resolves into Maxwell's Electric and Magnetic fields at right angles to each other with two equal and opposite charges. The energy of the spatiotemporal wave of the photon is given by

$$E = h\upsilon \tag{1.12}$$

Where h is the momentum of the area as it expands or contracts over time and υ is the frequency of the spatiotemporal wave. The Laplace force on a Planck unit charge moving perpendicular to a magnetic field is given by

$$\vec{F}_p = q_p \vec{v} \times \vec{B} \tag{1.13}$$

Substituting for the Planck unit charge in terms of the unit of area of space-time into the equation for the Laplace force to find the magnitude of the Planck force we get

$$\left|\vec{F}_p\right| = q_p vB = \left(t_p l_p\right)\left(\frac{l_p}{t_p}\right)\left(\frac{F_p}{l_p^{\,2}}\right) = F_p \tag{1.14}$$

The volumetric acceleration of space-time on the spherical surface of an object of Planck mass may be expressed in terms of Planck units of mass, length, time, and unit charge as follows:

$$\hbar c = \frac{q_p^{\,2}}{4\pi\varepsilon_0} = m_p \ddot{a} \tag{1.15}$$

$$\ddot{a} = \frac{q_p^{\,2}}{4\pi\varepsilon_0 m_p} = \frac{t_p^{\,2} l_p^{\,2}}{4\pi\left(\frac{t_p^{\,2} l_p^{\,2}}{F_q l_p^{\,2}}\right) m_p} = \frac{F_q l_p^{\,2}}{4\pi m_p} = \frac{m_p\left(\frac{\partial^2 l_p}{\partial t_p^{\,2}}\right)}{4\pi\left(\frac{\partial^2 m_p}{\partial l_p^{\,2}}\right)} \tag{1.16}$$

Thus, the acceleration of space-time about a spherical mass of a charged object, with a radius of Planck length, is directly proportional to the Planck force and indirectly proportional to the radial acceleration of the mass with respect to its radius about the geometry of the mass of the object.

§ 2. Electromagnetic Tubes of Force

The Planck unit force discussed above is acting normal to a unit of area of space or time for either a magnetic tube of force or an electric tube of force. A tube of force is the projection of the unit area of space or time in the direction of linear frequency of the wave of charge. As the distance from the center of gravity increases, the temporal wave contracts and the spatial wave expands. The electric tubes of force contract since they are made of temporal volume with an angular frequency that determines the type of charge, positive or negative, of the tube, in the direction of linear frequency. Similarly, as the distance from the center of gravity increases, the magnetic tube expands as the spatial wave expands with an angular frequency that determines type of charge, in the direction of linear frequency. There is pressure, normal force per unit of area, at right angles to tubes of force of one-half the product of the dielectric and diamagnetic density. The dielectric or diamagnetic density is equivalent to the number of electric or magnetic tubes of force per unit area of surface.

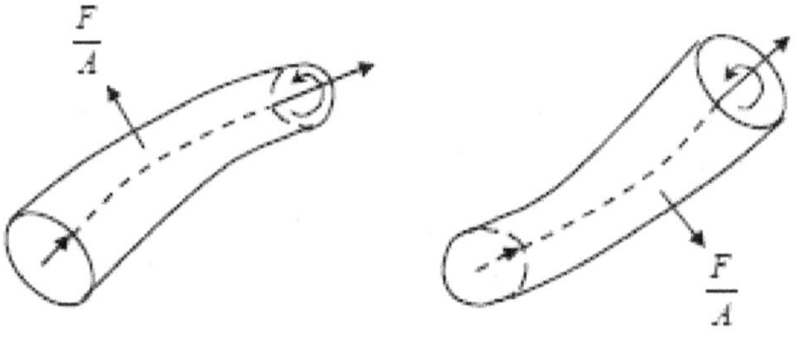

Figure 6.

The total pressure, or force per unit area, at right angles to tubes of force is given by

$$\frac{F_{Total}}{A} = \frac{1}{A}\left(F_{\varepsilon_0} + F_{\mu_0}\right) \tag{2.1}$$

Counter-rotating tubes of force attract and co-rotating tubes of force repel. By convention, all positive tubes of force have counterclockwise rotation and negative tubes of force have clockwise rotation. Spatial or temporal tubes may be either positive-counterclockwise rotation or negative-clockwise rotation in the direction of linear frequency. Tubes traveling in the same direction and with the same rotation repel. Parallel spatial or magnetic tubes of force traveling in opposite directions with opposite rotation tend to unite with each other and form into a single tube traveling in a direction determined by the magnetic poles creating the tubes of force.

§ 3. The Photon or Light Quantum

The photons that consist of the spatial or temporal unit of area of space-time are virtual, the difference between their energy and momentum is equivalent to zero rest mass, $E = pc$. If there is a difference in the energy and momentum of the photon, when the angular frequency is conserved and the linear frequency increases during compression of its wavelength, the photon would gain relativistic mass or virtual mass.

$$m_{rel}^{\ 2}c^4 = E^2 - p^2c^2 \tag{3.1}$$

As a virtual photon, or an elementary particle, gains relativistic mass, the spatiotemporal curvature about the photon, or particle, increases. Photons, or particles, consist of gluons, or comparatively, in very simple terms, points of energy suspended in space-time.

Thus, the energy of relativistic mass, E_{rel}, may be much greater than the energy of rest mass, or rest point(s) of energy, ∂E. The relativistic mass may be expressed as the product of the rest mass to the ratio of the relativistic curvature, R_{rel}, produced by motion, to the rest curvature, R.

$$m_{rel} = \frac{\partial E}{c^2} \cdot \frac{R_{rel}}{R} \tag{3.2}$$

$$E_{rel} = m_{rel} \cdot c^2 = \partial E \cdot \frac{R_{rel}}{R} \tag{3.3}$$

$$\frac{E_{rel}}{\partial E} = \frac{R_{rel}}{R} \tag{3.4}$$

Virtual photons are constantly emitted and reabsorbed by their source. A charged object of mass with an electric, and a magnetic field, is surrounded by photons, constantly being emitted and reabsorbed by their source.

Photons, real and virtual, are emitted and absorbed by charged particles, even though Photons are neutral. Photons are the carriers of the spatiotemporal or electromagnetic field only interacting electromagnetically with charged particles or objects, not with other photons. Photons do not interact with magnetic fields. Photons which make up electric or magnetic fields are not charged so other photons do not interact with them. The neutrality of a photon is the result of the zero-resultant force of the complex spatiotemporal surface.

A free and ordinary photon has two polarization states that correspond to the positive and negative charge orientation in space-time as can be easily demonstrated by experiment. An electron may interact with a free photon regardless of its source. The electron would scatter. If the photon is virtual, such as when two close electrons experience Coulomb-like forces, the photon has an additional longitudinal state of polarization in the linear frequency direction related to the sign of the charge of its source. Thus, an interacting electron receiving the photon is able to attract or repel

The approaching charged particle or object and has advanced information to warn off an incident and opposite charge.

The spatiotemporal force of a photon may be expressed as

$$\left|\vec{F}_{ST}\right| = \left|t_p l_p \left(\vec{E} + \vec{v} \times \vec{B}\right)\right| = \left| t_p l_p \left[\frac{F_q}{t_p l_p} + \left(\frac{l_p}{t_p} \times \frac{F_q}{l_p^{\,2}} \right) \right] \right| = 2\left|F_q\right| \qquad (3.5)$$

There are two equal and opposite forces related to the state of a free photon in space-time. A stationary charged object of mass experiences a temporal force but a moving object of mass experiences both a temporal and spatial force from the spatiotemporal (electromagnetic) field. Disturbances in the wave medium may unbalance the equilibrium of the spatiotemporal unit of area which results in pair production of free charges and potential energy.

§ 4. The unity field potential: the scalar, electric, magnetic, and gravitational fields

The spatiotemporal wave underlies the electromagnetic field and the gravitational field as the infrastructure of the wave medium of isotropic and homogeneous space-time. The spatiotemporal wave in its pure form is a scalar wave devoid of electromagnetic waves. All other field potentials or fields of force, electric, magnetic, and gravitational fields, originate from perturbations (such as curvature) of the state of the spatiotemporal wave that emerges in isotropic and homogeneous space-time.

The scalar wave medium is the spatiotemporal wave medium of all fields of forces and field potentials. The purely spatiotemporal or scalar wave medium is associated with a unique field measurement value that represents the conditions, attributes, including but not limited to phase, and magnitude of space and time at every point of measurement. The degree and type of perturbation or disturbance (such as curvature) of isotropic and homogeneous space-time and the presence of matter and charges, manifests the fields of force or field potentials that emerge in the spatiotemporal infrastructure of the medium.

Thus, if an electric or a magnetic potential exists in the scalar (spatiotemporal) wave medium, the gradient of that disturbance represents the electric field of force or the magnetic field of force. The gradient points in the direction where the field of force increases or decreases most with field potential. The potential energy of the field potential may be converted to kinetic energy as a charged object of mass or particle enters the field potential and moves toward or away from a source or sink. The kinetic energy of the moving charge is able to exert a force as its moves through a distance; thus, in this manner, the field potential can contribute to a force. However, if the field potential diverges or converges from its source as a function of distance, it originates a field of force such as an electric field, a magnetic field, or a gravitational field.

The field potential is underlain by, or superimposed upon, the scalar field, which diverges or converges as a function of distance. *The unity field potential of the scalar wave field unifies all fields of force, field potentials, and the spatiotemporal or scalar wave medium.*

Field Function	*Electromagnetism*	*Gravitation*	*Weak Interaction*	*Strong Interaction*	*Scalar Wave Medium*	*Unity Field*
Fields of Force	$F^{\varepsilon\beta}$	$F^{\mu\nu}$	$W^{\varepsilon\beta}$	$S^{\varepsilon\beta}$	$\Psi^{\mu\nu}$	$F^{\alpha\omega}$
Field Potentials	V^{00}	U^{00}	W^{00}	S^{00}	Ψ^{00}	P^{00}

Figure 7.

The Unity Field Force may be expressed as

$$F^{\alpha\omega} = F^{\varepsilon\beta} + F^{\mu\nu} + W^{\varepsilon\beta} + S^{\varepsilon\beta} + \Psi^{\mu\nu} \tag{4.1}$$

and the Unity Field Potential is given by

$$P^{00} = V^{00} + U^{00} + W^{00} + S^{00} + \Psi^{00} \tag{4.2}$$

The electric field has a temporal nature, so if there is a time-variant cyclic disturbance in the scalar wave medium about either a time variant or time invariant continuous charge distribution, then there would be a time-variant field potential also, and a resulting time variant electric field would emerge. The time-variant electric field of the time-variant charge distribution would increase or decrease more rapidly as a function of distance than the electric field of the time invariant charge distribution, but the direction of either time variant electric field would be the direction of the time variant scalar field.

The gravitational field and the electromagnetic field share a similar time variant relationship of divergence and convergence with space and time because the infrastructure of all fields of force is the scalar wave field. The scalar wave field of a gravitational field of a system of mass is time variant because a body of matter, as well as a continuous charge distribution, may change over time, through radiation, accretion, or depletion of matter or charge.

The direction of the electric field would depend on the sign of the charge: positive or negative. For a positive continuous charge distribution, the direction of the electric field is outward, away from the source, and for a negative continuous charge distribution the direction of the electric field is inward, toward the sink. The negative charge distribution, or accretion of electrons, constitute an increase in mass and in the gravitational field coming from the conductor, which in turn, dilate the temporal scalar waves about the conductor.

The temporal dilation of the scalar waves diverges the scalar field potential, and the voltage potential, toward the sink, so the electric field points inward toward the sink. For a positive charge distribution, or depletion of electrons, the opposite effect takes place.

The divergence of the temporal scalar field waves toward the sink also implies an accretion of negative charges on the conductor that would produce a divergent voltage potential toward the sink. As the temporal scalar waves diverge, the spatial scalar waves converge and the negative charge distribution density on the conductor converges to strengthen the electric field.

The Maxwell–Faraday equation states that a non-conservative

spatially-varying electric field is always accompanied by a time-varying magnetic field, and vice versa. We can express this relation as the relation between the temporal scalar wave field (the electric field) varying over space with the spatial scalar wave field (the magnetic field) varying over time in the scalar wave medium.

$$\nabla \times \vec{E} = -\frac{\partial \vec{B}}{\partial t} \tag{4.3}$$

$$\frac{\partial\left(\nabla \times \vec{E}\right)}{\partial t} = -\frac{\partial^2\left(\nabla \times \vec{A}\right)}{\partial t^2} \tag{4.4}$$

Moreover, the time-varying curl of the temporal scalar wave field (the electric field) equals the decelerative curl of the spatial scalar wave vector potential (the magnetic vector potential). The spatial scalar wave vector potential is the phase shift exerted by a spatial scalar wave on a unit charge moving through space. By the Aharonov–Bohm effect, if a particle with charge Q moves along a path from one point to another, the spatial scalar wave vector potential causes the phase of its quantum-mechanical wave function to shift. Therefore, the phase of the wave function depends upon the magnetic vector potential, and it has been observed experimentally that it also depends on the scalar electric potential as a particle moves through time.

Thus, the velocity of the time-varying curl of the temporal scalar wave field (the electric field) equals the deceleration of the phase shift on a unit charge moving through space-time. The temporal scalar wave potential times charge is scalar potential energy, and the Schrödinger equation tells us that the energy of the particle is the rate at which its phase evolves with time.

The temporal scalar potential (scalar electric potential) is the phase shift exerted by a temporal scalar wave on a unit charge moving through time. Thus, the spatial scalar wave vector potential (the magnetic vector potential) bears the same relationship to space as the temporal scalar potential (the electric scalar potential) bears to time.

For the near-field close to antennae, or through electromagnetic induction, we obtain:

$$\nabla \times \vec{B} = \mu_0 \varepsilon_0 \frac{\partial \vec{E}}{\partial t} \tag{4.5}$$

$$\nabla \times \vec{E} = \nabla \times v\vec{B} = v(\nabla \times \vec{B}) = v\left(\mu_0 \varepsilon_0 \frac{\partial \vec{E}}{\partial t}\right) = -\frac{\partial \vec{B}}{\partial t} \tag{4.6}$$

Since the velocity of the electromagnetic field in the far-field is the velocity of light, we obtain

$$c\left(\mu_0 \varepsilon_0 \frac{\partial E}{\partial t}\right) = c\left(\frac{1}{c^2}\frac{\partial E}{\partial t}\right) = \left(\frac{1}{c}\frac{\partial E}{\partial t}\right) = \frac{\partial B}{\partial t} \tag{4.7}$$

$$\frac{\partial E}{\partial t} = \left(\frac{\partial s}{\partial t}\right)\frac{\partial B}{\partial t} \tag{4.8}$$

$$\left(\frac{\partial s}{\partial t}\right) = \frac{\partial E}{\partial B} \tag{4.9}$$

In the equation above we can clearly see the magnitude relation in the far-field between the scalar wave medium (spatiotemporal wave medium) and the emerging electromagnetic field. The relation of electromagnetism and six-dimensional space-time wave medium is shown below.

$$\frac{\partial \vec{B}}{\partial t} = \frac{\partial \vec{E}}{\partial s} \tag{4.10}$$

$$\frac{\partial \vec{B}_X}{\partial t_X} + \frac{\partial \vec{B}_Y}{\partial t_Y} + \frac{\partial \vec{B}_Z}{\partial t_Z} = \frac{\partial \vec{E}_X}{\partial x} + \frac{\partial \vec{E}_Y}{\partial y} + \frac{\partial \vec{E}_Z}{\partial z} \tag{4.11}$$

In terms of the temporal and spatial divergence using the Tem Operator and the Del operator we may express the above equation as

$$\vec{\Re}_{\tau} \cdot \vec{B} = \nabla \cdot \vec{E} \tag{4.12}$$

Therefore, there is reason to believe, from the phenomena of light, that space-time is the surrounding wave medium and exists in the plenum of massive objects, capable of expansion and exerting motion to mass and energy through displacement of its spatiotemporal dimensions.

Thus, the attributes of potential energy and related field of force propagate and undulate as the spatiotemporal wave medium, at every point, expands in all its dimensions, as measured by our instruments or evidenced by our senses. The spatiotemporal (scalar) wave medium is highly malleable, adjustable, and underlies perturbations of the potential energy fields and fields of forces of homogeneous charge distributions that may be sustained or altered as electromagnetic or scalar conditions change. (Maxwell, 1865)

The forces acting between masses, or between homogeneous charge distributions, may be considered as directly related to the scalar wave medium and not treated with reference only to the condition of the bodies, their relative positions, and momenta.

In a sense, the scalar wave medium has been bereft of all mechanical and kinetic properties, but unequivocally has a very significant share in determining mechanical and electromagnetic occurrences, as it is in a constant state of motion. For a force, time is cause, motion is effect.

Time is the first thing to happen and the last. The spatiotemporal wave medium is the understructure and the prime mover for work that traditionally has been considered done by mass or energy in the dimensions of the medium. Time creates the space to allow things to happen elsewhere. Expanding space-time is that which makes motion possible.

If there are two nearby oppositely charged distributions, then an electric dipole would emerge.

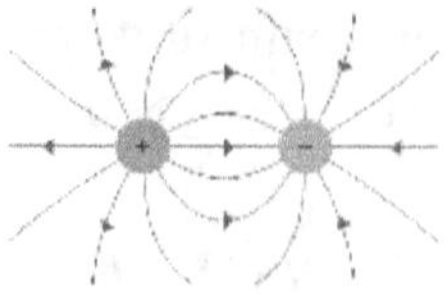

Figure 8.

If the scalar wave potential, and as a result the field potential is omnipresent, time-invariant or uniform, then in such field potential, or voltage potential, an electric field of force would not emerge, ceteris paribus. The voltage potential difference would be zero between any two points of the field potential that are measured.

The magnetic field has a spatial nature, so if there is a circular rotating and cyclic disturbance, in the scalar medium, then there would be a curl in the field potential also, and a resulting magnetic curl would emerge. The direction of the magnetic curl would follow Fleming's rules. The intensity of the magnetic field would increase in a converging scalar field as a function of distance, and it would decrease in a diverging scalar field. The mass of a permanent magnet has a gravitational field that diverges as a function of distance from the mass. A permanent magnet is a magnetic dipole, having a north pole and a south pole, because of the resultant effect of all its aligned and spinning internal magnetic domains that exists in the plenum of the mass of the magnet. These magnetic domains are polarized with spatial scalar wave quanta.

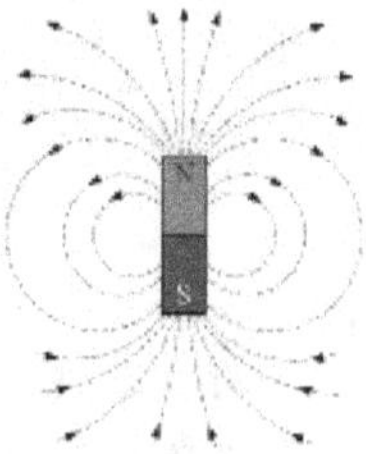

Figure 9.

A permanent magnet has a north pole and a south pole, magnetic tubes of force are scalar in nature and move from the north pole to the south pole of the magnet. There is a divergent scalar field at the north pole of the magnet originating a diverging magnetic field of force, and there is a diverging scalar field at the south pole that resembles a converging field from the perspective of the tubes of force from the north pole. Thus, tubes of force from the north pole and the south pole of the same permanent magnet have opposite spin and are chiral, from the perspective of each pole, so they are able to connect in the scalar wave medium, to complete the magnetic circuit.

In the gravitational field of a celestial body, the spatial scalar wave diverges from the mass as a function of distance, while the temporal scalar wave converges as a function of distance. Then, an electromagnetic wave propagating radially away from the celestial body of mass would have a gradually increasing intensity of its electric field due to the convergence of the temporal scalar wave as a function of distance and a gradually decreasing intensity of the magnetic field as a function of distance due to the divergence of the spatial scalar wave in a gravitational field. The overall momentum of the combined scalar wave, the spatiotemporal wave, is conserved.

In the diagram below, each circle represents a particular value of scalar potential for the spatial scalar field (green) and the temporal scalar field (blue). Circles of different sizes indicate divergence or convergence from the surface of mass. The gradient points in the direction where the scalar field of force increases or decreases most with scalar field potential. The diagram is only an approximate visualization aid.

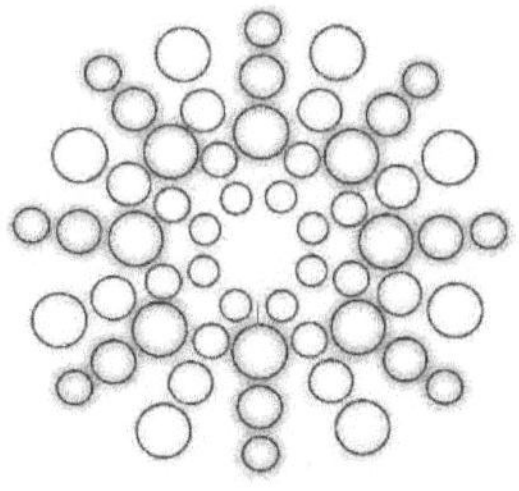

Figure 10.

The gradient of the gravitational field is the gradient of the temporal scalar field near a massive object. The scalar (spatiotemporal) wave medium expands at every point in space-time. Time is a form of compressed scalar wave field that expands through compression and rarefaction of space and time near massive bodies. Space and time create field potential that may convert a form of energy into another.

In that sense, space or time are energy producers. The combined scalar wave of the spatiotemporal wave medium sets the speed of light to "*c*" when and where the temporal scalar wave and the spatial scalar wave act as conjugate waves. The temporal scalar wave dilates as it nears a massive object decreasing its frequency as a function of distance (space). The spatial scalar wave compresses as it nears a massive object increasing its frequency as a function of time.

This effect results in a gravitational field near and about the massive object. In other words, this spatiotemporal effect results in curvature. However, space and time exist even in the absence of mass. Thus, *time is emergent and creates more space, which in turn allows the emergence of more time.*

The flow of charges, such as electrons, through a conducting medium, a conductor, is slower than the speed of light. Sometimes a lot slower than the speed of light due to electron drift, physical characteristics of the conducting material, and other attributes and conditions that may be involved.

Thus, if the combined scalar wave moves at the speed of light, the speed of flow of current, not necessarily the flow of individual electrons, lags the spatiotemporal (scalar) wave.

Consequently, the flow of charges shortens the scalar wave length at less than the speed of light as massive charges may be moving at relativistic speeds, and in that sense, compress space-time in the direction of propagation.

The combined scalar wave is longitudinal because both of its components (spatial and temporal) are also longitudinal. This is also the reason why gravitational waves are longitudinal.

A spherical antenna (a Tesla antenna), as in figure 11, is a longitudinal wave antenna that can transmit longitudinal electric scalar waves.

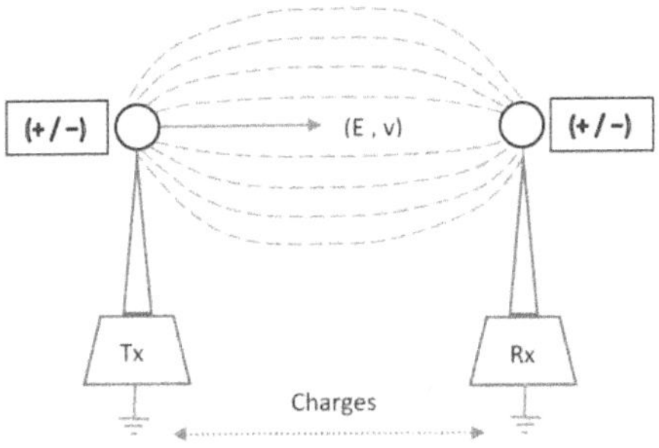

Figure 11.

A spherical antenna transmits oscillating longitudinal scalar waves as it oscillates the temporal scalar field potential around the geometry of the antenna of the transmitter at spatiotemporal resonance with the oscillating temporal scalar field potential at the antenna of the receiver at the speed and critical frequency of the spatiotemporal (scalar) wave. The charges would flow through the ground conducting medium and oscillate between antennae. At larger distances, the Earth ground conductor may work as a vast reservoir of negative charges that may be pumped up or drained down proportionally by the oscillating scalar field potential (voltage potential) of the system as the energy is conserved during every cycle. The entire electroscalar system may involve more than one receiver with a single transmitter.

Since the scalar fields are decelerative or accelerative flows of space and time, a diverging or converging scalar field may have within it an expansive or compressive, electric or magnetic, component. The combined scalar field manifests an electromagnetic wave. The spatial scalar pressure influences the density of continuous charge distributions increasing the intensity of a field of force. The scalar wave medium interacts with matter at the boundary and within the plenum of matter. *Space and time are conjugates in space-time. Space exists from time, and time exists from space.* As a temporal scalar wave dilates to a maximum, a spatial scalar wave compresses to a minimum. The event horizon of supermassive black hole is an ideal example of a fully dilated temporal wave in the scalar wave medium, or spatiotemporal wave medium, of the black hole.

§ 5. The wave medium of space-time as a field of force

As matter is converted to pure energy in free space-time, the energy of that matter returns to its source as it disperses in the wave medium of space-time. Mass and energy are physical forms of space-time. Space-time is the quintessence of physical forms. All forms of energy and momentum in six-dimensional space-time may be represented as forces.

$$\Psi(r,t)=\Psi e^{i\left(\frac{pr}{\hbar}-\frac{Et}{\hbar}\right)}=\Psi e^{i\left(\frac{pr}{\hbar}-\omega t\right)} \tag{5.1}$$

$$\frac{\partial\psi}{\partial t}=-i\omega\Psi \tag{5.2}$$

$$-\frac{1}{i}\frac{\partial\psi}{\partial t}=\omega\Psi \tag{5.3}$$

If the coefficient of the growth factor in the time-dependent equation of the wave function equals zero, the wave function is massless and results in the spatiotemporal wave itself. Furthermore, the change in the wave function with respect to time is entirely in the dimensions of time, and equal to the product of the amplitude and the angular frequency, in the direction of the linear frequency of the temporal wave function. Let us express the variable, Ψ, of the function as the temporal volume, a_t, as it expands or contracts.

$$i\frac{\partial\Psi}{\partial t}=i\dot{a}_t=\omega\Psi \tag{5.4}$$

Thus, the wave function embodies the three-dimensional property of time in the medium of space-time in the direction of the linear frequency axis.

Figure 12.

The expansion or contraction of space-time may be expressed three-dimensionally in space and time.

$$\frac{\partial^2 \Psi}{\partial t^2} = c^2 \frac{\partial^2 \Psi}{\partial r^2} \tag{5.5}$$

$$\frac{1}{c^2}\left[\frac{\partial^2 \Psi}{\partial t_x^{\ 2}} + \frac{\partial^2 \Psi}{\partial t_y^{\ 2}} + \frac{\partial^2 \Psi}{\partial t_z^{\ 2}}\right] = \frac{\partial^2 \Psi}{\partial x^2} + \frac{\partial^2 \Psi}{\partial y^2} + \frac{\partial^2 \Psi}{\partial z^2} \tag{5.6}$$

$$\ddot{a}_t = \frac{E}{m}\ddot{a}_S \tag{5.7}$$

The three-dimensional temporal acceleration is equivalent to the product of the energy-to-mass ratio of an object and the three-dimensional spatial acceleration about the object in space-time. In case of a charged object moving in the direction of the temporal (electric) field, the momentum vector of the charged object follows the direction of the linear frequency of the field with a magnitude proportional to the magnitude of the field. The spatial (magnetic) field rotates the charged object of mass and its momentum vector in the direction of rotation of the axis of angular frequency. A combination of a rotation and a Lorentz boost on a spatiotemporal coordinate system is a proper homogeneous Lorentz transformation. The spatial and temporal fields transform into each other with the change of reference frame as the Lorentz transformation is frame-of-reference dependent. A Lorentz boost in one frame of reference may be represented as a combination of a rotation and a Lorentz boost in another frame of reference.

A charged massive object experiences a force in a spatial field when it is moving, perpendicular to the spatial field and to the velocity of the tubes of force, while it would experience a force in the direction of a temporal field even when it is stationary. Thus, co-moving charged particles or co-moving charged massive objects would not experience a Lorentz force between them because of their spatial (magnetic) field, but there still is pressure at right angles to their tubes of force.

The curvature of space about an object of mass is the direct result of the expansion of the spatial wave as the temporal wave contracts as the distance increases outward from the center of gravitational acceleration of the object of mass. The wave medium acts on the physical form of the object of mass through the spatial and temporal wave actions. The spatial wave expands giving the space about the object the physical curvature about its form. Curvature of space does not create gravity.

The gravitational acceleration is the result of the spatiotemporal wave action upon the form of the object of mass as the spatiotemporal wave contracts and decelerates. Space-time is in motion as an isotropic and homogeneous spatiotemporal wave medium that expands or contracts, in which matter, particles, charges, and tubes of force, exist.

§ 6. The wave function equation

The wave function of space-time embodies the properties of wave motion in the spatiotemporal medium. The wave function equation is a very important differential equation that describes the propagation of a wave through its medium at a velocity equal to, or less than, the speed of light.

The eminent and contemporary physicists Jean le Rond d'Alembert and Leonhard Euler discovered the one-dimensional and three-dimensional spatial wave equations. Based upon their concepts, let us describe the role of energy in the wave equation of a three-dimensional spatial or temporal wave traveling through a six-dimensional wave medium.

Let us imagine a particle traveling through the expanding medium of space-time where the particle of mass has a predicted probability distribution given by its position and time for measurement results, within a homogeneous and isotropic spherical space-time wave function, $\Psi(r,t)$, emerging from a point source. The wave function embodies the momentarily attained sum of theoretically based future expectation of the probability distribution of a particle in the volume of space-time.

$$\Psi(r,t) = \Psi e^{i(pr/\hbar - Et/\hbar)} \tag{6.1}$$

Differentiating the space-time wave function with respect to time we get

$$\frac{\partial \Psi}{\partial t} = -i\omega\,\Psi \tag{6.2}$$

$$\frac{\partial^2 \Psi}{\partial t^2} = -i\omega\frac{\partial \Psi}{\partial t} = i^2\omega^2\Psi = -\omega^2\Psi \tag{6.3}$$

Differentiating the space-time wave function with respect to space we get

$$\frac{\partial \Psi}{\partial r} = i\frac{p}{\hbar}\Psi \tag{6.4}$$

$$\frac{\partial^2 \Psi}{\partial r^2} = i\frac{p}{\hbar}\frac{\partial \Psi}{\partial r} = i^2\frac{p^2}{\hbar^2}\Psi = -\frac{p^2}{\hbar^2}\Psi \tag{6.5}$$

Let us assume the total energy of the particle to be equal to the sum of its kinetic energy and its potential energy given by

$$\hbar\omega = \frac{1}{2}mv^2 + \frac{1}{2}kx^2 = \frac{p^2}{2m} + U(r,t) \tag{6.6}$$

The potential energy describes the forces acting on the moving particle. Multiplying both sides of the energy equation by a volume of space-time, Ψ , which represents the amplitude of the spherical wave function we obtain

$$\hbar\omega\Psi = \frac{p^2\Psi}{2m} + U(r,t)\Psi = \frac{\hbar^2}{i^2 2m}\frac{\partial^2\Psi}{\partial r^2} + U(r,t)\Psi \qquad (6.7)$$

$$i\hbar\frac{\partial\Psi}{\partial t} = -\frac{\hbar^2}{2m}\frac{\partial^2\Psi}{\partial r^2} + U(r,t)\Psi \qquad (6.8)$$

$$\frac{\partial^2\Psi}{\partial r^2} = -\frac{8\pi^2 m}{h^2}\left(i\hbar\frac{\partial\Psi}{\partial t} - U(r,t)\Psi\right) = -\frac{8\pi^2 m}{h^2}(E\Psi - U\Psi) \qquad (6.9)$$

Expanding the acceleration of the volume of space-time, first, in three spatial dimensions, and secondly, in three temporal dimensions we have

$$\frac{\partial^2\Psi}{\partial x^2} + \frac{\partial^2\Psi}{\partial y^2} + \frac{\partial^2\Psi}{\partial z^2} = \frac{1}{c^2}\left(\frac{\partial^2\Psi}{\partial t_X^{\ 2}} + \frac{\partial^2\Psi}{\partial t_Y^{\ 2}} + \frac{\partial^2\Psi}{\partial t_Z^{\ 2}}\right) = -\frac{8\pi^2 m}{h^2}(E-U)\Psi \qquad (6.10)$$

Note that the three-dimensional wave function may be expressed as a spatial wave or as a temporal wave in six-dimensional space-time. The three-dimensional temporal wave function represents the emerging probability distribution for a moving particle of mass as a function of energy.

$$\frac{\partial^2\Psi}{\partial t_X^{\ 2}} + \frac{\partial^2\Psi}{\partial t_Y^{\ 2}} + \frac{\partial^2\Psi}{\partial t_Z^{\ 2}} = -\frac{8\pi^2 mc^2}{h^2}(E-U)\Psi = -\frac{2}{\hbar^2}\left(E^2 - EU\right)\Psi \qquad (6.11)$$

The three-dimensional temporal wave is equivalent to the product of the square of the speed of light and the three-dimensional spatial wave in homogeneous and isotropic space-time. Collapsing the acceleration of the volume of the three-dimensional temporal wave to simplify the wave function equation in resultant coordinate time we obtain

$$\frac{\partial^2\Psi}{\partial t^2} = -\frac{2}{\hbar^2}\left(E^2 - EU\right)\Psi \tag{6.12}$$

$$-\hbar^2\omega^2\Psi = -2\left(E^2 - EU\right)\Psi \tag{6.13}$$

$$-E\Psi = -2\left(E\Psi - U\Psi\right) \tag{6.14}$$

$$\hat{H}\Psi = \left(2U\Psi\right) \tag{6.15}$$

Quantum-mechanical phase invariance, known more properly as gauge invariance, is a symmetry that underlies all modern quantum theories. It is basically a statement that action Lagrangians are invariant with respect to the replacement $e^{i\omega\tau}$.

$$\Psi\left(c\tau\right) \rightarrow e^{i\omega\tau}\Psi\left(c\tau\right) \tag{6.16}$$

$$e^{i\omega\tau}\Psi\left(c\tau\right) \approx \left(1+\omega\tau\right)\Psi\left(c\tau\right) \tag{6.17}$$

Where $\Psi\left(c\tau\right)$ is a wave function, $\Psi\left(\omega t\right)$ is an arbitrary function of space-time, and the coefficient $i\omega\tau$ is the growth factor of the temporal volume representing the emerging temporal wave for the expanding spatiotemporal medium.

At any instant of time, the volume of the emerging temporal wave is the realm of the predicted probability distribution for a particle of mass, as it moves in that region of space-time. Collapsing the spatiotemporal volume $\Psi\left(r\right)$ of the three-dimensional spatial wave, to simplify the wave function equation in resultant coordinate space, we obtain

$$\frac{\partial^2\Psi}{\partial r^2}=-\frac{2m}{\hbar^2}(E-U)\Psi \tag{6.18}$$

$$\hat{H}\Psi=(E-U)\Psi \tag{6.19}$$

6.1. The six-dimensional electrogravitic Dirac equations.

Let us examine the relativistic energy equation for the energy of fermions as proposed by the eminent physicist Paul Dirac,

$$E=\left(\sqrt[2]{1-\frac{v^2}{c^2}}\right)m'c^2+\vec{v}\cdot\vec{p} \tag{6.20}$$

Where E is the energy of the system, m' is the relativistic mass, c is the speed of light, v is velocity, and p is momentum.

Let us express a similar equation in six-dimensional space-time,

$$E\cdot\vec{\Psi}(r,t)=\left[\left(\sqrt[2]{1-\frac{v^2}{c^2}}\right)m'c^2+c\hat{p}\right]\vec{\Psi}(r,t)=i\hbar\left(\vec{\mathfrak{R}}\cdot\vec{\Psi}(r,t)\right) \tag{6.21}$$

$$\left[\beta m'c^2+c\left(\sum_{n=1}^{3}\alpha_n\hat{p}_n\right)\right]\vec{\Psi}(r,t)=i\hbar\left(\vec{\mathfrak{R}}\cdot\vec{\Psi}(r,t)\right) \tag{6.22}$$

Where α_n is a set of spatial six-component Dirac-Pauli matrices, β is a temporal six-component Dirac-Lorentz matrix, $\hbar$ is the reduced Planck constant, i is $\sqrt[2]{-1}$, $\vec{\mathfrak{R}}$ is the Robertonian six-dimensional operator, and $\vec{\Psi}(r,t)$ is the six-dimensional spatiotemporal wave function vector, or spinor, of the system. A spinor describes rotation at a specific spatiotemporal point independently of rotation at any other point in space-time.

$$\vec{\Psi}(r,t)=\sum_{n=1}^{6}\vec{\Psi}_n(r,t)=\begin{vmatrix}\vec{\Psi}_1(t)\\ \vec{\Psi}_2(r)\\ \vec{\Psi}_3(t)\\ \vec{\Psi}_4(r)\\ \vec{\Psi}_5(t)\\ \vec{\Psi}_6(r)\end{vmatrix} \tag{6.23}$$

$$\vec{\Re}=-\frac{1}{c}\frac{\partial}{\partial t_x}\vec{a}_{t_x}+\frac{\partial}{\partial x}\vec{a}_x-\frac{1}{c}\frac{\partial}{\partial t_y}\vec{a}_{t_y}+\frac{\partial}{\partial y}\vec{a}_y-\frac{1}{c}\frac{\partial}{\partial t_z}\vec{a}_{t_z}+\frac{\partial}{\partial z}\vec{a}_z \tag{6.24}$$

Moving all terms to one side of the equation we obtain

$$\left\{i\hbar\vec{\Re}-c\left(\sum_{n=1}^{3}\alpha_n\hat{p}_n\right)-\beta m'c^2\right\}\vec{\Psi}(r,t)=0 \tag{6.25}$$

We may express the energy difference of the above two left terms as a single energy expression, with six-component Dirac-Pauli temporal-spin matrices, or gamma matrices, γ^{ε} , and a three-dimensional spatial momentum matrix, $\hat{p}_n$, with six components.

$$\gamma^{\varepsilon}=\sum_{\varepsilon=1}^{3}\beta\alpha_{\varepsilon}+\sum_{\varepsilon=4}^{6}\beta \tag{6.26}$$

$$\gamma^1=\begin{vmatrix}0&0&0&0&0&1\\0&0&0&0&1&0\\0&0&0&-1&0&0\\0&0&-1&0&0&0\\0&1&0&0&0&0\\1&0&0&0&0&0\end{vmatrix} \tag{6.27}$$

$$\gamma^2=\begin{vmatrix}0&0&0&0&0&-i\\0&0&0&0&i&0\\0&0&0&i&0&0\\0&0&-i&0&0&0\\0&-i&0&0&0&0\\i&0&0&0&0&0\end{vmatrix} \tag{6.28}$$

$$\gamma^3 = \begin{vmatrix} 0 & 0 & 0 & 0 & 1 & 0 \\ 0 & 0 & 0 & 0 & 0 & -1 \\ 0 & 0 & -1 & 0 & 0 & 0 \\ 0 & 0 & 0 & 1 & 0 & 0 \\ 1 & 0 & 0 & 0 & 0 & 0 \\ 0 & -1 & 0 & 0 & 0 & 0 \end{vmatrix} \tag{6.29}$$

$$\hat{p}_n = \begin{vmatrix} 0 & 0 & 0 & 0 & 0 & \hat{p}_6 \\ 0 & 0 & 0 & 0 & \hat{p}_5 & 0 \\ 0 & 0 & 0 & \hat{p}_4 & 0 & 0 \\ 0 & 0 & \hat{p}_3 & 0 & 0 & 0 \\ 0 & \hat{p}_2 & 0 & 0 & 0 & 0 \\ \hat{p}_1 & 0 & 0 & 0 & 0 & 0 \end{vmatrix} \tag{6.30}$$

Thus, $\hat{\beta}^2 = \frac{v^2}{c^2}$, $\beta' = \sqrt[2]{1-\hat{\beta}^2}$, where we may substitute $v_n \to c\alpha_n$, and $\beta' \to \beta$. Since particles and anti-particles are luxons of mass, traveling at *c*, the temporal components of β are ± 1. The temporal six-component Dirac-Lorentz matrix is given by

$$\beta = \gamma^4 = \gamma^5 = \gamma^6 = \begin{vmatrix} 1 & 0 & 0 & 0 & 0 & 0 \\ 0 & 1 & 0 & 0 & 0 & 0 \\ 0 & 0 & -1 & 0 & 0 & 0 \\ 0 & 0 & 0 & -1 & 0 & 0 \\ 0 & 0 & 0 & 0 & 1 & 0 \\ 0 & 0 & 0 & 0 & 0 & 1 \end{vmatrix} \tag{6.31}$$

Simplifying the relativistic energy equation, we have,

$$i\hbar c\gamma^{\varepsilon}\vec{\Re} = i\hbar\vec{\Re} - c\left(\sum_{n=1}^{3}\alpha_n\hat{p}_n\right) \tag{6.32}$$

$$\left(i\hbar c\gamma^{\varepsilon}\vec{\Re} - \beta m'c^2\right)\vec{\Psi}(r,t) = 0 \tag{6.33}$$

The relativistic six-dimensional quantum mechanical wave equation, including electromagnetic interactions, describes all spin-½ massive particles for fermions (all quarks and leptons), that are symmetric under parity, or symmetric if the sign of one spatial coordinate is flipped. This equation is consistent with the Special Theory of Relativity and the Principles of Quantum Mechanics and includes the evolution of three-dimensional time. The equation encompasses six wave equations of motion for an electron, a positron, an electron neutrino, and their anti-particles, submerged in an external electromagnetic field in six-dimensional space-time.

The relativistic six-dimensional equation has six components or states, or six degrees of freedom, for particles and antiparticles, each component is a direction of spin or anti-spin. As predicted by Dirac, each particle, or antiparticle, is always moving at "c" with a trembling motion, $\langle v \rangle = \pm c$, due to weaker Coulombic forces near protons at Compton wavelength distances, which makes the motion appear slower, even though the motion abides by the Special Theory of Relativity.

The single relativistic six-dimensional equation unfolds into six coupled linear first-order partial differential equations for the six components that make up the six-dimensional quantum mechanical wave function. Let us consider the probability flux equation, or probability current equation, an equation describing the flow of probability in terms of probability per unit time per unit area, or linear probability per unit charge, or probability density per unit charge. The probability current is the rate of linear flow of probability per unit of charge. The probability current vector whose component normal to a surface gives the probability that a particle will cross the unit area of a charge on a surface during a unit time. Motion in quantum mechanics is probabilistic. Hence, the motion is how the probability for finding a particle moves around with time. Thus, it is useful to find a probability current, or current density, that relates to how the probability for locating a fermion might be changing with respect to space and time.

The four-dimensional probability current equation for a charged fermion is given by

$$\partial^{\varepsilon} F_{\varepsilon\beta} = J_{\varepsilon} = i\phi \overline{\Psi} \gamma_{\varepsilon} \Psi \tag{6.34}$$

Applying ∂_{ε} to both sides of the above equation,

$$\partial_{\varepsilon} \partial^{\varepsilon} F_{\varepsilon\beta} = \partial_{\varepsilon} \left(i\phi \overline{\Psi} \gamma_{\varepsilon} \Psi \right) \tag{6.35}$$

All matrices are four-by-four components, $\partial_{\varepsilon} \partial^{\varepsilon}$ is the d'Alembert operator, and Ψ has four components.

Let us convert the above four-dimensional equation to the six-dimensional Electromagnetic Dirac Equation for fermions, using redefined elements.

$$\vec{\Re}^2 F_{\varepsilon\beta} = \vec{\Re} \cdot \left(i\phi \vec{\Psi}^* \gamma_{\varepsilon} \vec{\Psi} \right) \tag{6.36}$$

All matrices are six-by-six matrices, $\vec{\Re}^2$ is the Double Robertonian six-dimensional operator, $F_{\varepsilon\beta}$ is the six-dimensional electromagnetic field strength, ϕ is the relative magnitude of the change of phase caused by gauge transformation, $\vec{\Psi}$ is the spatiotemporal spinor field vector with six components, $\vec{\Psi}^*$ is the conjugate spinor field vector with six components, *i* is $\sqrt[2]{-1}$, and γ^{ε} is the six-component Dirac-Pauli temporal-spin matrices, or gamma matrices.

The six-dimensional covariant electromagnetic field strength is

$$F_{\varepsilon\beta} = \begin{vmatrix} 0 & E_{t_y t_x} & -E_{t_z t_x} & -\frac{E_{xt_x}}{c} & -\frac{E_{yt_x}}{c} & -\frac{E_{zt_x}}{c} \\ -E_{t_x t_y} & 0 & E_{t_z t_y} & -\frac{E_{xt_y}}{c} & -\frac{E_{yt_y}}{c} & -\frac{E_{zt_y}}{c} \\ E_{t_x t_z} & -E_{t_y t_z} & 0 & -\frac{E_{xt_z}}{c} & -\frac{E_{yt_z}}{c} & -\frac{E_{zt_z}}{c} \\ \frac{E_{t_x x}}{c} & \frac{E_{t_y x}}{c} & \frac{E_{t_z x}}{c} & 0 & B_{yx} & -B_{zx} \\ \frac{E_{t_x y}}{c} & \frac{E_{t_y y}}{c} & \frac{E_{t_z y}}{c} & -B_{xy} & 0 & B_{zy} \\ \frac{E_{t_x z}}{c} & \frac{E_{t_y z}}{c} & \frac{E_{t_z z}}{c} & B_{xz} & -B_{yz} & 0 \end{vmatrix} \tag{6.37}$$

The six-dimensional electrogravitic tensor, $\Phi_{\varepsilon\beta}$, may be expressed in terms of the product of the six-dimensional Lorentz force, that includes the six-dimensional covariant electromagnetic field strength, $F_{\varepsilon\beta}$, and the six-dimensional electrogravitic curvature tensor, $\widetilde{R}_{\varepsilon\beta}$.

$$F_{\varepsilon\beta} u_{\varepsilon\beta} q_{\chi\chi} \widetilde{R}_{\varepsilon\beta} = 8\pi\Phi_{\varepsilon\beta} \tag{6.38}$$

$$F_{\varepsilon\beta} u_{\varepsilon\beta} q_{\chi\chi} \left(E_{\varepsilon\beta} + iB_{\varepsilon\beta}\right) = 8\pi\Phi_{\varepsilon\beta} \tag{6.39}$$

The six-dimensional Electrogravitic Dirac Equations are given by

$$\vec{\Re}^2 F_{\varepsilon\beta} u_{\varepsilon\beta} q_{\chi\chi} \widetilde{R}_{\varepsilon\beta} = 8\pi\, \vec{\Re}^2 \Phi_{\varepsilon\beta} \tag{6.40}$$

$$\vec{\Re} \cdot \left(i\phi\, \vec{\Psi}^* \gamma_\varepsilon \vec{\Psi}\right) u_{\varepsilon\beta}\, q_{\chi\chi} \widetilde{R}_{\varepsilon\beta} = 8\pi\, \vec{\Re}^2 \Phi_{\varepsilon\beta} \tag{6.41}$$

$$\vec{\Re} \cdot \left(i\phi\, \vec{\Psi}^* \gamma_\varepsilon \vec{\Psi}\right) u_{\varepsilon\beta} \widetilde{R}_{\varepsilon\beta} = \frac{\ddot{a}\, \vec{\Re}^2 \Phi_{\varepsilon\beta}}{q_{\chi\chi} r} \tag{6.42}$$

Where $\ddot{a}$ is the acceleration of space-time, "r" is a radial distance, $\widetilde{R}_{\varepsilon\beta}$ is the six-dimensional electrogravitic curvature tensor, $E_{\varepsilon\beta}$ and $B_{\varepsilon\beta}$ are the electric part and magnetic part of the complex electrogravitic Riemann tensor, $\widetilde{R}_{\mu\nu\varepsilon\beta}$, $q_{\chi\chi}$ is the charge of a particle, and $u_{\varepsilon\beta}$ is a covariant six-velocity of a particle with radial symmetry in spatial and temporal components.

Let us consider the six-dimensional Dirac equations that are consistent with the General Theory of Relativity and the Principles of Quantum Mechanics, that includes the evolution of three-dimensional time. These equations will also retain the invariance under Lorentz transformations.

Let us define six-dimensional metric, $g_{\mu\nu}$, in terms of the sextet, $\widetilde{e}_{\mu\nu}{}^{(a)(b)}$, and the six-dimensional Minkowski metric, η_{ab}, to satisfy six-dimensional Dirac equations,

$$g_{\mu\nu} = \widetilde{e}_{\mu\nu}{}^{(a)(b)} \eta_{(a)(b)} \tag{6.43}$$

In the notation, $\widetilde{e}_{\mu\nu}{}^{(a)(b)}$, (a) and (b) denote which components with respect to an orthonormal basis, and (μ) and (ν) denote which components of $\widetilde{e}^{(a)(b)}$ with respect to a coordinate basis. A sextet is a set of six-dimensional bases chosen for the tangent bundle that is not the one six-dimensional basis that naturally arises from the available coordinate system.

The benefit of applying a sextet field in terms of the General Theory of Relativity is in the application of components of the six-dimensional metric tensor with respect to a six-dimensional orthonormal basis instead of applying components of the same six-dimensional metric tensor with respect to a six-dimensional coordinate basis. The components of the six-dimensional metric tensor, $g_{\mu\nu}$, are usually with respect to a six-dimensional coordinate basis, while the components of $g_{\mu\nu}$ with respect to a six-dimensional orthonormal basis are in the six-dimensional Minkowski metric, $\eta_{(a)(b)}$.

The incentives to introduce sextet terminology begin with how natural observers would measure or express values anywhere in six-dimensional space-time. Moreover, the application of six-dimensional physics improves on spin-½ particles such as fermions, simpler solutions are possible on a more restricted six-dimensional metric, simplified mathematical representations in local inertial reference frames, and the inclusion of curvature, and acceleration, in six-dimensional space-time.

It is interesting to note that a sextet may be reduced to a tetrad. A four-dimensional tetrad is a folded six-dimensional sextet. By applying the summation convention to a pair of dummy indices, λ, that are bound to each other in the expression of a sextet, a basis contraction operation may be performed on a basis that arises from the natural pairing of a six-dimensional basis space and its dual. The indices $[\mu,\nu,(a),(b),c,\sigma,\gamma,\delta]$ are six-dimensional, such that any index value $\in \{1,2,3,4,5,6\}$, or $\in \{-t_x,-t_y,-t_z,x,y,z\}$.

$$e_{\mu}{}^{(a)} = \widetilde{e}_{\mu\lambda}{}^{(a)(\lambda)} \quad \text{where } b = \nu = \lambda \tag{6.44}$$

$$e_{\mu}{}^{(b)} = \widetilde{e}_{\mu\lambda}{}^{(\lambda)(b)} \quad \text{where } a = \nu = \lambda \tag{6.45}$$

A second-order curvature tensor, $R_{\mu\nu}$, on a six-dimensional Lorentzian manifold can be projected onto a six-dimensional Minkowski flat surface, $R_{(a)(b)}$, or vice versa, with the following expressions,

$$R_{\mu\nu} = \tilde{e}_{\mu\nu}{}^{(a)(b)} R_{(a)(b)} \tag{6.46}$$

$$R_{(a)(b)} = \tilde{e}_{(a)(b)}{}^{\mu\nu} R_{\mu\nu} \tag{6.47}$$

The projection is between two different six-dimensional spatiotemporal geometries.

A local Lorentz transformation, $\Lambda^{(a)}{}_{(b)}$, from a flat six-dimensional spatiotemporal spinor, Ψ , to a non-flat six-dimensional spinor, $\tilde{\Psi}$, may be expressed as

$$\tilde{\Psi} = \rho(\Lambda)\Psi \tag{6.48}$$

Where the spinorial expression $\rho(\Lambda)$ for Λ is given by

$$\rho(\Lambda) = 1 + \frac{1}{2} i\varepsilon_{\mu\nu}{}^{(a)(b)} \Sigma^{\mu\nu}{}_{(a)(b)} \tag{6.49}$$

Where $\Sigma^{\mu\nu}{}_{(a)(b)}$ is the six-dimensional spinorial representation of the generators of the six-dimensional Lorentzian transformation in terms of $\gamma_{\sigma}{}^{(c)}$ matrices. (Nakahara, 2003)

$$\Sigma^{\mu\nu}{}_{(a)(b)} = \frac{1}{4} i\left[\gamma^{\mu}{}_{(a)}, \gamma^{\nu}{}_{(b)}\right] \tag{6.50}$$

Let us define a six-dimensional covariant derivative, which is a locally Lorentz invariant vector that transforms as a spinor, with the following transformation restriction.

$$\nabla_{(a)}\Psi \to \rho(\Lambda)\left(\Lambda_{(a)}{}^{(b)}\right)\nabla_{(b)}\Psi \tag{6.51}$$

The covariant derivative of the spinor may be obtained through the following combination,

$$\nabla_{(a)}\Psi = e_{(a)}{}^{\mu}\left(\Diamond_{\mu} + \Omega_{\mu}\right)\Psi \tag{6.52}$$

Where the d'Robertonian Operator $\Diamond$ for six-dimensional space-time is given by

$$\Diamond = -\frac{1}{c}\frac{\partial}{\partial t_x} + \frac{\partial}{\partial x} - \frac{1}{c}\frac{\partial}{\partial t_y} + \frac{\partial}{\partial y} - \frac{1}{c}\frac{\partial}{\partial t_z} + \frac{\partial}{\partial z} \tag{6.53}$$

The transformation operator Ω_{μ} satisfies

$$\Omega_{\mu} \to \rho(\Lambda)\Omega_{\mu}\rho(\Lambda)^{-1} + \Diamond_{\mu}\rho(\Lambda)\rho(\Lambda)^{-1} \tag{6.54}$$

Let us consider an infinitesimal local Lorentz transformation at a point p,

$$\Lambda_{(a)}{}^{(b)}(p) = \delta_{(a)}{}^{(b)} + \varepsilon_{(a)}{}^{(b)}(p) \tag{6.55}$$

to find the explicit form of Ω_{μ}.

By combining terms, the spinor transforms as

$$\Psi \to e^{\left[\frac{1}{2}i\varepsilon^{(a)(b)}\Sigma_{(a)(b)}\right]\Psi} \approx \left(1 + \frac{1}{2}i\varepsilon^{(a)(b)}\Sigma_{(a)(b)}\right)\Psi \tag{6.56}$$

$$i\left(\Sigma_{(a)(b)}, \Sigma_{\gamma\delta}\right) = \eta_{\gamma(b)}\Sigma_{(a)\delta} - \eta_{\gamma(a)}\Sigma_{(b)\delta} + \eta_{\delta(b)}\Sigma_{\gamma(a)} - \eta_{\delta(a)}\Sigma_{\gamma(b)} \tag{6.57}$$

Under the same transformation, the operator transforms as

$$\Omega_{\mu} \to \left(1 + \frac{1}{2}i\varepsilon^{(a)(b)}\Sigma_{(a)(b)}\right)\Omega_{\mu}\left(1 - \frac{1}{2}i\varepsilon^{\gamma\delta}\Sigma_{\gamma\delta}\right) - \frac{1}{2}i\Diamond_{\mu}\varepsilon^{(a)(b)}\Sigma_{(a)(b)}\left(1 - \frac{1}{2}i\varepsilon^{\gamma\delta}\Sigma_{\gamma\delta}\right) \tag{6.58}$$

$$\Omega_\mu \rightarrow \Omega_\mu + \frac{1}{2} i\varepsilon^{(a)(b)} \left[\Sigma_{(a)(b)}, \Omega_\mu \right] - \frac{1}{2} i\lozenge_\mu \varepsilon^{(a)(b)} \Sigma_{(a)(b)} \qquad (6.59)$$

or in components we obtain

$$\Gamma^{(a)}{}_{\mu(b)} \rightarrow \Gamma^{(a)}{}_{\mu(b)} + \varepsilon^{(a)}{}_\gamma \Gamma^\gamma{}_{\mu(b)} - \Gamma^{(a)}{}_{\mu\gamma} \varepsilon^\gamma{}_{(b)} - \lozenge_\mu \varepsilon^{(a)}{}_{(b)} \qquad (6.60)$$

By combining the above equations,

$$\Omega_\mu \equiv \frac{1}{2} i\Gamma^{(a)}{}_\mu{}^{(b)} \Sigma_{(a)(b)} \equiv \frac{1}{2} i\Gamma_{(a)\mu(b)} \Sigma^{(a)(b)} = \frac{1}{2} ie^{(a)}{}_\nu \nabla_\mu e^{(b)\nu} \Sigma_{(a)(b)} \qquad (6.61)$$

$$\Gamma_{(a)\mu(b)} = e_{(a)\nu} \left(\lozenge_\mu e^\nu_{(b)} + \Gamma^\nu_{\mu\lambda} e^\lambda_{(b)} \right) \qquad (6.62)$$

Hence, the covariant operator is given by the expression,

$$\nabla_{(c)} \Psi = e^\mu{}_{(c)} \left(\lozenge_\mu + \frac{1}{2} ie^{(a)}{}_\nu \nabla_\mu e^{(b)\nu} \Sigma_{(a)(b)} \right) \Psi \qquad (6.63)$$

Let us reconsider the relativistic six-dimensional quantum mechanical wave equation for six-dimensional curved space-time,

$$\left(i\hbar c \tilde{e}^{\mu\nu}{}_{(a)(b)} \gamma^\varepsilon \left(\lozenge_\mu + \Omega_\mu \right) - \beta m' c^2 \right) \vec{\Psi}(r,t) = 0 \qquad (6.64)$$

If a particle is submerged in an electromagnetic field, minimal coupling, $ieA_{\mu\nu}$, may be included in the above equation, where $A_{\mu\nu}$ is the six-potential. The six-potential combines both an electric scalar potential and a magnetic vector potential into a single six-vector. Minimal coupling provides coupling between fields which involves only the charge distribution and not higher multipole moments of the charge distribution. Multipoles describe how much something behaves like another system that we can predict easily. It is possible to anticipate, based on the multipole data, a reaction to the field without knowing what its true shape is, and to gather hints about what that shape might be. Moreover, for instance, the multipole moment for a dipole is something like the center of charge, giving us a clue to the distance between the dipole located at the origin of a coordinate system and its center of charge.

$$\left(i\hbar c\tilde{e}^{\mu\nu}{}_{(a)(b)}\,\gamma^{\varepsilon}\left(\Diamond_{\mu}+\Omega_{\mu}\right)-\beta m'c^{2}+ieA_{\mu\nu}\right)\vec{\Psi}(r,t)=0 \quad (6.65)$$

These relativistic equations describe how particles and antiparticles are distinct according to the General Theory of Relativity, but the same particles, or antiparticles*, may travel as tardyons slower than "c", or hypothetically travel faster than c as tachyons, $\langle v\rangle \geq \pm c$, backward in linear, non-linear, or three-dimensional time to become distinct antiparticles, or particles*, from the perspective of forward three-dimensional time, or the retarded spatiotemporal wave function.

The scalar action that leads to the above equation may be expressed as

$$S=\int dx^{6}\left(\sqrt{-g}\right)\Psi^{*}\left[i\gamma^{(c)}e^{\mu}{}_{(c)}\times\left(\Diamond_{\mu}+\frac{1}{2}ie_{\nu}{}^{(a)}\,\nabla_{\mu}e^{(b)\nu}\Sigma_{(a)(b)}+ieA_{\mu\nu}\right)-m\right]\Psi \quad (6.66)$$

Where the term, $\gamma^{\varepsilon}=e^{\varepsilon}{}_{(a)}\gamma^{(a)}$, may be defined as six-component Dirac-Pauli temporal-spin matrices that satisfies the Clifford algebra $\gamma^{\varepsilon}\gamma^{\beta}+\gamma^{\beta}\gamma^{\varepsilon}=\gamma^{\varepsilon\beta}$.

§ 7. The collapse of the wave function

The probability of the wave function collapses to unity, or certainty, at a distinct spatial position in space-time and is null elsewhere. Probability is neither mass nor energy; it is just intangible information, nothing physical. As the probability of the wave function collapses, abstract information is not restrained by the speed of light between two distant spatial positions. Since the probability of the spatiotemporal wave function has complex amplitude, consisting of a dimensional magnitude and phase angle, conjugate temporal and spatial waves may interfere out of phase and collapse the superposition of the wave function.

A classical signal of information needs mass to encode its message and energy to transmit its waveform across a substantial spatial distance over time. The physical phenomenon of entanglement emerges as the outcomes of expectation values of at least a pair of random and correlated quantum mechanical systems are measured in more than one basis of measurement. Entanglement between particles is incorporeal instantaneous transmission of abstract information across space. Entangled quantum states may be spatially distant while may or may not coexist temporally which allows instantaneously (possibly orders of magnitude over *c*) appropriate quantum state correlation. Quantum states may be given for the quantum system. Hence, entangled particles exist in split spatial wave function instants but remain connected causally by a circuit of entanglement of the temporal wave function of the quantum system.

A tachyon $i(m)$ may be regarded as the quantum of entanglement, a particle with imaginary mass $\sqrt[2]{-1}$, of instantaneous quantum abstract information exchange between entangled quantum states of a system.

Thus, entangled quantum states of particles may share wavelength phase or spin information through the exchange of tachyons. Tachyons are the quantum objects of entanglement, emitted by faster-than-light particles through a specific medium, and with quantum states information in an electromagnetic bundle of energy with no effective charge.

A photon that is traveling at the speed of light travels at the speed of the retarded spatiotemporal wave function, or at the speed of the advanced spatiotemporal wave function. In a sense, the photon is at an equilibrium point in the spatiotemporal wave function. As a photon travels, backward in time in the direction of the advanced wave function as a tachyon, it appears as if the photon (tachyon) is traveling at a speed faster than light in the opposite direction to the forward or positive temporal polarity. In fact, the photon may still be traveling at the speed of light but in the direction of the advanced spatiotemporal wave function or negative spatiotemporal polarity, but from the positive spatiotemporal perspective it appears as a tachyon. If the same photon were to reverse direction in the positive spatiotemporal polarity, it may still appear to be traveling faster than light from the past to the future. Accordingly, *a photon may travel in the direction of the arrow of time but not faster than the speed of light with respect to an inertial frame of reference in the direction of the corresponding spatiotemporal wave function of its relative motion.*

Moreover, a photon traveling in a spatiotemporal direction may be absorbed and re-emitted by a particle in its past or future, or redirected by spatiotemporal curvature, which may cause the photon to reverse direction of travel in space-time. A photon has relativistic mass that is considered positive in the direction of positive spatiotemporal polarity, but the mass of a tachyon may be considered negative in the direction of negative spatiotemporal polarity.

The reinterpretation principle asserts that a tachyon traveling back in time can always be reinterpreted as a tachyon traveling forward in time, because observers cannot distinguish between the emission and absorption of tachyons.

The properties of photons (tachyons) are uncertain or fuzzy which endows tachyons wiggle room to avoid temporal inconsistencies. (Deutsch, 1991) Quantum systems, unlike classical systems with well-defined states, exist in superposition and permutations of quantum states. These assertions are supported by both the General Theory of Relativity and The Uncertainty Principle. (Ralph, 2014)

Tachyonic manifestations of photons are found through quantum tunneling of a photon through a quantum barrier, in laser media or inverted atomic populations, and in the Einstein-Podolsky-Rosen phenomenon, in which two distantly separated photons can apparently influence each other's behaviors at two distantly separated detectors, for example, through the entanglement of a quantum state of energy. Consequently, tachyon-type solutions are derived from Maxwell equations when coupled to any of the above conditions or environments. (Franson, 1989)

In the Laws of Physics, any information exchange that is not forbidden is possible under the principle of free information exchange. A handshake is the process by which two or more particles initiate the exchange of quantum states through tachyons.

Handshaking begins when a sending particle (source) emits a tachyon to a receiving particle (sink) indicating its quantum states and establishing a tachyonic channel. The sending and receiving particles exchange tachyons back and forth that enable them to agree on a protocol of quantum states of entanglement.

A tachyon may be regarded as an instant of an advanced photon, or luxon, with quantum state information, traveling backward in time at a speed greater than the speed of light. A tachyon may travel backward in time, forward in time, then backward in time again, as a quantum of entanglement between tardyons, or photons, that is, between a source and a sink of quantum state information that may or may not coexist at the same instant of time. As an advanced photon travels, backward in time as a tachyon $i(m)$, it is traveling faster than the speed of light through its wave medium as a current of entanglement from the perspective of the positive temporal polarity.

A quantum entanglement between tardyons, a quark and an antiquark pair production, through the Heisenberg-Schwinger effect in a very strong electric field, rushing away from each other approaching the speed of light, may be theorized to simultaneously create a spatiotemporal bridge connection in space-time between the pair.

The spatiotemporal bridge, as predicted in the General Theory of Relativity, establishes the wave medium, and circuit of entanglement, for tachyons that are quanta of entanglement.

During entanglement between particles, the tachyon would be emitted by the source to travel backward in time where the particles are sharing a circuit of entanglement, as the source and sink particles separate into different spatial paths, the tachyon would follow the temporal circuit of entanglement to the destination of the sink on which the tachyon exchanges quantum information or handshakes with the receiving particle.

Any entangled particle may be a source or a sink for a tachyon. This tachyonic action has created the illusion of spooky-action-at-a-distance. Since light and space-time are incorporeal, perhaps a supplementary axiom of General Theory of Relativity may be that photons as tachyons or space-time as the wave medium can travel faster than light.

A particle or particle system may be speculated to exist at the intersection of its advanced and retarded wave functions $\langle \Psi^- \| \Psi^+ \rangle$, a point of equilibrium of space-time in a time-symmetric quantum mechanical model.

A particle incorporates an expression of itself in the advanced and retarded wave functions of its existence. The advanced corporeal expression is antimatter and the retarded corporeal expression is matter.

Under the law of causal contingency: the retarded wave function Ψ^+ follows the law of forward-causality and the advanced wave function Ψ^- follows the law of retro-causality.

The following diagram illustrates the principle of causal contingency for a tachyon $i(m)$.

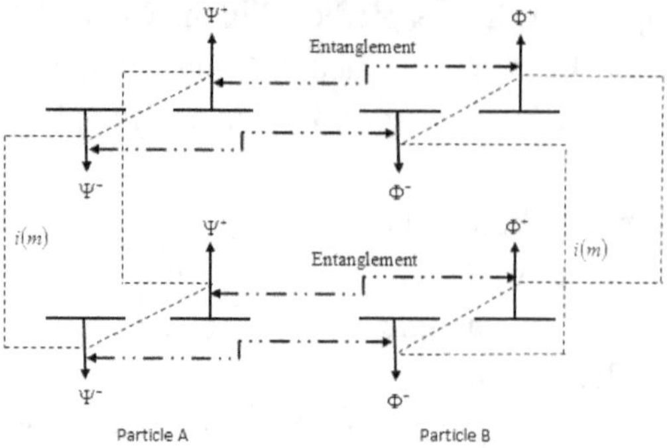

Figure 13.

Incidentally, the advanced photon solutions are a second set of solutions to Maxwell's electromagnetic equations. Electrons emit ordinary photons when they decelerate in response to colliding with a tachyon, or an advanced photon, that has traveled backward in time. Once the retarded photon or tachyon is emitted, it travels forward through the temporal volume until it completes the circuit, by striking and accelerating the entangled sink of the tachyon, or electron in the future, which in turn emits an advanced photon, or tachyon, without violating the law of causal contingency. (Feynman, 1964)

The General Theory of Relativity, the complex wave function of space-time, Maxwell's Equations of Electromagnetism, and the principle of causal contingency, affirm the probabilistic certainty of bidirectional temporal travel of photons as tachyons.

An electron in an excited state traveling faster than the speed of light through a material medium behaves as a luxon emitting electromagnetic radiation in the form of an advanced photon that travels faster than light backward in time as a tachyon. As the electron or luxon emits an advanced photon or tachyon, the electron decreases its energy level by an equivalent amount. When an inverted population of electrons travels faster than light through its material medium, multiple advanced photons, or tachyons, are emitted coherently.

An electron and a photon are different particles. Nonetheless, they both have wave functions that make them interfere in their respective double-slit experiments. Their wave functions obey the same laws describing their probabilistic propagation in space-time. Either a photon, or an electron, may interfere as a particle-wave duality in a double-slit experiment. There is interaction between the advanced and retarded waves of particles in space-time as the probabilities for particles to interfere. In some sense, waves are real and particles are not, but there is a real physical mechanism that causes the waves to collapse in certain circumstances and behave like particles. The existence or measurement of a quantum particle is a probabilistic function of its medium.

Let us express a tachyon as a particle of quantum abstract information in qubits and/or qutrits given by

$$i(m) = -\log_2\langle p(m)\rangle \tag{7.1}$$

Where $p(m)$ is the probability distribution that the quantum abstract information content of the imaginary mass "m" of an arbitrary tachyon is chosen from all possible tachyons "M" available in the background space-time.

The probability distribution $p(m)$ is given by

$$p(m) = 2^n \tag{7.2}$$

The tachyonic quantum of abstract information simplifies to

$$i(m) = \frac{1}{\log_2\langle p(m)\rangle} = \frac{1}{\log_2\langle 2^n\rangle} = \frac{1}{n} \tag{7.3}$$

Where the number 2^n is the number of ways in which the qubits quantity of tachyonic information can be arranged. A tachyonic information content unit is a qubit or qutrit; the qubit is the smallest possible unit of quantum abstract information. The qubit is a quantum of abstract information content of two possible states. The polarization of a photon, or the spin of a particle, are examples of two quantum state systems that can be used as qubits. Qubits may be entangled. A qutrit is a unit of quantum abstract information content that exists as a superposition of three orthogonal quantum states. A string of "n" qutrits represents 3^n different quantum states simultaneously. Qutrits may be entangled. Thus, it is possible for a qubit and a qutrit to be entangled. An arbitrary qubit or qutrit in a tachyon can neither be copied, deleted, nor destroyed. Quantum abstract information in a tachyon has finite size in qubits and/or qutrits.

The abstract information content of a tachyon i may be quantized in terms of the minimum number "n" of qubits and/or qutrits needed to encode the abstract information. Such abstract information content may be encoded with "n" qubits and b^2 classical bits that describe the relative arrangement of the "n" qubits.

Moreover, if an arbitrary tachyon has a probability distribution $|\Psi(i)|^N$ of 1 or certainty, where N is the number of quantum states, then its imaginary square mass is "-1" which implies the quantum wave function of the tachyon has collapsed to certainty and the quantum states exchange with the receiving particle has been made. If two distinct tachyons, or more, are exchanged between quantum particles, the total amount of abstract information exchanged is the sum of the measures of abstract information of all the distinct tachyons being exchanged.

A tachyon that conveys abstract information of certainty, which has already been received by the receiving particle, contains negligible abstract information. Infrequently received tachyonic quanta of abstract information contain more valuable abstract information for the receiving particle than the more frequently received tachyonic quanta of abstract information. (Shannon, 1948)

The imaginary mass of the tachyon may be expressed as

$$m^2 = -|\Psi(i)|^{-N} \tag{7.4}$$

Thus, if the probability distribution of a tachyon $|\Psi(i)|^N$ is "1", or "0", its square imaginary mass would respectively be "−1", or "0". The uncertainty of a tachyon may be measured by the number of possible arrangements of the quantum states of its sending particle or source. When the uncertainty of the tachyon is low, there are fewer arrangements of quantum states possible, and then less abstract 0information is delivered by the tachyon to the receiving particle or sink. The imaginary energy and momentum of a tachyon are given below. (Feinberg, 1967)

$$E_i = ipc = \frac{i^2\left\langle \sqrt[2]{|\Psi(i)|^{-N}} \right\rangle vc^2}{\sqrt[2]{v^2 - c^2}} \tag{7.5}$$

$$p_i = \frac{i\left\langle \sqrt[2]{-|\Psi(i)|^{-N}} \right\rangle (vc)}{\sqrt[2]{v^2 - c^2}} \tag{7.6}$$

The superluminal speed of a tachyon, from the perspective of the positive spatiotemporal polarity, is defined as the partial differential change in its energy over the partial differential change in its momentum.

$$v_i = \frac{\partial E}{\partial p} \tag{7.7}$$

The effect of the momentum of a photon was described by the eminent mathematician and astronomer Johannes Kepler to explain the observation that a tail of a comet always points away from the Sun. Tachyonic radiation may be analyzed as interactions between incident photons or electromagnetic waves upon a surface area of a particle or particle system. Photons or their waves have the property of momentum, which allows an advanced photon or its wave to be interchanged under classical conditions. The force exerted by a tachyonic wave that is completely absorbed by the surface of a receiving particle is given by

$$F(i) = \frac{\partial p}{\partial t} = \dot{p} \tag{7.8}$$

The direction of the space-time momentum vector of the tachyon is in the direction of the arrow of time which indicates the tachyon may travel forward or backward in time. The radiation pressure of an incident tachyon, when the pressure is completely exerted on the receiving surface of a particle (sink) after traveling through a distance *d*, may be expressed as

$$P(i) = \frac{1}{d}\frac{\partial^2 m}{\partial t^2} \tag{7.9}$$

Tachyons accelerate when they lose energy or imaginary mass to conserve energy and momentum. As a tachyon loses energy or imaginary mass, its abstract information on the quantum nature of the system outflows into its receiving particle, this outflow is compensated by the increase in kinetic energy or momentum. Tachyons flow in closed circuits of entanglement between particles. Neither the principle of conservation of energy nor the principle of causality is violated since entangled particles or systems form part of an encapsulated and inherited information structure for quantum state information types in the same history world tube of a probabilistic event. Thus, quantum information exchange is probabilistic in nature to conserve the causality of quantum events.

From the present perspective of a local inertial frame of reference, it seems like the tachyon was absorbed before it was emitted.
As a tachyon travels to its sink, it may share the quantum state

charge information of its source. As the tachyon accelerates, it would spontaneously radiate electromagnetic waves that would dissipate the electromagnetic energy of the tachyon on its path to its sink, the greater the tachyonic acceleration, the greater the electromagnetic radiation emission of the tachyon. The direction of propagation of the electromagnetic wave radiation is relative to the path of the tachyon and its relative superluminal velocity. A decrease in the wavelength of electromagnetic radiation emitted by an approaching tachyon, because of the Doppler Effect, would be displayed as a blueshift which is the displacement of the spectral lines to shorter wavelengths in the light coming from distant tachyons moving toward the observer. Since tachyons are rapidly propagating backward in time (negative spatiotemporal polarity) they would not be detected by a measuring instrument detecting particles propagating forward in time (positive spatiotemporal polarity).

As first predicted by Oliver Heaviside long ago, and detected experimentally by Cherenkov later, the radiation in the water surrounding a pool-type nuclear reactor, may be a good example for a tachyon field emitted by beta particles (fast high energy electrons) released as fission products decay. The beta particles radiate tachyons as they move through the water medium at a speed faster than light, though not faster than light in free space-time. (Heaviside, 1971)

Hence, the abstract information of a tachyon may be represented as quantum state events originated with specific probability distributions, as measures of uncertainty, that may be quantified as qubits and/or qutrits of abstract information that may represent quantum state sub-events whose probability distributions are additive.

A class of tachyons shares the same abstract information.
If the sending particle and the receiving particle are exchanging the same class of tachyons, there might be more than one tachyon with the same quantum abstract information content in the background *M*. (Simpson, 1949)

How much is the tachyon diversity in background M?

The tachyon diversity index is given by

$$D(i) = \frac{1}{\sum_{i=m}^{M} p_i^2} \tag{7.10}$$

Where $D(i)$ is a mathematical measure of uncertainty that characterizes tachyon diversity in background M, when the dataset in M is large and sampling is done with replacement. Thus, p_i is the proportion of a specific tachyon class $i(\hat{m})$ relative to the total number of tachyon classes in the background M. Then, the squared proportions for all the tachyon classes are added, and the reciprocal is taken. The diversity index would be at a maximum when all tachyon classes in the background M are distinct. If all classes of tachyons in background M are equally abundant, then $D(i) \geq 1/M$.

What is a measurement in quantum mechanics?

A quantum mechanical measurement may be performed by an object of mass or energy, or a measuring instrument, on the states of the quantum wave function of a particle, particle system, or another object.measurement is a physical interaction between two or more physical objects, allowing both Objects to be more certain about the states they are in as their quantum wave functions collapse. Quantum objects may measure each other as their physical fields interact and their spatiotemporal waves interfere. A single incident photon is a form of quantum measurement.

If a tree falls in the forest, and no one is around to see it, is it being measured?

Classical objects measure each other, such as braille instructions on embossed paper on a table, even in a dark room when people are not observing or using a measuring instrument, neither the braille instructions nor table, disappears into incorporeal probability distributions. A quantum mechanical measurement exchanges information (tachyons) between particles or particle systems which are the building blocks of macroscopic systems. Any interaction that conveys information is a form of quantum state measurement. Abstract information from quantum measurement reduces uncertainty, allows quantum state detection, and diminishes probabilities.

Active measurements of the wave function collapse its superposition, the combined two states of the temporal and spatial waves, changing to just one of the states of the quantum wave superposition. The collapse of the superposition in photons, that propagate through the wave function of light, has been widely researched. The quantum wave function leads the probabilistic path of the particle system through the wave medium of space-time while the system evolves along its deterministic path until measured.

Let us express the wave function of space-time as the superposition of the temporal wave and its conjugate the spatial wave given by

$$\Psi(r,t)=\Psi e^{i\left(\frac{p}{\hbar}-\omega t\right)} \tag{7.11}$$

Let is imagine that the wave function is measured, which collapses the wave to its probability $|\Psi|^2$ to

$$|\Psi|e^{i\frac{p}{\hbar}}|\Psi|e^{-i\omega t}=|\Psi|^2\frac{e^{i\frac{p}{\hbar}}}{e^{i\omega t}}=|\Psi|^2e^{i0}=|\Psi|^2 \tag{7.12}$$

The coefficient on the right side of the quantum wave function equals zero. The momentum "p" is the momentum of the spatial wave and ω is the angular frequency of the temporal wave of the quantum wave function. Then, under those conditions, the predictable probability distribution of the spatial wave is directly proportional to the predictable probability distribution of the temporal wave of the quantum wave function.

$$|\Psi|^2 e^{i\frac{p}{\hbar}} \equiv |\Psi|^2 e^{i\omega t} \qquad (7.13)$$

If the coefficient of the spatial wave is equal to the coefficient of the temporal wave, the quantum wave function collapses to the square of its amplitude which determines the probability distribution for either the spatial or temporal quantum wave function of the system in space-time.

$$i\frac{p}{\hbar} - i\omega t = 0 \qquad (7.14)$$

$$\frac{p}{\hbar} = \omega t \qquad (7.15)$$

The momentum "*p*" of the spatial wave function is associated with matter and energy of particles or systems of particles while the angular frequency "ω" is a temporal characteristic.

$$p = \hbar\omega t = Et \qquad (7.16)$$

The momentum of an isolated particle of mass in its quantum wave function of homogeneous and isotropic space-time, ceteris paribus, may be expressed as the mass of the particle times the velocity of the area of the amplitude, which is equivalent to the product of the mass of the particle, the angular frequency of the temporal wave function, and the predictable probability distribution of the particle.

$$m\frac{\partial\Psi^2}{\partial t} = mv_A = mc^2 t = m\omega|\Psi^2| \qquad (7.17)$$

$$\frac{\partial \Psi^2}{\partial t} = \omega |\Psi|^2 \tag{7.18}$$

The area of amplitude has a phase angle with respect to an inertial quantum frame of reference as the spatial and temporal waves travel through their medium. If the phase angle ϕ_S of the spatial wave function shifts with respect to the phase angle of the temporal wave function θ_t , or vice versa, the predictable probability distribution begins to collapse as the quantum wave function begins to collapse.

$$\Psi^2 e^{i(\phi_S - \theta_t)} \propto \Psi^2 e^{i\left(\frac{p}{\hbar} - \omega t\right)} \tag{7.19}$$

As the phase angles of the spatial and temporal waves shift apart, quantum decoherence of the wave function begins between the spatiotemporal states of superposition of the quantum wave of space-time. As the predictable probability distributions of the spatiotemporal waves begin to add constructively or destructively, the spatiotemporal waves dephase. As the spatiotemporal waves decohere and dephase the predictable probability distribution of the wave function splits into a temporal nondeterministic probability distribution and a spatial deterministic probability distribution which localizes the particle in space-time. Wherefore, the predictable spatiotemporal probability distributions may decohere into distinct wave functions in the wave medium of space-time that expands per the transcendental acceleration of the surface area, π , of each wave.

Decoherence may occur in a thermodynamically irreversible way when a particle system interacts with its spatiotemporal environment. Decoherence of the wave function of a particle system is indirectly proportional to absolute temperature, and directly proportional to the electromagnetic charge, in the volume of the wave function of space-time, since the electromagnetic stiffening of the spatiotemporal volume, or the lowering of the absolute temperature of the wave medium, influences motion which tends to decelerate the expansion of spatiotemporal waves. From a systemic perspective, time seems to be slowing down in the medium of the quantum wave.

$$\Psi^2 e^{i(\phi_S - \theta_t)} \propto \Psi^2 \frac{e^{iq}}{e^{iT}} \tag{7.20}$$

$$\Psi^2 e^{i(\phi_S - \theta_t)} \propto \Psi^2 e^{i(q-T)} \tag{7.21}$$

Decoherence is not the cause of the actual effect of wave function collapse. Decoherence provides a gradually observable measurement of the collapse of the wave function, as particle system abstract information on the quantum nature of the system outflows into the spatiotemporal background. Thus, the conjugates of the quantum wave function decouple from a coherent particle system, to obtain phases from their spatiotemporal backgrounds, as system information outflows from the system, and the quantum states of the superposition are distinctly conserved.

As the quantum wave states decouple into distinct spatiotemporal states, a comprehensive or systemic superposition still exists, until each conjugate spatiotemporal wave resumes expansion in its respective deterministic or nondeterministic quantum domain. The measurement problem represents the dichotomy of the quantum wave function which may be detected, if not observable in the present, by the existence of the distinct superposition states for the system in its future. Any future evolution of the system is based upon the entanglement of an active measurement with the state the system was discovered to be in when the measurement took place. After a nondemolition measurement, a non-entangled quantum state may become entangled and persist into its causal future, produced by the passive measurement itself.

As time passes freely, a present quantum state may be observed classically, recognized, and noted in the past. A present quantum state observation, which does not persist into its causal future, is unavailable to the present observer, but may be observable in the past. The present inconstant quantum state is always an unknown to our present consciousness through our senses, regardless of our measuring instrument. The quantum wave function endows a single photon with a probability distribution, that instead of determining events, provides the spatiotemporal locality when and where an event may happen. As time continues to pass freely, the collapse of the quantum wave function during an active measurement of a present inconstant quantum state does not require our present recognition since by the time our consciousness is aware of the observation the event may be over. An event which causes the creation of a photon precedes our recognition of the effect as a second event.

The temporal wave function may determine the future states of the quantum system, while a measuring device may entangle with the system's deterministic spatial wave function to predict precise results for active measurements. A predictable probability distribution of a quantum system embodies future results in the temporal wave function that may be nondeterministic in the present space, or deterministic in the future space of the system. Causality provides the deterministic premises for predictable effects in the direction of the arrow of time.

The internal interactions, or the external, with the spatiotemporal background, not necessarily only the immediate background, of a quantum system, change the predictable probability distribution of the system into an actual, distinctly well-defined outcome.

The above quantum state interactions of a system underlie the correspondence principle between a quantum system and a classical system in the same realm of space-time.

As stated by the correspondence principle, *the laws of quantum physics asymptotically approach the laws of classical physics in the limit of large quantum numbers and large numbers of particles*. Quantum mechanics can be used to describe large macroscopic systems.

The momenta and positions of macroscopic systems, such as measuring instruments, are uncertain even though the macroscopic systems are classical systems. Moreover, the uncertainty of the macroscopic systems is not zero even when the classical systems become vanishingly small. On the other hand, the uncertainty of a particle of mass describes its predictable future potential to be anywhere at any time within its quantum wave function.

Are the laws of Quantum Mechanics contradictory?

Quantum Mechanics has two different laws to describe how a system changes as time passes. The first law acts most of the time, and describes how the spatiotemporal waves expand, or contract, as they flow smoothly through spacetime.

Rule I:

"Except during a measurement of position or momentum, the spatiotemporal wavelet expands or contracts smoothly, and deterministically."

Hence, this law endows the quantum system with the characteristic to explore simultaneously distinct possible outcomes and the alternate histories of alternate realities of alternate universes for all possible outcomes available to the smooth flow of the spatiotemporal wavelet. This is an interdimensional law for all possible realities and outcomes of an arbitrary event in spacetime.

The second law involves the distinctive circumstance of measurement. During a measurement, a microscopic system interacts with a quantum system, to allow a single outcome to be manifested, but disallowing the probabilities of distinct outcomes to take place in a distinctive spacetime.

Rule II:

"During a measurement of position or momentum, the spatiotemporal wavelet collapses around the position, or momentum, where is being measured, with a probability that is approximately equal to the square of the height of the wavelet, or more accurately with a probability that is equal to the area of the hemisphere facing the measurement for a spherical spatiotemporal wavelet."

The second law involves probabilities which leads to contradiction with the first law under the present paradigm of Quantum Mechanics. This contradiction may be superseded by considering the first law as interdimensional, or multiversal, a superposition of states for the distinct possible outcomes and alternate histories of alternate realities of alternate universes. Each state with probability one, as the measuring device is able to observe distinct outcomes of each alternate history.

The second law treats the quantum system as being in a definite outcome, and the measuring device has observed only that outcome, with each distinct outcome having some definite probability.

Hence, the second law is a universal law, not an interdimensional law, or not a multiversal law, for the interaction of a microscopic system with a quantum system, in reference to a single universe of the possible multiverse. As in the law of Max Born.

Therefore, Quantum Mechanics has been very successful in its prediction and applications. By shifting the paradigm of the laws of Quantum Mechanics, the measurement problem may be solved. The outcome of that realization and its confirmation is that Quantum Mechanics will be compatible with realism.

§ 8. A fortunate legacy of notions and ideas from eminent predecessors

The following notions resemble those proposed or shared by, but not limited to: Faraday, J.J. Thomson, Heaviside, Helmholtz, Hertz, Maxwell, Minkowski, Poincare, Poynting and Tesla, long ago, which shed light on the concept of space-time tubes of force, which lead to the evolution of novel ideas.

- The wave medium fills all space-time. Swirling tubes of force exist in the wave medium.

- For a long time, studies of radioactivity have shown that the empty vacuum of space-time has spectroscopic structure similar to that of ordinary quantum solids and fluids. Thus, space-time is not empty, but a relative wave medium that has physical substance of its own.

- An element of force emerges from a polarized unit area of space-time with the passage of time. Energy may be conceptualized as a future ability of the element of force. The temporal momentum is the underlying infrastructure of the physical medium for the energy of work.

- Infinitesimal helices or swirls of space-time manifest electromagnetic waves or instances of mass. The transmutation of the space-time wave medium manifests mass or energy.

- When the force of space-time subsides or ceases, the manifestations of waves, charges, or instances of mass revert to the space-time wave medium. Space-time is the quintessence of physical reality. Measurements and observations of physical objects and forces describe interactions, deterministic occurrences and possible and feasible forms of the quintessence of the space-time medium.

- We understand walls in terms of bricks, bricks in terms of crystals, crystals in terms of molecules, molecules in terms of atoms, and so on, for matter is discrete, from one level to the next lower level. Nonetheless, the natural grain of the space-time wave medium that is applicable at each level of reality decreases by an order of amplitude of the spatiotemporal wave function when we shift our attention from one level to the next lower level. The mathematical limit of such pursuit is the quintessential point source of the spatiotemporal wave function in the medium of space-time.

- The wave medium is space-time and the most probable medium filling are the carriers of the force.

- Electrostatic force is spatiotemporal force that may produce physical motion.

- Tubes of force stiffen the wave medium.

- If the frequency and intensity of the wave increases, the exchange of spatiotemporal charges is slower, compression of the wave medium would result. Spatiotemporal (Electromagnetic) momentum would be imparted when very high frequency and high intensity opposite tubes of force retract and dissolve at the sources of emission.

- Physical motion, or propulsion, can result if the wave medium is thinned, through slower divergence or convergence, allowing an object of mass to move through such medium while the free wave medium behind the object expands or contracts about the volume of the wave medium.

- Tubes of force create movement of charged particles in the direction of linear frequency (direction of influence). Tubes of force are polarized.

- The wave medium is itself a vehicle of physical momentum for charged or uncharged mass. Electromagnetism and gravity are physical instances related to wave medium momenta.

- Tubes of force may impact momentum upon the area of the receiving end or at right angles to the tubes of force.

- The spatial or magnetic tubes of force are manifested by the movement and frequency, of the temporal or electric charges through the wave medium of space-time.

- Tubes of force may exist as closed loops in space-time, or begin and end as closed surfaces on charged objects of mass.

- Temporal (electric) tubes of force increase their influence or total force as they increase in number across the same cross-sectional area of space-time in the direction of linear frequency.

- Spatial (magnetic) tubes of force may move at the same velocity of light in opposite directions.

- A light ray is a spatiotemporal wave moving diagonally at "*c*", in the orthogonal spatiotemporal plane, in the direction of linear frequency of the wave, which may or may not be polarized, in the direction of rotation of the axis of the angular frequency.

- Momentum may be stored in a unit of spatiotemporal volume by the actions of spatial or temporal force vectors of the spatiotemporal (electromagnetic) field.

- The charge element of space-time has a resultant force of zero when both spatial and temporal unit areas of charge are conjugate, equal, or opposite, but may impact motion in the wave medium of space-time when either element of charge is acting alone.

§ 9. Conclusion

The spatiotemporal wave medium is the quintessence of reality; matter, energy, charge, motion and the physical states of particles or particle systems, are all manifestations of the same spatiotemporal wave medium. The dynamic spatiotemporal medium embodies our conceptualizations of space and time as well as the physical states we perceive through our senses. Nothing is what it seems, but everything is of the quintessence of reality, even the physical observer, outside of the observer's belief of his or her own nonphysical nature and higher order of existence.

The wave function of reality incorporates the probabilistic existence of physical states and encompasses the tenses of time, for particles or particle systems. The expansion or contraction of space-time is the realization and embodiment of the wave function. The wave function of space-time may be expressed as six-dimensional in our current understanding of the three-dimensional spatial wave function and the three-dimensional temporal wave function under the General Theory of Relativity. The six-dimensionality of the wave function is the probabilistic domain for the existence of all possible outcomes in space-time and the background for six-dimensional force fields.

References

Adloff, C. et al. (H1 collaboration) (1999). *"Charged particle cross sections in the photoproduction and extraction of the gluon density in the photon"*. European Physical Journal C. 10: 363–372.

Alcubierre, Miguel. (1994) *The warp drive: hyper-fast travel within general relativity*. Class. Quantum Grav. 11-5, L73-L77.

Arnowitt, R., Deser, S., Misner, C. (1959). *Dynamical Structure and Definition of Energy in General Relativity*. Physical Review. 116 (5): 1322–1330.

Atkins, Peter. *The Laws of Thermodynamics: A Very Short Introduction* (2010). Oxford University Press.

Baez, J. (1996). *Why are there eight gluons and not nine?* (http://math.ucr.edu/home/baez/physics/ParticleAndNuclear/gluons.html).

Baker, Bevan B., and Copson, E.T. (1987). *The Mathematical Theory of Huygens' Principle (Third edition)*. Chelsea Publishing Company, AMS, New York, NY.

Barbour, Julian, *The End of Time: The Next Revolution in Physics*, Oxford University Press, 1999.

Baumgarte, Thomas W., Shapiro, Stuart L. (2010). *Numerical Relativity, Solving Einstein Equations on the Computer*, Cambridge University Press, 40 West 20th Street, New York, NY 10011.

Bilson-Thompson, S.O., Leinweber, D.B., Williams, A.G. (2003). *Highly improved lattice field-strength tensor*. Annals of Physics. Adelaide, Australia: Elsevier 304 (1): 1-21.

Bohr, Niels (1958). *Atomic Physics and Human Knowledge* (1958). John Wiley and Sons, 111 River Street, Hoboken, NJ 07030-5774.

Born, M and Wolf, E (1999). *Principles of Optics: Electromagnetic Theory of Propagation, Interference and Diffraction of Light (7th edition)*. Cambridge University Press.

Bottema, O. and Roth, B. *Theoretical Kinematics* (1990). Dover Publications, Inc.

Brackenridge, J Bruce (1995). *The Key to Newton's Dynamics, the Kepler problem and the Principia*, The University of California Press, 2120 Berkeley Way, Berkeley, CA 94704.

Bronnikov, K.A. (1973). *Scalar-tensor theory and scalar charge.* Acta Physica Polonica. B4: 251–266.

Caldirola, P (1980). *The introduction of the chronon in the electron theory and a charged lepton mass formula.* Lett. Nuovo Cim. 27, pp. 225–228.

Carroll S. M., *The cosmological constant*, astro-ph/0004075, (2000).

Carroll S. M., *Quintessence and the rest of the world*, Phys. Rev. Lett. 81 (1998) 3067–3070, arXiv:astro-ph/9806099.

Cartan, Élie (1922). *"Sur une généralisation de la notion de courbure de Riemann et les espaces à torsion"* C. R. Acad. Sci. (Paris) 174 593–595.

Cartan, Élie (1923). *"Sur les variétés à connexion affine et la théorie de la relativité généralisée Part I"* Ann. Éc. Norm. 40: 325–412 and 41 1–25; Part II: 42 17–88.

Casimir, Hendrik B.G. (1948). *On the attraction between two perfectly conducting plates*. Communicated at the meeting of May 29, 1948.

Cengel, Yunus A., and Boles, Michael A. *Thermodynamics: An Engineering Approach* (2001). Dover Publications, Inc.

Ciufolini, I. and Wheeler, J. A. (1995) *Gravitation and Inertia*, Princeton University Press, Princeton, New Jersey.

Coan, Thomas, Liu, Tiankuan, and Ye, Jingbo (2005). *A Compact Apparatus for Muon Lifetime Measurement and Time Dilation Demonstration*, American Journal of Physics 74 (2006) 161-164.

Courant, R. (1962). *Methods of Mathematical Physics*, Vol. II: Partial Differential Equations. Interscience, New York.

Craig, W. and Weinstein, S. (2008). *On determinism and well-posedness in multiple time dimensions*. ArXiv.org: 0812.0210.

de Broglie, Louis (1953). *The revolution in physics; a non-mathematical survey of quanta*. Translated by Ralph W. Niemeyer. Noonday Press, 19 Union Sq. W, New York, NY 10003, pp. 47, 117, 178–186.

Deutsch, David, *Quantum Mechanics near Closed Time-like Lines*. Physical Review D 44, 3197–3217 (1991).

Dicke, R. H. (1957) Gravitation without a principle of equivalence, Rev. Mod. Phys. 29, 363-376.

Dorling, J. American Journal of Physics, Volume 38, Issue 4, pp. 539-540, (1970).

Dugas, Rene. *A History of Mechanics* (1988). Dover Publications, Inc.

Dziewonski, Adam M.; Anderson, Don L. (June 1981). *Preliminary reference Earth model*. Physics of the Earth and Planetary Interiors 25 (4): 297–356.

Eberhard, Phillippe H; Ross, Ronald R (1989). *Quantum theory cannot provide faster-than-light communication*. Foundations of Physics Letters 2 (2): pp. 127–149.

Einstein, Albert (1952) . *Relativity, The Special and the General Theory*, Crown Publishers Inc., One Park Avenue, New York, NY 10016.

Eldemuller, M., Dosch, H.G., Jamin, M. (1999). *The field strength correlator from QCD sum rules*. Nucl. Phys. Proc. Suppl. 86:421-425, 2000. Heidelberg, Germany.

Ellis, G. F. R. (1971) *General Relativity and Cosmology, International School of Physics, Enrico*

Ellis, H.G. (1973). *Ether flow through a drainhole: A particle model in general relativity*. Journal of Mathematical Physics. 14: 104–118. Bibcode:1973JMP...14...104E. doi:10.1063/1.1666161.

Fedosin S.G. *About the cosmological constant, acceleration field, pressure field and energy*. Jordan Journal of Physics. Vol. 9 (No. 1), pp. 1-30 (2016).

Feinberg, G. *Possibility of Faster-Than-Light Particles*, Physical Review, Volume 159, Number 5, (1967).

Fermi–Course XLVII, Academic Press, New York.

Feynman, Richard P., *The Feynman Lectures on Physics*, Volume II, Addison-Wesley, 1964.

Feynman, Richard P (1988). *QED: The Strange Theory of Light and Matter*. Princeton University Press, 32 Avenue of the Americas, New York, NY 10013.

Franson, J.D., Physical Review Letters, 62, 2205, (1989).

García-Parrado, Alfonso, Valiente Kroon, J.A. (2007). *Initial data sets for the Schwarzschild spacetime*, Phys. Rev. D 75, 024027.

Greiner, W., Schafer, G., (1994). "4". *Quantum Chromodynamics*. Springer.

Greene, Brian (1999). *The Elegant Universe: Superstrings, Hidden Dimensions, and the Quest for the Ultimate Theory*. W.W. Norton and Company, Inc., 500 Fifth Avenue, New York, NY 10110, pp. 97–109.

Griffiths, David. (1987). *Introduction to Elementary Particles.* John Wiley & Sons. pp. 280–281.

Gross, David J., and Wilczek, Frank, (1973). *Ultraviolet Behavior of Non-Abelian Gauge Theories*. Physical Review Letters, Vol. 30, No. 26, pages 1343–1346; June 25, 1973.

Hafele, J C (1971). *Performance and Results of Portable Clocks in Aircraft*, Washington University, St. Louis, Missouri.

Hawking S.W. and Ellis G. F. R., *The large-scale structure of space-time*. Cambridge University Press, Cambridge, England, (1973).

Heaviside, Oliver, *Electromagnetic Theory*, volume III. Chelsea Publishing Company, New York, 3rd edition, 1971.

Heaviside, Oliver (1888). *Electromagnetic waves, the propagation of potential, and the electromagnetic effects of a moving charge*. The Electrician.

Heisenberg, Werner (1930, repr. 1949). *The Physical Principles of the Quantum Theory*. Dover Publications Inc., 31 East 2nd Street, Mineola, NY 11501.

Hume, David. (1738). *A Treatise of Human Nature*. Longmans, Green, and Company, London, England. Edited in 1874.

Jackson, John D (1999). *Classical Electrodynamics* (3rd ed.), John Wiley and Sons, 111 River Street, Hoboken, NJ 07030-5774.

Jammer, Max. *Concepts of Mass in Contemporary Physics and Philosophy* (Princeton, NJ: Princeton University Press, 2000) pp.162–163.

Jönsson, Claus (1961). Zeitschrift für Physik 161. Reprinted in English as *Electron diffraction at multiple slits*. Am. J. Phys. 42 (1974), pp. 4-11.

Lanczos, Cornelius, *Lagrangian Multiplier and Riemannian Spaces*, Rev. Mod. Phys., 21 (1949) pp. 497–502.

Larmor, Joseph (1900). *Aether and Matter*, Pg. 174 Lorentz Transformation, Cambridge University Press, 32 Avenue of the Americas, New York, NY 10013.

Lense, J. and Thirring, H. *On the Influence of the Proper Rotation of Central Bodies on the Motions of Planets and Moons According to Einstein's Theory of Gravitation*. Physikalische Zeitschrift, 1918.

Lorentz, Hendrik Antoon (1920). *The Einstein Theory of Relativity: A Concise Statement*, (April 2009) Kessinger Publishing, LLC, P.O. Box 1404, Whitefish, MT 59937.

Lorentz, Hendrik Antoon (1909). *The Theory of Electrons and its Applications to the Phenomena of Light and Radiant Heat*, (January 2007) Cosimo Classics Publishing Inc., P.O. Box 416, Old Chelsea Station, New York, NY 10011.

Ludvigsen, Malcolm (1999). *General Relativity, A Geometric Approach*, Cambridge University Press, 40 West 20th Street, New York, NY 10011.

Maxwell, James Clerk, *A Dynamical Theory of the Electromagnetic Field*, Philosophical Transactions, Royal Society London, Published 1 January 1865.

McDonald, Kirk T., *The relation between expressions for time-dependent electromagnetic fields given by Jefimenko and by Panofsky and Phillips*, American Journal of Physics, 65 (11): 1074–1076, (2016).

Melia, Fulvio (2009). *Cracking the Einstein Code: Relativity and the Birth of Black Hole Physics.* The University of Chicago Press, 1427 E. 60th Street, Chicago, IL 60637.

Melia, Fulvio (2007). *The Galactic Supermassive Black Hole.* Princeton University Press, 41 William Street, Princeton, New Jersey 08540.

Mittelstaedt, P; Prieur, A; and Schieder, R (1987). *Unsharp particle-wave duality in a photon split-beam experiment*. Foundations of Physics 17, pp. 891-903.

Naber, Gregory L (1992). *The Geometry of Minkowski Spacetime*, Springer-Verlag New York, Inc. Publishers, New York.

Nakahara, Mikio, Geometry, *Topology and Physics*, IOP Publishing Ltd. Bristol and Philadelphia, 2003.

Newman, E T, Couch, R, Chinnapared, K, Exton, A, Prakash, A, and Torrence, R (1965). *Metric of a Rotating, Charged Mass*, J. Math. Phys. 6, pp. 918-919.

Newton, Isaac (1999). *The Principia: Mathematical Principles of Natural Philosophy*, The University of California Press, 2120 Berkeley Way, Berkeley, CA 94704.

Partovi, M. Hossein (1994). *QED corrections to Planck's radiation law and photon thermodynamics*, Phys. Rev. D 50, No. 2, 1118-1124.

Politzer, H. David, (1973). *Reliable Perturbative Results for Strong Interactions?* Review Letters, Vol. 30, No. 26, pages 1346–1349; June 25, 1973.

Pound, R. V., and Rebka Jr. G. A. (1959). *Gravitational Red-Shift in Nuclear Resonance,* Physical Review Letters. 3 (9): 439–441.

Puthoff, H.E., (2002) *Polarizable Vacuum (PV) Approach to General Relativity*, Found. Phys. 32, 927-943.

Ralph, Timothy C, et Al, *Experimental Simulation of Closed Timelike Curves*, Nature Communications 5, Article Number: 4145, Published 19 June 2014.

Rao, Achintya (2 July 2012). *"Why would I care about the Higgs boson?"*. CMS Public Website. CERN.

Raychaudhuri, A. K. (1955). *Relativistic cosmology I.* Phys. Rev. 98 (4): 1123.

Renker, D. (2006). *Geiger-mode avalanche photodiodes, history, properties and problems, Nuclear Instruments and Methods* in Physics Research, Section A, Volume 567, pp. 48-56.

Reuleaux, Franz. (2012). *The Kinematics of Machinery: Outlines of a Theory of Machines*. Dover Publications, Inc.

Riess et al., Supernova Search Team Collaboration, *Observational Evidence from Supernovae for an Accelerating Universe and a Cosmological Constant*, Astron. J. 116 (1998) 1009–1038, astro-ph/9805201.

Rucker, Rudolf v. B (1977). *Geometry, Relativity and the Fourth Dimension*, Dover Publications, Inc., New York, NY 10014.

Sauer, Tilman, and Majer Ulrich. (2009). *David Hilbert's Lectures on the Foundations of Physics 1915-1927*, Springer, Heidelberg, Germany.

Scully, Marlan O; Yoon-Ho Kim; Yu, R; Kulik, S P; and Shih, Y H (2000). *A Delayed Choice Quantum Eraser*. Physical Review Letters 84, pp. 1–5.

Shannon, Claude E., *A Mathematical Theory of Communication*, Bell System Technical Journal, in July and October of 1948.

Shifman, M., (2012). *Advanced Topics in Quantum Field Theory: A Lecture Course.* Cambridge University Press.

Simpson, E. H. (1949). *Measurement of diversity*. Nature 163: 688, 30 April 1949.

Skalsey M., Conti, R.S., Engbrecht, J.J., Gidley, D.W., Vallery, R.S., Zitzewitz, P.W. *A Viable Superluminal Hypothesis: Tachyon Emission from Orthopositronium.* Space Technology and Applications International Forum (2000). American Institute of Physics 1-56396-9 19-X.

Takeno, Hyoitiro, *On the spintensor of Lanczos*, Tensor, 15 (1964) pp. 103–119.

Taylor, Edwin F, Wheeler, John Archibald (1966). *Spacetime Physics*, W.H. Freeman and Company, 41 Madison Ave. E 26th, New York, NY 10010.

Tegmark, Max. *On the dimensionality of space-time*. Class. Quantum Grav. 14 (1997) L69–L75. IOP Publishing Limited. PII: S0264-9381(97)81824-2.

Tolman, Richard C. *Relativity, Thermodynamics, and Cosmology*. Oxford: Clarendon Press. 1934. LCCN 340-32023. Reissued (1987) New York: Dover ISBN 0-486-65383-8.

Tonomura, A; Endo, J; Matsuda, T; Kawasaki, T and Ezawa, H (1989). *Demonstration of single-electron buildup of an interference pattern*. Am. J. Phys. 57, pp. 117-120.

Wald, Robert M. (1984). *General Relativity*. The University of Chicago Press, 1427 E. 60th Street, Chicago, IL 60637.

Wald, Robert M (1977). *Space, Time and Gravity, The Theory of the Big Bang and Black Holes*, The University of Chicago Press, Chicago 60637.

Weinberg, Steven (2003). *The discovery of subatomic particles (Revised edition)*. Cambridge University Press.

Wheeler, J.A. (1955). *Geons*, Phys. Rev. 97:511–36.

Wilson, H. A. (1921) An electromagnetic theory of gravitation, Phys. Rev. 17, 54-59.

Yagi, K., Hatsuda, T., Miake, Y. (2005). *Quark-Gluon Plasma: From Big Bang to Little Bang*. Cambridge monographs on particle physics, nuclear physics, and cosmology. **23**. Cambridge University Press. Pp. 17-18.

Yang, Chen Ning "Frank", (2006). *An interview by Bill Zimmerman on (May 18, 2006)*. Stony Brook Masters Series.

Yang, C.N., and Mills, R.L. (1954) *Conservation of Isotopic Spin and Isotopic Gauge Invariance*. Physical Review, Volume 96, Number 1, October 1, 1954.

www.ingramcontent.com/pod-product-compliance
Lightning Source LLC
Chambersburg PA
CBHW071347210526
45465CB00001B/2
* 9 7 9 8 6 6 7 2 7 6 2 8 9 *